MECHANICS OF MATERIALS 2

An Introduction to the Mechanics of Elastic and

Plastic Deformation of Solids and Structural Materials

Also of interest

ASHBY
Materials Selection in Mechanical Design

ASHBY & JONES
Engineering Materials 1
Engineering Materials 2

BRANDES & BROOK
Smithells Metals Reference Book, 7th Edition

BRYDSON
Plastics Materials, 6th Edition

CAMPBELL
Castings

CHARLES, CRANE & FURNESS
Selection and Use of Engineering Materials, 2nd Edition

CRAWFORD
Plastics Engineering, 2nd Edition

HEARN
Mechanics of Materials 1

HULL & BACON
Introduction to Dislocations, 3rd Edition

JONES
Engineering Materials 3

LLEWELLYN
Steels: Metallurgy & Applications

SMALLMAN & BISHOP
Metals and Materials

MECHANICS OF MATERIALS 2

*An Introduction to the Mechanics of Elastic and
Plastic Deformation of Solids and Structural Materials*

THIRD EDITION

E. J. HEARN

PhD; BSc(Eng) Hons; CEng; FIMechE; FIProdE; FIDiagE

*University of Warwick
United Kingdom*

Butterworth-Heinemann
Linacre House, Jordan Hill, Oxford OX2 8DP
A division of Reed Educational and Professional Publishing Ltd

A member of the Reed Elsevier plc group

OXFORD BOSTON JOHANNESBURG
MELBOURNE NEW DELHI SINGAPORE

First published 1977
Reprinted with corrections 1980, 1981, 1982
Second edition 1985
Reprinted with corrections 1989
Reprinted 1992, 1995, 1996
Third edition 1997

British Library Cataloguing in Publication Data
Hearn, E. J. (Edwin John)
  Mechanics of materials. – 3rd ed.
  1. An introduction to the mechanics of elastic and plastic deformation of solids and structural
  materials
  1. Strength of materials 2. Strains and stress
  3. Deformations (Mechanics) 4. Elasticity
  I. Title
  620.1′12

ISBN 0 7506 3266 6

Library of Congress Cataloguing in Publication Data
Hearn, E. J. (Edwin John)
  Mechanics of materials 1: an introduction to the mechanics of elastic and plastic deformation of
  solids and structural materials/E. J. Hearn. – 3rd ed.
  p.   cm.
  Includes bibliographical references and index.
  ISBN 0 7506 3266 6
  1. Strength of materials.   I. Title
  TA405.H3
  620.1′123-dc21   96-49967
                                                                                   CIP

Typeset by Laser Words, Madras, India
Printed and bound in Great Britain by Scotprint Ltd, Musselburgh, Scotland.

CONTENTS

INTRODUCTION

This text is a revised and extended third edition of the highly successful text initially published in 1977 intended to cover the material normally contained in degree and honours degree courses in mechanics of materials and in courses leading to exemption from the academic requirements of the Engineering Council. It should also serve as a valuable reference medium for industry and for post-graduate courses. Published in two volumes, the text should also prove valuable for students studying mechanical science, stress analysis, solid mechanics or similar modules on Higher Certificate, Higher Diploma or equivalent courses in the UK or overseas and for appropriate NVQ* programmes.

The study of mechanics of materials is the study of the behaviour of solid bodies under load. The way in which they react to applied forces, the deflections resulting and the stresses and strains set up within the bodies, are all considered in an attempt to provide sufficient knowledge to enable any component to be designed such that it will not fail within its service life.

Typical components considered in detail in the first volume, *Mechanics of Materials 1*, include beams, shafts, cylinders, struts, diaphragms and springs and, in most simple loading cases, theoretical expressions are derived to cover the mechanical behaviour of these components. Because of the reliance of such expressions or certain basic assumptions, the text also includes a chapter devoted to the important experimental stress and strain measurement techniques in use today with recommendations for further reading.

Building upon the fundamentals established in *Mechanics of Materials 1*, this book extends the scope of material covered into more complex areas such as unsymmetrical bending, loading and deflection of struts, rings, discs, cylinders plates, diaphragms and thin walled sections. There is a new treatment of the Finite Element Method of analysis, and more advanced topics such as contact and residual stresses, stress concentrations, fatigue, creep and fracture are also covered.

Each chapter of both books contains a summary of essential formulae which are developed within the chapter and a large number of worked examples. The examples have been selected to provide progression in terms of complexity of problem and to illustrate the logical way in which the solution to a difficult problem can be developed. Graphical solutions have been introduced where appropriate. In order to provide clarity of working in the worked examples there is inevitably more detailed explanation of individual steps than would be expected in the model answer to an examination problem.

All chapters conclude with an extensive list of problems for solution by students together with answers. These have been collected from various sources and include questions from past examination papers in imperial units which have been converted to the equivalent SI values. Each problem is graded according to its degree of difficulty as follows:

* National Vocational Qualifications.

A   Relatively easy problem of an introductory nature.

A/B   Generally suitable for first-year studies.

B   Generally suitable for second or third-year studies.

C   More difficult problems generally suitable for third-year studies.

Gratitude is expressed to the following examination boards, universities and colleges who have kindly given permission for questions to be reproduced:

City University	C.U.
East Midland Educational Union	E.M.E.U.
Engineering Institutions Examination	E.I.E. and C.E.I.
Institution of Mechanical Engineers	I.Mech.E.
Institution of Structural Engineers	I.Struct.E.
Union of Educational Institutions	U.E.I.
Union of Lancashire and Cheshire Institutes	U.L.C.I.
University of Birmingham	U.Birm.
University of London	U.L.

Both volumes of the text together contain 150 worked examples and more than 500 problems for solution, and whilst it is hoped that no errors are present it is perhaps inevitable that some errors will be detected. In this event any comment, criticism or correction will be gratefully acknowledged.

The symbols and abbreviations throughout the text are in accordance with the latest recommendations of BS 1991 and PD 5686†

As mentioned above, graphical methods of solution have been introduced where appropriate since it is the author's experience that these are more readily accepted and understood by students than some of the more involved analytical procedures; substantial time saving can also result. Extensive use has also been made of diagrams throughout the text since in the words of the old adage "a single diagram is worth 1000 words".

Finally, the author is indebted to all those who have assisted in the production of this text; to Professor H. G. Hopkins, Mr R. Brettell, Mr R. J. Phelps for their work associated with the first edition, to Dr A. S. Tooth[1], Dr N. Walker[2], Mr R. Winters[2] for their contributions to the second edition and to Dr M. Daniels[3] for the extended treatment of the Finite Element Method which is the major change in this third edition. Thanks also go to the publishers for their advice and assistance, especially in the preparation of the diagrams and editing and to Dr. C. C. Perry (USA) for his most valuable critique of the first edition.

E. J. HEARN

---

† Relevant Standards for use in Great Britain: BS 1991; PD 5686: Other useful SI Guides: *The International System of Units*, N.P.L. Ministry of Technology, H.M.S.O. (Britain). Mechty, *The International System of Units (Physical Constants and Conversion Factors)*, NASA, No SP-7012, 3rd edn. 1973 (U.S.A.) *Metric Practice Guide*, A.S.T.M.Standard E380-72 (U.S.A.).
1. §23.27.   Dr. A. S. Tooth, University of Strathclyde, Glasgow.
2. §26   D. N. Walker and Mr. R. Winters, City of Birmingham Polytechnic.
3. §24.4   Dr M. Daniels, University of Central England.

NOTATION

Quantity	Symbol	SI Unit
Angle	$\alpha, \beta, \theta, \gamma, \phi$	rad (radian)
Length	L, s	m (metre)
		mm (millimetre)
Area	A	m^2
Volume	V	m^3
Time	t	s (second)
Angular velocity	ω	rad/s
Velocity	v	m/s
Weight	W	N (newton)
Mass	m	kg (kilogram)
Density	ρ	kg/m^3
Force	F or P or W	N
Moment	M	Nm
Pressure	P	Pa (Pascal)
		N/m^2
		bar (= 10^5 N/m^2)
Stress	σ	N/m^2
Strain	ε	–
Shear stress	τ	N/m^2
Shear strain	γ	–
Young's modulus	E	N/m^2
Shear modulus	G	N/m^2
Bulk modulus	K	N/m^2
Poisson's ratio	v	–
Modular ratio	m	–
Power	–	W (watt)
Coefficient of linear expansion	α	m/m°C
Coefficient of friction	μ	–
Second moment of area	I	m^4
Polar moment of area	J	m^4
Product moment of area	I_{xy}	m^4
Temperature	T	°C
Direction cosines	l, m, n	–
Principal stresses	$\sigma_1, \sigma_2, \sigma_3$	N/m^2
Principal strains	$\varepsilon_1, \varepsilon_2, \varepsilon_3$	–
Maximum shear stress	τ_{max}	N/m^2
Octahedral stress	σ_{oct}	N/m^2

Quantity	*Symbol*	*SI Unit*
Deviatoric stress	σ'	N/m^2
Deviatoric strain	ε'	–
Hydrostatic or mean stress	$\bar{\sigma}$	N/m^2
Volumetric strain	Δ	–
Stress concentration factor	K	–
Strain energy	U	J
Displacement	δ	m
Deflection	δ or y	m
Radius of curvature	ρ	m
Photoelastic material fringe value	f	N/m^2/fringe/m
Number of fringes	n	–
Body force stress	X, Y, Z	N/m^3
	F_R, F_θ, F_Z	
Radius of gyration	k	m
Slenderness ratio	L/k	–
Gravitational acceleration	g	m/s^2
Cartesian coordinates	x, y, z	–
Cylindrical coordinates	r, θ, z	–
Eccentricity	e	m
Number of coils or leaves of spring	n	–
Equivalent J or effective polar moment of area	J_{eq} or J_E	m^4
Autofrettage pressure	P_A	N/m^2 or bar
Radius of elastic-plastic interface	R_p	m
Thick cylinder radius ratio R_2/R_1	K	–
Ratio elastic–plastic interface radius to internal radius of thick cylinder R_p/R_1	m	–
Resultant stress on oblique plane	p_n	N/m^2
Normal stress on oblique plane	σ_n	N/m^2
Shear stress on oblique plane	τ_n	N/m^2
Direction cosines of plane	l, m, n	–
Direction cosines of line of action of resultant stress	l', m', n'	–
Direction cosines of line of action of shear stress	l_s, m_s, n_s	–
Components of resultant stress on oblique plane	p_{xn}, p_{yn}, p_{zn}	N/m^2
Shear stress in any direction ϕ on oblique plane	τ_ϕ	N/m^2
Invariants of stress	$\left\{ \begin{array}{l} I_1 \\ I_2 \\ I_3 \end{array} \right.$	N/m^2 $(N/m^2)^2$ $(N/m^2)^3$
Invariants of reduced stresses	J_1, J_2, J_3	
Airy stress function	ϕ	–

Quantity	Symbol	SI Unit
'Operator' for Airy stress function biharmonic equation	∇	–
Strain rate	$\dot{\varepsilon}$	s^{-1}
Coefficient of viscosity	η	
Retardation time (creep strain recovery)	t'	s
Relaxation time (creep stress relaxation)	t''	s
Creep contraction or lateral strain ratio	$J(t)$	–
Maximum contact pressure (Hertz)	p_0	N/m^2
Contact formulae constant	Δ	$(N/m^2)^{-1}$
Contact area semi-axes	a, b	m
Maximum contact stress	$\sigma_c = -p_0$	N/m^2
Spur gear contact formula constant	K	N/m^2
Helical gear profile contact ratio	m_p	–
Elastic stress concentration factor	K_t	–
Fatigue stress concentration factor	K_f	–
Plastic flow stress concentration factor	K_p	–
Shear stress concentration factor	Kt_s	–
Endurance limit for n cycles of load	S_n	N/m^2
Notch sensitivity factor	q	–
Fatigue notch factor	K_f	–
Strain concentration factor	K_ε	–
Griffith's critical strain energy release	G_c	
Surface energy of crack face	γ	Nm
Plate thickness	B	m
Strain energy	U	Nm
Compliance	C	mN^{-1}
Fracture stress	σ_f	N/m^2
Stress Intensity Factor	K or K_1	$N/m^{3/2}$
Compliance function	Y	–
Plastic zone dimension	r_p	m
Critical stress intensity factor	K_{IC}	$N/m^{3/2}$
"J" Integral	J	
Fatigue crack dimension	a	m
Coefficients of Paris Erdogan law	C, m	–
Fatigue stress range	σ_r	N/m^2
Fatigue mean stress	σ_m	N/m^2
Fatigue stress amplitude	σ_a	N/m^2
Fatigue stress ratio	R_s	–
Cycles to failure	N_f	–
Fatigue strength for N cycles	σ_N	N/m^2
Tensile strength	σ_{TS}	N/m^2
Factor of safety	F	–

Quantity	*Symbol*	*SI Unit*
Elastic strain range	$\Delta\varepsilon_e$	–
Plastic strain range	$\Delta\varepsilon_p$	–
Total strain range	$\Delta\varepsilon_t$	–
Ductility	D	
Secondary creep rate	ε_s^0	s^{-1}
Activation energy	H	Nm
Universal Gas Constant	R	J/kgK
Absolute temperature	T	°K
Arrhenius equation constant	A	–
Larson–Miller creep parameter	P_1	
Sherby–Dorn creep parameter	P_2	
Manson–Haford creep parameter	P_3	
Initial stress	σ_i	N/m^2
Time to rupture	t_r	s
Constants of power law equation	β, n	–

CHAPTER 1

UNSYMMETRICAL BENDING

Summary

The second moments of area of a section are given by

$$I_{xx} = \int y^2 \, dA \quad \text{and} \quad I_{yy} = \int x^2 \, dA$$

The product second moment of area of a section is defined as

$$I_{xy} = \int xy \, dA$$

which reduces to $I_{xy} = Ahk$ for a rectangle of area A and centroid distance h and k from the X and Y axes.

The *principal second moments of area* are the maximum and minimum values for a section and they occur about the principal axes. *Product second moments of area about principal axes are zero.*

With a knowledge of I_{xx}, I_{yy} and I_{xy} for a given section, the principal values may be determined using either Mohr's or Land's circle construction.

The following relationships apply between the second moments of area about different axes:

$$I_u = \tfrac{1}{2}(I_{xx} + I_{yy}) + \tfrac{1}{2}(I_{xx} - I_{yy}) \sec 2\theta$$

$$I_v = \tfrac{1}{2}(I_{xx} + I_{yy}) - \tfrac{1}{2}(I_{xx} - I_{yy}) \sec 2\theta$$

where θ is the angle between the U and X axes, and is given by

$$\tan 2\theta = \frac{2I_{xy}}{(I_{yy} - I_{xx})}$$

Then

$$I_u + I_v = I_{xx} + I_{yy}$$

The second moment of area about the neutral axis is given by

$$I_{\text{N.A.}} = \tfrac{1}{2}(I_u + I_v) + \tfrac{1}{2}(I_u - I_v) \cos 2\alpha_u$$

where α_u is the angle between the neutral axis (N.A.) and the U axis.

Also
$$I_{xx} = I_u \cos^2 \theta + I_v \sin^2 \theta$$

$$I_{yy} = I_v \cos^2 \theta + I_u \sin^2 \theta$$

$$I_{xy} = \tfrac{1}{2}(I_v - I_u) \sin 2\theta$$

$$I_{xx} - I_{yy} = (I_u - I_v) \cos 2\theta$$

1

Stress determination

For skew loading and other forms of bending about principal axes

$$\sigma = \frac{M_u v}{I_u} + \frac{M_v u}{I_v}$$

where M_u and M_v are the components of the applied moment about the U and V axes.
Alternatively, with $\sigma = Px + Qy$

$$M_{xx} = PI_{xy} + QI_{xx}$$
$$M_{yy} = -PI_{yy} - QI_{xy}$$

Then the inclination of the N.A. to the X axis is given by

$$\tan \alpha = -\frac{P}{Q}$$

As a further alternative,
$$\sigma = \frac{M'n}{I_{\text{N.A.}}}$$

where M' is the component of the applied moment about the N.A., $I_{N.A.}$ is determined either from the momental ellipse or from the Mohr or Land constructions, and n is the perpendicular distance from the point in question to the N.A.

Deflections of unsymmetrical members are found by applying standard deflection formulae to bending about either the principal axes or the N.A. taking care to use the correct component of load and the correct second moment of area value.

Introduction

It has been shown in Chapter 4 of *Mechanics of Materials 1*[†] that the simple bending theory applies when bending takes place about an axis which is perpendicular to a plane of symmetry. If such an axis is drawn through the centroid of a section, and another mutually perpendicular to it also through the centroid, then these axes are principal axes. Thus a plane of symmetry is automatically a principal axis. Second moments of area of a cross-section about its principal axes are found to be maximum and minimum values, while the product second moment of area, $\int xy\,dA$, is found to be zero. All plane sections, whether they have an axis of symmetry or not, have two perpendicular axes about which the product second moment of area is zero. *Principal axes are thus defined as the axes about which the product second moment of area is zero.* Simple bending can then be taken as bending which takes place about a principal axis, moments being applied in a plane parallel to one such axis.

In general, however, moments are applied about a convenient axis in the cross-section; the plane containing the applied moment may not then be parallel to a principal axis. Such cases are termed "unsymmetrical" or "asymmetrical" bending.

The most simple type of unsymmetrical bending problem is that of "skew" loading of sections containing at least one axis of symmetry, as in Fig. 1.1. This axis and the axis

[†] E.J. Hearn, *Mechanics of Materials 1*, Butterworth-Heinemann, 1997.

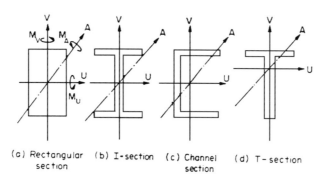

(a) Rectangular section (b) I-section (c) Channel section (d) T-section

Fig. 1.1. Skew loading of sections containing one axis of symmetry.

perpendicular to it are then principal axes and the term skew loading implies load applied at some angle to these principal axes. The method of solution in this case is to resolve the applied moment M_A about some axis A into its components about the principal axes. Bending is then assumed to take place simultaneously about the two principal axes, the total stress being given by

$$\sigma = \frac{M_u v}{I_u} + \frac{M_v u}{I_v}$$

With at least one of the principal axes being an axis of symmetry the second moments of area about the principal axes I_u and I_v can easily be determined.

With unsymmetrical sections (e.g. angle-sections, Z-sections, etc.) the principal axes are not easily recognized and the second moments of area about the principal axes are not easily found except by the use of special techniques to be introduced in §§1.3 and 1.4. In such cases an easier solution is obtained as will be shown in §1.8. Before proceeding with the various methods of solution of unsymmetrical bending problems, however, it is advisable to consider in some detail the concept of principal and product second moments of area.

1.1. Product second moment of area

Consider a small element of area in a plane surface with a centroid having coordinates (x, y) relative to the X and Y axes (Fig. 1.2). The second moments of area of the surface about the X and Y axes are defined as

$$I_{xx} = \int y^2 \, dA \quad \text{and} \quad I_{yy} = \int x^2 \, dA \qquad (1.1)$$

Similarly, the product second moment of area of the section is defined as follows:

$$I_{xy} = \int xy \, dA \qquad (1.2)$$

Since the cross-section of most structural members used in bending applications consists of a combination of rectangles the value of the product second moment of area for such sections is determined by the addition of the I_{xy} value for each rectangle (Fig. 1.3),

i.e. $$I_{xy} = Ahk \qquad (1.3)$$

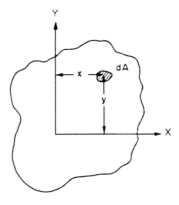

Fig. 1.2.

where h and k are the distances of the centroid of each rectangle from the X and Y axes respectively (taking account of the normal sign convention for x and y) and A is the area of the rectangle.

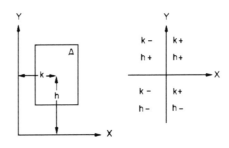

Fig. 1.3.

1.2. Principal second moments of area

The principal axes of a section have been defined in the introduction to this chapter. Second moments of area about these axes are then termed principal values and these may be related to the standard values about the conventional X and Y axes as follows.

Consider Fig. 1.4 in which GX and GY are any two mutually perpendicular axes inclined at θ to the principal axes GV and GU. A small element of area A will then have coordinates (u, v) to the principal axes and (x, y) referred to the axes GX and GY. The area will thus have a product second moment of area about the principal axes given by $uv\,dA$.

$\therefore$ total product second moment of area of a cross-section

$$I_{uv} = \int uv\,dA$$

$$= \int (x\cos\theta + y\sin\theta)(y\cos\theta - x\sin\theta)\,dA$$

$$= \int (xy \cos^2 \theta + y^2 \sin \theta \cos \theta - x^2 \cos \theta \sin \theta - xy \sin^2 \theta) \, dA$$

$$= (\cos^2 \theta - \sin^2 \theta) \int xy \, dA + \sin \theta \cos \theta \left[\int y^2 \, dA - \int x^2 \, dA \right]$$

$$= I_{xy} \cos 2\theta + \tfrac{1}{2}(I_{xx} - I_{yy}) \sin 2\theta$$

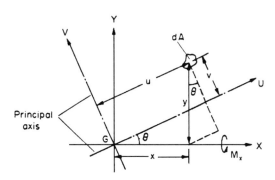

Fig. 1.4.

Now for principal axes the product second moment of area is zero.

$$\therefore \qquad 0 = I_{xy} \cos 2\theta + \tfrac{1}{2}(I_{xx} - I_{yy}) \sin 2\theta$$

$$\mathbf{\tan 2\theta} = \frac{-2I_{xy}}{(I_{xx} - I_{yy})} = \frac{2I_{xy}}{(I_{yy} - I_{xx})} \tag{1.4}$$

This equation, therefore, gives the direction of the principal axes.
To determine the second moments of area about these axes,

$$I_u = \int v^2 \, dA = \int (y \cos \theta - x \sin \theta)^2 \, dA$$

$$= \cos^2 \theta \int y^2 \, dA + \sin^2 \theta \int x^2 \, dA - 2 \cos \theta \sin \theta \int xy \, dA$$

$$= I_{xx} \cos^2 \theta + I_{yy} \sin^2 \theta - I_{xy} \sin 2\theta \tag{1.5}$$

Substituting for I_{xy} from eqn. (1.4),

$$I_u = \tfrac{1}{2}(1 + \cos 2\theta)I_{xx} + \tfrac{1}{2}(1 - \cos 2\theta)I_{yy} - \tfrac{1}{2}\frac{\sin^2 2\theta}{\cos 2\theta}(I_{yy} - I_{xx})$$

$$= \tfrac{1}{2}(1 + \cos 2\theta)I_{xx} + \tfrac{1}{2}(1 - \cos 2\theta)I_{yy} - \tfrac{1}{2}\left[\frac{(1 - \cos^2 2\theta)}{\cos 2\theta}(I_{yy} - I_{xx}) \right]$$

$$= \tfrac{1}{2}(1 + \cos 2\theta)I_{xx} + \tfrac{1}{2}(1 - \cos 2\theta)I_{yy} - \tfrac{1}{2}\sec 2\theta(I_{yy} - I_{xx}) + \tfrac{1}{2}\cos 2\theta(I_{yy} - I_{xx})$$

$$= \tfrac{1}{2}(I_{xx} + I_{yy}) + (I_{xx} - I_{yy})\cos 2\theta - (I_{yy} - I_{xx})\sec 2\theta + (I_{yy} - I_{xx})\cos 2\theta$$

i.e.

$$I_u = \tfrac{1}{2}(I_{xx} + I_{yy}) + \tfrac{1}{2}(I_{xx} - I_{yy})\sec 2\theta \qquad (1.6)$$

Similarly,

$$I_v = \int u^2\, dA = \int (x\cos\theta + y\sin\theta)^2\, dA$$

$$= \tfrac{1}{2}(I_{xx} + I_{yy}) - \tfrac{1}{2}(I_{xx} - I_{yy})\sec 2\theta \qquad (1.7)$$

N.B.–Adding the above expressions,

$$I_u + I_v = I_{xx} + I_{yy}$$

Also from eqn. (1.5),

$$I_u = I_{xx}\cos^2\theta + I_{yy}\sin^2\theta - I_{xy}\sin 2\theta$$

$$= \tfrac{1}{2}(1 + \cos 2\theta)I_{xx} + \tfrac{1}{2}(1 - \cos 2\theta)I_{yy} - I_{xy}\sin 2\theta$$

$$I_u = \tfrac{1}{2}(I_{xx} + I_{yy}) + \tfrac{1}{2}(I_{xx} - I_{yy})\cos 2\theta - I_{xy}\sin 2\theta \qquad (1.8)$$

Similarly,

$$I_v = \tfrac{1}{2}(I_{xx} + I_{yy}) - \tfrac{1}{2}(I_{xx} - I_{yy})\cos 2\theta + I_{xy}\sin 2\theta \qquad (1.9)$$

These equations are then identical in form with the complex-stress eqns. (13.8) and (13.9)[†] with I_{xx}, I_{yy}, and I_{xy} replacing σ_x, σ_y and τ_{xy} and Mohr's circle can be drawn to represent I values in exactly the same way as Mohr's stress circle represents stress values.

1.3. Mohr's circle of second moments of area

The construction is as follows (Fig. 1.5):

(1) Set up axes for second moments of area (horizontal) and product second moments of area (vertical).

(2) Plot the points A and B represented by (I_{xx}, I_{xy}) and $(I_{yy}, -I_{xy})$.

(3) Join AB and construct a circle with this as diameter. *This is then the Mohr's circle.*

(4) Since the principal moments of area are those about the axes with a zero product second moment of area they are given by the points where the circle cuts the horizontal axis.

Thus OC and OD are the principal second moments of area I_v and I_u.

The point A represents values on the X axis and B those for the Y axis. Thus, in order to determine the second moment of area about some other axis, e.g. the N.A., at some angle α counterclockwise to the X axis, construct a line from G at an angle 2α counterclockwise to GA on the Mohr construction to cut the circle in point N. The horizontal coordinate of N then gives the value of $I_{\text{N.A.}}$

[†] E.J. Hearn, *Mechanics of Materials 1*, Butterworth-Heinemann, 1997.

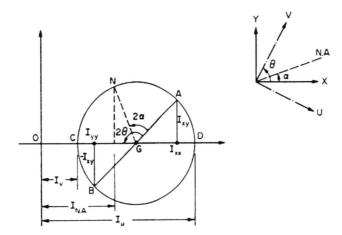

Fig. 1.5. Mohr's circle of second moments of area.

The procedure is therefore identical to that for determining the direct stress on some plane inclined at α to the plane on which σ_x acts in Mohr's stress circle construction, i.e. **angles are DOUBLED** on Mohr's circle.

1.4. Land's circle of second moments of area

An alternative graphical solution to the Mohr procedure has been developed by Land as follows (Fig. 1.6):

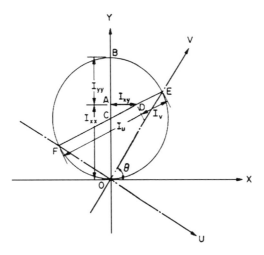

Fig. 1.6. Land's circle of second moments of area.

(1) From O as origin of the given XY axes mark off lengths $OA = I_{xx}$ and $AB = I_{yy}$ on the vertical axis.

(2) Draw a circle with *OB* as diameter and centre *C*. *This is then Land's circle of second moment of area.*

(3) From point *A* mark off $AD = I_{xy}$ parallel with the *X* axis.

(4) Join the centre of the circle *C* to *D*, and produce, to cut the circle in *E* and *F*. Then $ED = I_v$ and $DF = I_u$ are the principal moments of area about the principal axes *OV* and *OU* the positions of which are found by joining *OE* and *OF*. The principal axes are thus inclined at an angle θ to the *OX* and *OY* axes.

1.5. Rotation of axes: determination of moments of area in terms of the principal values

Figure 1.7 shows any plane section having coordinate axes *XX* and *YY* and principal axes *UU* and *VV*, each passing through the centroid *O*. Any element of area *dA* will then have coordinates (x, y) and (u, v), respectively, for the two sets of axes.

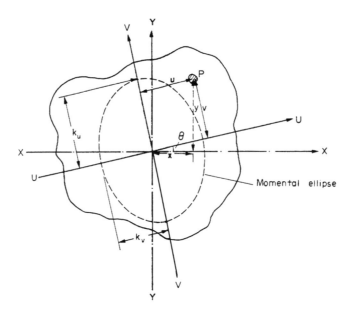

Fig. 1.7. The momental ellipse.

Now

$$I_{xx} = \int y^2 \, dA = \int (v \cos \theta + u \sin \theta)^2 \, dA$$

$$= \int v^2 \cos^2 \theta \, dA + \int 2uv \sin \theta \cos \theta \, dA + \int u^2 \sin^2 \theta \, dA$$

But *UU* and *VV* are the principal axes so that $I_{uv} = \int uv \, dA$ is zero.

$$\therefore \qquad\qquad I_{xx} = I_u \cos^2 \theta + I_v \sin^2 \theta \qquad\qquad (1.10)$$

Similarly,

$$I_{yy} = \int x^2 \, dA = \int (u\cos\theta - v\sin\theta)^2 \, dA$$

$$= \int u^2 \cos^2\theta \, dA - \int 2uv\sin\theta\cos\theta \, dA + \int v^2 \sin^2\theta \, dA$$

and with $\int uv \, dA = 0$

$$I_{yy} = I_v \cos^2\theta + I_u \sin^2\theta \tag{1.11}$$

Also

$$I_{xy} = \int xy \, dA = \int (u\cos\theta - v\sin\theta)(v\cos\theta + u\sin\theta) \, dA$$

$$= \int [uv(\cos^2\theta - \sin^2\theta) + (u^2 - v^2)\sin\theta\cos\theta] \, dA$$

$$= I_{uv} \cos 2\theta + \tfrac{1}{2}(I_v - I_u)\sin 2\theta \quad \text{and} \quad I_{uv} = 0$$

$$\therefore \qquad I_{xy} = \tfrac{1}{2}(I_v - I_u)\sin 2\theta \tag{1.12}$$

From eqns. (1.10) and (1.11)

$$I_{xx} - I_{yy} = I_u \cos^2\theta + I_v \sin^2\theta - I_v \cos^2\theta - I_u \sin^2\theta$$

$$= (I_u - I_v)\cos^2\theta - (I_u - I_v)\sin^2\theta$$

$$I_{xx} - I_{yy} = (I_u - I_v)\cos 2\theta \tag{1.13}$$

Combining eqns. (1.12) and (1.13) gives

$$\tan 2\theta = \frac{2I_{xy}}{I_{yy} - I_{xx}} \tag{1.14}$$

and combining eqns. (1.10) and (1.11) gives

$$I_{xx} + I_{yy} = I_u + I_v \tag{1.15}$$

Substitution into eqns. (1.10) and (1.11) then yields

$$I_u = \tfrac{1}{2}[(I_{xx} + I_{yy}) + (I_{xx} - I_{yy})\sec 2\theta] \qquad \text{(1.16) as (1.6)}$$

$$I_v = \tfrac{1}{2}[(I_{xx} + I_{yy}) - (I_{xx} - I_{yy})\sec 2\theta] \qquad \text{(1.17) as (1.7)}$$

1.6. The ellipse of second moments of area

The above relationships can be used as the basis for construction of the moment of area ellipse proceeding as follows:

(1) Plot the values of I_u and I_v on two mutually perpendicular axes and draw concentric circles with centres at the origin, and radii equal to I_u and I_v (Fig. 1.8).

(2) Plot the point with coordinates $x = I_u \cos\theta$ and $y = I_v \sin\theta$, the value of θ being given by eqn. (1.14).

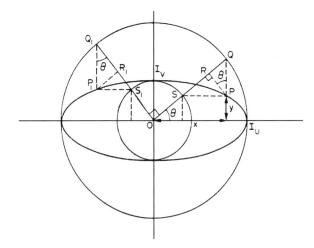

Fig. 1.8. The ellipse of second moments of area.

It then follows that

$$\frac{x^2}{(I_u)^2} + \frac{y^2}{(I_v)^2} = 1$$

This equation is the locus of the point P and represents the equation of an ellipse – the *ellipse of second moments of area*.

(3) Draw OQ at an angle θ to the I_u axis, cutting the circle through I_v in point S and join SP which is then parallel to the I_u axis. Construct a perpendicular to OQ through P to meet OQ in R.

Then

$$OR = OQ - RQ$$

$$= I_u - (I_u \sin\theta - I_v \sin\theta)\sin\theta$$

$$= I_u - (I_u - I_v)\sin^2\theta$$

$$= I_u \cos^2\theta + I_v \sin^2\theta$$

$$= I_{xx}$$

Similarly, repeating the process with OQ_1 perpendicular to OQ gives the result

$$OR_1 = I_{yy}$$

Further,

$$PR = PQ \cos\theta$$

$$= (I_u \sin\theta - I_v \sin\theta)\cos\theta$$

$$= \tfrac{1}{2}(I_u - I_v)\sin 2\theta = I_{xy}$$

Thus the construction shown in Fig. 1.8 can be used to determine the second moments of area and the product second moment of area about any set of perpendicular axis at a known orientation to the principal axes.

1.7. Momental ellipse

Consider again the general plane surface of Fig. 1.7 having radii of gyration k_u and k_v about the U and V axes respectively. An ellipse can be constructed on the principal axes with semi-major and semi-minor axes k_u and k_v, respectively, as shown.

Thus the perpendicular distance between the axis UU and a tangent to the ellipse which is parallel to UU is equal to the radius of gyration of the surface about UU. Similarly, the radius of gyration k_v is the perpendicular distance between the tangent to the ellipse which is parallel to the VV axis and the axis itself. Thus if the radius of gyration of the surface is required about any other axis, e.g. the N.A., then it is given by the distance between the N.A. and the tangent AA which is parallel to the N.A. (see Fig. 1.11). Thus

$$k_{\text{N.A.}} = h$$

The ellipse is then termed the *momental ellipse* and is extremely useful in the solution of unsymmetrical bending problems as described in §1.10.

1.8. Stress determination

Having determined both the values of the principal second moments of area I_u and I_v and the inclination of the principal axes U and V from the equations listed below,

$$I_u = \tfrac{1}{2}(I_{xx} + I_{yy}) + \tfrac{1}{2}(I_{xx} - I_{yy})\sec 2\theta \tag{1.16}$$

$$I_v = \tfrac{1}{2}(I_{xx} + I_{yy}) - \tfrac{1}{2}(I_{xx} - I_{yy})\sec 2\theta \tag{1.17}$$

and

$$\tan 2\theta = \frac{2I_{xy}}{(I_{yy} - I_{xx})} \tag{1.14}$$

the stress at any point is found by application of the simple bending theory simultaneously about the principal axes,

i.e.
$$\sigma = \frac{M_v u}{I_v} + \frac{M_u v}{I_u} \tag{1.18}$$

where M_v and M_u are the moments of the applied loads about the V and U axes, e.g. if loads are applied to produce a bending moment M_x about the X axis (see Fig. 1.14), then

$$M_v = M_x \sin\theta$$

$$M_u = M_x \cos\theta$$

the maximum value of M_x, and hence M_u and M_v, for cantilevers such as that shown in Fig. 1.10, being found at the root of the cantilever. The maximum stress due to bending will then occur at this position.

1.9. Alternative procedure for stress determination

Consider any unsymmetrical section, represented by Fig. 1.9. The assumption is made initially that the stress at any point on the unsymmetrical section is given by

$$\sigma = Px + Qy \tag{1.19}$$

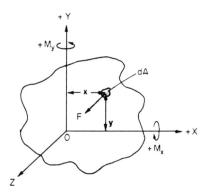

Fig. 1.9. Alternative procedure for stress determination.

where P and Q are constants; in other words it is assumed that bending takes place about the X and Y axes at the same time, stresses resulting from each effect being proportional to the distance from the respective axis of bending.

Now let there be a tensile stress σ on the element of area dA. Then

$$\text{force } F \text{ on the element} = \sigma \, dA$$

the direction of the force being parallel to the Z axis. The moment of this force about the X axis is then $\sigma \, dA \, y$.

$$\therefore \qquad \text{total moment} = M_x = \int \sigma \, dA \, y$$

$$= \int (Px + Qy)y \, dA = \int Pxy \, dA + \int Qy^2 \, dA$$

Now, by definition,

$$I_{xx} = \int y^2 \, dA, \quad I_{yy} = \int x^2 \, dA \quad \text{and} \quad I_{xy} = \int xy \, dA$$

the latter being termed the product second moment of area (see §1.1):

$$\therefore \qquad\qquad\qquad M_x = PI_{xy} + QI_{xx} \qquad\qquad\qquad (1.20)$$

Similarly, considering moments about the Y axis,

$$\therefore \qquad\qquad M_y = -\int \sigma \, dA \, x = -\int (Px + Qy)x \, dA$$

$$\therefore \qquad\qquad\qquad M_y = -PI_{yy} - QI_{xy} \qquad\qquad\qquad (1.21)$$

The sign convention used above for bending moments is the *corkscrew rule*. A positive moment is the direction in which a corkscrew or screwdriver has to be turned in order to produce motion of a screw in the direction of positive X or Y, as shown in Fig. 1.9. Thus with a knowledge of the applied moments and the second moments of area about any two perpendicular axes, P and Q can be found from eqns. (1.20) and (1.21) and hence the stress at any point (x, y) from eqn. (1.19).

Since stresses resulting from bending are zero on the N.A. the equation of the N.A. is

$$Px + Qy = 0$$

$$\frac{y}{x} = -\frac{P}{Q} = \tan \alpha_{\text{N.A.}} \tag{1.22}$$

where $\alpha_{\text{N.A.}}$ is the inclination of the N.A. to the X axis.

If the unsymmetrical member is drawn to scale and the N.A. is inserted through the centroid of the section at the above angle, the points of maximum stress can be determined quickly by inspection as the points most distant from the N.A., e.g. for the angle section of Fig. 1.10, subjected to the load shown, the maximum tensile stress occurs at R while the maximum compressive stress will arise at either S or T depending on the value of α.

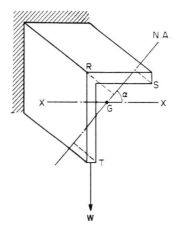

Fig. 1.10.

1.10. Alternative procedure using the momental ellipse

Consider the unsymmetrical section shown in Fig. 1.11 with principal axes UU and VV. Any moment applied to the section can be resolved into its components about the principal axes and the stress at any point found by application of eqn. (1.18).

For example, if vertical loads only are applied to the section to produce moments about the OX axis, then the components will be $M \cos \theta$ about UU and $M \sin \theta$ about VV. Then

$$\text{stress at P} = \frac{M \cos \theta}{I_u} v - \frac{M \sin \theta}{I_v} u \tag{1.23}$$

the value of θ having been obtained from eqn. (1.14).

Alternatively, however, the problem may be solved by realising that the N.A. and the plane of the external bending moment are conjugate diameters of an ellipse[†] – the *momental*

[†] *Conjugate diameters of an ellipse:* two diameters of an ellipse are conjugate when each bisects all chords parallel to the other diameter.

Two diameters $y = m_1 x$ and $y = m_2 x$ are conjugate diameters of the ellipse $\dfrac{x^2}{a^2} + \dfrac{y^2}{b^2} = 1$ if $m_1 m_2 = -\dfrac{b^2}{a^2}$.

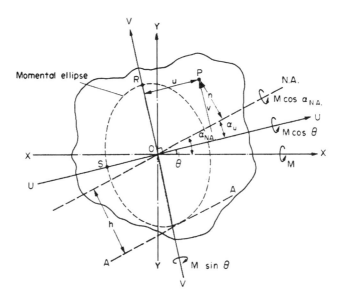

Fig. 1.11. Determination of stresses using the momental ellipse.

ellipse. The actual plane of resultant bending will then be perpendicular to the N.A., the inclination of which, relative to the U axis (α_u), is obtained by equating the above formula for stress at P to zero,

i.e.

$$\frac{M\cos\theta}{I_u}v = \frac{M\sin\theta}{I_v}u$$

so that

$$\tan\alpha_u = \frac{v}{u} = \frac{I_u}{I_v}\tan\theta$$

$$= \frac{k_u^2}{k_v^2}\tan\theta \qquad (1.24)$$

where k_u and k_v are the radii of gyration about the principal axes and hence the semi-axes of the momental ellipse.

The N.A. can now be added to the diagram to scale. The second moment of area of the section about the N.A. is then given by Ah^2, where h is the perpendicular distance between the N.A. and a tangent AA to the ellipse drawn parallel to the N.A. (see Fig. 1.11 and §1.7).

The bending moment about the N.A. is $M\cos\alpha_{N.A.}$ where $\alpha_{N.A.}$ is the angle between the N.A. and the axis XX about which the moment is applied.

The stress at P is now given by the simple bending formula

$$\sigma = \frac{M\cos\alpha_{N.A.}}{I_{N.A.}}n \qquad (1.25)$$

the distance n being measured perpendicularly from the N.A. to the point P in question.

As for the procedure introduced in §1.7, this method has the advantage of immediate indication of the points of maximum stress once the N.A. has been drawn. The solution does, however, involve the use of principal moments of area which must be obtained by calculation or graphically using Mohr's or Land's circle.

1.11. Deflections

The deflections of unsymmetrical members in the directions of the principal axes may always be determined by application of the standard deflection formulae of §5.7.[†]

For example, the deflection at the free end of a cantilever carrying an end-point-load is

$$\frac{WL^3}{3EI}$$

With the appropriate value of I and the correct component of the load perpendicular to the principal axis used, the required deflection is obtained.

Thus
$$\delta_v = \frac{W_u L^3}{3EI_u} \quad \text{and} \quad \delta_u = \frac{W_v L^3}{3EI_v} \qquad (1.26)$$

where W_u and W_v are the components of the load *perpendicular* to the U and V principal axes respectively.

The total resultant deflection is then given by combining the above values vectorially as shown in Fig. 1.12,

i.e.
$$\delta = \sqrt{(\delta_u^2 + \delta_v^2)}$$

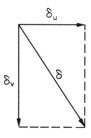

Fig. 1.12.

Alternatively, since bending always occurs about the N.A., the deflection equation can be written in the form

$$\delta = \frac{W'L^3}{3EI_{\text{N.A.}}} \qquad (1.27)$$

where $I_{\text{N.A.}}$ is the second moment of area about the N.A. and W' is the component of the load perpendicular to the N.A. The value of $I_{\text{N.A.}}$ may be found either graphically using Mohr's circle or the momental ellipse, or by calculation using

$$I_{\text{N.A.}} = \tfrac{1}{2}[(I_u + I_v) + (I_u - I_v)\cos 2\alpha_u] \qquad (1.28)$$

where α_u is the angle between the N.A. and the principal U axis.

[†] E.J. Hearn, *Mechanics of Materials 1*, Butterworth-Heinemann, 1997.

Examples

Example 1.1

A rectangular-section beam 80 mm × 50 mm is arranged as a cantilever 1.3 m long and loaded at its free end with a load of 5 kN inclined at an angle of 30° to the vertical as shown in Fig. 1.13. Determine the position and magnitude of the greatest tensile stress in the section. What will be the vertical deflection at the end? $E = 210$ GN/m².

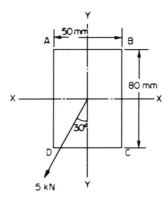

Fig. 1.13.

Solution

In the case of symmetrical sections such as this, subjected to skew loading, a solution is obtained by resolving the load into its components parallel to the two major axes and applying the bending theory simultaneously to both axes, i.e.

$$\sigma = \frac{M_{xx}y}{I_{xx}} \pm \frac{M_{yy}x}{I_{yy}}$$

Now the most highly stressed areas of the cantilever will be those at the built-in end where

$$M_{xx} = 5000\cos 30° \times 1.3 = 5629 \text{ Nm}$$

$$M_{yy} = 5000\sin 30° \times 1.3 = 3250 \text{ Nm}$$

The stresses on the short edges *AB* and *DC* resulting from bending about *XX* are then

$$\frac{M_{xx}}{I_{xx}}y = \frac{5629 \times 40 \times 10^{-3} \times 12}{50 \times 80^3 \times 10^{-12}} = 105.5 \text{ MN/m}^2$$

tensile on *AB* and compressive on *DC*.

The stresses on the long edges *AD* and *BC* resulting from bending about *YY* are

$$\frac{M_{yy}}{I_{yy}}x = \frac{3250 \times 25 \times 10^{-3} \times 12}{80 \times 50^3 \times 10^{-12}} = 97.5 \text{ MN/m}^2$$

tensile on *BC* and compressive on *AD*.

The maximum tensile stress will therefore occur at point *B* where the two tensile stresses add, i.e.

$$\text{maximum tensile stress} = 105.5 + 97.5 = \textbf{203 MN/m}^2$$

The deflection at the free end of the cantilever is then given by

$$\delta = \frac{WL^3}{3EI}$$

Therefore deflection vertically (i.e. along the YY axis) is

$$\delta_v = \frac{(W\cos 30°)L^3}{3EI_{xx}} = \frac{5000 \times 0.866 \times 1.3^3 \times 12}{3 \times 210 \times 10^9 \times 50 \times 80^3 \times 10^{-12}}$$

$$= 0.0071 = \textbf{7.1 mm}$$

Example 1.2

A cantilever of length 1.2 m and of the cross section shown in Fig. 1.14 carries a vertical load of 10 kN at its outer end, the line of action being parallel with the longer leg and arranged to pass through the shear centre of the section (i.e. there is no twisting of the section, see §7.5[†]). Working from first principles, find the stress set up in the section at points A, B and C, given that the centroid is located as shown. Determine also the angle of inclination of the N.A.

$$I_{xx} = 4 \times 10^{-6} \text{ m}^4, \quad I_{yy} = 1.08 \times 10^{-6} \text{ m}^4$$

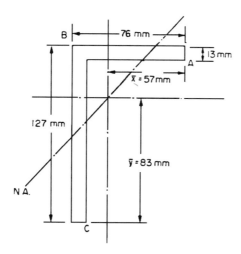

Fig. 1.14.

Solution

The product second moment of area of the section is given by eqn. (1.3).

$$I_{xy} = \Sigma Ahk$$

$$= \{76 \times 13(\tfrac{1}{2} \times 76 - 19)(44 - \tfrac{1}{2} \times 13)$$

$$+ 114 \times 13[-(83 - \tfrac{1}{2} \times 114)][-(19 - \tfrac{1}{2} \times 13)]\}10^{-12}$$

[†] E.J. Hearn, *Mechanics of Materials 1*, Butterworth-Heinemann, 1997.

$$= (0.704 + 0.482)10^{-6} = 1.186 \times 10^{-6} \text{ m}^4$$

From eqn. (1.20) $M_x = PI_{xy} + QI_{xx} = 10\,000 \times 1.2 = 12\,000$

i.e. $1.186P + 4Q = 12\,000 \times 10^6$ (1)

Since the load is vertical there will be no moment about the Y axis and eqn. (1.21) gives

$$M_y = -PI_{yy} - QI_{xy} = 0$$

∴ $-1.08P - 1.186Q = 0$

∴ $\dfrac{P}{Q} = -\dfrac{1.186}{1.08} = -1.098$

But the angle of inclination of the N.A. is given by eqn. (1.22) as

$$\tan \alpha_{\text{N.A.}} = -\frac{P}{Q} = 1.098$$

i.e. $\alpha_{\text{N.A.}} = \mathbf{47°41'}$

Substituting $P = -1.098Q$ in eqn. (1),

$$1.186(-1.098Q) + 4Q = 12\,000 \times 10^6$$

∴ $Q = \dfrac{12\,000 \times 10^6}{2.69} = 4460 \times 10^6$

∴ $P = -4897 \times 10^6$

If the N.A. is drawn as shown in Fig. 1.14 at an angle of 47°41′ to the XX axis through the centroid of the section, then this is the axis about which bending takes place. The points of maximum stress are then obtained by inspection as the points which are the maximum perpendicular distance from the N.A.

Thus B is the point of maximum tensile stress and C the point of maximum compressive stress.

Now from eqn (1.19) the stress at any point is given by

$$\sigma = Px + Qy$$

∴   stress at $A = -4897 \times 10^6(57 \times 10^{-3}) + 4460 \times 10^6(31 \times 10^{-3})$

$$= \mathbf{-141\ MN/m^2\ (compressive)}$$

stress at $B = -4897 \times 10^6(-19 \times 10^{-3}) + 4460 \times 10^6(44 \times 10^{-3})$

$$= \mathbf{289\ MN/m^2\ (tensile)}$$

stress at $C = -4897 \times 10^6(-6 \times 10^{-3}) + 4460 \times 10^6(-83 \times 10^{-3})$

$$= \mathbf{-341\ MN/m\ (compressive)}$$

Example 1.3

(a) A horizontal cantilever 2 m long is constructed from the Z-section shown in Fig. 1.15. A load of 10 kN is applied to the end of the cantilever at an angle of 60° to the horizontal as shown. Assuming that no twisting moment is applied to the section, determine the stresses at points A and B. ($I_{xx} \times 48.3 \times 10^{-6}$ m^4, $I_{yy} = 4.4 \times 10^{-6}$ m^4.)

(b) Determine the principal second moments of area of the section and hence, by applying the simple bending theory about each principal axis, check the answers obtained in part (a).

(c) What will be the deflection of the end of the cantilever? $E = 200$ GN/m^2.

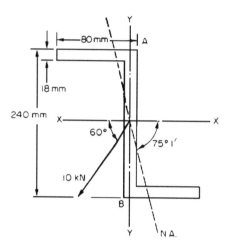

Fig. 1.15.

Solution

(a) For this section I_{xy} for the web is zero since its centroid lies on both axes and hence h and k are both zero. The contributions to I_{xy} of the other two portions will be negative since in both cases either h or k is negative.

$\therefore$
$$I_{xy} = -2(80 \times 18)(40 - 9)(120 - 9)10^{-12}$$
$$= -9.91 \times 10^{-6} \text{ m}^4$$

Now, at the built-in end,

$$M_x = +10\,000 \sin 60° \times 2 = +17\,320 \text{ Nm}$$
$$M_y = -10\,000 \cos 60° \times 2 = -10\,000 \text{ Nm}$$

Substituting in eqns. (1.20) and (1.21),

$$17\,320 = PI_{xy} + QI_{xx} = (-9.91P + 48.3Q)10^{-6}$$
$$-10\,000 = -PI_{yy} - QI_{xy} = (-4.4P + 9.91Q)10^{-6}$$

$\therefore$
$$1.732 \times 10^{10} = -9.91P + 48.3Q \qquad (1)$$
$$-1 \times 10^{10} = -4.4P + 9.91Q \qquad (2)$$

(1) $\times \dfrac{4.4}{9.91}$,

$$0.769 \times 10^{10} = -4.4P + 21.45Q \qquad (3)$$

(3) − (2),

$$1.769 \times 10^{10} = 11.54Q$$

∴ $$Q = 1533 \times 10^6$$

and substituting in (2) gives

$$P = 5725 \times 10^6$$

The inclination of the N.A. relative to the X axis is then given by

$$\tan \alpha_{\text{N.A.}} = -\frac{P}{Q} = -\frac{5725}{1533} = -3.735$$

$$\alpha_{\text{N.A.}} = \mathbf{-75°1'}$$

This has been added to Fig. 1.15 and indicates that the points A and B are on either side of the N.A. and equidistant from it. Stresses at A and B are therefore of equal magnitude but opposite sign.

Now

$$\sigma = Px + Qy$$

∴   stress at A $= 5725 \times 10^6 \times 9 \times 10^{-3} + 1533 \times 10^6 \times 120 \times 10^{-3}$

$$= \mathbf{235 \ \ MN/m^2 \ (tensile)}$$

Similarly,

stress at B $= \mathbf{235 \ MN/m^2 \ (compressive)}$

(b) The principal second moments of area may be found from Mohr's circle as shown in Fig. 1.16 or from eqns. (1.6) and (1.7),

i.e. $$I_u, I_v = \tfrac{1}{2}(I_{xx} + I_{yy}) \pm \tfrac{1}{2}(I_{xx} - I_{yy}) \sec 2\theta$$

with $$\tan 2\theta = \frac{2I_{xy}}{I_{yy} - I_{xx}} = \frac{-2 \times 9.91 \times 10^{-6}}{(4.4 - 48.3)10^{-6}}$$

$$= 0.451$$

∴ $$2\theta = 24°18', \ \ \theta = 12°9'$$

∴ $$I_u, I_v = \tfrac{1}{2}[(48.3 + 4.4) \pm (48.3 - 4.4)1.0972]10^{-6}$$

$$= \tfrac{1}{2}[52.7 \pm 48.17]10^{-6}$$

∴ $$I_u = 50.43 \times 10^{-6} \ \text{m}^4$$

$$I_v = 2.27 \times 10^{-6} \ \text{m}^4$$

The required stresses can now be obtained from eqn. (1.18).

$$\sigma = \frac{M_v u}{I_v} + \frac{M_u v}{I_u}$$

Now $$M_u = 10\,000 \sin(60° - 12°9') \times 2$$

$$= 10\,000 \sin 47°51' \times 2 = 14\,828 \ \text{Nm}$$

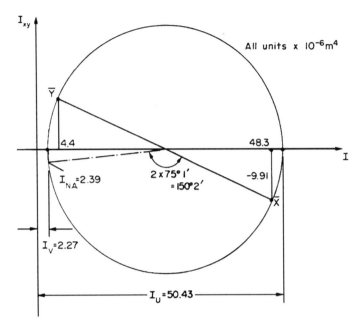

Fig. 1.16.

and
$$M_v = 10\,000 \cos 47°51' \times 2 = 13\,422 \text{ Nm}$$

and, for A,

$$u = x \cos\theta + y \sin\theta = (9 \times 0.9776) + (120 \times 0.2105)$$
$$= 34.05 \text{ mm}$$

$$v = y \cos\theta - x \sin\theta = (120 \times 0.9776) - (9 \times 0.2105)$$
$$= 115.4 \text{ mm}$$

$$\therefore \quad \sigma = \frac{14\,828 \times 115.4 \times 10^{-3}}{50.43 \times 10^{-6}} + \frac{13\,422 \times 34.05 \times 10^{-3}}{2.27 \times 10^{-6}}$$

$$= \mathbf{235 \ MN/m^2} \text{ as before.}$$

(c) The deflection at the free end of a cantilever is given by

$$\delta = \frac{WL^3}{3EI}$$

Therefore component of deflection perpendicular to the V axis

$$\delta_v = \frac{W_v L^3}{3EI_v} = \frac{10\,000 \cos 47°51' \times 2^3}{3 \times 200 \times 10^9 \times 2.27 \times 10^{-6}}$$

$$= 39.4 \times 10^{-3} = 39.4 \text{ mm}$$

and component of deflection perpendicular to the U axis

$$\delta_u = \frac{W_u L^3}{3EI_u} = \frac{10\,000 \sin 47°51' \times 2^3}{3 \times 200 \times 10^9 \times 50.43 \times 10^{-6}}$$

$$= 1.96 \times 10^{-3} = 1.96 \text{ mm}$$

The total deflection is then given by

$$= \sqrt{(\delta_u^2 + \delta_v^2)} = 10^{-3} \sqrt{(39.4^2 + 1.96^2)} = 39.45 \times 10^{-3}$$

$$= \mathbf{39.45 \text{ mm}}$$

Alternatively, since bending actually occurs about the N.A., the deflection can be found from

$$\delta = \frac{W_{\text{N.A.}} L^3}{3EI_{\text{N.A.}}}$$

its direction being normal to the N.A.

From Mohr's circle of Fig. 1.16, $I_{\text{N.A.}} = 2.39 \times 10^{-6}$ m^4

$$\therefore \qquad \delta = \frac{10\,000 \sin(30° + 14°59') \times 2^3}{3 \times 200 \times 10^9 \times 2.39 \times 10^{-6}} = 39.44 \times 10^{-3}$$

$$= \mathbf{39.44 \text{ mm}}$$

Example 1.4

Check the answer obtained for the stress at point B on the angle section of Example 1.2 using the momental ellipse procedure.

Solution

The semi-axes of the momental ellipse are given by

$$k_u = \sqrt{\frac{I_u}{A}} \quad \text{and} \quad k_v = \sqrt{\frac{I_v}{A}}$$

The ellipse can then be constructed by setting off the above dimensions on the principal axes as shown in Fig. 1.17 (The inclination of the N.A. can be determined as in Example 1.2 or from eqn. (1.24).) The second moment of area of the section about the N.A. is then obtained from the momental ellipse as

$$I_{\text{N.A.}} = Ah^2$$

Thus for the angle section of Fig. 1.14

$$I_{xy} = 1.186 \times 10^{-6} \text{ m}^4, \quad I_{xx} = 4 \times 10^{-6} \text{ m}^4, \quad I_{yy} = 1.08 \times 10^{-6} \text{ m}^4$$

The principal second moments of area are then given by Mohr's circle of Fig. 1.18 or from the equation

$$I_u, I_v = \tfrac{1}{2}[(I_{xx} + I_{yy}) \pm (I_{xx} - I_{yy}) \sec 2\theta]$$

where

$$\tan 2\theta = \frac{2I_{xy}}{(I_{yy} - I_{xx})} = \frac{2 \times 1.186 \times 10^{-6}}{(1.08 - 4)10^{-6}} = -0.8123$$

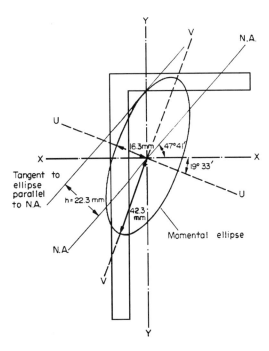

Fig. 1.17.

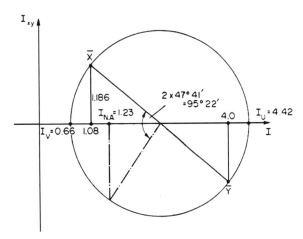

Fig. 1.18.

$$\therefore \qquad 2\theta = -39°5', \ \theta = -19°33'$$

and $$\sec 2\theta = -1.2883$$

$$\therefore \qquad I_u, I_v = \tfrac{1}{2}[(4 + 1.08) \pm (4 - 1.08)(-1.2883)]10^{-6}$$
$$= \tfrac{1}{2}[5.08 \pm 3.762]10^{-6}$$

$$I_u = 4.421 \times 10^{-6}, \ I_v = 0.659 \times 10^{-6} \ \text{m}^4$$

and $\qquad\qquad A = [(76 \times 13) + (114 \times 13)]10^{-6} = 2.47 \times 10^{-3} \ \text{m}^2$

$\therefore \qquad\qquad\qquad k_u = \sqrt{\left(\dfrac{4.421 \times 10^{-6}}{2.47 \times 10^{-3}} \right)} = 0.0423 = 42.3 \ \text{mm}$

$$k_v = \sqrt{\left(\dfrac{0.659 \times 10^{-6}}{2.47 \times 10^{-3}} \right)} = 0.0163 = 16.3 \ \text{mm}$$

The momental ellipse can now be constructed as described above and drawn in Fig. 1.17 and by measurement

$$h = 22.3 \ \text{mm}$$

Then $\qquad\qquad I_{\text{N.A.}} = Ah^2 = 2.47 \times 10^{-3} \times 22.3^2 \times 10^{-6}$

$$= 1.23 \times 10^{-6} \ \text{m}^4$$

(This value may also be obtained from Mohr's circle of Fig. 1.18.)
The stress at B is then given by

$$\sigma = \frac{M_{\text{N.A.}} n}{I_{\text{N.A.}}}$$

where

$$n = \text{perpendicular distance from } B \text{ to the N.A.}$$

$$= 44 \ \text{mm}$$

and $\qquad\qquad M_{\text{N.A.}} = 10\,000 \cos 47°41' \times 1.2 = 8079 \ \text{Nm}$

$\therefore \qquad\qquad \text{stress at } B = \dfrac{8079 \times 44 \times 10^{-3}}{1.23 \times 10^{-6}} = \mathbf{289 \ MN/m^2}$

This confirms the result obtained with the alternative procedure of Example 1.2.

Problems

1.1 (B). A rectangular-sectioned beam of 75 mm × 50 mm cross-section is used as a simply supported beam and carries a uniformly distributed load of 500 N/m over a span of 3 m. The beam is supported in such a way that its long edges are inclined at 20° to the vertical. Determine:
(a) the maximum stress set up in the cross-section;
(b) the vertical deflection at mid-span.
$E = 208 \ \text{GN/m}^2$. $\hfill$ [17.4 MN/m^2; 1.76 mm.]

1.2 (B). An I-section girder 1.3 m long is rigidly built in at one end and loaded at the other with a load of 1.5 kN inclined at 30° to the web. If the load passes through the centroid of the section and the girder dimensions are: flanges 100 mm × 20 mm, web 200 mm × 12 mm, determine the maximum stress set up in the cross-section. How does this compare with the maximum stress set up if the load is vertical?
$\hfill$ [18.1, 4.14 MN/m^2.]

1.3 (B). A 75 mm × 75 mm × 12 mm angle is used as a cantilever with the face AB horizontal, as shown in Fig. 1.19. A vertical load of 3 kN is applied at the tip of the cantilever which is 1 m long. Determine the stress at A, B and C.
$\hfill$ [196.37, −207 MN/m^2.]

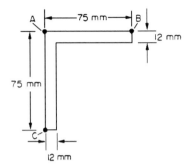

Fig. 1.19.

1.4 (B). A cantilever of length 2 m is constructed from 150 mm × 100 mm by 12 mm angle and arranged with its 150 mm leg vertical. If a vertical load of 5 kN is applied at the free end, passing through the shear centre of the section, determine the maximum tensile and compressive stresses set up across the section.

[B.P.] [169, − 204 MN/m².]

1.5 (B). A 180 mm × 130 mm × 13 mm unequal angle section is arranged with the long leg vertical and simply supported over a span of 4 m. Determine the maximum central load which the beam can carry if the maximum stress in the section is limited to 90 MN/m². Determine also the angle of inclination of the neutral axis.

$$I_{xx} = 12.8 \times 10^{-6} \text{ m}^4, I_{yy} = 5.7 \times 10^{-6} \text{ m}^4.$$

What will be the vertical deflection of the beam at mid-span? $E = 210$ GN/m². [8.73 kN, 41.6°, 7.74 mm.]

1.6 (B). The unequal-leg angle section shown in Fig. 1.20 is used as a cantilever with the 130 mm leg vertical. The length of the cantilever is 1.3 m. A vertical point load of 4.5 kN is applied at the free end, its line of action passing through the shear centre.

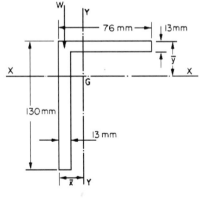

Fig. 1.20.

The properties of the section are as follows:

$$\bar{x} = 19 \text{ mm}, \bar{y} = 45 \text{ mm}, I_{xx} = 4 \times 10^{-6} \text{ m}^4, I_{yy} = 1.1 \times 10^{-6} \text{ m}^4, I_{xy} = 1.2 \times 10^{-6} \text{ m}^4.$$

Determine:
(a) the magnitude of the principal second moments of area together with the inclination of their axes relative to *XX*;
(b) the position of the neutral plane (*N*−*N*) and the magnitude of I_{NN};
(c) the end deflection of the centroid *G* in magnitude, direction and sense.

Take $E = 207$ GN/m^2 (2.07 Mbar).

[444×10^{-8} m^4, 66×10^{-8} m^4, $-19°51'$ to XX, $47°42'$ to XX, 121×10^{-8} m^4, 8.85 mm at $-42°18'$ to XX.]

1.7 (B). An extruded aluminium alloy section having the cross-section shown in Fig. 1.21 will be used as a cantilever as indicated and loaded with a single concentrated load at the free end. This load F acts in the plane of the cross-section but may have any orientation within the cross-section. Given that $I_{xx} = 101.2 \times 10^{-8}$ m^4 and $I_{yy} = 29.2 \times 10^{-8}$ m^4:

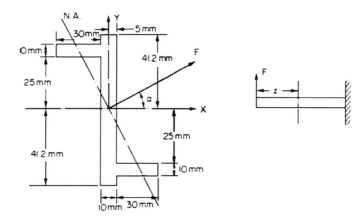

Fig. 1.21.

(a) determine the values of the principal second moments of area and the orientation of the principal axes;
(b) for such a case that the neutral axis is orientated at $-45°$ to the X-axis, as shown, find the angle α of the line of action of F to the X-axis and hence determine the numerical constant K in the expression $\sigma = KFz$, which expresses the magnitude of the greatest bending stress at any distance z from the free end.

[City U.] [116.1×10^{-8}, 14.3×10^{-8}, $22.5°$, $-84°$, 0.71×10^5.]

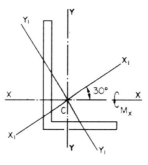

Fig. 1.22.

1.8 (B). A beam of length 2 m has the unequal-leg angle section shown in Fig. 1.22 for which $I_{xx} = 0.8 \times 10^{-6}$ m^4, $I_{yy} = 0.4 \times 10^{-6}$ m^4 and the angle between $X - X$ and the principal second moment of area axis $X_1 - X_1$ is 30°. The beam is subjected to a constant bending moment (M_x) of magnitude 1000 Nm about the $X - X$ axis as shown.
Determine:
(a) the values of the principal second moments of area Ix_1 and Iy_1 respectively;
(b) the inclination of the N.A., or line of zero stress ($N - N$) relative to the axis $X_1 - X_1$ and the value of the second moment of area of the section about $N - N$, that is I_N;

(c) the magnitude, direction and sense of the resultant maximum deflection of the centroid C.

For the beam material, Young's modulus $E = 200$ GN/m^2. For a beam subjected to a constant bending moment M, the maximum deflection δ is given by the formula

$$\delta = \frac{ML^2}{8EI}$$

$[1 \times 10^{-6}, 0.2 \times 10^{-6}$ m^4, $-70°54'$ to X_1X_1, 0.2847×10^{-6} m^4, 6.62 mm, $90°$ to N.A.$]$

CHAPTER 2

STRUTS

Summary

The allowable stresses and end loads given by Euler's theory for struts with varying end conditions are given in Table 2.1.

Table 2.1.

End condition	Fixed–free	Pinned–pinned (or rounded)	Fixed–pinned	Fixed–fixed
Euler load P_e	$\dfrac{\pi^2 EI}{4L^2}$	$\dfrac{\pi^2 EI}{L^2}$	$\dfrac{2\pi^2 EI}{L^2}$	$\dfrac{4\pi^2 EI}{L^2}$
	or, writing $I = Ak^2$, where $k =$ radius of gyration			
	$\dfrac{\pi^2 EA}{4(L/k)^2}$	$\dfrac{\pi^2 EA}{(L/k)^2}$	$\dfrac{2\pi^2 EA}{(L/k)^2}$	$\dfrac{4\pi^2 EA}{(L/k)^2}$
Euler stress σ_e	$\dfrac{\pi^2 E}{4(L/k)^2}$	$\dfrac{\pi^2 E}{(L/k)^2}$	$\dfrac{2\pi^2 E}{(L/k)^2}$	$\dfrac{4\pi^2 E}{(L/k)^2}$

Here L is the length of the strut and the term L/k is known as the *slenderness ratio*.

Validity limit for Euler formulae

$$L/k = \sqrt{\left(\frac{C\pi^2 E}{\sigma_y}\right)}$$

where C is a constant depending on the end condition of the strut.

Rankine–Gordon Formula

$$\sigma = \frac{\sigma_y}{1 + a(L/k)^2}$$

where $a = (\sigma_y/\pi^2 E)$ theoretically but is usually found by experiment. Typical values are given in Table 2.2.

Table 2.2.

Material	Compressive yield stress (MN/m^2)	a	
		Pinned ends	Fixed ends
Mild steel	315	1/7500	1/30 000
Cast iron	540	1/1600	1/64 000
Timber	35	1/3000	1/12 000

N.B. The value of a for pinned ends is always four times that for fixed ends

Perry–Robertson Formula

$$N\sigma = \frac{[\sigma_y + (\eta + 1)\sigma_e]}{2} - \sqrt{\left\{\left[\frac{\sigma_y + (\eta + 1)\sigma_e}{2}\right]^2 - \sigma_y\sigma_e\right\}}$$

where η is a constant depending on the material.

For a brittle material

$$\eta = 0.015L/k$$

For a ductile material

$$\eta = 0.3\left(\frac{L}{100k}\right)^2$$

These values will be modified for eccentric loading conditions. The Perry–Robertson formula is the basis of BS 449 as shown in §2.7.

Struts with initial curvature

$$\text{Maximum deflection } \delta_{\max} = \left[\frac{P_e}{(P_e - P)}\right]C_0$$

$$\text{Maximum stress } \sigma_{\max} = \frac{P}{A} \pm \left[\frac{PP_e}{(P_e - P)}\right]\frac{C_0h}{I}$$

where C_0 is the initial central deflection and h is the distance of the highest strained fibre from the neutral axis (N.A.).

Smith–Southwell formula for eccentrically loaded struts

With pinned ends the maximum stress reached in the strut is given by

$$\sigma_{\max} = \sigma\left[1 + \frac{eh}{k^2}\sec\frac{L}{2}\sqrt{\left(\frac{\sigma}{Ek^2}\right)}\right]$$

or

$$\sigma_{\max} = \sigma\left[1 + \frac{eh}{k^2}\sec\frac{1}{2}\pi\sqrt{\left(\frac{\sigma}{\sigma_E}\right)}\right]$$

where e is the eccentricity of loading, h is the distance of the highest strained fibre from the N.A., k is the minimum radius of gyration of the cross-section, and σ is the applied load/cross-sectional area.

Since the required allowable stress σ cannot be obtained directly from this equation a solution is obtained graphically or by trial and error.

With other end conditions the value L in the above formula should be replaced by the appropriate *equivalent strut* length (see §2.2).

Webb's approximation for the Smith–Southwell formula

$$\sigma_{\max} = \frac{P}{A}\left[1 + \frac{eh}{k^2}\left(\frac{P_e + 0.26P}{P_e - P}\right)\right]$$

Laterally loaded struts

(a) Central concentrated load

$$\text{Maximum deflection} = \frac{W}{2nP}\left[\tan\frac{nL}{2} - \frac{nL}{2}\right]$$

$$\text{maximum bending moment (B.M.)} = \frac{W}{2n}\tan\frac{nL}{2}$$

(b) Uniformly distributed load

$$\text{Maximum deflection} = \frac{w}{n^2 P}\left[\left(\sec\frac{nL}{2} - 1\right) - \frac{n^2 L^2}{8}\right]$$

$$\text{maximum B.M.} = \frac{w}{n^2}\left(\sec\frac{nL}{2} - 1\right)$$

Introduction

Structural members which carry compressive loads may be divided into two broad categories depending on their relative lengths and cross-sectional dimensions. Short, thick members are generally termed *columns* and these usually fail by *crushing* when the yield stress of the material in compression is exceeded. Long, slender columns or *struts*, however, fail by *buckling* some time before the yield stress in compression is reached. The buckling occurs owing to one or more of the following reasons:

(a) the strut may not be perfectly straight initially;
(b) the load may not be applied exactly along the axis of the strut;
(c) one part of the material may yield in compression more readily than others owing to some lack of uniformity in the material properties throughout the strut.

At values of load below the buckling load a strut will be in stable eqilibrium where the displacement caused by any lateral disturbance will be totally recovered when the disturbance is removed. At the buckling load the strut is said to be in a state of neutral equilibrium, and theoretically it should then be possible to gently deflect the strut into a simple sine wave provided that the amplitude of the wave is kept small. This can be demonstrated quite simply using long thin strips of metal, e.g. a metal rule, and gentle application of compressive loads.

Theoretically, it is possible for struts to achieve a condition of unstable equilibrium with loads exceeding the buckling load, any slight lateral disturbance then causing failure by buckling; this condition is never achieved in practice under static load conditions. Buckling occurs immediately at the point where the buckling load is reached owing to the reasons stated earlier.

The above comments and the contents of this chapter refer to the *elastic* stability of struts only. It must also be remembered that struts can also fail plastically, and in this case the failure is irreversible.

2.1. Euler's theory

(a) Strut with pinned ends

Consider the axially loaded strut shown in Fig. 2.1 subjected to the crippling load P_e producing a deflection y at a distance x from one end. Assume that the ends are either pin-jointed or rounded so that there is no moment at either end.

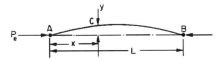

Fig. 2.1. Strut with axial load and pinned ends.

$$\text{B.M. at } C = EI\frac{d^2y}{dx^2} = -P_e y$$

$$\therefore \quad EI\frac{d^2y}{dx^2} + P_e y = 0$$

$$\therefore \quad \frac{d^2y}{dx^2} + \frac{P_e}{EI}y = 0$$

i.e. in operator form, with $D \equiv d/dx$,

$$(D^2 + n^2)y = 0, \quad \text{where } n^2 = P_e/EI$$

This is a second-order differential equation which has a solution of the form

$$y = A\cos nx + B\sin nx$$

i.e.
$$y = A\cos\sqrt{\left(\frac{P_e}{EI}\right)}x + B\sin\sqrt{\left(\frac{P_e}{EI}\right)}x$$

Now at $x = 0, y = 0$ $\qquad\qquad\qquad \therefore A = 0$

and at $x = L, y = 0$ $\qquad\qquad\qquad \therefore B\sin L\sqrt{(P_e/EI)} = 0$

$$\therefore \quad \text{either } B = 0 \text{ or } \sin L\sqrt{\left(\frac{P_e}{EI}\right)} = 0$$

If $B = 0$ then $y = 0$ and the strut has not yet buckled. Thus the solution required is

$$\sin L\sqrt{\left(\frac{P_e}{EI}\right)} = 0 \quad \therefore L\sqrt{\left(\frac{P_e}{EI}\right)} = \pi$$

$$\therefore \qquad P_e = \frac{\pi^2 EI}{L^2} \qquad\qquad\qquad\qquad (2.1)$$

It should be noted that other solutions exist for the equation

$$\sin L \sqrt{\left(\frac{P}{EI}\right)} = 0 \quad \text{i.e.} \quad \sin nL = 0$$

The solution chosen of $nL = \pi$ is just one particular solution; the solutions $nL = 2\pi$, 3π, 5π, etc., are equally as valid mathematically and they do, in fact, produce values of P_e which are equally valid for modes of buckling of the strut different from that of the simple bow of Fig. 2.1. Theoretically, therefore, there are an infinite number of values of P_e, each corresponding with a different mode of buckling. The value selected above is the so-called *fundamental* mode value and is the lowest critical load producing the single-bow buckling condition. The solution $nL = 2\pi$ produces buckling in two half-waves, 3π in three half-waves, etc., as shown in Fig. 2.2. If load is applied sufficiently quickly to the strut, it is possible to pass through the fundamental mode and to achieve at least one of the other modes which are theoretically possible. In practical loading situations, however, this is rarely achieved since the high stress associated with the first critical condition generally ensures immediate collapse. The buckling load of a strut with pinned ends is, therefore, for all practical purposes, given by eqn. (2.1).

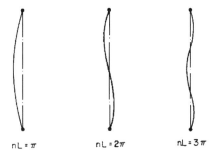

Fig. 2.2. Strut failure modes.

(b) One end fixed, the other free

Consider now the strut of Fig. 2.3 with the origin at the fixed end.

$$\text{B.M. at } C = EI\frac{d^2y}{dx^2} = +P(a - y)$$

$$\therefore \qquad \frac{d^2y}{dx^2} + \frac{Py}{EI} = \frac{Pa}{EI}$$

$$\therefore \qquad (D^2 + n^2)y = n^2 a \qquad\qquad (2.2)$$

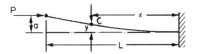

Fig. 2.3. Fixed–free strut.

N.B.–It is always convenient to arrange the diagram and origin such that the differential equation is achieved in the above form since the solution will then always be of the form

$$y = A \cos nx + B \sin nx + \text{(particular solution)}$$

The *particular solution* is a particular value of y which satisfies eqn. (2.2), and in this case can be shown to be $y = a$.

$\therefore$
$$y = A \cos nx + B \sin nx + a$$

Now when $x = 0$, $y = 0$

$\therefore$
$$A = -a$$

when $x = 0$, $dy/dx = 0$

$\therefore$
$$B = 0$$

$\therefore$
$$y = -a \cos nx + a$$

But when $x = L$, $y = a$

$\therefore$
$$a = -a \cos nL + a$$
$$0 = \cos nL$$

The fundamental mode of buckling in this case therefore is given when $nL = \frac{1}{2}\pi$.

$\therefore$
$$L\sqrt{\left(\frac{P}{EI}\right)} = \frac{\pi}{2}$$

or
$$P_e = \frac{\pi^2 EI}{4L^2} \tag{2.3}$$

(c) Fixed ends

Consider the strut of Fig. 2.4 *with the origin at the centre.*

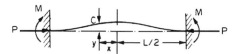

Fig. 2.4. Strut with fixed ends.

In this case the B.M. at C is given by

$$EI\frac{d^2y}{dx^2} = M - Py$$

$$\frac{d^2y}{dx^2} + \frac{P}{EI}y = \frac{M}{EI}$$

$$(D^2 + n^2)y = M/EI$$

Here the particular solution is

$$y = \frac{M}{n^2 EI} = \frac{M}{P}$$

$$\therefore \qquad y = A \cos nx + B \sin nx + M/P$$

Now when $x = 0$, $dy/dx = 0$ $\therefore B = 0$

and when $x = \frac{1}{2}L$, $y = 0$ $\qquad \therefore A = \frac{-M}{P} \sec \frac{nL}{2}$

$$\therefore \qquad y = -\frac{M}{P} \sec \frac{nL}{2} \cos nx + \frac{M}{P}$$

But when $x = \frac{1}{2}L$, dy/dx is also zero,

$$\therefore \qquad 0 = \frac{nM}{P} \sec \frac{nL}{2} \sin \frac{nL}{2}$$

$$0 = \frac{nM}{P} \tan \frac{nL}{2}$$

The fundamental buckling mode is then given when $nL/2 = \pi$

$$\therefore \qquad \frac{L}{2} \sqrt{\left(\frac{P}{EI}\right)} = \pi$$

or $$P_e = \frac{4\pi^2 EI}{L^2} \qquad\qquad (2.4)$$

(d) One end fixed, the other pinned

In order to maintain the pin-joint on the horizontal axis of the unloaded strut, it is necessary in this case to introduce a vertical load F at the pin (Fig. 2.5). The moment of F about the built-in end then balances the fixing moment.

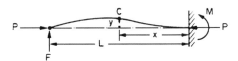

Fig. 2.5. Strut with one end pinned, the other fixed.

With the origin at the built-in end the B.M. at C is

$$EI\frac{d^2y}{dx^2} = -Py + F(L - x)$$

$$\frac{d^2y}{dx^2} + \frac{P}{EI}y = \frac{F}{EI}(L - x)$$

$$(D^2 + n^2)y = \frac{F}{EI}(L - x)$$

The particular solution is

$$y = \frac{F}{n^2 EI}(L - x) = \frac{F}{P}(L - x)$$

The full solution is therefore

$$y = A \cos nx + B \sin nx + \frac{F}{P}(L - x)$$

When $x = 0$, $y = 0$, $\quad \therefore A = -\frac{FL}{P}$

When $x = 0$, $dy/dx = 0$, $\therefore B = \frac{F}{nP}$

$$y = -\frac{FL}{P} \cos nx + \frac{F}{nP} \sin nx + \frac{F}{P}(L - x)$$

$$= \frac{F}{nP}[-nL \cos nx + \sin nx + n(L - x)]$$

But when $x = L$, $y = 0$

$$\therefore \quad\quad\quad\quad\quad nL \cos nL = \sin nL$$

$$\tan nL = nL$$

The lowest value of nL (neglecting zero) which satisfies this condition and which therefore produces the fundamental buckling condition is $nL = 4.5$ radians.

$$\therefore \quad\quad\quad\quad\quad L\sqrt{\left(\frac{P}{EI}\right)} = 4.5$$

or

$$P_e = \frac{20.25EI}{L^2} \quad\quad\quad\quad (2.5)$$

or, approximately

$$P_e = \frac{2\pi^2 EI}{L^2} \quad\quad\quad\quad (2.6)$$

2.2. Equivalent strut length

Having derived the result for the buckling load of a strut with pinned ends the Euler loads for other end conditions may all be written in the same form,

i.e.

$$P_e = \frac{\pi^2 EI}{l^2} \quad\quad\quad\quad (2.7)$$

where l is the *equivalent length* of the strut and can be related to the actual length of the strut depending on the end conditions. The equivalent length is found to be the length of a simple bow (half sine-wave) in each of the strut deflection curves shown in Fig. 2.6. The buckling load for each end condition shown is then readily obtained.

The use of the equivalent length is not restricted to the Euler theory and it will be used in other derivations later.

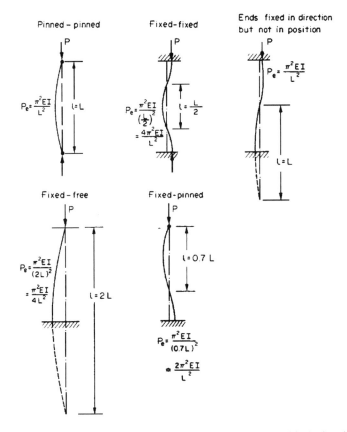

Fig. 2.6. "Equivalent length" of struts with different end conditions. In each case *l* is the length of a single bow.

2.3. Comparison of Euler theory with experimental results (see Fig. 2.7)

Between $L/k = 40$ and $L/k = 100$ neither the Euler results nor the yield stress are close to the experimental values, each suggesting a critical load which is in excess of that which is actually required for failure—a very unsafe situation! Other formulae have therefore been derived to attempt to obtain closer agreement between the actual failing load and the predicted value in this particular range of slenderness ratio.

(a) Straight-line formula

$$P = \sigma_y A[1 - n(L/k)] \tag{2.8}$$

the value of *n* depending on the material used and the end condition.

(b) Johnson parabolic formula

$$P = \sigma_y A[1 - b(L/k)^2] \tag{2.9}$$

the value of *b* depending also on the end condition.

Neither of the above formulae proved to be very successful, and they were replaced by:

(c) Rankine–Gordon formula

$$\frac{1}{P_R} = \frac{1}{P_e} + \frac{1}{P_c} \tag{2.10}$$

where P_e is the Euler buckling load and P_c is the crushing (compressive yield) load $= \sigma_y A$. This formula has been widely used and is discussed fully in §2.5.

2.4. Euler "validity limit"

From the graph of Fig. 2.7 and the comments above, it is evident that the Euler theory is unsafe for small L/k ratios. It is useful, therefore, to determine the limiting value of L/k below which the Euler theory should not be applied; this is termed the *validity limit*.

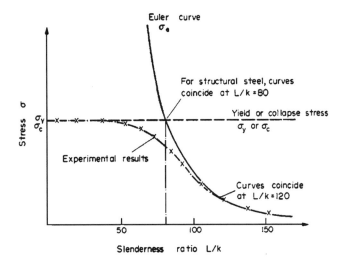

Fig. 2.7. Comparison of experimental results with Euler curve.

The validity limit is taken to be the point where the Euler σ_e equals the yield or crushing stress σ_y, i.e. the point where the strut load

$$P = \sigma_y A$$

Now the Euler load can be written in the form

$$P_e = C\frac{\pi^2 EI}{L^2} = C\frac{\pi^2 EAk^2}{L^2}$$

where C is a constant depending on the end condition of the strut.
Therefore in the limiting condition

$$\sigma_y A = C\frac{\pi^2 EAk^2}{L^2}$$

$$\frac{L}{k} = \sqrt{\left(\frac{C\pi^2 E}{\sigma_y}\right)}$$

The value of this expression will vary with the type of end condition; as an example, low carbon steel struts with pinned ends give $L/k \simeq 80$.

2.5. Rankine or Rankine–Gordon formula

As stated above, the Rankine formula is a combination of the Euler and crushing loads for a strut

$$\frac{1}{P_R} = \frac{1}{P_e} + \frac{1}{P_c}$$

For very short struts P_e is very large; $1/P_e$ can therefore be neglected and $P_R = P_c$. For very long struts P_e is very small and $1/P_e$ is very large so that $1/P_c$ can be neglected. Thus $P_R = P_e$.

The Rankine formula is therefore valid for extreme values of L/k. It is also found to be fairly accurate for the intermediate values in the range under consideration. Thus, re-writing the formula in terms of stresses,

$$\frac{1}{\sigma A} = \frac{1}{\sigma_e A} + \frac{1}{\sigma_y A}$$

i.e.

$$\frac{1}{\sigma} = \frac{1}{\sigma_e} + \frac{1}{\sigma_y} = \frac{\sigma_e + \sigma_y}{\sigma_e \sigma_y}$$

$$\sigma = \frac{\sigma_e \sigma_y}{\sigma_e + \sigma_y} = \frac{\sigma_y}{[1 + (\sigma_y/\sigma_e)]}$$

For a strut with both ends pinned

$$\sigma_e = \frac{\pi^2 E}{(L/k)^2}$$

$\therefore$

$$\sigma = \frac{\sigma_y}{1 + \dfrac{\sigma_y}{\pi^2 E}\left(\dfrac{L}{k}\right)^2}$$

i.e. Rankine stress

$$\sigma_R = \frac{\sigma_y}{1 + a(L/k)^2} \qquad (2.11)$$

where $a = \sigma_y/\pi^2 E$, theoretically, but having a value normally found by experiment for various materials. This will take into account other types of end condition.

Therefore Rankine load

$$P_R = \frac{\sigma_y A}{1 + a(L/k)^2} \qquad (2.12)$$

Typical values of a for use in the Rankine formula are given in Table 2.3.

However, since the values of a are not exactly equal to the theoretical values, the Rankine loads for long struts will not be identical to those estimated by the Euler theory as suggested earlier.

Table 2.3.

Material	σ_y or σ_c (MN/m²)	a	
		Pinned ends	Fixed ends
Low carbon steel	315	1/7500	1/30 000
Cast iron	540	1/1600	1/64 000
Timber	35	1/3000	1/12 000

N.B. *a* for pinned ends = 4 × (*a* for fixed ends)

2.6. Perry–Robertson formula

The Perry–Robertson proof is based on the assumption that any imperfections in the strut, through faulty workmanship or material or eccentricity of loading, can be allowed for by giving the strut an initial curvature. For ease of calculation this is assumed to be a cosine curve, although the actual shape assumed has very little effect on the result.

Consider, therefore, the strut AB of Fig. 2.8, of length L and pin-jointed at the ends. The initial curvature y_0 at any distance x from the centre is then given by

$$y_0 = C_0 \cos \frac{\pi x}{L}$$

Fig. 2.8. Strut with initial curvature.

If a load P is now applied at the ends, this deflection will be increased to $y + y_0$.

$$\therefore \qquad BM_c = EI\frac{d^2 y}{dx^2} = -P\left(y + C_0 \cos \frac{\pi x}{L}\right)$$

$$\therefore \qquad \frac{d^2 y}{dx^2} + \frac{P}{EI}\left(y + C_0 \cos \frac{\pi x}{L}\right) = 0$$

the solution of which is

$$y = A \sin \sqrt{\left(\frac{P}{EI}\right)} x + B \cos \sqrt{\left(\frac{P}{EI}\right)} x + \left[\left(\frac{PC_0}{EI} \cos \frac{\pi x}{L}\right) \bigg/ \left(\frac{\pi^2}{L^2} - \frac{P}{EI}\right)\right]$$

where A and B are the constants of integration.
Now when $x = \pm L/2$, $y = 0$

$$\therefore \qquad A = B = 0$$

$$\therefore \qquad y = \left[\left(\frac{PC_0}{EI} \cos \frac{\pi x}{L}\right) \bigg/ \left(\frac{\pi^2}{L^2} - \frac{P}{EI}\right)\right] = \left[\left(PC_0 \cos \frac{\pi x}{L}\right) \bigg/ \left(\frac{\pi^2 EI}{L^2} - P\right)\right]$$

Therefore dividing through, top and bottom, by A,

$$y = \left[\left(\frac{P}{A} C_0 \cos \frac{\pi x}{L} \right) \bigg/ \left(\frac{\pi^2 EI}{L^2 A} - \frac{P}{A} \right) \right]$$

But $P/A = \sigma$ and $(\pi^2 EI)/(L^2 A) = \sigma_e$ (the Euler stress for pin-ended struts)

$$\therefore \qquad\qquad y = \frac{\sigma}{(\sigma_e - \sigma)} C_0 \cos \frac{\pi x}{L}$$

Therefore total deflection at any point is given by

$$y + y_0 = \left[\frac{\sigma}{(\sigma_e - \sigma)} \right] C_0 \cos \frac{\pi x}{L} + C_0 \cos \frac{\pi x}{L}$$

$$= \left[\frac{\sigma_e}{(\sigma_e - \sigma)} \right] C_0 \cos \frac{\pi x}{L} \qquad\qquad (2.13)$$

$$\therefore \qquad \text{Maximum deflection (when } x = 0) = \left[\frac{\sigma_e}{(\sigma_e - \sigma)} \right] C_0 \qquad\qquad (2.14)$$

$$\therefore \qquad\qquad \text{maximum B.M.} = P \left[\frac{\sigma_e}{(\sigma_e - \sigma)} \right] C_0 \qquad\qquad (2.15)$$

$$\therefore \qquad \text{maximum stress owing to bending} = \frac{My}{I} = \frac{P}{I} \left[\frac{\sigma_e}{(\sigma_e - \sigma)} \right] C_0 h$$

where h is the distance of the outside fibre from the N.A. of the strut.

Therefore the maximum stress owing to combined bending and thrust is given by

$$\sigma_{\max} = \frac{P}{I} \left[\frac{\sigma_e}{(\sigma_e - \sigma)} \right] C_0 h + \frac{P}{A} \qquad\qquad (2.16)$$

$$= \frac{P}{Ak^2} \left[\frac{\sigma_e}{(\sigma_e - \sigma)} \right] C_0 h + \frac{P}{A}$$

$$= \sigma \left[\frac{\eta \sigma_e}{(\sigma_e - \sigma)} + 1 \right] \quad \text{where } \eta = \frac{C_0 h}{k^2}$$

If $\sigma_{\max} = \sigma_y$, the compressive yield stress for the material of the strut, the above equation when solved for σ gives

$$\sigma = \frac{[\sigma_y + (\eta + 1)\sigma_e]}{2} - \sqrt{\left\{ \left[\frac{\sigma_y + (\eta + 1)\sigma_e}{2} \right]^2 - \sigma_y \sigma_e \right\}} \qquad\qquad (2.17)$$

This is the Perry–Robertson formula required. If the material is brittle, however, and failure is likely to occur in tension, then the sign between the two square-bracketed terms becomes positive and σ_y is the *tensile* yield strength.

2.7. British Standard procedure (BS 449)

With a *load factor N* applied, the Perry–Robertson equation becomes

$$N\sigma = \frac{[\sigma_y + (\eta + 1)\sigma_e]}{2} - \sqrt{\left\{\left[\frac{\sigma_y + (\eta + 1)\sigma_e}{2}\right]^2 - \sigma_y\sigma_e\right\}} \qquad (2.18)$$

With values for steel of $\sigma_y = 225$ MN/m^2, $E = 200$ GN/m^2, $N = 1.7$ and $\eta = 0.3(L/100k)^2$, the above equation gives the graph shown in Fig. 2.9. This graph then indicates the basis of design using BS449: 1959 (amended 1964). Allowable values are provided in the standard, however, in tabular form.

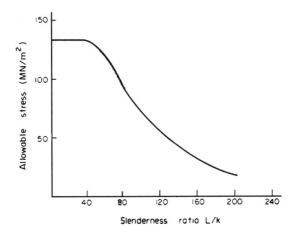

Fig. 2.9. Graph of allowable stress as given in BS 449: 1964 (in tabulated form) against slenderness ratio.

If, however, design is based on the *safety factor* method instead of the *load factor* method, then N is omitted and σ_y/n replaces σ_y in the formula, where n is the safety factor.

2.8. Struts with initial curvature

In §2.6 the Perry–Robertson equation was derived on the assumption that strut imperfections could be allowed for by giving the strut an initial curvature. This proof applies equally well, of course, for struts which have genuine initial curvatures and, provided the curvature is small, the precise shape of the curve has little effect on the end result.

Thus for an initial curvature with a central deflection C_0,

$$\text{maximum deflection} = \left[\frac{\sigma_e}{(\sigma_e - \sigma)}\right] C_0 = \left[\frac{P_e}{(P_e - P)}\right] C_0 \qquad (2.19)$$

$$\text{maximum B.M.} = P\left[\frac{\sigma_e}{(\sigma_e - \sigma)}\right] C_0 = \left[\frac{PP_e}{(P_e - P)}\right] C_0 \qquad (2.20)$$

and

$$\sigma_{\max} = \frac{P}{A} \pm \left[\frac{P\sigma_e}{(\sigma_e - \sigma)}\right] \frac{hC_0}{I}$$

$$= \frac{P}{A} \pm \left[\frac{PP_e}{(P_e - P)}\right] \frac{hC_0}{I} \qquad (2.21)$$

where h is the distance from the N.A. to the outside fibres of the strut.

2.9. Struts with eccentric load

For eccentric loading at the ends of a strut Ayrton and Perry suggest that the Perry–Robertson formula can be modified by replacing C_0 by $(C_0 + 1.2e)$ where e is the eccentricity.
Then

$$\eta' = \eta + 1.2\frac{eh}{k^2} \qquad (2.22)$$

and η' replaces η in the original Perry–Robertson equation.

(a) Pinned ends – the Smith–Southwell formula

For a more fundamental treatment consider the strut loaded as shown in Fig. 2.10 carrying a load P at an eccentricity e on one principal axis. In this case there is strictly no 'buckling" load as previously described since the strut will bend immediately load is applied, bending taking place about the other principal axis.

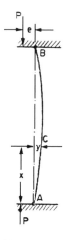

Fig. 2.10. Strut with eccentric load (pinned ends)

Applying a similar procedure to that used previously

$$\text{B.M. at } C = -P(y + e)$$

∴

$$EI\frac{d^2y}{dx^2} = -P(y + e)$$

∴

$$\frac{d^2y}{dx^2} + n^2(y + e) = 0$$

where $$n = \sqrt{(P/EI)}$$

This is a second-order differential equation, the solution of which is as follows:

$$y = A \sin nx + B \cos nx - e$$

Now when $x = 0$, $y = 0$

$\therefore$ $$B = e$$

and when $x = \dfrac{L}{2}, \dfrac{dy}{dx} = 0$

$\therefore$ $$0 = nA \cos n\frac{L}{2} - ne \sin n\frac{L}{2}$$

$\therefore$ $$A = e \tan \frac{nL}{2}$$

$\therefore$ $$y + e = e \tan \frac{nL}{2} \sin nx + e \cos nx$$

$\therefore$ maximum deflection, when $$x = L/2 \text{ and } y = \delta, \text{ is}$$

$$\delta + e = e \frac{\sin^2 \dfrac{nL}{2}}{\cos \dfrac{nL}{2}} + e \cos \frac{nL}{2}$$

$$= e \frac{\left(\sin^2 \dfrac{nL}{2} + \cos^2 \dfrac{nL}{2} \right)}{\cos \dfrac{nL}{2}} = e \sec \frac{nL}{2} \qquad (2.23)$$

$\therefore$ $$\text{maximum B.M.} = P(\delta + e) = Pe \sec \frac{nL}{2} \qquad (2.24)$$

$\therefore$ $$\text{maximum stress owing to bending} = \frac{My}{I} = Pe \sec \frac{nL}{2} \times \frac{h}{I}$$

where h is the distance from the N.A. to the highest stressed fibre.

Therefore the total maximum compressive stress owing to combined bending and thrust, assuming a *ductile* material[†], is given by

$$\sigma_{max} = \frac{P}{A} + \left(Pe \sec \frac{nL}{2} \right) \frac{h}{I}$$

$$= \sigma \left[1 + \frac{eh}{k^2} \sec \frac{nL}{2} \right] \qquad (2.25)$$

$$= \sigma \left[1 + \frac{eh}{k^2} \sec \frac{L}{2} \sqrt{\left(\frac{P}{EI} \right)} \right]$$

[†] For a brittle material which is relatively weak in tension it is the maximum tensile stress which becomes the criterion of failure and the bending and direct stress components are opposite in sign.

i.e.
$$\sigma_{max} = \sigma \left[1 + \frac{eh}{k^2} \sec \frac{L}{2} \sqrt{\left(\frac{\sigma}{Ek^2}\right)} \right] \qquad (2.26)$$

This formula is known as the Smith–Southwell formula.

Unfortunately, since $\sigma = P/A$, the above equation represents a function of P (the required unknown) which can only be solved by trial and error or graphically. A good approximation however, is obtained as shown below:

Webb's approximation

From above
$$\sigma_{max} = \sigma \left[1 + \frac{eh}{k^2} \sec \frac{nL}{2} \right] \qquad (2.25)(bis)$$

Let
$$\frac{nL}{2} = \theta$$

Then
$$\theta = \frac{L}{2} \sqrt{\left(\frac{P}{EI}\right)} = \frac{\pi}{2} \sqrt{\left(\frac{L^2}{\pi^2} \frac{P}{EI}\right)} = \frac{\pi}{2} \sqrt{\left(\frac{P}{P_e}\right)}$$

Now for θ between 0 and $\pi/2$,

$$\sec \theta \simeq \frac{1 + 0.26 \left(\frac{2\theta}{\pi}\right)^2}{1 - \left(\frac{2\theta}{\pi}\right)^2} = \frac{1 + 0.26 \dfrac{P}{P_e}}{1 - \dfrac{P}{P_e}} = \frac{P_e + 0.26P}{P_e - P}$$

Therefore substituting in eqn. (2.25)

$$\sigma_{max} = \sigma \left[1 + \frac{eh}{k^2} \left(\frac{P_e + 0.26P}{P_e - P}\right) \right]$$

$$= \frac{P}{A} \left[1 + \frac{eh}{k^2} \left(\frac{P_e + 0.26P}{P_e - P}\right) \right] \qquad (2.27)$$

where σ_{max} is the maximum allowable stress in the strut material, P_e is the Euler buckling load for axial loading, and P is the maximum allowable value of the eccentric load.

The above equation can be re-written into a more readily observed quadratic equation in P, thus:

$$P^2 \left[1 - 0.26 \frac{eh}{k^2} \right] - P \left[P_e \left(1 + \frac{eh}{k^2}\right) + \sigma_{max}A \right] + \sigma_{max}AP_e = 0 \qquad (2.28)$$

For any given eccentric load condition P is the only unknown and the equation can be readily solved.

(b) One end fixed, the other free

Consider the strut shown in Fig. 2.11.

$$BM_c = EI \frac{d^2y}{dx^2} = P(e_0 - y)$$

$$\therefore \qquad \frac{d^2 y}{dx^2} + n^2 y = n^2 e_0$$

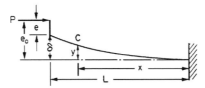

Fig. 2.11. Strut with eccentric load (one end fixed, the other free)

The solution of the expression is

$$y = A \cos nx + B \sin xn + e_0$$

At $x = 0$, $y = 0$ $\qquad \therefore A + e_0 = 0$ or $A = -e_0$

At $x = 0$, $dy/dx = 0$ $\qquad \therefore B = 0$

$$\therefore \qquad y = -e_0 \cos nx + e_0$$

Now at $x = L$, $y = \delta$

$$\therefore \qquad \delta = -e_0 \cos nL + e_0$$

$$= e_0 (1 - \cos nL)$$

$$= (\delta + e)(1 - \cos nL)$$

$$= \delta - \delta \cos nL + e - e \cos nL$$

$$\therefore \qquad \delta \cos nL = e - e \cos nL$$

$$\therefore \qquad \delta = e(\sec nL - 1)$$

or $\qquad \delta + e = e \sec nL$

This is the same form of solution as that obtained previously for pinned ends with L replaced by $2L$, i.e. the Smith–Southwell formula will apply in this case provided that the equivalent length of the strut ($l = 2L$) is used in place of L.

Thus the Smith–Southwell formula can be written in the form

$$\sigma_{\max} = \sigma \left[1 + \frac{eh}{k^2} \sec \frac{l}{2} \sqrt{\left(\frac{\sigma}{Ek^2} \right)} \right] \qquad (2.29)$$

the value of the equivalent length l to be used for any given end condition being given by the diagrams of Fig. 2.6, §2.2.

The exception to this rule, however, is the case of fixed ends where the only effect of eccentricity of loading is to increase the fixing moments within the supports at each end; there will be no effect on the deflection or stress in the strut itself. Thus, eccentricity of loading can be neglected in the case of fixed-ended struts – an important factor since most practical struts can be considered to be of this type.

2.10. Laterally loaded struts

(a) Central concentrated load

With the origin at the centre of the strut as shown in Fig. 2.12,

$$\text{B.M. at } C = EI\frac{d^2y}{dx^2} = -Py - \frac{W}{2}\left(\frac{L}{2} - x\right)$$

$$\frac{d^2y}{dx^2} + n^2y = -\frac{W}{2EI}\left(\frac{L}{2} - x\right)$$

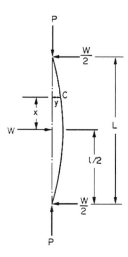

Fig. 2.12.

The solution of this equation is similar to that of §2.1(d),

i.e. $$y = A\cos nx + B\sin nx - \frac{W}{2P}\left(\frac{L}{2} - x\right)$$

Now when $x = 0$, $dy/dx = 0$ $\therefore B = -\dfrac{W}{2nP}$

and when $x = L/2$, $y = 0$ $\therefore A = \dfrac{W}{2nP}\tan\dfrac{nL}{2}$

$\therefore$ $$y = \frac{W}{2nP}\left[\tan\frac{nL}{2}\cos nx - \sin nx - n\left(\frac{L}{2} - x\right)\right]$$

The maximum deflection occurs where x is zero,

i.e. $$y_{\text{max}} = \frac{W}{2nP}\left[\tan\frac{nL}{2} - \frac{nL}{2}\right] \tag{2.30}$$

The maximum B.M. acting on the strut is at the same position and is given by

$$M_{\text{max}} = -Py_{\text{max}} - \frac{W}{2}\frac{L}{2}$$

$$= -\frac{W}{2n} \tan \frac{nL}{2} \tag{2.31}$$

(b) Uniformly distributed load

Consider now the uniformly loaded strut of Fig. 2.13 with the origin again selected at the centre but y measured from the maximum deflected position.

$$\text{B.M.}_{\cdot c} = EI\frac{d^2 y}{dx^2} = P(\delta - y) + \frac{wL}{2}\left(\frac{L}{2} - x\right) - \frac{w}{2}\left(\frac{L}{2} - x\right)^2$$

$$= P\delta - Py + \frac{w}{2}\left(\frac{L^2}{4} - x^2\right)$$

$$\therefore \qquad \frac{d^2 y}{dx^2} + n^2 y = \frac{w}{2EI}\left(\frac{L^2}{4} - x^2\right) + n^2\delta$$

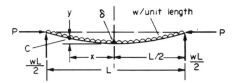

Fig. 2.13.

The solution of this equation is

$$y = A\cos nx + B\sin nx - \frac{w}{2P}\left(\frac{L^2}{4} - x^2\right) + \delta + \frac{2w}{2n^2 P}$$

i.e. $\qquad y - \delta = A\cos nx + B\sin nx - \frac{w}{2P}\left(\frac{L^2}{4} - x^2 - \frac{2}{n^2}\right)$

When $x = 0$, $dy/dx = 0$ $\therefore B = 0$

When $x = L/2$, $y = \delta$ $\therefore A = \frac{w}{n^2 P}\sec\frac{nL}{2}$

$$\therefore \qquad y - \delta = \frac{w}{n^2 P}\left[\left(\sec\frac{nL}{2}\cos nx - 1\right) - n^2\left(\frac{L^2}{8} - \frac{x^2}{2}\right)\right]$$

Thus the maximum deflection δ, when $y = 0$ and $x = 0$, is given by

$$\delta = y_{\max} = \frac{w}{n^2 p}\left[\left(\sec\frac{nL}{2} - 1\right) - \frac{n^2 L^2}{8}\right] \tag{2.32}$$

and the maximum B.M. is

$$M_{\max} = P\delta + \frac{wL^2}{8} = \frac{w}{n^2}\left(\sec\frac{nL}{2} - 1\right) \tag{2.33}$$

In the case of a member carrying a tensile load (i.e. a *tie*) together with a uniformly distributed load, the above procedure applies with the sign for P reversed. The relevant differential expression then becomes

$$\frac{d^2 y}{dx^2} - n^2 y = \frac{w}{2EI} \left[\frac{L^2}{4} - x^2 \right] + n^2 \delta$$

i.e. $(D^2 - n^2)y$ in place of $(D^2 + n^2)y$ as usual.

The solution of this equation involves hyperbolic functions but remains of identical form to that obtained previously,

i.e. $$M = A \cosh nx + B \sinh nx + \text{etc.}$$

giving $$M_{\text{max}} = \frac{w}{n^2} \left(\operatorname{sech} \frac{nL}{2} - 1 \right)$$

2.11. Alternative procedure for any strut-loading condition

If deflections are not the primary interest and only the B.M.'s and hence maximum stress are required, it is convenient to commence the analysis with a differential expression for the B.M. M.

This is most easily achieved by considering the moment divided into two parts:

(a) that due to the end load P;
(b) that due to any transverse load (M').

Thus total moment $M = -Py + M'$

Differentiating twice, $$\frac{d^2 M}{dx^2} + P\frac{d^2 y}{dx^2} = \frac{d^2 M'}{dx^2}$$

But $$P\frac{d^2 y}{dx^2} = \frac{P}{EI}\left(EI\frac{d^2 y}{dx^2}\right) = n^2 M$$

$\therefore$ $$\frac{d^2 M}{dx^2} + n^2 M = \frac{d^2 M'}{dx^2}$$

The general solution will be of the form

$$M = A \cos nx + B \sin nx + \text{ particular solution}$$

Now for zero transverse load (or for any concentrated load) $(d^2 M'/dx^2)$ is zero, the particular solution is also zero, and the solution for the above expression is in the form

$$M = A \cos nx + B \sin nx$$

Thus, for an **eccentrically loaded strut** (Smith–Southwell):

shear force $= \dfrac{dM}{dx} = 0$ when $x = 0$ $\therefore B = 0$

and $$M = Pe \qquad \text{when } x = \tfrac{1}{2}L \quad \therefore A = Pe \sec \frac{nL}{2}$$

Therefore substituting, $M = Pe \sec \dfrac{nL}{2} \cos nx$

and $\qquad\qquad \boldsymbol{M_{\text{max}} = Pe \sec \dfrac{nL}{2}} \quad$ as before

For a **central concentrated load** (see Fig. 2.12)

$$M' = \frac{W}{2}\left(\frac{L}{2} - x\right)$$

$\therefore \qquad\qquad \dfrac{d^2 M'}{dx^2} = 0 \quad$ and the particular solution $= 0$

$\therefore \qquad\qquad M = A\cos nx + B\sin nx$

$\qquad$ Shear force $= \dfrac{dM}{dx} = \dfrac{W}{2} \qquad$ when $x = 0 \qquad \therefore B = \dfrac{W}{2n}$

and $\qquad\qquad M = 0 \qquad$ when $x = \tfrac{1}{2}L \quad \therefore A = -\dfrac{W}{2n}\tan\dfrac{nL}{2}$

$\therefore \qquad\qquad M = -\dfrac{W}{2n}\left[\tan\dfrac{nL}{2}\cos nx + \sin nx\right]$

and $\qquad\qquad \boldsymbol{M_{\text{max}} = -\dfrac{W}{2n}\tan\dfrac{nL}{2}} \quad$ as before

For **a uniformly distributed lateral load** (see Fig. 2.13)

$$M' = \frac{w}{2}\left[\frac{L^2}{4} - x^2\right] \quad \text{(see page 47)}$$

$\therefore \qquad\qquad \dfrac{d^2 M'}{dx^2} = -w$

Hence $\qquad \dfrac{d^2 M}{dx^2} + n^2 M = -w \quad$ and the particular integral is $\dfrac{w}{n^2}$

$\therefore \qquad\qquad M = A\cos nx + B\sin nx - w/n^2$

Now when $x = 0$, $dM/dx = 0 \quad \therefore B = 0$

and when $x = L/2$, $M = 0 \quad \therefore A = \dfrac{w}{n^2}\sec\dfrac{nL}{2}$

$\therefore \qquad\qquad M = \dfrac{w}{n^2}\left[\sec\dfrac{nL}{2}\cos nx - 1\right]$

and $\qquad\qquad \boldsymbol{M_{\text{max}} = \dfrac{w}{n^2}\left[\sec\dfrac{nL}{2} - 1\right]} \quad$ as before

2.12. Struts with unsymmetrical cross-sections

The formulae derived in the preceding paragraphs have assumed that buckling takes place about an axis of symmetry. Loading is then normally applied to produce bending on the strongest or major principal axis (that about which I has a maximum value) so that buckling is assumed to occur about the minor axis. It is also assumed that the end conditions allow rotation in this direction and this is normally achieved by loading through ball ends.

For sections with only one axis of symmetry, e.g. channel or T-sections, the shear centre is not coincident with the centroid and torsional effects are often introduced. These may, in some cases, affect the failure condition of the strut. Certainly, in the case of totally unsymmetrical sections, the loading condition always involves considerable torsion and the theoretical buckling load has little relevance. One popular form of section which falls in this category is the unequal-leg angle section.

Some sections, e.g. cruciform sections, are subject to both flexural and torsional buckling and the reader is referred to more advanced texts for the methods of treatment is such cases.

A special form of failure is associated with hollow low carbon steel columns with small thickness to diameter ratios when the strut is found to *crinkle*, i.e. the material forms into folds when the direct stress is approximately equal to the yield stress. Southwell has investigated this problem and produced the formula

$$\sigma = E\frac{t}{R}\left[\frac{1}{3(1-v^2)}\right]^{1/2}$$

where σ is the stress causing yielding, R is the mean radius of the column and t is the thickness. It should be noted, however, that this type of failure is not common since very small t/R ratios of the order of 1/400 are required before crinkling can occur.

Examples

Example 2.1

Two 300 mm × 120 mm I-section joists are united by 12 mm thick plates as shown in Fig. 2.14 to form a 7 m long stanchion. Given a factor of safety of 3, a compressive yield stress of 300 MN/m^2 and a constant a of 1/7500, determine the allowable load which can be carried by the stanchion according to the Rankine–Gordon formulae.

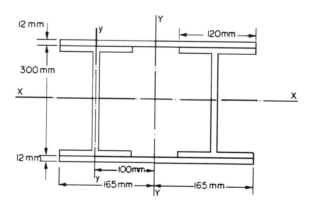

Fig. 2.14.

The relevant properties of each joist are:

$$I_{xx} = 96 \times 10^{-6} \text{ m}^4, \quad I_{yy} = 4.2 \times 10^{-6} \text{ m}^4, \quad A = 6 \times 10^{-3} \text{ m}^2$$

Solution

For the strut of Fig. 2.14:

$$I_{xx} \text{ for joists} = 2 \times 96 \times 10^{-6} = 192 \times 10^{-6} \text{ m}^4$$

$$I_{xx} \text{ for plates} = 0.33 \times \frac{0.324^3}{12} - \frac{0.33 \times 0.300^3}{12}$$

$$= \frac{0.33}{12}[0.034 - 0.027] = 192.5 \times 10^{-6} \text{ m}^4$$

$$\therefore \qquad \text{total } I_{xx} = (192 + 192.5)10^{-6} = 384.5 \times 10^{-6} \text{ m}^4$$

From the parallel axis theorem:

$$I_{yy} \text{ for joists} = 2(4.2 \times 10^{-6} + 6 \times 10^{-3} \times 0.1^2)$$

$$= 128.4 \times 10^{-6} \text{ m}^4$$

$$\text{and} \qquad I_{yy} \text{ for plates} = 2 \times 0.012 \times \frac{0.33^3}{12} = 71.9 \times 10^{-6} \text{ m}^4$$

$$\therefore \qquad \text{total } I_{yy} = 200.3 \times 10^{-6} \text{ m}^4$$

Now the smallest value of the Rankine–Gordon stress σ_R is given when k, and hence I, is a minimum.

$$\therefore \qquad \text{smallest } I = I_{yy} = 200.3 \times 10^{-6} = Ak^2$$

$$\text{total area } A = 2 \times 6 \times 10^{-3} + 2 \times 0.33 \times 12 \times 10^{-3} = 19.92 \times 10^{-3}$$

$$\therefore \qquad 19.92 \times 10^{-3}k^2 = 200.3 \times 10^{-6}$$

$$\therefore \qquad k^2 = \frac{200.3 \times 10^{-6}}{19.92 \times 10^{-3}} = 10.05 \times 10^{-3}$$

$$\therefore \qquad \left(\frac{L}{k}\right)^2 = \frac{7^2}{10.05 \times 10^{-3}} = 4.9 \times 10^3$$

$$\text{and} \qquad \sigma_R = \frac{\sigma_y}{1 + a\left(\dfrac{L}{k}\right)^2} = \frac{300 \times 10^6}{1 + \dfrac{4.9 \times 10^3}{7500}}$$

$$\therefore \qquad = \frac{300 \times 10^6}{1.653} = 181.45 \text{ MN/m}^2$$

$$\therefore \qquad \text{allowable load} = \sigma_R \times A = 181.45 \times 10^6 \times 19.92 \times 10^{-3} = 3.61 \text{ MN}$$

With a factor of safety of 3 the maximum permissible load therefore becomes

$$P_{\text{max}} = \frac{3.61 \times 10^6}{3} = \mathbf{1.203 \text{ MN}}$$

Example 2.2

An 8 m long column is constructed from two 400 mm × 250 mm I-section joists joined as shown in Fig. 2.15. One end of the column is arranged to be fixed and the other free and a load equal to one-third of the Euler load is applied. Determine the load factor provided if the Perry–Robertson formula is used as the basis for design.

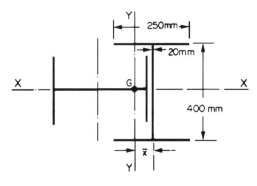

Fig. 2.15.

For each joist:

$$I_{max} = 213 \times 10^{-6} \text{ m}^4, \quad I_{min} = 9.6 \times 10^{-6} \text{ m}^4, \quad A = 8.4 \times 10^{-3} \text{ m}^2,$$

with web and flange thicknesses of 20 mm. For the material of the joist, $E = 208 \text{ GN/m}^2$ and $\sigma_y = 270 \text{ MN/m}^2$.

Solution

To find the position of the centroid G of the built-up section take moments of area about the centre line of the vertical joist.

$$2 \times 8.4 \times 10^{-3} \bar{x} = 8.4 \times 10^{-3}(200 + 10)10^{-3}$$

$$\bar{x} = \frac{210}{2} \times 10^{-3} = 105 \text{ mm}$$

Now $$I_{xx} = (213 + 9.6)10^{-6} = 222.6 \times 10^{-6} \text{ m}^4$$

and $$I_{yy} = [213 + 8.4(210 - 105)^2]10^{-6} + [9.6 + 8.4 \times 105^2]10^{-6}$$

i.e. greater than I_{xx}.

∴   least $I = 222.6 \times 10^{-6} \text{ m}^4$

∴   least $k^2 = \dfrac{222.6 \times 10^{-6}}{2 \times 8.4 \times 10^{-3}} = 13.25 \times 10^{-3}$

Now Euler load for fixed–free ends

$$= \frac{\pi^2 EI}{4L^2} = \frac{\pi^2 \times 208 \times 10^9 \times 222.6 \times 10^{-6}}{4 \times 8^2}$$

$$= 1786 \times 10^3 = 1.79 \text{ MN}$$

Therefore actual load applied to the column

$$= \frac{1.79}{3} = 0.6 \text{ MN}$$

i.e.
$$\text{actual stress} = \frac{\text{load}}{\text{area}} = \frac{0.6 \times 10^6}{2 \times 8.4 \times 10^{-3}}$$

$$= \mathbf{35.7 \ MN/m^2}$$

The Perry–Robertson constant is

$$\eta = 0.3 \left(\frac{L}{100k}\right)^2 = 0.3 \left(\frac{8^2}{10^4 \times 13.25 \times 10^{-3}}\right)$$

$$= 0.144$$

and
$$N\sigma = \frac{(\sigma_y + 1.144\sigma_e)}{2} - \sqrt{\left\{\left[\frac{(\sigma_y + 1.144\sigma_e)}{2}\right]^2 - \sigma_y\sigma_e\right\}}$$

But
$$\sigma_y = 270 \text{ MN/m}^2 \text{ and } \sigma_e = \frac{1.79 \times 10^6}{2 \times 8.4 \times 10^{-3}} = 106.5 \text{ MN/m}^2$$

i.e. in units of MN/m^2:

$$\therefore \qquad N\sigma = \frac{(270 + 121.8)}{2} - \sqrt{\left\{\left[\frac{270 + 121.8}{2}\right]^2 - 270 \times 106.5\right\}}$$

$$= 196 - 98 = 98$$

$$\therefore \qquad \text{load factor } N = \frac{98}{35.7} = \mathbf{2.75}$$

Example 2.3

Determine the maximum compressive stress set up in a 200 mm × 60 mm I-section girder carrying a load of 100 kN with an eccentricity of 6 mm from the critical axis of the section (see Fig. 2.16). Assume that the ends of the strut are pin-jointed and that the overall length is 4 m.

Take $I_{yy} = 3 \times 10^{-6} \text{ m}^4$, $A = 6 \times 10^{-3} \text{ m}^2$, $E = 207 \text{ GN/m}^2$.

Solution

Normal stress on the section

$$\sigma = \frac{P}{A} = \frac{100 \times 10^3}{6 \times 10^{-3}} = \frac{100}{6} \text{ MN/m}^2$$

$$I = Ak^2 = 3 \times 10^{-6} \text{ m}^4$$

$$\therefore \qquad k^2 = \frac{3 \times 10^{-6}}{6 \times 10^{-3}} = 5 \times 10^{-4} \text{ m}^2$$

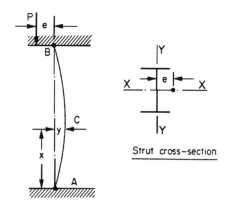

Fig. 2.16.

Now from eqn. (2.26)

$$\sigma_{max} = \sigma \left[1 + \frac{eh}{k^2} \sec \frac{L}{2} \sqrt{\left(\frac{\sigma}{Ek^2} \right)} \right]$$

∴   with $e = 6$ mm   and $h = 30$ mm

$$\sigma_{max} = \frac{100}{6} \left[1 + \frac{30 \times 6 \times 10^{-6}}{5 \times 10^{-4}} \sec 2 \sqrt{\left(\frac{100 \times 10^6 \times 10^4}{6 \times 207 \times 10^9 \times 5} \right)} \right]$$

$$= \frac{100}{6} [1 + 0.36 \sec 2 \sqrt{(0.161)}]$$

$$= \frac{100}{6} [1 + 0.36 \times 1.44] = \textbf{25.3 MN/m}^2$$

Example 2.4

A horizontal strut 2.5 m long is constructed from rectangular section steel, 50 mm wide by 100 mm deep, and mounted with pinned ends. The strut carries an axial load of 120 kN together with a uniformly distributed lateral load of 5 kN/m along its complete length. If $E = 200$ GN/m^2 determine the maximum stress set up in the strut.

Check the result using the approximate Perry method with

$$M_{max} = M_0 \left[\frac{P_e}{P_e - P} \right]$$

Solution

From eqn. (2.34)

$$M_{max} = \frac{w}{n^2} \left(\sec \frac{nL}{2} - 1 \right)$$

where $$n^2 = \frac{P}{EI} = \frac{120 \times 10^3 \times 12}{200 \times 10^9 \times 50 \times 100^3 \times 10^{-12}}$$

$$= 0.144$$

$$\therefore \qquad \frac{nL}{2} = \frac{2.5}{2}\sqrt{(0.144)} = 0.474 \text{ radian}$$

$$\therefore \qquad M_{max} = \frac{5 \times 10^3}{0.144}(\sec 0.474 - 1)$$

$$= 34.7 \times 10^3(1.124 - 1) = 4.3 \times 10^3 \text{ Nm}$$

The maximum stress due to the axial load and the eccentricity caused by bending is then given by

$$\sigma_{max} = \frac{P}{A} + \frac{My}{I}$$

$$= \frac{120 \times 10^3}{(0.1 \times 0.05)} + \frac{4.34 \times 10^3 \times 0.05 \times 12}{(50 \times 100^3)10^{-12}}$$

$$= 24 \times 10^6 + 51.6 \times 10^6$$

$$= \mathbf{75.6 \ MN/m^2}$$

Using the approximate Perry method,

$$M_{max} = M_0\left[\frac{P_e}{P_e - P}\right]$$

where

$$M_0 = \text{B.M. due to lateral load only} = \frac{wL^2}{8}$$

But

$$P_e = \frac{\pi^2 EI}{L^2} = \frac{\pi^2 \times 200 \times 10^9}{2.5^2} \times \frac{(50 \times 100^3)10^{-12}}{12}$$

$$= 1.316 \text{ MN}$$

$$\therefore \qquad M_{max} = \frac{wL^2}{8}\left[\frac{P_e}{P_e - P}\right]$$

$$= \frac{5 \times 10^3 \times 2.5^2}{8}\left[\frac{1316 \times 10^3}{(1316 - 120)10^3}\right]$$

$$= 4.3 \times 10^3 \text{ Nm}$$

In this case, therefore, the approximate method yields the same answer for maximum B.M. as the full solution. The maximum stress will then also be equal to that obtained above, i.e. **75.6 MN/m²**.

Example 2.5

A hollow circular steel strut with its ends fixed in position has a length of 2 m, an outside diameter of 100 mm and an inside diameter of 80 mm. Assuming that, before loading, there is an initial sinusoidal curvature of the strut with a maximum deflection of 5 mm, determine the maximum stress set up due to a compressive end load of 200 kN. $E = 208 \text{ GN/m}^2$.

Solution

The assumed sinusoidal initial curvature may be expressed alternatively in the complementary cosine form

$$y_0 = \delta_0 \cos \frac{\pi x}{L} \quad \text{(Fig. 2.17)}$$

Now when P is applied, y_0 increases to y and the central deflection increases from $\delta_0 = 5$ mm to δ.

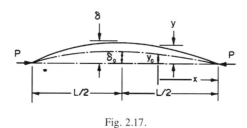

Fig. 2.17.

For the above initial curvature it can be shown that

$$\delta = \left[\frac{P_e}{P_e - P} \right] \delta_0$$

∴ $$\text{maximum B.M.} = P\delta_0 \left[\frac{P_e}{P_e - P} \right]$$

where P_e for ends fixed in direction only $= \dfrac{\pi^2 EI}{L^2}$

$$I = \frac{\pi}{64}(0.1^4 - 0.08^4) = \frac{\pi}{64}(1 - 0.41)10^{-4} = 2.896 \times 10^{-6} \text{m}^4$$

∴ $$P_e = \frac{\pi^2 \times 208 \times 10^9 \times 2.89 \times 10^{-6}}{4} = 1.486 \text{ MN}$$

∴ $$\text{maximum B.M.} = 200 \times 10^3 \times 5 \times 10^{-3} \left[\frac{1486 \times 10^3}{(1486 - 200)10^3} \right] = 1.16 \text{ kN m}$$

∴ $$\text{maximum stress} = \frac{P}{A} + \frac{My}{I} = \frac{200 \times 10^3 \times 4}{\pi(0.1^2 - 0.08^2)} + \frac{1.16 \times 10^3 \times 0.05}{2.89 \times 10^{-6}}$$

$$= 70.74 \times 10^6 + 20.07 \times 10^6$$

$$= \mathbf{90.8 \text{ MN/m}^2}$$

Problems

2.1 (A/B). Compare the crippling loads given by the Euler and Rankine–Gordon formulae for a pin-jointed cylindrical strut 1.75 m long and of 50 mm diameter. (For Rankine–Gordon use $\sigma_y = 315$ MN/m²; $a = 1/7500$; $E = 200$ GN/m².)   [197.7, 171 kN.]

2.2 (A/B). In an experiment an alloy rod 1 m long and of 6 mm diameter, when tested as a simply supported beam over a length of 750 mm, was found to have a maximum deflection of 5.8 mm under the action of a central load of 5 N.

(a) Find the Euler buckling load when this rod is tested as a strut, pin-jointed and guided at both ends.

(b) What will be the central deflection of this strut when the material reaches a yield stress of 240 MN/m²?

$$\text{(Clue: maximum stress} = \frac{P}{A} \pm \frac{My}{I} \text{ where } M = P \times \delta_{\max}.)$$ [74.8 N; 67 mm.]

2.3 (B) A steel strut is built up of two T-sections riveted back to back to form a cruciform section of overall dimensions 150 mm × 220 mm. The dimensions of each T-section are 150 mm × 15 mm × 110 mm high. The ends of the strut are rigidly secured and its effective length is 7 m. Find the maximum safe load that this strut can carry with a factor of safety of 5, given $\sigma_y = 315$ MN/m² and $a = 1/30000$ in the Rankine–Gordon formula.
[192 kN.]

2.4 (B). State the assumptions made when deriving the Euler formula for a strut with pin-jointed ends. Derive the Euler crippling load for such a strut–the general equation of bending and also the solution of the differential equation may be assumed.

A straight steel rod 350 mm long and of 6 mm diameter is loaded axially in compression until it buckles. Assuming that the ends are pin-jointed, find the critical load using the Euler formula. Also calculate the maximum central deflection when the material reaches a yield stress of 300 MN/m² compression. Take $E = 200$ GN/m².
[1.03 kN; 5.46 mm.]

2.5 (B). A steel stanchion 5 m long is to be built of two I-section rolled steel joists 200 mm deep and 150 mm wide flanges with a 350 mm wide × 20 mm thick plate riveted to the flanges as shown in Fig. 2.18. Find the spacing of the joists so that for an axially applied load the resistance to buckling may be the same about the axes XX and YY. Find the maximum allowable load for this condition with ends pin-jointed and guided, assuming $a = 1/7500$ and $\sigma_y = 315$ MN/m² in the Rankine formula.

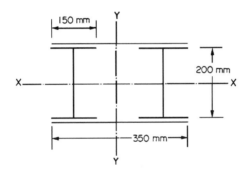

Fig. 2.18.

If the maximum working stress in compression σ for this strut is given by $\sigma = 135[1 - 0.005 \, L/k]$ MN/m², what factor of safety must be used with the Rankine formula to give the same result? For each R.S.J. $A = 6250$ mm², $k_x = 85$ mm, $k_y = 35$ mm. [180.6 mm, 6.23 MN; 2.32.]

2.6 (B). A stanchion is made from two 200 mm × 75 mm channels placed back to back, as shown in Fig. 2.19, with suitable diagonal bracing across the flanges. For each channel $I_{xx} = 20 \times 10^{-6} \text{m}^4$, $I_{yy} = 1.5 \times 10^{-6}$ m⁴, the cross-sectional area is 3.5×10^{-3} m² and the centroid is 21 mm from the back of the web.

At what value of p will the radius of gyration of the whole cross-section be the same about the X and Y axes? The strut is 6 m long and is pin-ended. Find the Euler load for the strut and discuss briefly the factors which cause the actual failure load of such a strut to be less than the Euler load. $E = 210$ GN/m². [163.6 mm; 2.3 MN.]

2.7 (B). In tests it was found that a tube 2 m long, 50 mm outside diameter and 2 mm thick when used as a pin-jointed strut failed at a load of 43 kN. In a compression test on a short length of this tube failure occurred at a load of 115 kN.

(a) Determine whether the value of the critical load obtained agrees with that given by the Euler theory.

(b) Find from the test results the value of the constant a in the Rankine–Gordon formula. Assume $E = 200$ GN/m².
[Yes; 1/7080.]

2.8 (B). Plot, on the same axes, graphs of the crippling stresses for pin-ended struts as given by the Euler and Rankine–Gordon formulae, showing the variation of stress with slenderness ratio

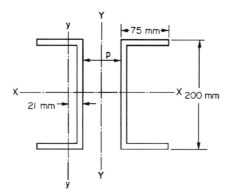

Fig. 2.19.

For the Euler formula use L/k values from 80 to 150, and for the Rankine formula L/k from 0 to 150, with $\sigma_y = 315$ MN/m^2 and $a = 1/7500$.

From the graphs determine the values of the stresses given by the two formulae when $L/k = 130$ and the slenderness ratio required by both formulae for a crippling stress of 135 MN/m^2. $E = 210$ GN/m^2.

[122.6 MN/m^2, 96.82 MN/m^2; 124,100.]

2.9 (B/C). A timber strut is 75 mm × 75 mm square-section and is 3 m high. The base is rigidly built-in and the top is unrestrained. A bracket at the top of the strut carries a vertical load of 1 kN which is offset 150 mm from the centre-line of the strut in the direction of one of the principal axes of the cross-section. Find the maximum stress in the strut at its base cross-section if $E = 9$ GN/m^2. [I.Mech.E.] [2.3 MN/m^2.]

2.10 (B/C). A slender column is built-in at one end and an eccentric load is applied at the free end. Working from first principles find the expression for the maximum length of column such that the deflection of the free end does not exceed the eccentricity of loading. [I.Mech.E.] [sec^{-1} $2/\sqrt{(P/EI)}$.]

2.11 (B/C). A slender column is built-in one end and an eccentric load of 600 kN is applied at the other (free) end. The column is made from a steel tube of 150 mm o.d. and 125 mm i.d. and it is 3 m long. Deduce the equation for the deflection of the free end of the beam and calculate the maximum permissible eccentricity of load if the maximum stress is not to exceed 225 MN/m^2. $E = 200$ GN/m^2. [I.Mech.E.] [4 mm.]

2.12 (B). A compound column is built up of two 300 mm × 125 mm R.S.J.s arranged as shown in Fig. 2.20. The joists are braced together; the effects of this bracing on the stiffness may, however, be neglected. Determine the safe height of the column if it is to carry an axial load of 1 MN. Properties of joists: $A = 6 \times 10^{-3}$ m^2; $k_{yy} = 27$ mm; $k_{xx} = 125$ mm.

The allowable stresses given by BS449: 1964 may be found from the graph of Fig. 2.9. [8.65 m.]

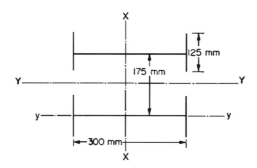

Fig. 2.20.

2.13 (B). A 10 mm long column is constructed from two 375 mm × 100 mm channels placed back to back with a distance *h* between their centroids and connected together by means of narrow batten plates, the effects of which may be ignored. Determine the value of *h* at which the section develops its maximum resistance to buckling.

Estimate the safe axial load on the column using the Perry–Robertson formula (a) with a load factor of 2, (b) with a factor of safety of 2. For each channel $I_{xx} = 175 \times 10^{-6}$ m^4, $I_{yy} = 7 \times 10^{-6}$ m^4, $A = 6.25 \times 10^{-3}$ m^2, $E = 210$ GN/m^2 and yield stress $= 300$ MN/m^2. Assume $\eta = 0.003\ L/k$ and that the ends of the column are effectively pinned. [328 mm; 1.46, 1.55 MN.]

2.14 (B). (a) Compare the buckling loads that would be obtained from the Rankine–Gordon formula for two identical steel columns, one having both ends fixed, the other having pin-jointed ends, if the slenderness ratio is 100.

(b) A steel column, 6 m high, of square section 120 mm × 120 mm, is designed using the Rankine–Gordon expression to be used as a strut with both ends pin-jointed.

The values of the constants used were $a = 1/7500$, and $\sigma_c = 300$ MN/m^2. If, in service, the load is applied axially but parallel to and a distance *x* from the vertical centroidal axis, calculate the maximum permissible value of *x*. Take $E = 200$ GN/m^2. [7.4; 0.756 m.]

2.15 (B). Determine the maximum compressive stress set up in a 200 mm × 60 mm I-section girder carrying a load of 100 kN with an eccentricity of 6 mm. Assume that the ends of the strut are pin-jointed and that the overall length is 4 m.

Take $I = 3 \times 10^{-6}$ m^4; $A = 6 \times 10^{-3}$ m^2 and $E = 207$ GN/m^2. [25.4 MN/m^2.]

2.16 (B). A slender strut, initially straight, is pinned at each end. It is to be subjected to an eccentric compressive load whose line of action is parallel to the original centre-line of the strut.

(a) Prove that the central deflection *y* of the strut, relative to its initial centre-line, is given by the expression

$$y = e \left[\sec \frac{1}{2} \sqrt{\left(\frac{PL^2}{EI} \right)} - 1 \right]$$

where *P* is the applied load, *L* is the effective length of the strut, *e* is the eccentricity of the line of action of the load from the initially straight strut axis and *EI* is the flexural rigidity of the strut cross-section.

(b) Using the above formula, and assuming that the strut is made of a ductile material, show that, for a maximum compressive stress, σ, the value of *P* is given by the expression

$$P = \frac{\sigma A}{\dfrac{he}{k^2} \sec \dfrac{1}{2} \sqrt{\left(\dfrac{PL^2}{EI} \right)} + 1}$$

the symbols *A*, *h* and *k* having their usual meanings.

(c) Such a strut, of constant tubular cross-section throughout, has an outside diameter of 64 mm, a principal second moment of area of 52×10^{-8} m^4 and a cross-sectional area of 12.56×10^{-4} m^2. The effective length of the strut is 2.5 m. If $P = 120$ kN and $\sigma = 300$ MN/m^2, determine the permissible value of *e*. Take $E = 200$ GN/m^2. [B.P.] [6.25 mm.]

2.17 (C). A strut of length *L* has each end fixed in an elastic material which can exert a restraining moment μ per radian. Prove that the critical load *P* is given by the equation

$$P + \mu \sqrt{\left(\frac{P}{EI} \right)} \tan \frac{L}{2} \sqrt{\left(\frac{P}{EI} \right)} = 0$$

The designed buckling load of a 1 m long strut, assuming the ends to be rigidly fixed, was 2.5 kN. If, during service, the ends were found to rotate with each mounting exerting a restraining moment of 1 kN m per radian, show that the buckling load decreases by 20%. [C.E.I.]

2.18 (C). A uniform elastic bar of circular cross-section and of length *L*, free at one end and rigidly built-in at the other end, is subjected to a single concentrated load *P* at the free end. In general the line of action of *P* may be at an angle θ to the axis of the bar ($0 < \theta < \pi/2$) so that the bar is simultaneously compressed and bent. For this general case:

(a) Show that the deflection at the free end is given by

$$\delta = \tan\theta \left\{ \left(\frac{\tan mL}{m} - L \right) \right\}$$

(b) Hence show that as $\theta \to \pi/2$, then $\delta \to PL^3/3EI$
(c) Show that when $\theta = 0$ no deflection unless P has certain particular values.

Note that in the above, m^2 denotes $P\cos\theta/EI$.
The following expression may be used in part (b) where appropriate:

$$\tan\alpha = \alpha + \frac{\alpha^3}{3} + \frac{2\alpha^5}{15} \hspace{3cm} \text{[City U.]}$$

2.19 (C). A slender strut of length L is encastré at one end and pin-jointed at the other. It carries an axial load P and a couple M at the pinned end. If its flexural rigidity is EI and $P/EI = n$, show that the magnitude of the couple at the fixed end is

$$M \left[\frac{nL - \sin nL}{nL\cos nL - \sin nL} \right]$$

What is the value of this couple when (a) P is one-quarter the Euler critical load and (b) P is zero?

[U.L.] [0.571 M, 0.5 M.]

2.20 (C). An initially straight strut of length L has lateral loading w per metre and a longitudinal load P applied with an eccentricity e at both ends.

If the strut has area A, second moment of area I, section modulus Z and the end moments and lateral loading have opposing effects, find an expression for the central bending moment and show that the maximum stress at the centre will be equal to

$$\frac{P}{A} + \frac{\left(Pe - \dfrac{wEI}{P} \right) \sec \dfrac{L}{2}\sqrt{\left(\dfrac{P}{EI} \right)} + \dfrac{wEI}{P}}{Z} \hspace{2cm} \text{[U.L.]}$$

CHAPTER 3

STRAINS BEYOND THE ELASTIC LIMIT

Summary

For rectangular-sectioned beams strained up to and beyond the elastic limit, i.e. for *plastic bending*, the bending moments (B.M.) which the beam can withstand at each particular stage are:

maximum elastic moment
$$M_E = \frac{BD^2}{6}\sigma_y$$

partially plastic moment
$$M_{PP} = \frac{B\sigma_y}{12}[3D^2 - d^2]$$

fully plastic moment
$$M_{FP} = \frac{BD^2}{4}\sigma_y$$

where σ_y is the stress at the elastic limit, or *yield stress*.

$$\text{Shape factor } \lambda = \frac{\text{fully plastic moment}}{\text{maximum elastic moment}}$$

For **I-section beams**:

$$M_E = \sigma_y \left[\frac{BD^3}{12} - \frac{bd^3}{12}\right]\frac{2}{D}$$

$$M_{FP} = \sigma_y \left[\frac{BD^2}{4} - \frac{bd^2}{4}\right]$$

The position of the neutral axis (N.A.) for fully plastic unsymmetrical sections is given by:

$$\text{area of section above or below N.A.} = \tfrac{1}{2} \times \text{total area of cross-section}$$

Deflections of partially plastic beams are calculated on the basis of the elastic areas only.

In plastic limit or ultimate collapse load procedures the normal elastic safety factor is replaced by a load factor as follows:

$$\text{load factor} = \frac{\text{collapse load}}{\text{allowable working load}}$$

For **solid shafts**, radius R, strained up to and beyond the elastic limit in shear, i.e. for *plastic torsion*, the torques which can be transmitted at each stage are

maximum elastic torque
$$T_E = \frac{\pi R^3}{2}\tau_y$$

partially plastic torque
$$T_{PP} = \frac{\pi \tau_y}{6}[4R^3 - R_1^3] \quad \text{(yielding to radius } R_1)$$

61

fully plastic torque $$T_{FP} = \frac{2\pi R^3}{3}\tau_y$$

where τ_y is the shear stress at the elastic limit, or shear yield stress. Angles of twist of partially plastic shafts are calculated on the basis of the elastic core only.

For **hollow shafts**, inside radius R_1, outside radius R yielded to radius R_2,

$$T_{PP} = \frac{\pi\tau_y}{6R_2}[4R^3R_2 - R_2^4 - 3R_1^4]$$

$$T_{FP} = \frac{2\pi\tau_y}{3}[R^3 - R_1^3]$$

For **eccentric loading** of rectangular sections the fully plastic moment is given by

$$M_{FP} = \frac{BD^2}{4}\sigma_y - \frac{P^2N^2}{4B\sigma_y}$$

where P is the axial load, N the load factor and B the width of the cross-section.

The maximum allowable moment is then given by

$$M = \frac{BD^2}{4N}\sigma_y - \frac{P^2N}{4B\sigma_y}$$

For a **solid rotating disc**, radius R, the collapse speed ω_p is given by

$$\omega_p^2 = \frac{3\sigma_y}{\rho R^2}$$

where ρ is the density of the disc material.

For **rotating hollow discs** the collapse speed is found from

$$\omega_p^2 = \frac{3\sigma_y}{\rho}\left[\frac{R_2 - R_1}{R_2^3 - R_1^3}\right]$$

Introduction

When the design of components is based upon the elastic theory, e.g. the simple bending or torsion theory, the dimensions of the components are arranged so that the maximum stresses which are likely to occur under service loading conditions do not exceed the allowable working stress for the material in either tension or compression. The allowable working stress is taken to be the yield stress of the material divided by a convenient safety factor (usually based on design codes or past experience) to account for unexpected increase in the level of service loads. If the maximum stress in the component is likely to exceed the allowable working stress, the component is considered unsafe, yet it is evident that complete failure of the component is unlikely to occur even if the yield stress is reached at the outer fibres provided that some portion of the component remains elastic and capable of carrying load, i.e. the strength of a component will normally be much greater than that assumed on the basis of initial yielding at any position. To take advantage of the inherent additional strength, therefore, a different design procedure is used which is often referred to as *plastic limit design*. The revised design procedures are based upon a number of basic assumptions about the material behaviour.

Figure 3.1 shows a typical stress–strain curve for annealed low carbon steel indicating the presence of both upper and lower yield points and strain-hardening characteristics.

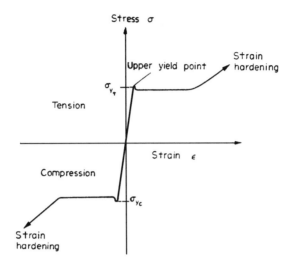

Fig. 3.1. Stress–strain curve for annealed low-carbon steel indicating upper and lower yield points and strain-hardening characteristics.

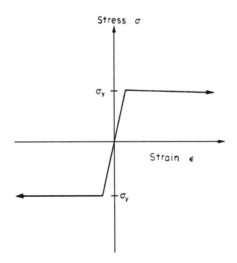

Fig. 3.2. Assumed stress–curve for plastic theory – no strain-hardening, equal yield points, $\sigma_{y_t} = \sigma_{y_c} = \sigma_y$.

Figure 3.2 shows the assumed material behaviour which:

(a) ignores the presence of upper and lower yields and suggests only a single yield point;
(b) takes the yield stress in tension and compression to be equal;

(c) assumes that yielding takes place at constant strain thereby ignoring any strain-hardening characteristics. Thus, once the material has yielded, stress is assumed to remain constant throughout any further deformation.

It is further assumed, despite assumption (c), that transverse sections of beams in bending remain plane throughout the loading process, i.e. strain is proportional to distance from the neutral axis.

   It is now possible on the basis of the above assumptions to determine the moment which must be applied to produce:

(a) the maximum or limiting elastic conditions in the beam material with yielding just initiated at the outer fibres;

(b) yielding to a specified depth;

(c) yielding across the complete section.

The latter situation is then termed a fully plastic state, or *"plastic hinge"*. Depending on the support and loading conditions, one or more plastic hinges may be required before complete collapse of the beam or structure occurs, the load required to produce this situation then being termed the *collapse load*. This will be considered in detail in §3.6.

3.1. Plastic bending of rectangular-sectioned beams

   Figure 3.3(a) shows a rectangular beam loaded until the yield stress has just been reached in the outer fibres. The beam is still completely elastic and the bending theory applies, i.e.

$$M = \frac{\sigma I}{y}$$

∴   maximum elastic moment $= \sigma_y \times \dfrac{BD^3}{12} \times \dfrac{2}{D}$

$$M_E = \frac{BD^2}{6}\sigma_y \tag{3.1}$$

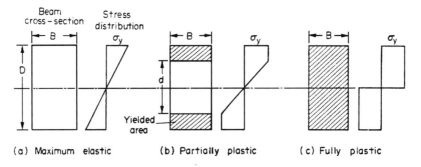

Fig. 3.3. Plastic bending of rectangular-section beam.

If loading is then increased, it is assumed that instead of the stress at the outside increasing still further, more and more of the section reaches the yield stress σ_y. Consider the stage shown in Fig. 3.3(b).

Partially plastic moment,

$$M_{PP} = \text{moment of elastic portion} + \text{total moment of plastic portion}$$

$$\therefore \qquad M_{PP} = \frac{Bd^2}{6}\sigma_y + 2\left\{\sigma_y \times B\left[\frac{D}{2} - \frac{d}{2}\right]\left[\frac{1}{2}\left(\frac{D}{2} - \frac{d}{2}\right) + \frac{d}{2}\right]\right\}$$

$$\underset{\text{stress}}{} \qquad \underset{\text{area}}{} \qquad \underset{\text{moment arm}}{}$$

$$M_{PP} = \sigma_y\left[\frac{Bd^2}{6} + \frac{B}{4}(D - d)(D + d)\right]$$

$$= \frac{B\sigma_y}{12}[2d^2 + 3(D^2 - d^2)] = \frac{B\sigma_y}{12}[\mathbf{3D^2 - d^2}] \qquad (3.2)$$

When loading has been continued until the stress distribution is as in Fig. 3.3(c) (assumed), the beam with collapse. The moment required to produce this fully plastic state can be obtained from eqn. (3.2), since d is then zero,
i.e.

$$\text{fully plastic moment, } M_{FP} = \frac{B\sigma_y}{12} \times 3D^2 = \frac{BD^2}{4}\sigma_y \qquad (3.3)$$

This is the moment therefore which produces a plastic hinge in a rectangular-section beam.

3.2. Shape factor – symmetrical sections

The shape factor is defined as the ratio of the moments required to produce fully plastic and maximum elastic states:

$$\text{shape factor } \lambda = \frac{\mathbf{M_{FP}}}{\mathbf{M_E}} \qquad (3.4)$$

It is a factor which gives a measure of the increase in strength or load-carrying capacity which is available beyond the normal elastic design limits for various shapes of section, e.g. for the *rectangular section* above,

$$\text{shape factor } = \frac{BD^2}{4}\sigma_y \bigg/ \frac{BD^2}{6}\sigma_y = \mathbf{1.5}$$

Thus rectangular-sectioned beams can carry 50% additional moment to that which is required to produce initial yielding at the edge of the beam section before a fully plastic hinge is formed. (It will be shown later that even greater strength is available beyond this stage depending on the support conditions used.) It must always be remembered, however, that should the stresses exceed the yield at any time during service there will be some associated *permanent set* or deflection when load is removed, and consideration should be given to whether or not this is acceptable. Bearing in mind, however, that normal design office practice involves the use of a safety factor to take account of abnormalities of loading, it should be evident that even at this stage considerable advantages are obtained by application of this factor to the fully plastic condition rather than the limiting elastic case. It is then possible to arrange for all normal loading situations to be associated with elastic stresses in the beam, the additional strength in the partially plastic condition being used as the safety margin to take account of unexpected load increases.

Figure 3.4 shows the way in which moments build up with increasing depth or penetration of yielding and associated radius of curvature as the beam bends.

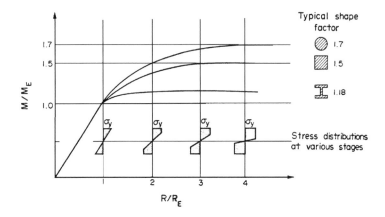

Fig. 3.4. Variation of moment of resistance of beams of various cross-section with depth of plastic penetration and associated radius of curvature.

Here the moment M carried by the beam at any particular stage and its associated radius of curvature R are considered as ratios of the values at the maximum elastic or initial yield condition. It will be noticed that at large curvature ratios, i.e. high plastic penetrations, the values of M/M_E approach the shape factor of the sections indicated, e.g. 1.5 for the rectangular section.

Shape factors of other symmetrical sections such as the I-section beam are found as follows (Fig. 3.5).

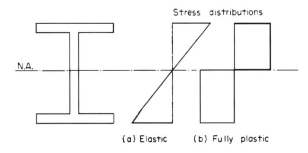

Fig. 3.5. Plastic bending of symmetrical (I-section) beam.

First determine the value of the maximum elastic moment M_E by applying the simple bending theory

$$\frac{M}{I} = \frac{\sigma}{y}$$

with y the maximum distance from the N.A. (the axis of symmetry passing through the centroid) to an outside fibre and $\sigma = \sigma_y$, the yield stress.

Then, in the fully plastic condition, the stress will be uniform across the section at σ_y and the section can be divided into any convenient number of rectangles of area A and centroid distance h from the neutral axis.

Then
$$M_{FP} = \sum (\sigma_y A) h \qquad (3.5)$$

The shape factor M_{FP}/M_E can then be determined.

3.3. Application to I-section beams

When the B.M. applied to an I-section beam is just sufficient to initiate yielding in the extreme fibres, the stress distribution is as shown in Fig. 3.5(a) and the value of the moment is obtained from the simple bending theory by subtraction of values for convenient rectangles.

i.e.
$$M_E = \frac{\sigma I}{y}$$

$$= \sigma_y \left[\frac{BD^3}{12} - \frac{bd^3}{12} \right] \frac{2}{D}$$

If the moment is then increased to produce full plasticity across the section, i.e. a plastic hinge, the stress distribution is as shown in Fig. 3.5(b) and the value of the moment is obtained by applying eqn. (3.3) to the same convenient rectangles considered above.

$$M_{FP} = \sigma_y \left[\frac{BD^2}{4} - \frac{bd^2}{4} \right]$$

The value of the shape factor can then be obtained as the ratio of the above equations M_{FP}/M_E. A typical value of shape factor for commercial rolled steel joists is 1.18, thus indicating only an 18% increase in "strength" capacity using plastic design procedures compared with the 50% of the simple rectangular section.

3.4. Partially plastic bending of unsymmetrical sections

Consider the T-section beam shown in Fig. 3.6. Whilst stresses remain within the elastic limit the position of the N.A. can be obtained in the usual way by taking moments of area

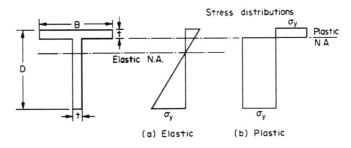

Fig. 3.6. Plastic bending of unsymmetrical (T-section) beam.

about some convenient axis as described in Chapter 4.[†] A typical position of the elastic N.A. is shown in the figure. Application of the simple blending theory about the N.A. will then yield the value of M_E as described in the previous paragraph.

Whatever the state of the section, be it elastic, partially plastic or fully plastic, equilibrium of forces must always be maintained, i.e. at any section the tensile forces on one side of the N.A. must equal the compressive forces on the other side.

$$\sum \text{stress} \times \text{area above N.A.} = \sum \text{stress} \times \text{area below N.A.}$$

In the *fully plastic* condition, therefore, when the stress is equal throughout the section, the above equation reduces to

$$\sum \textbf{areas above N.A.} = \sum \textbf{areas below N.A.} \qquad (3.6)$$

and in the special case shown in Fig. 3.5 the N.A. will have moved to a position coincident with the lower edge of the flange. Whilst this position is peculiar to the particular geometry chosen for this section it is true to say that for all unsymmetrical sections the N.A. will move from its normal position when the section is completely elastic as plastic penetration proceeds. In the ultimate stage when a plastic hinge has been formed the N.A. will be positioned such that eqn. (3.6) applies, or, often more conveniently,

$$\textbf{area above or below N.A.} = \tfrac{1}{2} \textbf{ total area} \qquad (3.7)$$

In the partially plastic state, as shown in Fig. 3.7, the N.A. position is again determined by applying equilibrium conditions to the forces above and below the N.A. The section is divided into convenient parts, each subjected to a force = average stress × area, as indicated, then

$$F_1 + F_2 = F_3 + F_4 \qquad (3.8)$$

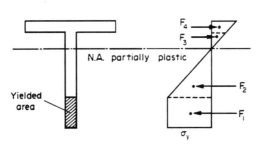

Fig. 3.7. Partially plastic bending of unsymmetrical section beam.

and this is an equation in terms of a single unknown $\bar{y}_p$, which can then be determined, as can the independent values of F_1, F_2, F_3 and F_4.

The sum of the moments of these forces about the N.A. then yields the value of the partially plastic moment M_{PP}. Example 3.2 describes the procedure in detail.

[†] E.J. Hearn, *Mechanics of Materials 1*, Butterworth-Heinemann, 1997.

3.5. Shape factor – unsymmetrical sections

Whereas with symmetrical sections the position of the N.A. remains constant as the axis of symmetry through the centroid, in the case of unsymmetrical sections additional work is required to take account of the movement of the N.A. position. However, having determined the position of the N.A. in the fully plastic condition using eqn. (3.6) or (3.7), the procedure outlined in §3.2 can then be followed to evaluate shape factors of unsymmetrical sections – see Example 3.2.

3.6. Deflections of partially plastic beams

Deflections of partially plastic beams are normally calculated on the assumption that the yielded areas, having yielded, offer no resistance to bending. Deflections are calculated therefore on the basis of the elastic core only, i.e. by application of simple bending theory and/or the standard deflection equations of Chapter 5[†] to the elastic material only. Because the second moment of area I of the central core is proportional to the fourth power of d, and I appears in the denominator of deflection formulae, deflections increase rapidly as d approaches zero, i.e. as full plasticity is approached.

If an experiment is carried out to measure the deflection of beams as loading, and hence B.M., is increased, the deflection graph for simply supported end conditions will appear as shown in Fig. 3.8. Whilst the beam is elastic the graph remains linear. The initiation of yielding in the outer fibres of the beam is indicated by a slight change in slope, and when plastic penetration approaches the centre of the section deflections increase rapidly for very small increases in load. For rectangular sections the ratio M_{FP}/M_E will be 1.5 as determined theoretically above.

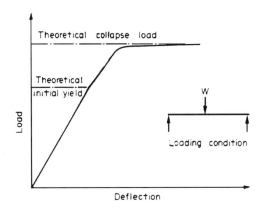

Fig. 3.8. Typical load-deflection curve for plastic bending.

3.7. Length of yielded area in beams

Consider a simply supported beam of rectangular section carrying a central concentrated load W. The B.M. diagram will be as shown in Fig. 3.9 with a maximum value of $WL/4$ at

[†] E.J. Hearn, *Mechanics of Materials 1*, Butterworth-Heinemann, 1997.

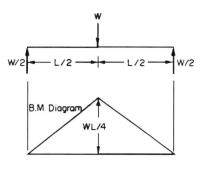

Fig. 3.9.

the centre. If loading is increased, yielding will commence therefore at the central section when $(WL/4) = (BD^2/6)\sigma_y$ and will gradually penetrate from the outside fibres towards the N.A. As this proceeds with further increase in loads, the B.M. at points away from the centre will also increase, and in some other positions near the centre it will also reach the value required to produce the initial yielding, namely $BD^2\sigma_y/6$. Thus, when full plasticity is achieved at the central section with a load W_p, there will be some other positions on either side of the centre, distance x from the supports, where yielding has just commenced at the outer fibres; between these two positions the beam will be in some elastic–plastic state. Now at distance x from the supports:

$$\text{B.M.} = W_p\frac{x}{2} = \frac{2}{3}M_{FP} = \frac{2}{3}\frac{W_pL}{4}$$

$$\therefore \qquad\qquad x = \frac{L}{3}$$

The central third of the beam span will be affected therefore by plastic yielding to some depth. At any general section within this part of the beam distance x' from the supports the B.M. will be given by

$$\text{B.M.} = W_p\frac{x'}{2} = \frac{B\sigma_y}{12}[3D^2 - d^2] \qquad\qquad (1)$$

Now since

$$\frac{BD^2}{4}\sigma_y = W_p\frac{L}{4} \qquad \sigma_y = \frac{W_pL}{BD^2}$$

Therefore substituting in (1),

$$W_p\frac{x'}{2} = \frac{B}{12}[3D^2 - d^2]\frac{W_pL}{BD^2}$$

$$x' = \frac{(3D^2 - d^2)}{6D^2}L$$

$$x' = \frac{L}{2}\left[1 - \frac{d^2}{3D^2}\right]$$

This is the equation of a parabola with

$$x' = L/2 \text{ when } d = 0 \text{ (i.e. fully plastic section)}$$

and $x' = L/3$ when $d = D$ (i.e. section elastic)

The yielded portion of the beam is thus as indicated in Fig. 3.10.

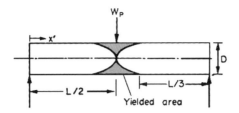

Fig. 3.10. Yielded area in beam carrying central point load.

Other beam support and loading cases may be treated similarly. That for a simply supported beam carrying a uniformly distributed load produces linear boundaries to the yielded areas as shown in Fig. 3.11.

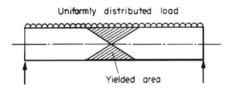

Fig. 3.11. Yielded area in beam carrying uniformly distributed load.

3.8. Collapse loads – plastic limit design

Having determined the moment required to produce a plastic hinge for the shape of beam cross-section used in any design it is then necessary to decide from a knowledge of the support and loading conditions how many such hinges are required before complete collapse of the beam or structure takes place, and to calculate the corresponding load. Here it is necessary to consider a plastic hinge as a pin-joint and to decide how many pin-joints are required to convert the structure into a "mechanism". If there are a number of points of "local" maximum B.M., i.e. peaks in the B.M. diagram, the first plastic hinge will form at the numerical maximum; if further plastic hinges are required these will occur successively at the next highest value of maximum B.M. etc. It is assumed that when a plastic hinge has developed at any cross-section the moment of resistance at that point remains constant until collapse of the whole structure takes place owing to the formation of the required number of further plastic hinges.

Consider, therefore, the following loading cases.

(a) Simply supported beam or cantilever

Whatever the loading system there will only be one point of maximum B.M. and plastic collapse will occur with **one** plastic hinge at this point (Fig. 3.12).

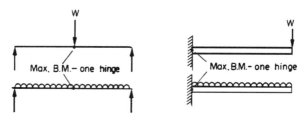

Fig. 3.12.

(b) Propped cantilever

In the case of propped cantilevers, i.e. cantilevers carrying opposing loads, the B.M. diagram is as shown in Fig. 3.13. The maximum B.M. then occurs at the built-in support and a plastic hinge forms first at this position. Due to the support of the prop, however, the beam does not collapse at this stage but requires another plastic hinge before complete failure or collapse occurs. This is formed at the other local position of maximum B.M., i.e. at the prop position, the moments at the support remaining constant until that at the prop also reaches the value required to form a plastic hinge.

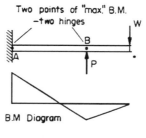

Fig. 3.13.

Collapse therefore occurs when $M_A = M_B = M_{FP}$, and thus **two** plastic hinges are required.

(c) Built-in beam

In this case there are three positions of local maximum **B.M.**, two of them being at the supports, and **three** plastic hinges are required for collapse (Fig. 3.14).

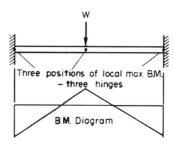

Fig. 3.14.

Other structures may require even more plastic hinges depending on their particular support conditions and degree of redundancy, but these need not be considered here. It should be evident, however, that there is now even more strength or load-carrying capacity available beyond that suggested by the shape factor, i.e. with a knowledge of the yield stress and hence the maximum elastic moment for any particular cross-section, the shape factor determines the increase in moment required to produce a fully plastic section or plastic hinge; depending on the support and loading conditions it may then be possible to increase the moment beyond this value until a sufficient number of plastic hinges have been formed to produce complete collapse. In order to describe the increased strength available using this "plastic limit" or "collapse load" procedure a *load factor* is introduced defined as

$$\text{load factor} = \frac{\text{collapse load}}{\text{allowable working load}} \tag{3.9}$$

This is completely different from, and must not be confused with, the safety factor, which is a factor to be applied to the yield stress in simple *elastic* design procedures.

3.9. Residual stresses after yielding: elastic-perfectly plastic material

Reference to the results of simple tensile or proof tests detailed in §1.7[†] shows that when materials are loaded beyond the yield point the resulting deformation does not disappear completely when load is removed and the material is subjected to permanent deformation or so-called *permanent set* (Fig. 3.15). In bending applications, therefore, when beams may be subjected to moments producing partial plasticity, i.e. part of the beam section remains elastic whilst the outer fibres yield, this permanent set associated with the yielded areas prevents those parts of the material which are elastically stressed from returning to their unstressed state when load is removed. *Residual stress* are therefore produced. In order to determine the magnitude of these residual stresses it is normally assumed that the unloading process, from either partially plastic or fully plastic states, is completely elastic (see Fig. 3.15). The

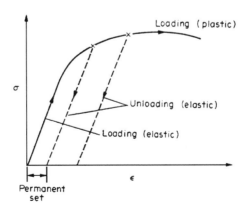

Fig. 3.15. Tensile test stress–strain curve showing elastic unloading process from any load condition.

[†] E.J. Hearn, *Mechanics of Materials 1*, Butterworth-Heinemann, 1997.

unloading stress distribution is therefore linear and it can be subtracted graphically from the stress distribution in the plastic or partially plastic state to obtain the residual stresses.

Consider, therefore, the rectangular beam shown in Fig. 3.16 which has been loaded to its fully plastic condition as represented by the stress distribution rectangles *oabc* and *odef*. The bending stresses which are then superimposed during the unloading process are given by the line *goh* and are opposite to sign. Subtracting the two distributions produces the shaded areas which then indicate the residual stresses which remain after unloading the plastically deformed beam. In order to quantify these areas, and the values of the residual stresses, it should be observed that the loading and unloading moments must be equal, i.e. the moment of the force due to the rectangular distribution *oabc* about the N.A. must equal the moment of the force due to the triangular distribution *oag*.

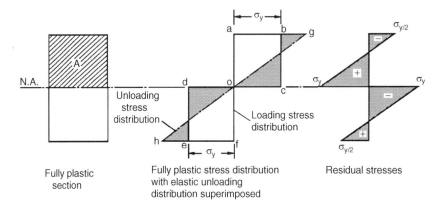

Fig. 3.16. Residual stresses produced after unloading a rectangular-section beam from a fully plastic state.

Now, moment due to *oabc*

$$= \text{stress} \times \text{area} \times \text{moment arm}$$

$$= ab \times A \times oa/2$$

and moment due to *oag*

$$= \text{average stress} \times \text{area} \times \text{moment arm}$$

$$= ag/2 \times A \times 2oa/3$$

Equating these values of moment yields

$$ag = \tfrac{3}{2}ab$$

Now $ab = \text{yield stress } \sigma_y \quad \therefore ag = 1\tfrac{1}{2}\sigma_y$

Thus the residual stresses at the outside surfaces of the beam $= \tfrac{1}{2}\sigma_y$. The maximum residual stresses occur at the N.A. and are equal to the yield stress. The complete residual stress distribution is shown in Fig. 3.16.

In loading cases where only partial plastic bending has occurred in the beam prior to unloading the stress distributions obtained, using a similar procedure to that outlined above, are shown in Fig. 3.17. Again, the unloading process is assumed elastic and the line *goh* in

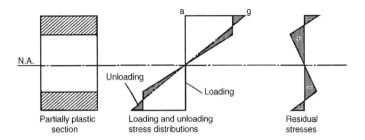

Fig. 3.17. Residual stress produced after unloading a rectangular-section beam from a partially plastic state.

this case is positioned such that the moments of the loading and unloading stress distributions are once more equal, i.e. the stress at the outside fibre ag is determined by considering the plastic moment M_{pp} applied to the beam assuming it to be elastic; thus

$$ag = \sigma = \frac{M_y}{I} = \frac{M_{PP}}{I} \frac{D}{2}$$

Whereas in the previous case the maximum residual stress occurs at the centre of the beam, in this case it may occur either at the outside or at the inner boundary of the yielded portion depending on the depth of plastic penetration. There is no residual stress at the centre of the beam.

Because of the permanent set mentioned above and the resulting stresses, beams which have been unloaded from plastic or partially plastic states will be deformed from their original shape. The straightening moment which is required at any section to return the beam to its original position is that which is required to remove the residual stresses from the elastic core (see Example 3.3).

The residual or permanent radius of curvature R after load is removed can be found from

$$\frac{1}{R} = \frac{1}{R_E} - \frac{1}{R_P} \tag{3.10}$$

where R_P is the radius of curvature in the plastic condition and R_E is the elastic spring-back, calculated by applying the simple bending theory to the complete section with a moment of M_{PP} or M_{FP} as the case may be.

3.10. Torsion of shafts beyond the elastic limit – plastic torsion

The method of treatment of shafts subjected to torques sufficient to initiate yielding of the material is similar to that used for plastic bending of beams (§3.1), i.e. it is usual to assume a stress–strain curve for the shaft material of the form shown in Fig. 3.2, the stress being proportional to strain up to the elastic limit and constant thereafter. It is also assumed that plane cross-sections remain plane and that any radial line across the section remains straight.

Consider, therefore, the cross-section of the shaft shown in Fig. 3.18(a) with its associated shear stress distribution. Whilst the shaft remains elastic the latter remains linear, and as the torque increases the shear stress in the outer fibres will eventually reach the yield stress in shear of the material τ_y. The torque at this point will be the maximum that the shaft can withstand whilst it is completely elastic.

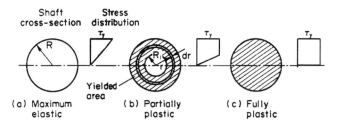

Fig. 3.18. Plastic torsion of a circular shaft.

From the torsion theory

$$\frac{T}{J} = \frac{\tau}{r}$$

Therefore maximum elastic torque

$$T_E = \frac{\tau_y J}{R} = \frac{\tau_y}{R}\frac{\pi R^4}{2}$$

$$= \frac{\pi R^3}{2}\tau_y \tag{3.11}$$

If the torque is now increased further it is assumed that, instead of the stress in the outer fibre increasing beyond τ_y, more and more of the material will yield and take up the stress τ_y, giving the stress distribution shown in Fig. 3.18(b). Consider the case where the material has yielded to a radius R_1, then:
Partially plastic torque

$$T_{PP} = \text{torque owing to elastic core} + \text{torque owing to plastic portion}$$

The first part is obtained directly from eqn. (3.11) with R_1 replacing R,

i.e.
$$\frac{\pi R_1^3}{2}\tau_y$$

For the second part consider an element of radius r and thickness dr, carrying a stress τ_y, (see Fig. 3.18(b)),

$$\text{force on element} = \tau_y \times 2\pi r\, dr$$

$$\text{contribution to torque} = \text{force} \times \text{radius}$$

$$= (\tau_y \times 2\pi r\, dr)r$$

$$= 2\pi r^2\, dr\tau_y$$

∴
$$\text{total contribution} = \int_{R_1}^{R} \tau_y 2\pi r^2\, dr$$

$$= 2\pi\tau_y \left[\frac{r^3}{3}\right]_{R_1}^{R}$$

$$= \frac{2\pi\tau_y}{3}[R^3 - R_1^3]$$

Therefore, partially plastic torque

$$T_{PP} = \frac{\pi R_1^3}{2}\tau_y + \frac{2\pi}{3}\tau_y[R^3 - R_1^3]$$

$$= \frac{\pi\tau_y}{6}[4R^3 - R_1^3] \tag{3.12}$$

In Fig. 3.18(c) the torque has now been increased until the whole cross-section has yielded, i.e. become plastic. The torque required to reach this situation is then easily determined from eqn. (3.12) since $R_1 = 0$.

∴   fully plastic torque $T_{FP} = \dfrac{\pi\tau_y}{6} \times 4R^3$

$$= \frac{2\pi}{3}R^3\tau_y \tag{3.13}$$

There is thus a considerable torque capacity beyond that required to produce initial yield, the ratio of fully plastic to maximum elastic torques being

$$\frac{T_{FP}}{T_E} = \frac{2\pi R^3}{3}\tau_y \times \frac{2}{\pi R^3}\tau_y$$

$$= \frac{4}{3}$$

The fully plastic torque for a solid shaft is therefore 33% greater than the maximum elastic torque. As in the case of beams this can be taken account of in design procedures to increase the allowable torque which can be carried by the shaft or it may be treated as an additional safety factor. In any event it must be remembered that should stresses in the shaft at any time exceed the yield point for the material, then some permanent deformation will occur.

3.11. Angles of twist of shafts strained beyond the elastic limit

Angles of twist of shafts in the partially plastic condition are calculated on the basis of the elastic core only, thus assuming that once the outer regions have yielded they no longer offer any resistance to torque. This is in agreement with the basic assumption listed earlier that radial lines remain straight throughout plastic torsion, i.e. $\theta_{PP} = \theta_E$ for the core.

For the elastic core, therefore,

$$\frac{\tau_y}{R_1} = \frac{G\theta}{L}$$

i.e. $$\theta_{PP} = \frac{\tau_y L}{R_1 G} \tag{3.14}$$

3.12. Plastic torsion of hollow tubes

Consider the hollow tube of Fig. 3.19 with internal radius R_1 and external radius R subjected to a torque sufficient to produce yielding to a radius R_2. The torque carried by the

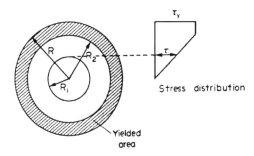

Fig. 3.19. Plastic torsion of a hollow shaft.

equivalent partially plastic solid shaft, i.e. ignoring the central hole, is given by eqn. (3.12) with R_2 replacing R_1 as

$$\frac{\pi \tau_y}{6}[4R^3 - R_2^3]$$

The torque carried by the hollow tube can then be determined by subtracting from the above the torque which would be carried by a solid shaft of diameter equal to the central hole and subjected to a shear stress at its outside fibre equal to τ.
i.e. from eqn. (3.11) torque on imaginary shaft

$$= \frac{\pi R_1^3}{2}\tau$$

but by proportions of the stress distribution diagram

$$\tau = \frac{R_1}{R_2}\tau_y$$

Therefore torque on imaginary shaft equal in diameter to the hollow core

$$= \frac{\pi R_1^4}{2R_2}\tau_y$$

Therefore, partially plastic torque for the hollow tube

$$T_{PP} = \frac{\pi \tau_y}{6}[4R^3 - R_2^3] - \frac{\pi}{2}\frac{R_1^4}{R_2}\tau_y$$

$$= \frac{\pi \tau_y}{6R_2}[4R^3 R_2 - R_2^4 - 3R_1^4] \qquad (3.15)$$

The fully plastic torque is then obtained when $R_2 = R_1$,

i.e. $$T_{FP} = \frac{\pi \tau_y}{6R_1}[4R^3 R_1 - 4R_1^4] = \frac{2\pi \tau_y}{3}[R^3 - R_1^3] \qquad (3.16)$$

This equation could also have been obtained by adaptation of eqn. (3.13), subtracting a fully plastic core of diameter equal to the central hole.

As an aid in visualising the stresses and torque capacities of members loaded to the fully plastic condition an analogy known as the *sand-heap analogy* has been introduced. Whilst full details have been given by Nadai[†] it is sufficient for the purpose of this text to note that

[†] A. Nadai, *Theory of Flow and Fracture of Solids*, Vol. 1, 2nd edn., McGraw-Hill, New York, 1950.

if dry sand is poured on to a raised flat surface having the same shape as the cross-section of the member under consideration, the sand heap will assume a constant slope, e.g. a cone on a circular disc and a pyramid on a square base. The volume of the sand heap, and hence its weight, is then found to be directly proportional to the fully plastic torque which would be carried by that particular shape of cross-section. Thus by calibration, i.e. with a knowledge of the fully plastic torque for a circular shaft, direct comparison of the weight of appropriate sand heaps yields an immediate indication of the fully plastic torque of some other more complicated section.

3.13. Plastic torsion of case-hardened shafts

Consider now the case-hardened shaft shown in Fig. 3.20. Whilst it is often assumed in such cases that the shear-modulus is the same for the material of the case and core, this is certainly not the case for the yield stresses; indeed, there is often a considerable difference, the value for the case being generally much larger than that for the core. Thus, when the shaft is subjected to a torque sufficient to initiate yielding at the outside fibres, the normal triangular elastic stress distribution required to maintain straight radii must be modified, since this would imply that some of the core material is stressed beyond its yield stress. Since the basic assumption used throughout this treatment is that stress remains constant at the yield stress for any increase in strain, it follows that the stress distribution must be as indicated in Fig. 3.20. The shaft thus contains at this stage a plastic region sandwiched between two elastic layers. Torques for each portion must be calculated separately, therefore, and combined to yield the partially plastic torque for the case-hardened shaft. (Example 3.5.)

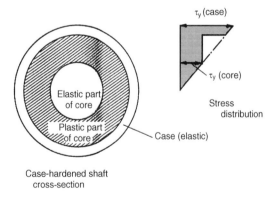

Fig. 3.20. Plastic torsion of a case-hardened shaft.

3.14. Residual stresses after yield in torsion

If shafts are stressed at any time beyond their elastic limit to a partially plastic state as described previously, a permanent deformation will remain when torque is removed. Associated with this plastic deformation will be a system of residual stresses which will affect the strength of the shaft in subsequent loading cycles. The magnitudes of the residual stresses are determined using the method described in detail for beams strained beyond the elastic limit on page 73, i.e. the removal of torque is assumed to be a completely elastic process so that the associated stress distribution is linear. The residual stresses are thus obtained by subtracting the elastic unloading stress distribution from that of the partially plastic loading condition. Now, from eqn. (3.12), partially plastic torque $= T_{PP}$.

Therefore elastic torque to be applied during unloading $= T_{PP}$.

The stress τ' at the outer fibre of the shaft which would be achieved by this torque, assuming elastic material, is given by the torsion theory

$$\frac{T}{J} = \frac{\tau}{R}, \quad \text{i.e. } \tau' = \frac{T_{PP}R}{J}$$

Thus, for a solid shaft the residual stress distribution is obtained as shown in Fig. 3.21.

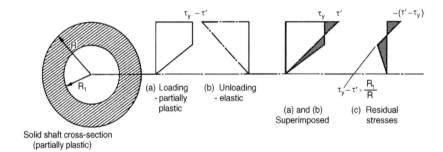

Fig. 3.21. Residual stresses produced in a solid shaft after unloading from a partially plastic state.

Similarly, for hollow shafts, the residual stress distribution will be as shown in Fig. 3.22.

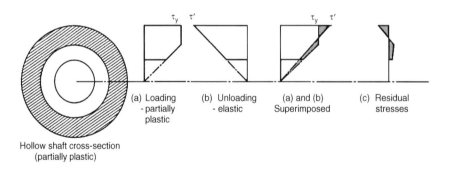

Fig. 3.22. Residual stresses produced in a hollow shaft after unloading from a partially plastic state.

3.15. Plastic bending and torsion of strain-hardening materials

(a) Inelastic bending

Whilst the material in this case no longer follows Hookes' law it is necessary to assume that cross-sections of the beam remain plane during bending so that strains remain proportional to distance from the neutral axis.

Consider, therefore, the rectangular section beam shown in Fig. 3.23(b) with its neutral axis positioned at a distance h_1 from the lower surface and h_2 from the upper surface. Bearing in mind the assumption made in the preceding paragraph we can now locate the neutral axis position by the usual equilibrium conditions.

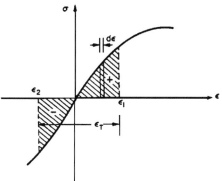

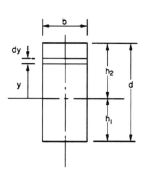

Fig. 3.23(a). Stress–strain curve for a beam in bending constructed from a strain–hardening material.

Fig. 3.23(b).

i.e. Since the sum of forces normal to any cross-section must always be zero then:

$$\int \sigma \, dA = \int_{-h_1}^{h_2} \sigma \cdot b \, dy = 0$$

But, from eqn (4.1)[†]

$$y = R\frac{\sigma}{E} = R\varepsilon \quad \therefore dy = R \, d\varepsilon$$

$$\therefore \qquad \int_{-h_1}^{h_2} \sigma b R \, d\varepsilon = 0$$

or

$$\int_{\varepsilon_1}^{\varepsilon_2} \sigma b R \, d\varepsilon = 0.$$

where ε_1 and ε_2 are the strains in the top and bottom surfaces of the beam, respectively. They are also indicated on Fig. 3.23(a).

Since b and R are constant then the position of the neutral axis must be such that:

$$\int_{\varepsilon_1}^{\varepsilon_2} \sigma \, d\varepsilon = 0 \tag{3.17}$$

i.e. the total area under the σ–ε curve between ε_1 and ε_2 must be zero. This is achieved by marking the length ε_T on the horizontal axis of Fig. 3.23(a) in such a way as to make the positive and negative areas of the diagram equal. This identifies the appropriate values for ε_1 and ε_2 with:

$$\varepsilon_T = |\varepsilon_1 + \varepsilon_2| = \frac{h_1}{R} + \frac{h_2}{R} = \frac{1}{R}(h_1 + h_2)$$

[†] E.J. Hearn, *Mechanics of Materials 1*, Butterworth-Heinemann, 1997.

i.e. $$\varepsilon_T = \frac{d}{R} \tag{3.18}$$

Because strains have been assumed linear with distance from the neutral axis the position of the N.A. is then obtained by simple proportions:

$$\frac{h_1}{h_2} = \frac{\varepsilon_1}{\varepsilon_2} \tag{3.19}$$

The value of the applied bending moment M is then given by the sum of the moments of forces above and below the neutral axis.

i.e. $$M = \int \sigma \, dA \cdot y = \int_{-h_1}^{h_2} \sigma \cdot b \, dy \cdot y$$

and, since $dy = R \cdot d\varepsilon$ and $y = R\varepsilon$.

$$M = \int_{\varepsilon_1}^{\varepsilon_2} \sigma b \cdot R^2 \varepsilon \, d\varepsilon = R^2 b \int_{\varepsilon_1}^{\varepsilon_2} \sigma \varepsilon \, d\varepsilon.$$

Substituting, from eqn. (3.18), $R = d/\varepsilon_T$:

$$M = \frac{bd^2}{\varepsilon_T^2} \int_{\varepsilon_1}^{\varepsilon_2} \sigma \varepsilon \, d\varepsilon. \tag{3.20}$$

The integral part of this expression is the first moment of area of the shaded parts of Fig. 3.23(a) about the vertical axis and evaluation of this integral allows the determination of M for any assumed value of ε_T.

An alternative form of the expression is obtained by multiplying the top and bottom of the expression by $12R$ using $R = d/\varepsilon_T$ for the numerator,

i.e. $$M = 12 \frac{(d/\varepsilon_T)}{12R} \left[\frac{bd^2}{\varepsilon_T^2} \int_{\varepsilon_1}^{\varepsilon_2} \sigma \varepsilon \, d\varepsilon \right] = \frac{1}{R} \cdot \frac{bd^3}{12} \cdot \frac{12}{\varepsilon_T^3} \int_{\varepsilon_1}^{\varepsilon_2} \sigma \varepsilon \, d\varepsilon.$$

which can be reduced to a form similar to the standard bending eqn. (4.3)[†] $M = EI/R$

i.e. $$M = \frac{E_r I}{R} \tag{3.21}$$

with E_r known as the *reduced modulus* and given by:

$$E_r = \frac{12}{\varepsilon_T^3} \int_{\varepsilon_1}^{\varepsilon_2} \sigma \varepsilon \, d\varepsilon. \tag{3.22}$$

The appropriate value of the reduced modulus E_r for any particular curvature is best obtained from a curve of E_r against ε_T. This is constructed rather laboriously by determining the relevant values of ε_1 and ε_2 for a set of assumed ε_T values using the condition of equal positive and negative areas for each ε_T value and then evaluating the integral of eqn. (3.22). Having found E_r, the value of the bending moment for any given curvature R is found from eqn. (3.21).

It is sometimes useful to remember that, because strains are linear with distance from the neutral axis, the distribution of bending stresses across the beam section will take exactly the same form as that of the stress–strain diagram of Fig. 3.23(a) turned through 90° with

[†] E.J. Hearn, *Mechanics of Materials 1*, Butterworth-Heinemann, 1977.

ε_T replaced by the beam depth d. The position of the neutral axis indicated by eqn. (3.19) is then readily observed.

(b) Inelastic torsion

A similar treatment can be applied to the torsion of shafts constructed from materials which exhibit strain hardening characteristics. Figure 3.24 shows the shear stress–shear strain curve for such a material.

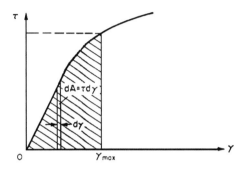

Fig. 3.24. Shear stress–shear strain curve for torsion of materials exhibiting strain-hardening characteristics.

Once again it is necessary to assume that cross-sections of the shaft remain plane and that the radii remain straight under torsion. The shear strain at any radius r is then given by eqn. (8.9)[†] as:

$$\gamma = \frac{r\theta}{L}$$

For a shaft of radius R the maximum shearing strain is thus

$$\gamma_{\max} = \frac{R\theta}{L}$$

the corresponding shear stress being given by the relevant ordinate of Fig. 3.24.

Now the torque T has been shown in §8.1[†] to be given by:

$$T = \int_0^R 2\pi r^2 \tau' \, dr$$

where τ' is the shear stress at any general radius r.

Now, since $\qquad \gamma = \dfrac{r\theta}{L} \quad$ then $\quad d\gamma = \dfrac{\theta}{L} \cdot dr$

and, substituting for r and dr, we have:

$$T = \int_0^{\gamma_{\max}} 2\pi \left(\frac{\gamma L}{\theta}\right)^2 \tau' \frac{L}{\theta} \cdot d\gamma$$

$$= \frac{2\pi L^3}{\theta^3} \int_0^{\gamma_{\max}} \tau' \gamma^2 \, d\gamma \tag{3.23}$$

[†] E.J. Hearn, *Mechanics of Materials 1*, Butterworth-Heinemann, 1997.

The integral part of the expression is the second moment of area of the shaded portion of Fig. 3.24 about the vertical axis. Thus, determination of this quantity for a given y_{max} value yields the corresponding value of the applied torque T.

As for the case of inelastic bending, the form of the shear stress–strain curve, Fig. 3.24, is identical to the shear stress distribution across the shaft section with the y axis replaced by radius r.

3.16. Residual stresses – strain-hardening materials

The procedure for determination of residual stresses arising after unloading from given stress states is identical to that described in §3.9 and §3.14.

For example, it has been shown previously that the stress distribution across a beam section in inelastic bending will be similar to that shown in Fig. 3.23(a) with the beam depth corresponding to the strain axis. Application of the elastic unloading stress distribution as described in §3.9 will then yield the residual stress distribution shown in Fig. 3.25. The same procedure should be adopted for residual stresses in torsion situations, reference being made to §3.14.

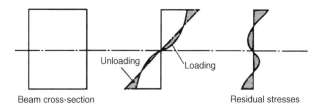

Fig. 3.25. Residual stresses produced in a beam constructed from a strain-hardening material.

3.17. Influence of residual stresses on bending and torsional strengths

The influence of residual stresses on the future loading of members has been summarised by Juvinall[†] into the following rule:

An overload causing yielding produces residual stresses which are favourable to future overloads in the same direction and unfavourable to future overloads in the opposite direction.

This suggests that the residual stresses represent a favourable stress distribution which has to be overcome by any further load system before any adverse stress can be introduced into the member of structure. This principle is taken advantage of by spring manufacturers, for example, who intentionally yield springs in the direction of anticipated service loads as part of the manufacturing process. A detailed discussion of residual stress can be found in the *Handbook of Experimental Stress Analysis* of Hetényi.[‡]

[†] R. C. Juvinall, *Engineering Considerations of Stress, Strain and Strength*, McGraw-Hill, 1967.
[‡] M. Hetényi, *Handbook of Experimental Stress Analysis*, John Wiley, 1966.

3.18. Plastic yielding in the eccentric loading of rectangular sections

When a column or beam is subjected to an axial load and a B.M., as in the application of eccentric loads, the elastic stress distribution is as shown in Fig. 3.25(a), the N.A. being displaced from the centroidal axis of the section. As the load increases the yield stress will be reached on one side of the section first as shown in Fig. 3.26(b) and, as in the case of the partially plastic bending of unsymmetrical sections in §3.4, the N.A. will move as plastic penetration proceeds. In the limiting case, when plasticity has spread across the complete section, the N.A. will be situated at a distance h from the centroidal axis (the axis through the centroid of the section) (Fig. 3.26(c)). The precise position of the N.A. is related to the excess of the total tensile force over the total compressive force, i.e. to the area shown shaded in Fig. 3.26(c). In simple bending, for example, there is no resultant force across the section and the shaded area reduces to zero. Thus, the magnitude of the axial load for full plasticity as given by the shaded area

$$= P_{FP} = 2h \times B \times \sigma_y$$

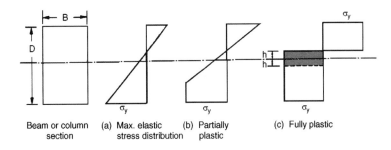

Fig. 3.26. Plastic yielding of eccentrically loaded rectangular section.

where B is the width of the section,

i.e.
$$h = \frac{P_{FP}}{2B\sigma_y} \tag{3.24}$$

The fully plastic load is sometimes written in terms of a load factor N defined as

$$\text{load factor } N = \frac{\text{fully plastic load}}{\text{axial load}} = \frac{P_{FP}}{P}$$

then
$$h = \frac{PN}{2B\sigma_y} \tag{3.25}$$

The fully plastic moment on the section is given by the difference in the moments produced by the stress distributions of Fig. 3.27,

i.e.
$$M_{FP} = \frac{BD^2}{4}\sigma_y - 2(Bh\sigma_y)\frac{h}{2}$$

$$M_{FP} = \frac{BD^2}{4}\sigma_y - Bh^2\sigma_y$$

$$= \frac{BD^2}{4}\sigma_y - \frac{P^2N^2}{4B\sigma_y} \tag{3.26}$$

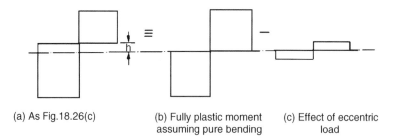

(a) As Fig.18.26(c)   (b) Fully plastic moment   (c) Effect of eccentric
                             assuming pure bending   load

Fig. 3.27.

The fully plastic moment required in eccentric load conditions is therefore reduced from that in the simple bending case by an amount depending on the values of the load, yield stress, section shape and load factor.

The maximum allowable working moment for a single plastic hinge in eccentric loading situations with a load factor N is therefore given by

$$M = \frac{M_{FP}}{N} = \frac{BD^2}{4N}\sigma_y - \frac{P^2N}{4B\sigma_y} \tag{3.27}$$

3.19. Plastic yielding and residual stresses under axial loading with stress concentrations

If a bar with uniform cross-section is loaded beyond its yield point in pure tension (or compression), the bar will experience permanent deformation when load is removed but no residual stresses will be created since all of the material in the cross-section will have yielded simultaneously and all will return to the same unloaded condition. If, however, stress concentrations such as notches, keyways, holes, etc., are present in the bar, these will result in local stress increases or *stress concentrations*, and the material will yield at these positions before the rest of the cross-section. If the local stress concentration factor is K then the maximum stress in the section with an axial load P is given by

$$\sigma_{max} = K(P/A)$$

where P/A is the mean stress across the section assuming no stress concentration is present.

When the load has been increased to a value P_y, just sufficient to initiate yielding at the root of the notch or other stress concentration, the stress distribution will be as shown in Fig. 3.28(a). Since equilibrium considerations require the mean stress across the section to equal P_y/A it follows that the stress at the centre of the section must be less than P_y/A.

If the load is now increased to P_2, yielding will continue at the root of the notch and plastic penetration will proceed towards the centre of the section. At some stage the stress distribution will appears as in Fig. 3.28(b) with a mean stress value of P_2/A. If the load is then removed the residual stresses may be obtained using the procedure of §§3.9 and 3.14, i.e. by superimposing an elastic stress distribution of opposite sign but equal moment value (shown dotted in Fig. 3.28(b)). The resulting residual stress distribution would then be similar to that shown in Fig. 3.28(c).

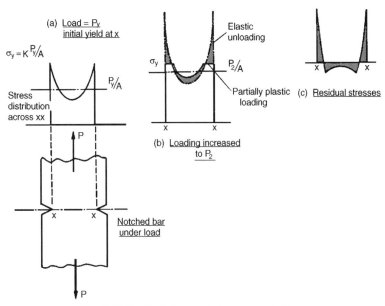

Fig. 3.28. Residual stresses at stress concentrations.

Whilst subsequent application of loads above the value of P_2 will cause further yielding, no yielding will be caused by the application of loads up to the value of P_2 however many times they are applied. With a sufficiently high value of stress concentration factor it is possible to produce a residual stress distribution which exceeds the compressive yield stress at the root of the notch, i.e. the material will be stressed from tensile yield to compressive yield throughout one cycle. Provided that further cycles remain within these limits, the component will not experience additional yielding, and it can be considered safe in, for example, high strain, low-cycle fatigue conditions.

3.20. Plastic yielding of axially symmetric components[†]

(a) Thick cylinders under internal pressure – collapse pressure

Consider the thick cylinder shown in Fig. 3.29 subjected to an internal pressure P_1 of sufficient magnitude to produce yielding to a radius R_p.

Now for ductile materials, from §10.17,[‡] yield is deemed to occur when

$$\sigma_y = \sigma_H - \sigma_r. \tag{3.28}$$

but from eqn. (10.2),[‡] the equilibrium equation,

$$\sigma_H - \sigma_r = r\frac{d\sigma_r}{dr}$$

[†] J. Heyman, *Proc. I.Mech.E.* **172** (1958). W.R.D. Manning, *High Pressure Engineering*, Bulleid Memorial Lecture, 1963, University of Nottingham.

[‡] E.J. Hearn, *Mechanics of Materials 1*, Butterworth-Heinemann, 1997.

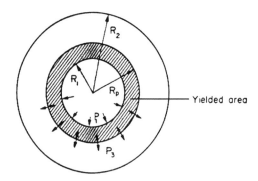

Fig. 3.29.

$\therefore$ $$\sigma_r = r\frac{d\sigma_r}{dr}$$

$\therefore$ $$\frac{d\sigma_r}{dr} = \frac{\sigma_y}{r}$$

Integrating: $$\sigma_r = \sigma_y \log_e r + \text{ constant}$$

Now $$\sigma_r = -P_3 \quad \text{at} \quad r = R_p$$

$\therefore$ $$\text{constant} = -P_3 - \sigma_y \log_e R_p$$

$\therefore$ $$\sigma_r = \sigma_y \log_e r - P_3 - \sigma_y \log_e R_p$$

i.e. $$\sigma_r = \sigma_y \log_e \frac{r}{R_p} - P_3 \tag{3.29}$$

and from eqn. (3.28) $$\sigma_H = \sigma_y + \sigma_r$$

$\therefore$ $$\sigma_H = \sigma_y \left(1 + \log_e \frac{r}{R_p}\right) - P_3 \tag{3.30}$$

These equations thus yield the hoop and radial stresses *throughout the plastic zone* in terms of the radial pressure at the elastic–plastic interface P_3. The *numerical* value of P_3 may be determined as follows (the sign has been allowed for in the derivation of eqn. (3.29)).

At the stage where plasticity has penetrated partly through the cylinder walls the cylinder may be considered as a compound cylinder with the inner tube plastic and the outer tube elastic, the latter being subjected to an internal pressure P_3. From eqns. (10.5) and (10.6)[†] the hoop and radial stresses *in the elastic portion* are therefore given by

$$\sigma_r = \frac{P_3 R_p^2}{(R_2^2 - R_p^2)}\left[\frac{R_p^2 - R_2^2}{R_p^2}\right]$$

and $$\sigma_H = \frac{P_3 R_p^2}{(R_2^2 - R_p^2)}\left[\frac{R_p^2 + R_2^2}{R_p^2}\right]$$

[†] E.J. Hearn, *Mechanics of Materials 1*, Butterworth-Heinemann, 1997.

i.e. the maximum shear stress is

$$\frac{\sigma_H - \sigma_r}{2} = \frac{P_3 R_p^2}{2(R_2^2 - R_p^2)} \left[\frac{2R_2^2}{R_p^2} \right] = \frac{P_3 R_2^2}{(R_2^2 - R_p^2)}$$

Again, applying the Tresca yield criterion,

$$\frac{\sigma_y}{2} = \frac{P_3 R_2^2}{(R_2^2 - R_p^2)}$$

i.e. the radial pressure at the elastic interface is

$$P_3 = \frac{\sigma_y}{2R_2^2} [R_2^2 - R_p^2] \tag{3.31}$$

Thus from eqns. (3.29) and (3.30) the stresses in the plastic zone are given by

$$\sigma_r = \sigma_y \left[\log_e \frac{r}{R_p} - \frac{1}{2R_2^2} (R_2^2 - R_p^2) \right] \tag{3.32}$$

and

$$\sigma_H = \sigma_y \left[\left(1 + \log_e \frac{r}{R_p} \right) - \frac{1}{2R_2^2} (R_2^2 - R_p^2) \right] \tag{3.33}$$

The pressure required for complete plastic "collapse" of the cylinder is given by eqn. (3.29) when $r = R_1$ and $R_p = R_2$ with $P_3 = P_2 = 0$ (at the outside edge).

For "collapse" $$\sigma_r = -P_1 = \sigma_y \log_e \frac{R_1}{R_2} \tag{3.34}$$

With a knowledge of this collapse pressure the design pressure can be determined by dividing it by a suitable load factor as described in §3.8.

The *pressure at initial yield* is found from eqn. (3.31) when $R_p = R_1$,

i.e.   initial yield pressure $$= \frac{\sigma_y}{2R_2^2} [R_2^2 - R_1^2] \tag{3.35}$$

Finally, the *internal pressure required to cause yielding to a radius* R_p is given by eqn. (3.32) when $r = R_1$,

i.e. $$\sigma_r = -P_1 = \sigma_y \left[\log_e \frac{R_1}{R_p} - \frac{1}{2R_2^2} (R_2^2 - R_p^2) \right] \tag{3.36}$$

(b) Thick cylinders under internal pressure ("auto-frettage")

When internal pressure is applied to thick cylinders it has been shown that maximum tensile stresses are set up at the inner surface of the bore. If the internal pressure is increased sufficiently, yielding of the cylinder material will take place at this position and the working safety factor n according to the Tresca theory will be given by

$$\sigma_H - \sigma_r = \sigma_y/n$$

where σ_H and σ_r are the hoop and radial stresses at the bore.

Fortunately, the condition is not too serious at this stage since there remains a considerable bulk of elastic material surrounding the yielded area which contains the resulting strains within reasonable limits. As the pressure is increased further, however, plastic penetration takes place deeper and deeper into the cylinder wall and eventually the whole cylinder will yield. Fatigue life of the cylinder will also be heavily dependent upon the value of the maximum tensile stress at the bore so that any measures which can be taken to reduce the level of this stress will be beneficial to successful operation of the cylinder. Such methods include the use of compound cylinders with force or shrink fits and/or external wire winding; the largest effect is obtained, however, with a process known as *"autofrettage"*.

If the pressure inside the cylinder is increased beyond the initial yield value so that plastic penetration occurs only partly into the cylinder wall then, on release of the pressure, the elastic zone attempts to return to its original dimensions but is prevented from doing so by the permanent deformation or "set" of the yielded material. The result is that residual stresses are introduced, the elastic material being held in a state of residual tension whilst the inside layers are brought into residual compression. On subsequent loading cycles, therefore, the cylinder is able to withstand a higher internal pressure since the compressive residual stress at the bore has to be overcome before this region begins to experience tensile stresses. The autofrettage process has the same effect as shrinking one tube over another without the complications of the shrinking process. With careful selection of cylinder dimensions and autofrettage pressure the resulting residual compressive stresses can significantly reduce or even totally eliminate tensile stresses which would otherwise be achieved at the bore under working conditions. As a result the fatigue life and the safety factor at the bore are considerably enhanced and for this reason gun barrels and other pressure vessels are often pre-stressed in this way prior to service.

Care must be taken in the design process, however, since the autofrettage process introduces a secondary critical stress region at the position of the elastic/plastic interface of the autofrettage pressure loading condition. This will be discussed further below.

The autofrettage pressure required for yielding to any radius R_p is given by the High Pressure Technology Association (HPTA) code of practice[†] as

$$P_A = \frac{\sigma_y}{2}\left[\frac{K^2 - m^2}{K^2}\right] + \sigma_y \log_e m \tag{3.37}$$

where $K = R_2/R_1$ and $m = R_p/R_1$, where R_1 the internal radius and R_2 the external radius.

This is simply a modified form of eqn. (3.36) developed in the preceding section.

The maximum allowable autofrettage pressure is then given as that which will produce yielding to the geometric mean radius $R_p = \sqrt{R_1 R_2}$.

Stress distribution under autofrettage pressure loading

From eqns. (3.32) and (3.33) the stresses in the plastic zone at any radius r are given by:

$$\sigma_r = \sigma_y\left[\log_e\left(\frac{r}{R_p}\right) - \frac{1}{2}\left(1 - \frac{R_p^2}{R_2^2}\right)\right] \tag{3.38}$$

$$\sigma_H = \sigma_y\left[1 + \log_e\left(\frac{r}{R_p}\right) - \frac{1}{2}\left(1 - \frac{R_p^2}{R_2^2}\right)\right] \tag{3.39}$$

[†] *High Pressure Safety Code.* High Pressure Technology Association, 1975.

Also in §3.20 (a) it has been shown that stresses at any radius r in the elastic zone are obtained in terms of the radial pressure $P_3 = P_p$ set up at the elastic–plastic interface with:

$$\sigma_r = \frac{P_p R_p^2}{(R_2^2 - R_p^2)}\left[1 - \frac{R_2^2}{r^2}\right] \qquad (3.40)$$

$$\sigma_H = \frac{P_p \cdot R_p^2}{(R_2^2 - R_p^2)}\left[1 + \frac{R_2^2}{r^2}\right] \qquad (3.41)$$

with

$$P_p = \frac{\sigma_y}{2R_2^2}[R_2^2 - R_p^2] \qquad (3.31)(bis)$$

The above equations yield hoop and radial stress distributions throughout the cylinder wall typically of the form shown in Fig. 3.30.

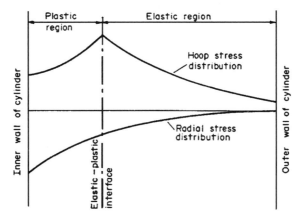

Fig. 3.30. Stress distributions under autofrettage pressure.

Residual stress distributions

Residual stress after unloading can then be obtained using the procedure introduced in §3.9 of elastic unloading, i.e. the autofrettage loading pressure is assumed to be removed (applied in a negative sense) elastically across the whole cylinder, the unloading elastic stress distribution being given by eqns. (10.5) and (10.6)[†] as:

$$\sigma_r = P_A\left[\frac{1 - \left(\dfrac{R_2}{r}\right)^2}{K^2 - 1}\right] \qquad (3.42)$$

$$\sigma_H = P_A\left[\frac{1 + \left(\dfrac{R_2}{r}\right)^2}{K^2 - 1}\right] \quad \text{with } K = R_2/R_1 \qquad (3.43)$$

[†] E.J. Hearn, *Mechanics of Materials 1*, Butterworth-Heinemann, 1997.

Superposition of these distributions on the previous loading distributions allows the two curves to be subtracted for both hoop and radial stresses and produces residual stresses of the form shown in Figs. 3.31 and 3.32.

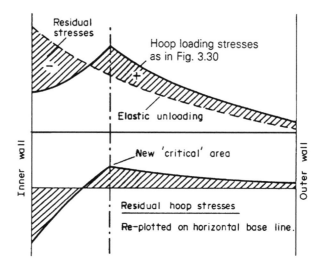

Fig. 3.31. Determination of residual hoop stresses by elastic unloading.

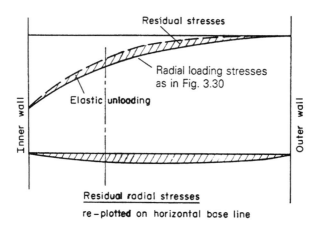

Fig. 3.32. Determination of residual radial stresses by elastic unloading.

Working stress distributions

Finally, if the stress distributions due to an elastic internal working pressure P_w are superimposed on the residual stress state then the final working stress state is produced as in Figs. 3.33 and 3.34.

The elastic working stresses are given by eqns. (3.42) and (3.43) with P_A replaced by P_w. Alternatively a Lamé line solution can be adopted. The final stress distributions show that

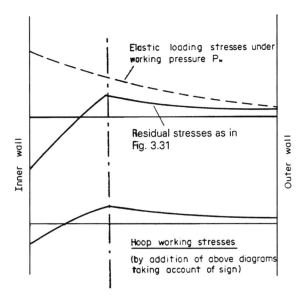

Fig. 3.33. Evaluation of hoop working stresses.

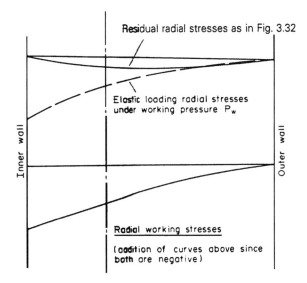

Fig. 3.34. Evaluation of radial working stresses.

the maximum tensile stress, instead of being at the bore as in the plain cylinder, is now at the elastic/plastic interface position. Application of the Tresca maximum shear stress failure criterion:

i.e.
$$\sigma_H - \sigma_r = \sigma_y/n$$

also indicates the elastic/plastic interface as now more critical than the internal bore – see Fig. 3.35.

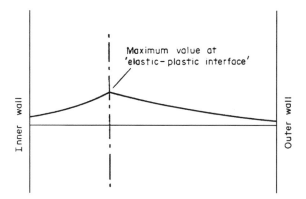

Fig. 3.35. Distribution of maximum shear stress = $\frac{1}{2}(\sigma_\theta - \sigma_r)$ by combination of Figs. 3.33 and 3.34.

Effect of axial stresses and end restraint

Depending on the end conditions which can be assumed for the cylinder during both the autofrettage process and its normal working condition a further complication can arise since the axial stresses σ_z which are produced can affect the application of the Tresca criterion.

Strictly, Tresca requires the use of the greatest difference in the principal stresses which, if σ_z is zero, $= \sigma_H - \sigma_r$. If, however, σ_z has a value it must be used in conjunction with σ_H and σ_r to produce the greatest difference.

The procedure used above to determine residual hoop and radial stresses and subsequent working stresses should therefore be repeated for axial stresses with values in the plastic region being found as suggested by Franklin and Morrison[†] from:

$$\sigma_z = P_A \frac{(1 - 2v)}{(K^2 - 1)} + v(\sigma_H - \sigma_r) \tag{3.44}$$

and axial stresses under elastic conditions being given by eqn. (10.7)[‡] with $P_2 = 0$ and $P_1 = P_A$ or P_W as required.

(c) Rotating discs

It will be shown in Chapter 4 that the centrifugal forces which act on rotating discs produce two-dimensional tensile stress systems. At any given radius the hoop or circumferential stress is always greater than, or equal to, the radial stress, the maximum values occurring at the inside radius. It follows, therefore, that yielding will first occur at the inside surface when the speed of rotation has increased sufficiently to make the circumferential stress equal to the tensile yield stress. With further increase of speed, plastic penetration will gradually proceed towards the centre of the disc and eventually complete plastic collapse will occur.

[†] G.J. Franklin and J.L.M. Morrison, Autofrettage of cylinders: reduction of pressure/external expansion curves and calculation of residual stresses. *Proc. J. Mech. E.* **174** (35) 1960.
[‡] E.J. Hearn, *Mechanics of Materials 1*, Butterworth-Heinemann, 1997.

Now for a *solid disc* the equilibrium eqn. (4.1) derived on page 120 is, with $\sigma_H = \sigma_y$,

$$\sigma_y - \sigma_r - r\frac{d\sigma_r}{dr} = \rho r^2 \omega^2$$

∴
$$\sigma_r + r\frac{d\sigma_r}{dr} = \sigma_y - \rho r^2 \omega^2$$

Integrating,
$$r\sigma_r = r\sigma_y - \rho\frac{r^3\omega^2}{3} + A \tag{1}$$

Now since the stress cannot be infinite at the centre where $r = 0$, then A must be zero.

∴
$$r\sigma_r = r\sigma_y - \rho\frac{r^3\omega^2}{3}$$

Now at $r = R$, i.e. at the outside of the disc, $\sigma_r = 0$.

∴
$$R\sigma_y = \rho\frac{R^3\omega^2}{3}$$

i.e. the collapse speed ω_p is given by

$$\omega_p^2 = \frac{3\sigma_y}{\rho R^2} \tag{3.45}$$

For a *disc with a central hole*, (1) still applies, but in this case the value of the constant A is determined from the condition

$$\sigma_r = 0 \text{ at } r = R_1 \text{ the inside radius}$$

i.e.
$$A = \rho\frac{R_1^3\omega^2}{3} - R_1\sigma_y$$

Again, $\sigma_r = 0$ at the outside surface where $r = R$.
  Substituting in (1),

$$0 = R\sigma_y - \rho\frac{R^3\omega^2}{3} + \rho\frac{R_1^3\omega^2}{3} - R_1\sigma_y$$

$$0 = \sigma_y(R - R_1) - \rho\frac{\omega^2}{3}(R^3 - R_1^3)$$

i.e. the collapse speed ω_p is given by

$$\omega_p^2 = \frac{3\sigma_y}{\rho}\frac{(R - R_1)}{(R^3 - R_1^3)} \tag{3.46}$$

  If a rotating disc is stopped after only partial penetration, residual stresses are set up similar to those discussed in the case of thick cylinders under internal pressure (auto-frettage). Their values may be determined in precisely the same manner as that described in earlier sections, namely, by calculating the elastic stress distribution at an appropriate higher speed and subtracting this from the partially plastic stress distribution. Once again, favourable compressive residual stresses are set up on the surface of the central hole which increases the stress range – and hence the speed limit – available on subsequent cycles. This process is sometimes referred to as *overspeeding*.

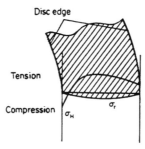

Fig. 3.36. Residual stresses produced after plastic yielding ("overspeeding") of rotating disc with a central hole.

A typical residual stress distribution is shown in Fig. 3.36.

Examples

Example 3.1

(a) A rectangular-section steel beam, 50 mm wide by 20 mm deep, is used as a simply supported beam over a span of 2 m with the 20 mm dimension vertical. Determine the value of the central concentrated load which will produce initiation of yield at the outer fibres of the beam.

(b) If the central load is then increased by 10% find the depth to which yielding will take place at the centre of the beam span.

(c) Over what length of beam will yielding then have taken place?

(d) What are the maximum deflections for each load case?

For steel σ_y in simple tension and compression $= 225$ MN/m^2 and $E = 206.8$ GN/m^2.

Solution

(a) From eqn. (3.1) the B.M. required to initiate yielding is

$$\frac{BD^2}{6}\sigma_y = \frac{50 \times 20^2 \times 10^{-9}}{6} \times 225 \times 10^6 = 750 \text{ N m}$$

But the maximum B.M. on a beam with a central point load is $WL/4$, at the centre.

$$\therefore \qquad \frac{W \times 2}{4} = 750$$

i.e. $$W = \mathbf{1500 \ N}$$

The load required to initiate yielding is 1500 N.

(b) If the load is increased by 10% the new load is

$$W' = 1500 + 150 = 1650 \text{ N}$$

The maximum B.M. is therefore increased to

$$M' = \frac{W'L}{4} = \frac{1650 \times 2}{4} = 825 \text{ Nm}$$

and this is sufficient to produce yielding to a depth d, and from eqn. (3.2),

$$M_{pp} = \frac{B\sigma_y}{12}[3D^2 - d^2] = 825 \text{ Nm}$$

$$\therefore \qquad 825 = \frac{50 \times 10^{-3} \times 225 \times 10^6}{12}[3 \times 2^2 - d^2]10^{-4}$$

where d is the depth of the elastic core in centimetres,

$$\therefore \qquad 8.8 = 12 - d^2$$

$$d^2 = 3.2 \text{ and } d = 1.79 \text{ cm}$$

$$\therefore \qquad \text{depth of yielding} = \tfrac{1}{2}(D - d) = \tfrac{1}{2}(20 - 17.9) = \textbf{1.05 mm}$$

(c) With the central load at 1650 N the yielding will have spread from the centre as shown in Fig. 3.37. At the extremity of the yielded region, a distance x from each end of the beam, the section will just have yielded at the extreme surface fibres, i.e. the moment carried at this section will be the maximum elastic moment and given by eqn. (3.1) – see part (a) above.

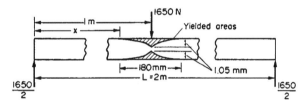

Fig. 3.37.

Now the B.M. at the distance x from the support is

$$\frac{1650x}{2} = \frac{BD^2}{6}\sigma_y = 750$$

$$\therefore \qquad x = \frac{2 \times 750}{1650} = 0.91 \text{ m}$$

Therefore length of beam over which yielding has occurred

$$= 2 - 2 \times 0.91 = 0.18 \text{ m} = \textbf{180 mm}$$

(d) For $W = 1500$ N the beam is completely elastic and the maximum deflection, at the centre, is given by the standard form of eqn. (5.15)[†]:

$$\delta = \frac{WL^3}{48EI} = \frac{1500 \times 2^3 \times 12}{48 \times 206.8 \times 10^9 \times 50 \times 20^3 \times 10^{-12}}$$

$$= 0.0363 \text{ m} = \textbf{36.3 mm}$$

With $W = 1650$ N and the beam partially plastic, deflections are calculated on the basis of the elastic core only,

$$\text{i.e.} \qquad \delta = \frac{W'L^3}{48EI'} = \frac{1650 \times 2^3 \times 12}{48 \times 206.8 \times 10^9 \times 50 \times 17.9^3 \times 10^{-12}}$$

$$= 0.0556 \text{ m} = \textbf{55.6 mm}$$

[†] E.J. Hearn, *Mechanics of Materials 1*, Butterworth-Heinemann, 1997.

Example 3.2

(a) Determine the "shape factor" of a T-section beam of dimensions 100 mm × 150 mm × 12 mm as shown in Fig. 3.38.

(b) A cantilever is to be constructed from a beam with the above section and is designed to carry a uniformly distributed load over its complete length of 2 m. Determine the maximum u.d.1. that the cantilever can carry if yielding is permitted over the lower part of the web to a depth of 25 mm. The yield stress of the material of the cantilever is 225 MN/m^2.

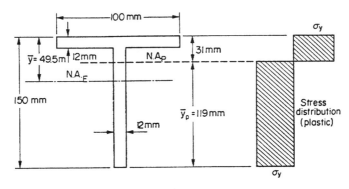

Fig. 3.38.

Solution

(a)   Shape factor $= \dfrac{\text{fully plastic moment}}{\text{maximum elastic moment}}$

To determine the maximum moment carried by the beam while completely elastic we must first determine the position of the N.A.

Take moments of area about the top edge (see Fig. 3.38):

$$(100 \times 12 \times 6) + (138 \times 12 \times 81) = [(100 \times 12) + (138 \times 12)]\ \bar{y}$$

$$7200 + 134136 = (1200 + 1656)\ \bar{y}$$

∴ $\bar{y} = 49.5$ mm

∴ $I_{NA} = \left[\dfrac{100 \times 49.5^3}{3} + \dfrac{12 \times 100.5^3}{3} - \dfrac{88 \times 37.5^3}{3} \right] 10^{-12}\ \text{m}^4$

$$= \tfrac{1}{3}[121.29 + 121.81 - 46.4]10^{-7}$$

$$= 6.56 \times 10^{-6}\ \text{m}^4$$

Now from the simple bending theory the moment required to produce the yield stress at the edge of the section (in this case the lower edge), i.e. the maximum elastic moment, is

$$M_E = \dfrac{\sigma I}{y_{\text{max}}} = \sigma_y \times \dfrac{6.56 \times 10^{-6}}{100.5 \times 10^{-3}} = 0.065 \times 10^{-3}\sigma_y$$

When the section becomes fully plastic the N.A. is positioned such that

$$\text{area below N.A.} = \text{half total area}$$

i.e. if the plastic N.A. is a distance $\bar{y}_p$ above the base, then

$$\bar{y}_p \times 12 = \tfrac{1}{2}(1200 + 1656)$$

$\therefore$
$$\bar{y}_p = 119 \text{ mm}$$

The fully plastic moment is then obtained by considering the moments of forces on convenient rectangular parts of the section, each being subjected to a uniform stress σ_y,

i.e.
$$M_{FP} = \left[\sigma_y(100 \times 12)(31 - 6) + \sigma_y(31 - 12) \times 12 \times \frac{(31 - 12)}{2} \right.$$

$$\left. + \sigma_y(119 \times 12)\frac{119}{2} \right] 10^{-9}$$

$$= \sigma_y(30\,000 + 2166 + 84\,966)10^{-9}$$

$$= 0.117 \times 10^{-3}\sigma_y$$

$\therefore \qquad$ shape factor $= \dfrac{M_{FP}}{M_E} = \dfrac{0.117 \times 10^{-3}}{0.065 \times 10^{-3}} = \mathbf{1.8}$

(b) For this part of the question the load on the cantilever is such that yielding has progressed to a depth of 25 mm over the lower part of the web. It has been shown in §3.4 that whilst plastic penetration proceeds, the N.A. of the section moves and is always positioned by the rule:

$$\text{compressive force above N.A.} = \text{tensile force below N.A.}$$

Thus if the partially plastic N.A. is positioned a distance y above the extremity of the yielded area as shown in Fig. 3.39, the forces exerted on the various parts of the section may be established (proportions of the stress distribution diagram being used to determine the various values of stress noted in the figure).

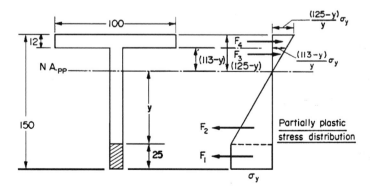

Fig. 3.39.

Force on yielded area $F_1 = $ stress $\times$ area

$$= 225 \times 10^6 (12 \times 25 \times 10^{-6})$$

$$= 67.5 \text{ kN}$$

Force on elastic portion of web below N.A.

$$F_2 = \text{ average stress } \times \text{ area}$$

$$= \frac{225 \times 10^6}{2} (12 \times y \times 10^{-6})$$

$$= 1.35 y \text{ kN}$$

where y is in millimetres.

Force in web above N.A.

$$F_3 = \text{ average stress } \times \text{ area}$$

$$= \frac{(113 - y)}{2y} (225 \times 10^6)(113 - y)12 \times 10^{-6}$$

$$= 1.35 \frac{(113 - y)^2}{y} \text{ kN}$$

Force in flange

$$F_4 = \text{ average stress } \times \text{ area}$$

$$= \frac{1}{2} \left[\frac{(113 - y)}{y} + \frac{(125 - y)}{y} \right] (225 \times 10^6) 100 \times 12 \times 10^{-6} \text{ approximately}$$

$$= \frac{(238 - 2y)}{2y} 225 \times 10^6 \times 100 \times 12 \times 10^{-6}$$

$$= 135 \frac{(238 - 2y)}{y} \text{ kN}$$

Now for the resultant force across the section to be zero,

$$F_1 + F_2 = F_3 + F_4$$

$$67.5 + 1.35 y = \frac{1.35(113 - y)^2}{y} + \frac{135(238 - 2y)}{y}$$

$\therefore \qquad 67.5y + 1.35y^2 = 17.24 \times 10^3 - 305y + 1.35y^2 + 32.13 \times 10^3 - 270y$

$$642.5 y = 49\,370$$

$$y = 76.8 \text{ mm}$$

Substituting back,

$$F_1 = 67.5 \text{ kN} \quad F_2 = 103.7 \text{ kN}$$

$$F_3 = 23 \text{ kN} \quad\;\; F_4 = 148.1 \text{ kN}$$

The moment of resistance of the beam can now be obtained by taking the moments of these forces about the N.A. Here, for ease of calculation, it is assumed that F_4 acts at the mid-point of the web. This, in most cases, is sufficiently accurate for practical purposes.

$$\text{Moment of resistance} = \left\{ F_1(y + 12.5) + F_2 \left(\frac{2y}{3} \right) + F_3 \left[\frac{2}{3}(113 - y) \right] \right.$$

$$\left. + F_4[(113 - y) + 6] \right\} 10^{-3} \text{ kNm}$$

$$= (6030 + 5312 + 554 + 6243)10^{-3} \text{ kNm}$$

$$= 18.14 \text{ kNm}$$

Now the maximum B.M. present on a cantilever carrying a u.d.l. is $wL^2/2$ at the support

$$\therefore \qquad \frac{wL^2}{2} = 18.15 \times 10^3$$

The maximum u.d.l. which can be carried by the cantilever is then

$$w = \frac{18.15 \times 10^3 \times 2}{4} = \mathbf{9.1 \ kN/m}$$

Example 3.3

(a) A steel beam of rectangular section, 80 mm deep by 30 mm wide, is simply supported over a span of 1.4 m and carries a u.d.l. w. If the yield stress of the material is 240 MN/m^2, determine the value of w when yielding of the beam material has penetrated to a depth of 20 mm from each surface of the beam.

(b) What will be the magnitudes of the residual stresses which remain when load is removed?

(c) What external moment must be applied to the unloaded beam in order to return it to its undeformed (straight) position?

Solution

(a) From eqn. (3.2) the partially plastic moment carried by a rectangular section is given by

$$M_{pp} = \frac{B\sigma_y}{12}[3D^2 - d^2]$$

Thus, for the simply supported beam carrying a u.d.l., the maximum B.M. will be at the centre of the span and given by

$$BM_{max} = \frac{wL^2}{8} = \frac{B\sigma_y}{12}[3D^2 - d^2]$$

$$\therefore \qquad w = \frac{8 \times 30 \times 10^{-3} \times 240 \times 10^6}{1.4^2 \times 12}[3 \times 80^2 - 40^2]10^{-6}$$

$$= \mathbf{43.1 \ kN/m}$$

(b) From the above working

$$M_{pp} = \frac{B\sigma_y}{12}[3D^2 - d^2] = \frac{wL^2}{8}$$

$$= 43.1 \times 10^3 \times \frac{1.4^2}{8} = 10.6 \text{ kNm}$$

During the unloading process a moment of equal value but opposite sense is applied to the beam assuming it to be completely elastic. Thus the equivalent maximum elastic stress σ' introduced at the outside surfaces of the beam by virtue of the unloading is given by the simple bending theory with $M = M_{pp} = 10.6$ kNm,

i.e.
$$\sigma' = \frac{My}{I} = \frac{10.6 \times 10^3 \times 40 \times 10^{-3} \times 12}{30 \times 80^3 \times 10^{-12}}$$

$$= 0.33 \times 10^9 = 330 \text{ MN/m}^2$$

The unloading, elastic stress distribution is then linear from zero at the N.A. to ± 330 MN/m^2 at the outside surfaces, and this may be subtracted from the partially plastic loading stress distribution to yield the residual stresses as shown in Fig. 3.40.

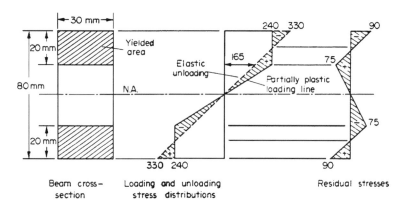

Fig. 3.40.

(c) The residual stress distribution of Fig. 3.40 indicates that the central portion of the beam, which remains elastic throughout the initial loading process, is subjected to a residual stress system when the beam is unloaded from the partially plastic state. The beam will therefore be in a deformed state. In order to remove this deformation an external moment must be applied of sufficient magnitude to return the elastic core to its unstressed state. The required moment must therefore introduce an elastic stress distribution producing stresses of ± 75 MN/m^2 at distances of 20 mm from the N.A. Thus, applying the bending theory,

$$M = \frac{\sigma I}{y} = \frac{75 \times 10^6}{20 \times 10^{-3}} \times \frac{30 \times 80^3 \times 10^{-12}}{12}$$

$$= \textbf{4.8 kNm}$$

Alternatively, since a moment of 10.6 kNm produces a stress of 165 MN/m² at 20 mm from the N.A., then, by proportion, the required moment is

$$M = 10.6 \times \frac{75}{165} = \textbf{4.8 kNm}$$

Example 3.4

A solid circular shaft, of diameter 50 mm and length 300 mm, is subjected to a gradually increasing torque T. The yield stress in shear for the shaft material is 120 MN/m² and, up to the yield point, the modulus of rigidity is 80 GN/m².

(a) Determine the value of T and the associated angle of twist when the shaft material first yields.
(b) If, after yielding, the stress is assumed to remain constant for any further increase in strain, determine the value of T when the angle of twist is increased to twice that at yield.

Solution

(a) For this part of the question the shaft is elastic and the simple torsion theory applies,

i.e.
$$T = \frac{\tau J}{R} = \frac{120 \times 10^6}{25 \times 10^{-3}} \times \frac{\pi(25 \times 10^{-3})^4}{2} = 2950$$
$$= \textbf{2.95 kNm}$$

$$\theta = \frac{\tau L}{GR} = \frac{120 \times 10^6 \times 300 \times 10^{-3}}{80 \times 10^9 \times 25 \times 10^{-3}} = 0.018 \text{ radian}$$
$$= \textbf{1.03}°$$

If the torque is now increased to double the angle of twist the shaft will yield to some radius R_1. Applying the torsion theory to the elastic core only,

$$\theta = \frac{\tau L}{GR}$$

i.e.
$$2 \times 0.018 = \frac{120 \times 10^6 \times 300 \times 10^{-3}}{80 \times 10^9 \times R_1}$$

∴
$$R_1 = \frac{120 \times 10^6 \times 300 \times 10^{-3}}{2 \times 0.018 \times 80 \times 10^9} = 0.0125 = 12.5 \text{ mm}$$

Therefore partially plastic torque, from eqn. (3.12),

$$= \frac{\pi \tau_y}{6}[4R^3 - R_1^3]$$

$$= \frac{\pi \times 120 \times 10^6}{6}[4 \times 25^3 - 12.5^3]10^{-9}$$

$$= \textbf{3.8 kNm}$$

Example 3.5

A 50 mm diameter steel shaft is case-hardened to a depth of 2 mm. Assuming that the inner core remains elastic up to a yield stress in shear of 180 MN/m² and that the case can also be assumed to remain elastic up to failure at the shear stress of 320 MN/m², calculate:

(a) the torque required to initiate yielding at the outside surface of the case;
(b) the angle of twist per metre length at this stage.

Take $G = 85$ GN/m² for both case and core whilst they remain elastic.

Solution

Since the modulus of rigidity G is assumed to be constant throughout the shaft whilst elastic, the angle of twist θ will be constant.

The stress distribution throughout the shaft cross-section at the instant of yielding of the outside surface of the case is then as shown in Fig. 3.41, and it is evident that whilst the failure stress of 320 MN/m² has only just been reached at the outside of the case, the yield stress of the core of 180 NM/m² has been exceeded beyond a radius r producing a fully plastic annulus and an elastic core.

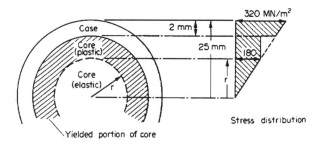

Fig. 3.41.

By proportions, since $G_{\text{case}} = G_{\text{core}}$, then

$$\left(\frac{\tau}{r}\right)_{\text{case}} = \left(\frac{\tau}{r}\right)_{\text{core}}$$

$$\frac{180}{r} = \frac{320}{25}$$

$$\therefore \qquad r = \frac{180}{320} \times 25 = 14.1 \text{ mm}$$

The shaft can now be considered in three parts:

(i) A solid elastic core of 14.1 mm external radius;
(ii) A fully plastic cylindrical region between $r = 14.1$ mm and $r = 23$ mm;
(iii) An elastic outer cylinder of external diameter 50 mm and thickness 2 mm.

Torque on elastic core $\qquad = \dfrac{\tau_y J}{R} = \dfrac{180 \times 10^6}{14.1 \times 10^{-3}} \times \dfrac{\pi (14.1 \times 10^{-3})^4}{2}$

$$= 793 \text{ Nm} = 0.793 \text{ kNm}$$

Torque on plastic section $\qquad = 2\pi\tau_y \int_{r_1}^{r_2} r^2\,dr$

$$= \frac{2\pi \times 180 \times 10^6}{3}[23^3 - 14.1^3]10^{-9}$$

$$= \frac{2\pi \times 180 \times 10^6 \times 9364 \times 10^{-9}}{3}$$

$$= 3.53 \text{ kNm}$$

Torque on elastic outer case $\quad = \dfrac{\tau_y J}{r} = \dfrac{320 \times 10^6}{25 \times 10^{-3}}\pi\left[\dfrac{25^4 - 23^4}{2}\right]10^{-12}$

$$= 2.23 \text{ kNm}$$

Therefore total torque required $= (0.793 + 3.53 + 2.23)10^3$

$$= \mathbf{6.55 \text{ kNm}}$$

Since the angle of twist is assumed constant across the whole shaft its value may be determined by application of the simple torsion theory to either the case or the elastic core.

For the case: $\qquad \dfrac{\theta}{L} = \dfrac{\tau}{GR} = \dfrac{320 \times 10^6}{85 \times 10^9 \times 25 \times 10^{-3}}$

$$= 0.15 \text{ rad} = \mathbf{8.6°}$$

Example 3.6

A hollow circular bar of 100 mm external diameter and 80 mm internal diameter (Fig. 3.42) is subjected to a gradually increasing torque T. Determine the value of T:

(a) when the material of the bar first yields;
(b) when plastic penetration has occurred to a depth of 5 mm;
(c) when the section is fully plastic.

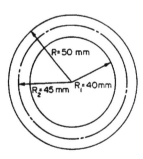

Fig. 3.42.

The yield stress in shear of the shaft material is 120 MN/m².

Determine the distribution of the residual stresses present in the shaft when unloaded from conditions (b) and (c).

Solution

(a) Maximum elastic torque from eqn. (3.11)

$$= \frac{\pi \tau_y}{2R}[R^4 - R_1^4] = \frac{\pi \times 120 \times 10^6}{2 \times 50 \times 10^{-3}}(625 - 256)10^{-8}$$

$$= 13900 \text{ Nm} = \textbf{13.9 kNm}$$

(b) Partially plastic torque, from eqns. (3.11) and (3.13),

$$= \frac{\pi \tau_y}{2R_2}[R_2^4 - R_1^4] + \frac{2\pi \tau_y}{3}[R^3 - R_2^3]$$

$$= \frac{\pi \times 120 \times 10^6}{2 \times 45 \times 10^{-3}}(4.5^4 - 256)10^{-8} + \frac{2\pi \times 120 \times 10^6}{3}(125 - 91)10^{-6}$$

$$= 6450 + 8550 = 15000 \text{ Nm} = \textbf{15 kNm}$$

(c) Fully plastic torque from eqn. (3.16) or eqn. (3.13)

$$= \frac{2\pi \tau_y}{3}[R^3 - R_1^3]$$

$$= \frac{2\pi \times 120 \times 10^6}{3}[125 - 64]10^{-6} = 15330 = \textbf{15.33 kNm}$$

In order to determine the residual stresses after unloading, the unloading process is assumed completely elastic.

Thus, unloading from condition (b) is equivalent to applying a moment of 15 kNm of opposite sense to the loading moment on a complete elastic bar. The effective stress introduced at the outer surface by this process is thus given by the simple torsion theory

$$\frac{T}{J} = \frac{\tau}{R}$$

i.e.

$$\tau = \frac{TR}{J} = \frac{15 \times 10^3 \times 50 \times 10^{-3} \times 2}{\pi \times (50^4 - 40^4)10^{-12}}$$

$$= \frac{15 \times 10^3 \times 50 \times 10^{-3} \times 2}{\pi(5^4 - 4^4)10^{-8}}$$

$$= 129 \text{ MN/m}^2$$

The unloading stress distribution is then linear, from zero at the centre of the bar to 129 MN/m^2 at the outside. This can be subtracted from the partially plastic loading stress distribution as shown in Fig. 3.43 to produce the residual stress distribution shown.

Similarly, unloading from the fully plastic state is equivalent to applying an elastic torque of 15.33 kNm of opposite sense. By proportion, from the above calculation,

$$\text{equivalent stress at outside of shaft on unloading} = \frac{15.33}{15} \times 129 = 132 \text{ MN/m}^2$$

Subtracting the resulting unloading distribution from the fully plastic loading one gives the residual stresses shown in Fig. 3.44.

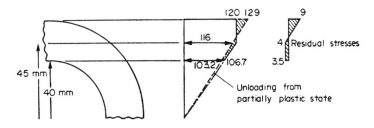

Fig. 3.43.

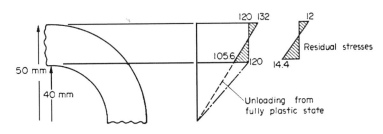

Fig. 3.44.

Example 3.7

(a) A thick cylinder, inside radius 62.5 mm and outside radius 190 mm, forms the pressure vessel of an isostatic compacting press used in the manufacture of ceramic components. Determine, using the Tresca theory of elastic failure, the safety factor on initial yield of the cylinder when an internal working pressure P_W of 240 MN/m² is applied.

(b) In view of the relatively low value of the safety factor which is achieved at this working pressure the cylinder is now subjected to an autofrettage pressure P_A of 580 MN/m². Determine the residual stresses produced at the bore of the cylinder when the autofrettage pressure is removed and hence determine the new value of the safety factor at the bore when the working pressure $P_W = 240$ MN/m² is applied.

The yield stress of the cylinder material is $\sigma_y = 850$ MN/m² and axial stresses may be ignored.

Solution

(a) Plain cylinder – working conditions $K = 190/62.5 = 9.24$
From eqn $(10.5)^{\dagger}$

$$\sigma_{rr} = -P\left[\frac{(R_2/r)^2 - 1}{K^2 - 1}\right] = \frac{-240}{8.24}\left[\frac{0.19^2}{r^2} - 1\right]$$

$$= -240 \text{ MN/m}^2 \text{ at the bore surface } (r = 0.0625 \text{ mm})$$

---

† E.J. Hearn, *Mechanics of Materials 1*, Butterworth-Heinemann, 1997.

and from eqn. (10.6)[†]

$$\sigma_{\theta\theta} = P \left[\frac{(R_2/r)^2 + 1}{K^2 - 1} \right] = \frac{240}{8.24} \left[\frac{0.19^2}{r^2} + 1 \right]$$

$$= 298.3 \text{ MN/m}^2 \text{ at the bore surface}$$

Thus, assuming axial stress will be the intermediate stress (σ_2) value, the critical stress conditions for the cylinder at the internal bore are $\sigma_1 = 298.3$ MN/m^2 and $\sigma_3 = -240$ MN/m^2.
∴ Applying the Tresca theory of failure ($\sigma_1 - \sigma_3 = \sigma_y/n$)

$$\text{Safety factor } n = \frac{850}{298.3 - (-240)} = \mathbf{1.58}$$

(b) Autofrettage conditions

From eqn 3.37 the radius R_p of the elastic/plastic interface under autofrettage pressure of 580 MN/m^2 will be given by:

$$P_A = \frac{\sigma_y}{2} \left[\frac{K^2 - m^2}{K^2} \right] + \sigma_y \log_e m$$

∴

$$580 \times 10^6 = \frac{850 \times 10^6}{2} \left[\frac{3.04^2 - m^2}{3.04^2} \right] + 850 \times 10^6 \log_e m$$

By trial and error:

m	$850 \log_e m$	$\dfrac{850}{2} \left[\dfrac{3.04^2 - m^2}{3.04^2} \right]$	P_A
1.6	399.5	307.3	706.8
1.4	286.0	334.8	620.8
1.3	223.0	347.3	570.3
1.33	242.4	343.6	585.6
1.325	239.2	344.2	583.4

∴ to a good approximation $m = 1.325 = R_p/R_1$

$$\therefore R_p = 1.325 \times 62.5 = 82.8 \text{ mm}$$

∴ From eqns 3.38 and 3.39 stresses in the plastic zone are:

$$\sigma_{rr} = 850 \times 10^6 \left[\log_e \left(\frac{r}{82.8} \right) - \frac{1}{2 \times 190^2} (190^2 - 82.8^2) \right]$$

$$= 850 \times 10^6 [\log_e (r/82.8) - 0.405]$$

and

$$\sigma_{\theta\theta} = 850 \times 10^6 [\log_e (r/82.8) + 0.595]$$

∴ At the bore surface where $r = 62.5$ mm the stresses due to autofrettage are:

$$\sigma_{rr} = -580 \text{ MN/m}^2 \text{ and } \sigma_{\theta\theta} = 266.7 \text{ MN/m}^2.$$

[†] E.J. Hearn, *Mechanics of Materials 1*, Butterworth-Heinemann, 1997.

Residual stresses are then obtained by *elastic* unloading of the autofrettage pressure, i.e. by applying $\sigma_{rr} = +580$ MN/m^2 at the bore in eqns (10.5) and (10.6)[†]; i.e. by proportions:

$$\sigma_{rr} = 580 \text{ MN/m}^2 \text{ and } \sigma_{\theta\theta} = -298.3 \times \frac{580}{240} = -721 \text{ MN/m}^2.$$

Giving residual stresses at the bore of:

$$\sigma'_{rr} = 580 - 580 = \mathbf{0}$$

$$\sigma'_{\theta\theta} = 266.7 - 721 = \mathbf{-453 \text{ MN/m}^2}$$

Working stresses are then obtained by the addition of elastic loading stresses due to an internal working pressure of 240 MN/m^2

i.e. from part (a) $\sigma_{rr} = -240$ MN/m^2, $\sigma_{\theta\theta} = 298.3$ MN/m^2

∴ final working stresses are:

$$\sigma_{rr_w} = 0 - 240 = -240 \text{ MN/m}^2$$

$$\sigma_{\theta\theta_w} = 298.3 - 454.3 = -156 \text{ MN/m}^2.$$

∴ New safety factor according to Tresca theory

$$n = \frac{850}{-156 - (-240)} = \mathbf{10.1}.$$

N.B. It is unlikely that the Tresca theory will give such a high value in practice since the axial working stress (ignored in this calculation) may well become the major principal stress σ_1 in the working condition and increase the magnitude of the denominator to reduce the resulting value of n.

Problems

3.1 (A/B). Determine the shape factors for the beam cross-sections shown in Fig. 3.45, in the case of section (c) treating the section both with and without the dotted area. [1.23, 1.81, 1.92, 1.82.]

All dimensions in mm

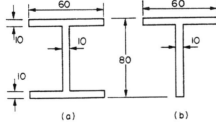
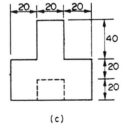

Fig. 3.45.

[†] E.J. Hearn, *Mechanics of Materials 1*, Butterworth-Heinemann, 1997.

3.2 (B). A 50 mm × 20 mm rectangular-section beam is used simply supported over a span of 2 m. If the beam is used with its long edges vertical, determine the value of the central concentrated load which must be applied to produce initial yielding of the beam material.

If this load is then increased by 10% determine the depth to which yielding will take place at the centre of the beam span.

Over what length of beam has yielding taken place?

What are the maximum deflections for each load case? Take $\sigma_y = 225$ MN/m^2 and $E = 206.8$ GN/m^2.

[1.5 kN; 1.05 mm; 180 mm; 36.3, 55.5 mm.]

3.3 (B). A steel bar of rectangular section 72 mm × 30 mm is used as a simply supported beam on a span of 1.2 m and loaded at mid-span. If the yield stress is 280 MN/m^2 and the long edges of the section are vertical, find the loading when yielding first occurs.

Assuming that a further increase in load causes yielding to spread inwards towards the neutral axis, with the stress in the yielded part remaining at 280 MN/m^2, find the load required to cause yielding for a depth of 12 mm at the top and bottom of the section at mid-span, and find the length of beam over which yielding has occurred.

[24.2 kN; 31 kN; 0.264 m.]

3.4 (B). A 300 mm × 125 mm I-beam has flanges 13 mm thick and web 8.5 mm thick. Calculate the shape factor and the moment of resistance in the fully plastic state. Take $\sigma_y = 250$ MN/m^2 and $I_{xx} = 85 \times 10^{-6}$ m^4.

[1.11, 141 kNm.]

3.5 (B). Find the shape factor for a 150 mm × 75 mm channel in pure bending with the plane of bending perpendicular to the web of the channel. The dimensions are shown in Fig. 3.46 and $Z = 21 \times 10^{-6}$ m^3.

[2.2.]

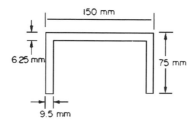

Fig. 3.46.

3.6 (B). A cantilever is to be constructed from a 40 mm × 60 mm T-section beam with a uniform thickness of 5 mm. The cantilever is to carry a u.d.l. over its complete length of 1 m. Determine the maximum u.d.l. that the cantilever can carry if yielding is permitted over the lower part of the web to a depth of 10 mm. $\sigma_y = 225$ MN/m^2.

[2433 N/m.]

3.7 (B). A 305 mm × 127 mm symmetrical I-section has flanges 13 mm thick and a web 5.4 mm thick. Treating the web and flanges as rectangles, calculate the bending moment of resistance of the cross-section (a) at initial yield, (b) for full plasticity of the flanges only, and (c) for full plasticity of the complete cross-section. Yield stress in simple tension and compression = 310 MN/m^2. What is the shape factor of the cross-section?

[167, 175.6, 188.7 kNm; 1.13.]

3.8 (B). A steel bar of rectangular section 80 mm by 40 mm is used as a simply supported beam on a span of 1.4 m and point-loaded at mid-span. If the yield stress of the steel is 300 MN/m^2 in simple tension and compression and the long edges of the section are vertical, find the load when yielding first occurs.

Assuming that a further increase in load causes yielding to spread in towards the neutral axis with the stress in the yielded part remaining constant at 300 MN/m^2, determine the load required to cause yielding for a depth of 10 mm at the top and bottom of the section at mid-span and find the length of beam over which yielding at the top and bottom faces will have occurred.   [U.L.] [36.57, 44.6 kN; 0.232 m.]

3.9 (B). A straight bar of steel of rectangular section, 76 mm wide by 25 mm deep, is simply supported at two points 0.61 m apart. It is subjected to a uniform bending moment of 3 kNm over the whole span. Determine the depth of beam over which yielding will occur and make a diagram showing the distribution of bending stress over the full depth of the beam. Yield stress of steel in tension and compression = 280 MN/m^2.

Estimate the deflection at mid-span assuming $E = 200$ GN/m^2 for elastic conditions.   [5.73, 44.4 mm.]

3.10 (B). A symmetrical I-section beam of length 6 m is simply supported at points 1.2 m from each end and is to carry a u.d.l. w kN/m run over its entire length. The second moment of area of the cross-section about the neutral axis parallel to the flanges is 6570 cm^4 and the beam cross-section dimensions are: flange width and thickness, 154 mm and 13 mm respectively, web thickness 10 mm, overall depth 254 mm.

(a) Determine the value of w to just cause initial yield, stating the position of the transverse section in the beam length at which it occurs.

(b) By how much must w be increased to ensure full plastic penetration of the flanges only, the web remaining elastic?

Take the yield stress of the beam material in simple tension and compression as 340 MN/m^2.

[B.P.] [195, 20 kN/m.]

3.11 (B). A steel beam of rectangular cross-section. 100 mm wide by 50 mm deep, is bent to the arc of a circle until the material just yields at the outer fibres, top and bottom. Bending takes place about the neutral axis parallel to the 100 mm side. If the yield stress for the steel is 330 MN/m^2 in simple tension and compression, determine the applied bending moment and the radius of curvature of the neutral layer. $E = 207$ GN/m^2.

Find how much the bending moment has to be increased so that the stress distribution is as shown in Fig. 3.47.

[I.Mech.E.] [13.75 kNm; 15.7 m; 16.23 kNm.]

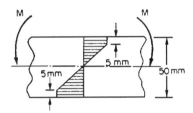

Fig. 3.47.

3.12 (B). A horizontal steel cantilever beam, 2.8 m long and of uniform I-section throughout, has the following cross-sectional dimensions: flanges 150 mm × 25 mm, web 13 mm thick, overall depth 305 mm. It is fixed at one end and free at the other.

(a) Determine the intensity of the u.d.l. which the beam has to carry across its entire length in order to produce fully developed plasticity of the cross-section.

(b) What is the value of the shape factor of the cross-section?

(c) Determine the length of the beam along the top and bottom faces, measured from the fixed end, over which yielding will occur due to the load found in (a).

Yield stress of steel = 330 MN/m^2.   [106.2 kN/m; 1.16; 0.2 m.]

3.13 (B). A rectangular steel beam, 60 mm deep by 30 mm wide, is supported on knife-edges 2m apart and loaded with two equal point loads at one-third of the span from each end. Find the load at which yielding just begins, the yield stress of the material in simple tension and compression being 300 MN/m^2.

If the loads are increased to 25% above this value, estimate how far the yielding penetrates towards the neutral axis, assuming that the yield stress remains constant.   [U.L.] [8.1 kN; 8.79 mm.]

3.14 (B). A steel bar of rectangular section, 72 mm deep by 30 mm wide, is used as a beam simply supported at each end over a span of 1.2 m and loaded at mid-span with a point load. The yield stress of the material is 280 MN/m^2. Determine the value of the load when yielding first occurs.

Find the load to cause an inward plastic penetration of 12 mm at the top and bottom of the section at mid-span. Also find the length, measured along the top and bottom faces, over which yielding has occurred, and the residual stresses present after unloading.   [U.L.] [24.2 kN; 31 kN; 0.26 m, $\mp$79, $\pm$40.7 MN/m^2.]

3.15 (B). A symmetrical I-section beam, 300 mm deep, has flanges 125 mm wide by 13 mm thick and a web 8.5 mm thick. Determine:

(a) the applied bending moment to cause initial yield;

(b) the applied bending moment to cause full plasticity of the cross-section;

(c) the shape factor of the cross-section.

Take the yield stress = 250 MN/m^2 and assume $I = 85 \times 10^6$ mm^4.

[141 × 10^6 N mm; 156 × 10^6 N mm; 1.11.]

3.16 (B). A rectangular steel beam AB, 20 mm wide by 10 mm deep, is placed symmetrically on two knife-edges C and D, 0.5 m apart, and loaded by applying equal loads at the ends A and B. The steel follows a linear stress/strain law ($E = 200$ GN/m^2) up to a yield stress of 300 MN/m^2; at this constant stress considerable plastic deformation occurs. It may be assumed that the properties of the steel are the same in tension and compression.

Calculate the bending moment on the central part of the beam *CD* when yielding commences and the deflection at the centre relative to the supports.

If the loads are increased until yielding penetrates half-way to the neutral axis, calculate the new value of the bending moment and the corresponding deflection.   [U.L.] [100 Nm, 9.375 mm; 137.5 Nm, 103 mm.]

3.17 (B). A steel bar of rectangular material, 75 mm × 25 mm, is used as a simply supported beam on a span of 2 m and is loaded at mid-span. The 75 mm dimension is placed vertically and the yield stress for the material is 240 MN/m². Find the load when yielding first occurs.

The load is further increased until the bending moment is 20% greater than that which would cause initial yield. Assuming that the increased load causes yielding to spread inwards towards the neutral axis, with the stress in the yielded part remaining at 240 MN/m², find the depth at the top and bottom of the section at mid-span to which the yielding will extend. Over what length of the beam has yielding occurred?

[B.P.] [11.25 kN; 8.45 mm; 0.33 m.]

3.18 (B). The cross-section of a beam is a channel, symmetrical about a vertical centre line. The overall width of the section is 150 mm and the overall depth 100 mm. The thickness of both the horizontal web and each of the vertical flanges is 12 mm. By comparing the behaviour in both the elastic and plastic range determine the shape factor of the section. Work from first principles in both cases.   [1.806.]

3.19 (B). The T-section beam shown in Fig. 3.48 is subjected to increased load so that yielding spreads to within 50 mm of the lower edge of the flange. Determine the bending moment required to produce this condition. $\sigma_y = 240$ MN m².   [44 kN m.]

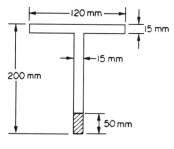

Fig. 3.48.

3.20 (B). A steel beam of I-section with overall depth 300 mm, flange width 125 mm and length 5 m, is simply supported at each end and carries a uniformly distributed load of 114 kN/m over the full span. Steel reinforcing plates 12 mm thick are welded symmetrically to the outside of the flanges producing a section of overall depth 324 mm. If the plate material is assumed to behave in an elastic-ideally plastic manner, determine the plate width necessary such that yielding has just spread through each reinforcing plate at mid-span under the given load.

Determine also the positions along the reinforcing plates at which the outer surfaces have just reached the yield point. At these sections what is the horizontal shearing stress at the interfaces of the reinforcing plates and the flanges?

Take the yield stress $\sigma_y = 300$ MN/m² and the second moment of area of the basic I-section to be 80×10^{-6} m⁴.

[C.E.I.] [175 mm; 1.926 m; 0.94 MN/m².]

3.21 (B). A horizontal cantilever is propped at the free end to the same level as the fixed end. It is required to carry a vertical concentrated load *W* at any position between the supports. Using the normal assumption of plastic limit design, determine the least favourable position of the load. (Note that the calculation of bending moments under elastic conditions is not required.)

Hence calculate the maximum permissible value of *W* which may be carried by a rectangular-section cantilever with depth *d* equal to twice the width over a span *L*. Assume a load factor of *n* and a yield stress for the beam material σ_y.   [0.586 *L* from built-in end; $d^3\sigma_y / 1.371 \, Ln$.]

3.22 (B). (a) Sketch the idealised stress–strain diagram which is used to establish a quantitative relationship between stress and strain in the plastic range of a ductile material. Include the effect of strain-hardening.

(b) Neglecting strain–hardening, sketch the idealised stress–strain diagram and state, in words, the significance of any alteration you make in the diagram shown for part (a) when calculations are made, say, for pure bending beyond the yield point.

(c) A steel beam of rectangular cross-section, 200 mm wide × 100 mm deep, is bent to the arc of a circle, bending taking place about the neutral axis parallel to the 200 mm side.

Determine the bending moment to be applied such that the stress distribution is as shown in (i) Fig. 3.49(a) and (ii) Fig. 3.49(b).

Take the yield stress of steel in tension and compression as 250 MN/m^2.   [B.P.] [98.3, 125 kN m.]

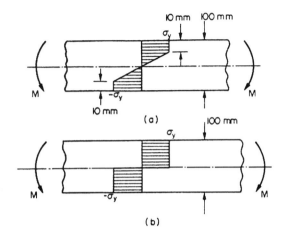

Fig. 3.49.

3.23 (B). (a) A rectangular section beam is 80 mm wide, 120 mm deep and is simply supported at each end over a span of 4 m. Determine the maximum uniformly distributed load that the beam can carry:

(i) if yielding of the beam material is permitted to a depth of 40 mm;
(ii) before complete collapse occurs.

(b) What residual stresses would be present in the beam after unloading from condition (a) (i)?

(c) What external moment must be applied to the beam to hold the deformed bar in a straight position after unloading from condition (a) (i)?

The yield stress of the material of the beam = 280 MN/m^2.

[B.P.] [38.8, 40.3 kN/m; ± 123, ± 146 MN/m^2; 84.3 kN m.]

3.24 (C). A rectangular beam 80 mm wide and 20 mm deep is constructed from a material with a yield stress in tension of 270 MN/m^2 and a yield stress in compression of 300 MN/m^2. If the beam is now subjected to a pure bending moment find the value required to produce:

(a) initial yield;
(b) initial yield on the compression edge;
(c) a fully plastic section.   [1.44; 1.59, 2.27 kN m.]

3.25 (C). Determine the load factor of a propped cantilever carrying a concentrated load W at the centre.

Allowable working stress = 150 MN/m^2, yield stress = 270 MN/m^2. The cantilever is of I-section with dimensions 300 mm × 80 mm × 8 mm.   [2.48.]

3.26 (C). A 300 mm × 100 mm beam is carried over a span of 7 m the ends being rigidly built in. Find the maximum point load which can be carried at 3 m from one end and the maximum working stress set up.

Take a load factor of 1.8 and σ_y = 240 MN/m^2.

$I = 85 \times 10^{-6}$ m^4 and the shape factor = 1.135.   [100 kN; 172 MN/m^2.]

3.27 (C). A 300 mm × 125 mm I-beam is carried over a span of 20 m the ends being rigidly built in. Find the maximum point load which can be carried at 8 m from one end and the maximum working stress set up. Take a load factor of 1.8 and σ_y = 250 MN/m^2; $Z = 56.6 \times 10^{-5}$ m^3 and shape factor $\lambda = 1.11$.

[36 kN; 183 MN/m^2.]

3.28 (C). Determine the maximum intensity of loading that can be sustained by a simply supported beam, 75 mm wide × 100 mm deep, assuming perfect elastic–plastic behaviour with a yield stress in tension and compression of 135 MN/m^2. The beam span is 2 m.

What will be the distribution of residual stresses in the beam after unloading?

[50.6 kN/m; 67, 135, −67 MN/m^2.]

3.29 (C). A short column of 0.05 m square cross-section is subjected to a compressive load of 0.5 MN parallel to but eccentric from the central axis. The column is made from elastic – perfectly plastic material which has a yield stress in tension or compression of 300 MN/m². Determine the value of the eccentricity which will result in the section becoming just fully plastic. Also calculate the residual stress at the outer surfaces after elastic unloading from the fully plastic state.   [10.4 mm; 250, 150 MN/m².]

3.30 (C). A rectangular beam 75 mm wide and 200 mm deep is constructed from a material with a yield stress in tension of 270 MN/m² and a yield stress in compression of 300 MN/m². If the beam is now subjected to a pure bending moment, determine the value of the moment required to produce (a) initial yield, (b) initial yield on the compression edge, (c) a fully plastic section.   [135, 149.2, 213.2 kN m.]

3.31 (C). Figure 3.50 shows the cross-section of a welded steel structure which forms the shell of a gimbal frame used to support the ship-to-shore transport platform of a dock installation. The section is symmetrical about the vertical centre-line with a uniform thickness of 25 mm throughout.

As a preliminary design study what would you assess as the maximum bending moment which the section can withstand in order to prevent:

(a) initial yielding at any point in the structure if the yield stress for the material is 240 MN/m²,
(b) complete collapse of the structure?

What would be the effect of adverse weather conditions which introduce instantaneous loads approaching, but not exceeding that predicted in (b). Quantify your answers where possible.

State briefly the factors which you would consider important in the selection of a suitable material for such a structure.   [309.3 kN m; 423 kN m; local yielding, residual stress max = 279 MN/m².]

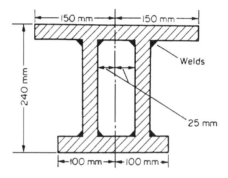

Fig. 3.50.

3.32 (B). A solid shaft 40 mm diameter is made of a steel the yield point of which in shear is 150 MN/m². After yielding, the stress remains constant for a very considerable increase in strain. Up to the yield point the modulus of rigidity $G = 80$ GN/m². If the length of the shaft is 600 mm calculate:
(a) the angle of twist and the twisting moment when the shaft material first yields;
(b) the twisting moment when the angle of twist is increased to twice that at yield.   [3.32°; 1888, 2435 Nm.]

3.33 (B). A solid steel shaft, 76 mm diameter and 1.53 m long, is subjected to pure torsion. Calculate the applied torque necessary to cause initial yielding if the material has a yield stress in pure tension of 310 MN/m². Adopt the Tresca criterion of elastic failure.

(b) If the torque is increased to 10% above that at first yield, determine the radial depth of plastic penetration. Also calculate the angle of twist of the shaft at this increased torque. Up to the yield point in shear, $G = 83$ GN/m².

(c) Calculate the torque to be applied to cause the cross-section to become fully plastic.
                          [13.36 kNm; 4.26 mm, 0.085 rad; 17.8 kNm.]

3.34 (B). A hollow steel shaft having outside and inside diameters of 32 mm and 18 mm respectively is subjected to a gradually increasing axial torque. The yield stress in shear is reached at the surface when the torque is 1 kNm, the angle of twist per metre length then being 7.3°. Find the magnitude of the yield shear stress.

If the torque is increased to 1.1 kN m, calculate (a) the depth to which yielding will have penetrated, and (b) the angle of twist per metre length.

State any assumptions made and prove any special formulae used.

[U.L.] [172.7 MN/m²; 1.8 mm; 8.22°.]

3.35 (B). A hollow shaft, 50 mm diameter and 25 mm bore, is made of steel with a yield stress in shear of 150 MN/m^2 and a modulus of rigidity of 83 GN/m^2. Calculate the torque and the angle of twist when the material first yields, if the shaft has a length of 2 m.

On the assumption that the yield stress, after initial yield, then remains constant for a considerable increase in strain, calculate the depth of penetration of plastic yield for an increase in torque of 10% above that at initial yield. Determine also the angle of twist of the shaft at the increased torque.

[U.L.] [3.45 kN m; 8.29°; 2.3 mm, 9.15°.]

3.36 (C). A steel shaft of length 1.25 m has internal and external diameters of 25 mm and 50 mm respectively. The shear stress at yield of the steel is 125 MN/m^2. The shear modulus of the steel is 80 GN/m^2. Determine the torque and overall twist when (a) yield first occurs, (b) the material has yielded outside a circle of diameter 40 mm, and (c) the whole section has just yielded. What will be the residual stresses after unloading from (b) and (c)?

[2.88, 3.33, 3.58 kN m; 0.0781, 0.0975, 0.1562 rad, (a) 19.7, −9.2, −5.5 MN/m^2, (b) 30.6, −46.75 MN/m^2.]

3.37 (B). A shaft having a diameter of 90 mm is turned down to 87 mm for part of its length. If a torque is applied to the shaft of sufficient magnitude just to produce yielding at the surface of the shaft in the unturned part, determine the depth of yielding which would occur in the turned part. Find also the angle of twist per unit length in the turned part to that in the unturned part of the shaft. [U.L.] [5.3 mm; 1.18.]

3.38 (B). A steel shaft, 90 mm diameter, is solid for a certain distance from one end but hollow for the remainder of its length with an inside diameter of 38 mm. If a pure torque is transmitted from one end of the shaft to the other of such a magnitude that yielding just occurs at the surface of the solid part of the shaft, find the depth of yielding in the hollow part of the shaft and the ratio of the angles of twist per unit length for the two parts of the shaft. [U.L.] [1.5 mm; 1.0345:1.]

3.39 (B). A steel shaft of solid circular cross-section is subjected to a gradually increasing torque. The diameter of the shaft is 76 mm and it is 1.22 m long. Determine for initial yield conditions in the outside surface of the shaft (a) the angle of twist of one end relative to the other, (b) the applied torque, and (c) the total resilience stored.

Assume a yield in shear of 155 MN/m^2 and a shear modulus of 85 GN/m^2. If the torque is increased to a value 10% greater than that at initial yield, estimate (d) the depth of penetration of plastic yielding and (e) the new angle of twist. [B.P.] [3.35°; 13.4 kN m; 391 J; 4.3 mm; 3.8°.]

3.40 (B). A solid steel shaft, 50 mm diameter and 1.22 m long, is transmitting power at 10 rev/s.

(a) Determine the power to be transmitted at this speed to cause yielding of the outer fibres of the shaft if the yield stress in shear is 170 MN/m^2.

(b) Determine the increase in power required to cause plastic penetration to a radial depth of 6.5 mm, the speed of rotation remaining at 10 rev/s. What would be the angle of twist of the shaft in this case? G for the steel is 82 GN/m^2. [B.P.] [262 kW, 52 kW, 7.83°.]

3.41 (B). A marine propulsion shaft of length 6 m and external diameter 300 mm is initially constructed from solid steel bar with a shear stress at yield of 150 MN/m^2.

In order to increase its power/weight ratio the shaft is machined to convert it into a hollow shaft with internal diameter 260 mm, the outer diameter remaining unchanged.

Compare the torques which may be transmitted by the shaft in both its initial and machined states:

(a) when yielding first occurs,

(b) when the complete cross-section has yielded.

If, in service, the hollow shaft is subjected to an unexpected overload during which condition (b) is achieved, what will be the distribution of the residual stresses remaining in the shaft after torque has been removed?

[795 kN m, 346 kN m, 1060 kN m, 370 kN m; −10.2, +11.2 MN/m^2.]

3.42 (C). A solid circular shaft 100 mm diameter is in an elastic–plastic condition under the action of a pure torque of 24 kN m. If the shaft is of steel with a yield stress in shear of 120 MN/m^2 determine the depth of the plastic zone in the shaft and the angle of twist over a 3 m length. Sketch the residual shear stress distribution on unloading. $G = 85$ GN/m^2. [0.95 mm; 4.95°.]

3.43 (C). A column is constructed from elastic – perfectly plastic material and has a cross-section 60 mm square. It is subjected to a compressive load of 0.8 MN parallel to the central longitudinal axis of the beam but eccentric from it. Determine the value of the eccentricity which will produce a fully plastic section if the yield stress of the column material is 280 MN/m^2.

What will be the values of the residual stresses at the outer surfaces of the column after unloading from this condition? [7 mm; 213, −97 MN/m^2.]

3.44 (C). A beam of rectangular cross-section with depth d is constructed from a material having a stress–strain diagram consisting of two linear portions producing moduli of elasticity E_1 in tension and E_2 in compression.

Assuming that the beam is subjected to a positive bending moment M and that cross-sections remain plane, show that the strain on the outer surfaces of the beam can be written in the form

$$\varepsilon_1 = \frac{d}{R}\left[\frac{\sqrt{E_2}}{\sqrt{E_1} + \sqrt{E_2}}\right]$$

where R is the radius of curvature.

Hence derive an expression for the bending moment M in terms of the elastic moduli, the second moment of area I of the beam section and R the radius of curvature.

$$\left[M = \frac{4E_1 E_2 I}{R(\sqrt{E_1} + \sqrt{E_2})^2}\right]$$

3.45 (C). Explain what is meant by the term "autofrettage" as applied to thick cylinder design. What benefits are obtained from autofrettage and what precautions should be taken in its application?

(b) A thick cylinder, inside radius 62.5 mm and outside radius 190 mm, forms the pressure vessel of an isostatic compacting press used in the manufacture of sparking plug components. Determine, using the Tresca theory of elastic failure, the safety factor on initial yield of the cylinder when an internal working pressure P_w of 240 MN/m^2 is applied.

(c) In view of the relatively low value of safety factor which is achieved at this working pressure, the cylinder is now subjected to an autofrettage pressure of $P_A = 580$ MN/m^2.

Determine the residual stresses produced at the bore of the cylinder when the autofrettage pressure is removed and hence determine the new value of the safety factor at the bore when the working pressure P_w is applied.

The yield stress of the cylinder material $\sigma_y = 850$ MN/m^2 and axial stresses may be ignored.

3.46 (C). A thick cylinder of outer radius 190 mm and radius ratio $K = 3.04$ is constructed from material with a yield stress of 850 MN/m^2 and tensile strength 1 GN/m^2. In order to prepare it for operation at a working pressure of 248 MN/m^2 it is subjected to an initial autofrettage pressure of 584 MN/m^2.

Ignoring axial stresses, compare the safety factors against initial yielding of the bore of the cylinder obtained with and without the autofrettage process. [1.53, 8.95.]

3.47 (C). What is the maximum autofrettage pressure which should be applied to a thick cylinder of the dimensions given in problem 3.46 in order to achieve yielding to the geometric mean radius?

Determine the maximum hoop and radial residual stresses produced by the application and release of this pressure and plot the distributions of hoop and radial residual stress across the cylinder wall.

[758 MN/m^2; -55.2 MN/m^2; -8.5 MN/m^2.]

CHAPTER 4

RINGS, DISCS AND CYLINDERS SUBJECTED TO ROTATION AND THERMAL GRADIENTS

Summary

For *thin rotating rings and cylinders* of mean radius R, the tensile hoop stress set up is given by

$$\sigma_H = \rho \omega^2 R^2$$

The radial and hoop stresses at any radius r in a *disc of uniform thickness* rotating with an angular velocity ω rad/s are given by

$$\sigma_r = A - \frac{B}{r^2} - (3 + v)\frac{\rho\omega^2 r^2}{8}$$

$$\sigma_H = A + \frac{B}{r^2} - (1 + 3v)\frac{\rho\omega^2 r^2}{8}$$

where A and B are constants, ρ is the density of the disc material and v is Poisson's ratio.

For a *solid disc* of radius R these equations give

$$\sigma_r = (3 + v)\frac{\rho\omega^2}{8}(R^2 - r^2)$$

$$\sigma_H = \frac{\rho\omega^2}{8}\left[(3 + v)R^2 - (1 + 3v)r^2\right]$$

At the centre of the solid disc these equations yield the maximum stress values

$$\sigma_{H_{max}} = \sigma_{r_{max}} = (3 + v)\frac{\rho\omega^2 R^2}{8}$$

At the outside radius,

$$\sigma_r = 0$$

$$\sigma_H = (1 - v)\frac{\rho\omega^2 R^2}{4}$$

For a *disc with a central hole*,

$$\sigma_r = (3 + v)\frac{\rho\omega^2}{8}\left[R_1^2 + R_2^2 - \frac{R_1^2 R_2^2}{r^2} - r^2\right]$$

$$\sigma_H = \frac{\rho\omega^2}{8}\left[(3 + v)\left(R_1^2 + R_2^2 + \frac{R_1^2 R_2^2}{r^2}\right) - (1 + 3v)r^2\right]$$

117

the maximum stresses being

$$\sigma_{H_{max}} = \frac{\rho\omega^2}{4}\left[(3+v)R_2^2 + (1-v)R_1^2\right] \quad \text{at the centre}$$

and
$$\sigma_{r_{max}} = (3+v)\frac{\rho\omega^2}{8}\left[R_2 - R_1\right]^2 \quad \text{at } r = \sqrt{(R_1R_2)}$$

For *thick cylinders* or *solid shafts* the results can be obtained from those of the corresponding disc by replacing

$$v \text{ by } v/(1-v),$$

e.g. hoop stress at the centre of a rotating solid shaft is

$$\sigma_H = \left[3 + \frac{v}{(1-v)}\right]\frac{\rho\omega^2 r^2}{8}$$

Rotating thin disc of uniform strength

For uniform strength, i.e. $\sigma_H = \sigma_r = \sigma$ (constant over plane of disc), the disc thickness must vary according to the following equation:

$$t = t_0 e^{(-\rho\omega^2 r^2)/(2\sigma)}$$

4.1. Thin rotating ring or cylinder

Consider a thin ring or cylinder as shown in Fig. 4.1 subjected to a radial pressure p caused by the centrifugal effect of its own mass when rotating. The centrifugal effect on a unit length of the circumference is

$$p = m\omega^2 r$$

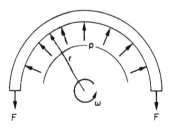

Fig. 4.1. Thin ring rotating with constant angular velocity ω.

Thus, considering the equilibrium of half the ring shown in the figure,

$$2F = p \times 2r \quad \text{(assuming unit length)}$$

$$F = pr$$

where F is the hoop tension set up owing to rotation.

The cylinder wall is assumed to be so thin that the centrifugal effect can be assumed constant across the wall thickness.

$$\therefore \qquad F = \text{ mass} \times \text{acceleration} = m\omega^2 r^2 \times r$$

This tension is transmitted through the complete circumference and therefore is resisted by the complete cross-sectional area.

$$\therefore \qquad \text{hoop stress} = \frac{F}{A} = \frac{m\omega^2 r^2}{A}$$

where A is the cross-sectional area of the ring.

Now with unit length assumed, m/A is the mass of the material per unit volume, i.e. the density ρ.

$$\therefore \qquad \text{hoop stress} = \rho\omega^2 r^2$$

4.2. Rotating solid disc

(a) General equations

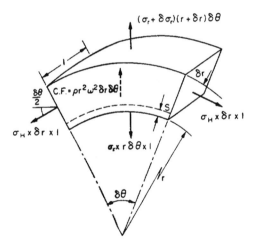

Fig. 4.2. Forces acting on a general element in a rotating solid disc.

Consider an element of a disc at radius r as shown in Fig. 4.2. Assuming unit thickness:

$$\text{volume of element} = r\,\delta\theta \times \delta r \times 1 = r\,\delta\theta\delta r$$

$$\text{mass of element} = \rho r\,\delta\theta\delta r$$

Therefore centrifugal force acting on the element

$$= m\omega^2 r$$

$$= \rho r\delta\theta\delta r\omega^2 r = \rho r^2\omega^2\delta\theta\,\delta r$$

Now for equilibrium of the element radially

$$2\sigma_H \, \delta r \sin \frac{\delta\theta}{2} + \sigma_r r \delta\theta - (\sigma_r + \delta\sigma_r)(r + \delta r)\delta\theta = \rho r^2 \omega^2 \delta\theta \, \delta r$$

If $\delta\theta$ is small,

$$\sin \frac{\delta\theta}{2} = \frac{\delta\theta}{2} \text{ radian}$$

Therefore in the limit, as $\delta r \to 0$ (and therefore $\delta\sigma_r \to 0$) the above equation reduces to

$$\sigma_H - \sigma_r - r\frac{d\sigma_r}{dr} = \rho r^2 \omega^2 \tag{4.1}$$

If there is a radial movement or "shift" of the element by an amount s as the disc rotates, the radial strain is given by

$$\varepsilon_r = \frac{ds}{dr} = \frac{1}{E}(\sigma_r - v\sigma_H) \tag{4.2}$$

Now it has been shown in §9.1.3(a)[†] that the diametral strain is equal to the circumferential strain.

$$\therefore \qquad \frac{s}{r} = \frac{1}{E}(\sigma_H - v\sigma_r) \tag{4.3}$$

$$s = \frac{1}{E}(\sigma_H - v\sigma_r)$$

Differentiating, $\qquad \dfrac{ds}{dr} = \dfrac{1}{E}(\sigma_H - v\sigma_r) + \dfrac{r}{E}\left[\dfrac{d\sigma_H}{dr} - \dfrac{v d\sigma_r}{dr}\right] \tag{4.4}$

Equating eqns. (4.2) and (4.4) and simplifying,

$$(\sigma_H - \sigma_r)(1 + v) + r\frac{d\sigma_H}{dr} - vr\frac{d\sigma_r}{dr} = 0 \tag{4.5}$$

Substituting for $(\sigma_H - \sigma_r)$ from eqn. (4.1),

$$\left(r\frac{d\sigma_r}{dr} + \rho r^2 \omega^2\right)(1 + v) + r\frac{d\sigma_H}{dr} - vr\frac{d\sigma_r}{dr} = 0$$

$$\therefore \qquad \frac{d\sigma_H}{dr} + \frac{d\sigma_r}{dr} = -\rho r\omega^2(1 + v)$$

Integrating,

$$\sigma_H + \sigma_r = -\frac{\rho r^2 \omega^2}{2}(1 + v) + 2A \tag{4.6}$$

where $2A$ is a convenient constant of integration.
Subtracting eqn. (4.1),

$$2\sigma_r + r\frac{d\sigma_r}{dr} = -\frac{\rho r^2 \omega^2}{2}(3 + v) + 2A$$

But $\qquad 2\sigma_r + r\dfrac{d\sigma_r}{dr} = \dfrac{d}{dr}\left[(r^2\sigma_r)\right] \times \dfrac{1}{r}$

[†] E.J. Hearn, *Mechanics of Materials 1*, Butterworth-Heinemann, 1997.

$$\frac{d}{dr}(r^2\sigma_r) = r\left[-\frac{\rho r^2\omega^2}{2}(3+v) + 2A\right]$$

$$r^2\sigma_r = -\frac{\rho r^4\omega^2}{8}(3+v) + \frac{2Ar^2}{2} - B$$

where $-B$ is a second convenient constant of integration,

$$\sigma_r = A - \frac{B}{r^2} - (3+v)\frac{\rho\omega^2 r^2}{8} \tag{4.7}$$

and from eqn. (4.5),

$$\sigma_H = A + \frac{B}{r^2} - (1+3v)\frac{\rho\omega^2 r^2}{8} \tag{4.8}$$

For a solid disc the stress at the centre is given when $r = 0$. With r equal to zero the above equations will yield infinite stresses whatever the speed of rotation unless B is also zero,
i.e. $B = 0$ and hence $B/r^2 = 0$ gives the only finite solution.

Now at the outside radius R the radial stress must be zero since there are no external forces to provide the necessary balance of equilibrium if σ_r were not zero.

Therefore from eqn. (4.7),

$$\sigma_r = 0 = A - (3+v)\frac{\rho\omega^2 R^2}{8}$$

$$\therefore \qquad A = (3+v)\frac{\rho\omega^2 R^2}{8}$$

Substituting in eqns. (4.7) and (4.8) the hoop and radial stresses at any radius r in a solid disc are given by

$$\sigma_H = (3+v)\frac{\rho\omega^2 R^2}{8} - (1+3v)\frac{\rho\omega^2 r^2}{8}$$

$$= \frac{\rho\omega^2}{8}\left[(3+v)R^2 - (1+3v)r^2\right] \tag{4.9}$$

$$\sigma_r = (3+v)\frac{\rho\omega^2 R^2}{8} - (3+v)\frac{\rho\omega^2 r^2}{8}$$

$$= (3+v)\frac{\rho\omega^2}{8}[R^2 - r^2] \tag{4.10}$$

(b) Maximum stresses

At the *centre* of the disc, where $r = 0$, the above equations yield equal values of hoop and radial stress which may also be seen to be the maximum stresses in the disc, i.e. maximum hoop and radial stress (at the centre)

$$= (3+v)\frac{\rho\omega^2 R^2}{8} \tag{4.11}$$

At the *outside* of the disc, at $r = R$, the equations give

$$\sigma_r = 0 \quad \text{and} \quad \sigma_H = (1 - v)\frac{\rho\omega^2 R^2}{4} \tag{4.12}$$

The complete distributions of radial and hoop stress across the radius of the disc are shown in Fig. 4.3.

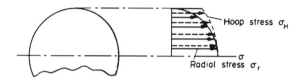

Fig. 4.3. Hoop and radial stress distributions in a rotating solid disc.

4.3. Rotating disc with a central hole

(a) General equations

The general equations for the stresses in a rotating hollow disc may be obtained in precisely the same way as those for the solid disc of the previous section,

i.e.
$$\sigma_r = A - \frac{B}{r^2} - (3 + v)\frac{\rho\omega^2 r^2}{8}$$

$$\sigma_H = A + \frac{B}{r^2} - (1 + 3v)\frac{\rho\omega^2 r^2}{8}$$

The only difference to the previous treatment is the conditions which are required to evaluate the constants A and B since, in this case, B is not zero.

The above equations are similar in form to the Lamé equations for pressurised thick rings or cylinders with modifying terms added. Indeed, should the condition arise in service where a rotating ring or cylinder is also pressurised, then the pressure and rotation boundary conditions may be substituted simultaneously to determine appropriate values of the constants A and B.

However, returning to the rotation only case, the required boundary conditions are zero radial stress at both the inside and outside radius,

i.e. at $r = R_1$, $\qquad\qquad \sigma_r = 0$

$\therefore \qquad\qquad 0 = A - \frac{B}{R_1^2} - (3 + v)\frac{\rho\omega^2 R_1^2}{8}$

and at $r = R_2$, $\qquad\qquad \sigma_r = 0$

$\therefore \qquad\qquad 0 = A - \frac{B}{R_2^2} - (3 + v)\frac{\rho\omega^2 R_2^2}{8}$

Subtracting and simplifying,

$$B = (3 + v)\frac{\rho\omega^2 R_1^2 R_2^2}{8}$$

and

$$A = (3 + v)\frac{\rho\omega^2(R_1^2 + R_2^2)}{8}$$

Substituting in eqns. (4.7) and (4.8) yields the final equation for the stresses

$$\sigma_r = (3 + v)\frac{\rho\omega^2}{8}\left[R_1^2 + R_2^2 - \frac{R_1^2 R_2^2}{r^2} - r^2\right] \tag{4.13}$$

$$\sigma_H = \frac{\rho\omega^2}{8}\left[(3 + v)\left(R_1^2 + R_2^2 + \frac{R_1^2 R_2^2}{r^2}\right) - (1 + 3v)r^2\right] \tag{4.14}$$

(b) Maximum stresses

The *maximum hoop stress* occurs at the inside radius where $r = R_1$,

i.e.

$$\sigma_{H\,max} = \frac{\rho\omega^2}{8}\left[(3 + v)(R_1^2 + R_2^2 + R_2^2) - (1 + 3v)R_1^2\right]$$

$$= \frac{\rho\omega^2}{4}\left[(3 + v)R_2^2 + (1 - v)R_1^2\right] \tag{4.15}$$

As the value of the inside radius approaches zero the maximum hoop stress value approaches

$$\frac{\rho\omega^2}{4}(3 + v)R_2^2$$

This is **twice** the value obtained at the centre of a solid disc rotating at the same speed. Thus the drilling of even a very small hole at the centre of a solid disc will double the maximum hoop stress set up owing to rotation.

At the outside of the disc when $r = R_2$

$$\sigma_{H\,min} = \frac{\rho\omega^2}{4}\left[(3 + v)R_1^2 + (1 - v)R_2^2\right]$$

The *maximum radial stress* is found by consideration of the equation

$$\sigma_r = (3 + v)\frac{\rho\omega^2}{8}\left[R_1^2 + R_2^2 - \frac{R_1^2 R_2^2}{r^2} - r^2\right] \tag{4.13}(bis)$$

This will be a maximum when $\dfrac{d\sigma_r}{dr} = 0$,

i.e.

$$\text{when } 0 = \frac{d}{dr}\left[R_1^2 + R_2^2 - \frac{R_1^2 R_2^2}{r^2} - r^2\right]$$

$$0 = R_1^2 R_2^2 \frac{2}{r^3} - 2r$$

$$r^4 = R_1^2 R_2^2$$

$$r = \sqrt{(R_1 R_2)} \tag{4.16}$$

Substituting for r in eqn. (4.13).

$$\sigma_{r_{max}} = (3 + v)\frac{\rho\omega^2}{8}\left[R_1^2 + R_2^2 - R_1R_2 - R_1R_2\right]$$

$$= (3 + v)\frac{\rho\omega^2}{8}\left[R_2 - R_1\right]^2 \tag{4.17}$$

The complete radial and hoop stress distributions are indicated in Fig. 4.4.

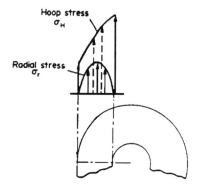

Fig. 4.4. Hoop and radial stress distribution in a rotating hollow disc.

4.4. Rotating thick cylinders or solid shafts

In the case of rotating thick cylinders the longitudinal stress σ_L must be taken into account and the longitudinal strain is assumed to be constant. Thus, writing the equations for the strain in three mutually perpendicular directions (see §4.2),

$$\varepsilon_L = \frac{1}{E}(\sigma_L - v\sigma_H - v\sigma_r) \tag{4.18}$$

$$\varepsilon_r = \frac{1}{E}(\sigma_r - v\sigma_H - v\sigma_L) = \frac{ds}{dr} \tag{4.19}$$

$$\varepsilon_H = \frac{1}{E}(\sigma_H - v\sigma_r - v\sigma_L) = \frac{s}{r} \tag{4.20}$$

From eqn. (4.20)

$$Es = r[\sigma_H - v(\sigma_r + \sigma_L)]$$

Differentiating,

$$E\frac{ds}{dr} = r\left[\frac{d\sigma_H}{dr} - v\frac{d\sigma_r}{dr} - v\frac{d\sigma_L}{dr}\right] + 1\left[\sigma_H - v\sigma_r - v\sigma_L\right]$$

Substituting for $E(ds/dr)$ in eqn. (4.19),

$$\sigma_r - v\sigma_H - v\sigma_L = r\left[\frac{d\sigma_H}{dr} - v\frac{d\sigma_r}{dr} - v\frac{d\sigma_L}{dr}\right] + \sigma_H - v\sigma_r - v\sigma_L$$

$$\therefore \qquad 0 = (\sigma_H - \sigma_r)(1 + v) + r\frac{d\sigma_H}{dr} - vr\frac{d\sigma_r}{dr} - vr\frac{d\sigma_L}{dr}$$

Now, since ε_L is constant, differentiating eqn. (4.18),

$$\frac{d\sigma_L}{dr} = v\left[\frac{d\sigma_H}{dr} + \frac{d\sigma_r}{dr}\right]$$

$\therefore$
$$0 = (\sigma_H - \sigma_r)(1 + v) + r(1 - v^2)\frac{d\sigma_H}{dr} - vr(1 + v)\frac{d\sigma_r}{dr}$$

Dividing through by $(1 + v)$,

$$0 = (\sigma_H - \sigma_r) + r(1 - v)\frac{d\sigma_H}{dr} - vr\frac{d\sigma_r}{dr}$$

But the general equilibrium equation will be the same as that obtained in §4.2, eqn. (4.1),

i.e.
$$\sigma_H - \sigma_r - r\frac{d\sigma_r}{dr} = \rho\omega^2 r^2$$

Therefore substituting for $(\sigma_H - \sigma_r)$,

$$0 = \rho\omega^2 r^2 + r\frac{d\sigma_r}{dr} + r(1 - v)\frac{d\sigma_H}{dr} - vr\frac{d\sigma_r}{dr}$$

$$0 = \rho\omega^2 r^2 + r(1 - v)\left[\frac{d\sigma_H}{dr} + \frac{d\sigma_r}{dr}\right]$$

$\therefore$
$$\frac{d\sigma_H}{dr} + \frac{d\sigma_r}{dr} = -\frac{\rho\omega^2 r}{(1 - v)}$$

Integrating,

$$\sigma_H + \sigma_r = -\frac{\rho\omega^2 r^2}{2(1 - v)} + 2A$$

where $2A$ is a convenient constant of integration. This equation can now be compared with the equivalent equation of §4.2, when it is evident that similar results for σ_H and σ_r can be obtained if $(1 + v)$ is replaced by $1/(1 - v)$ or, alternatively, if v is replaced by $v/(1 - v)$, see §8.14.2. **Thus hoop and radial stresses in rotating thick cylinders can be obtained from the equations for rotating discs provided that Poisson's ratio v is replaced by $v/(1 - v)$**, e.g. the stress at the centre of a rotating solid shaft will be given by eqn. (4.11) for a solid disc modified as stated above,

i.e.
$$\sigma_H = \left[3 + \frac{v}{(1 - v)}\right]\frac{\rho\omega^2 R^2}{8} \tag{4.21}$$

4.5. Rotating disc of uniform strength

In applications such as turbine blades rotating at high speeds it is often desirable to design for constant stress conditions under the action of the high centrifugal forces to which they are subjected.

Consider, therefore, an element of a disc subjected to equal hoop and radial stresses,

i.e.
$$\sigma_H = \sigma_r = \sigma \quad \text{(Fig. 4.5)}$$

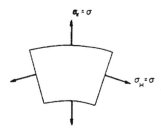

Fig. 4.5. Stress acting on an element in a rotating disc of uniform strength.

The condition of equal stress can only be achieved, as in the case of uniform strength canti-levers, by varying the thickness. Let the thickness be t at radius r and $(t + \delta t)$ at radius $(r + \delta r)$.

Then centrifugal force on the element

$$= \text{mass} \ \times \ \text{acceleration}$$

$$= (\rho t r \delta\theta \delta r)\omega^2 r$$

$$= \rho t \omega^2 r^2 \delta\theta \delta r$$

The equilibrium equation is then

$$\rho t \omega^2 r^2 \delta\theta \delta r + \sigma(r + \delta r)\delta\theta(t + \delta t) = 2\sigma t \delta r \sin \tfrac{1}{2}\delta\theta + \sigma_r t \delta\theta$$

i.e. in the limit

$$\sigma t \, dr = \rho \omega^2 r^2 t \, dr + \sigma t \, dr + \sigma r \, dt$$

$$\therefore \qquad\qquad \sigma r \, dt = -\rho \omega^2 r^2 t \, dr$$

$$\therefore \qquad\qquad \frac{dt}{dr} = -\frac{\rho\omega^2 r t}{\sigma}$$

Integrating,

$$\log_e t = -\frac{\rho\omega^2 r^2}{2\sigma} + \log_e A$$

where $\log_e A$ is a convenient constant.

$$\therefore \qquad\qquad t = A e^{(-\rho\omega^2 r^2)/(2\sigma)}$$

where $r = 0$ $\qquad\qquad t = A = t_0$

i.e. *for uniform strength* the thickness of the disc must vary according to the following equation,

$$t = t_0 e^{(-\rho\omega^2 r^2)/(2\sigma)} \qquad\qquad (4.22)$$

4.6. Combined rotational and thermal stresses in uniform discs and thick cylinders

If the temperature of any component is raised *uniformly* then, provided that the material is free to expand, expansion takes place without the introduction of any so-called thermal or temperature stresses. In cases where components, e.g. discs, are subjected to *thermal gradients*, however, one part of the material attempts to expand at a faster rate than another owing to the difference in temperature experienced by each part, and as a result stresses are developed. These are analogous to the differential expansion stresses experienced in compound bars of different materials and treated in §2.3.[†]

Consider, therefore, a disc initially unstressed and subjected to a temperature rise T. Then, for a radial movement s of any element, eqns. (4.2) and (4.3) may be modified to account for the strains due to temperature thus:

$$\frac{ds}{dr} = \frac{1}{E}(\sigma_r - v\sigma_H + E\alpha T) \tag{4.23}$$

and

$$\frac{s}{r} = \frac{1}{E}(\sigma_H - v\sigma_r + E\alpha T) \tag{4.24}$$

where α is the coefficient of expansion of the disc material (see §2.3)[†]

From eqn. (4.24),

$$\frac{ds}{dr} = \frac{1}{E}\left[(\sigma_H - v\sigma_r + E\alpha T) + r\left(\frac{d\sigma_H}{dr} - v\frac{d\sigma_r}{dr} + E\alpha\frac{dT}{dr}\right)\right]$$

Therefore from eqn. (4.23),

$$\frac{1}{E}(\sigma_r - v\sigma_H + E\alpha T) = \frac{1}{E}\left[(\sigma_H - v\sigma_r + E\alpha T) - r\left(\frac{d\sigma_H}{dr} - v\frac{d\sigma_r}{dr} + E\alpha\frac{dT}{dr}\right)\right]$$

∴

$$(\sigma_H - \sigma_r)(1 + v) + r\frac{d\sigma_H}{dr} - vr\frac{d\sigma_r}{dr} + E\alpha r\frac{dT}{dr} = 0 \tag{4.25}$$

but, from the equilibrium eqn. (4.1),

$$\sigma_H - \sigma_r - r\frac{d\sigma_r}{dr} = \rho r^2 \omega^2$$

Therefore substituting for $(\sigma_H - \sigma_r)$ in eqn. (4.25),

$$(1 + v)\left(\rho r^2\omega^2 + r\frac{d\sigma_r}{dr}\right) + r\frac{d\sigma_H}{dr} - vr\frac{d\sigma_r}{dr} + E\alpha r\frac{dT}{dr} = 0$$

$$(1 + v)\rho r^2\omega^2 + r\frac{d\sigma_r}{dr} + r\frac{d\sigma_H}{dr} + E\alpha r\frac{dT}{dr} = 0$$

$$\frac{d\sigma_H}{dr} + \frac{d\sigma_r}{dr} = -(1 + v)\rho r\omega^2 - E\alpha\frac{dT}{dr}$$

Integrating,

$$\sigma_H + \sigma_r = -(1 + v)\frac{\rho r^2\omega^2}{2} - E\alpha T + 2A \tag{4.26}$$

where, again, $2A$ is a convenient constant.

Subtracting eqn. (4.1),

$$2\sigma_r + r\frac{d\sigma_r}{dr} = -\frac{\rho r^2\omega^2}{2}(3 + v) - E\alpha T + 2A$$

[†] E.J. Hearn, *Mechanics of Materials 1*, Butterworth-Heinemann, 1997.

But $\qquad 2\sigma_r + r\dfrac{d\sigma_r}{dr} = \dfrac{d}{dr}\left[(r^2\sigma_r) \times \dfrac{1}{r}\right]$

$\therefore \qquad \dfrac{d}{dr}(r^2\sigma_r) = r\left[-\dfrac{\rho r^2\omega^2}{2}(3+v) - E\alpha T + 2A\right]$

Integrating, $\qquad r^2\sigma_r = -\dfrac{\rho r^4\omega^2}{8}(3+v) - E\alpha\int Tr\,dr + \dfrac{2Ar^2}{2} - B$

where, as in eqn. (4.7), $-B$ is a second convenient constant of integration.

$\therefore \qquad \sigma_r = A - \dfrac{B}{r^2} - \dfrac{\rho r^2\omega^2}{8}(3+v) - \dfrac{Ea}{r^2}\int Tr\,dr \qquad (4.27)$

Then, from eqn. (4.26),

$$\sigma_H = A + \dfrac{B}{r^2} - (1+3v)\dfrac{\rho r^2\omega^2}{8} - EaT + \dfrac{Ea}{r^2}\int Tr\,dr \qquad (4.28)$$

i.e. the expressions obtained for the hoop and radial stresses are those of the standard Lamé equations for simple pressurisation with (a) modifying terms for rotational effects as obtained in previous sections of this chapter, and (b) modifying terms for thermal effects.

A solution to eqns. (4.27) and (4.28) *for discs* may thus be obtained provided that the way in which T varies with r is known. Because of the form of the equations it is clear that, if required, pressure, rotational and thermal effects can be considered simultaneously and the appropriate values of A and B determined.

For *thick cylinders* **with an axial length several times the outside diameter the above plane stress equations may be modified to the equivalent plane strain equations (see §8.14.2) by replacing v by $v/(1-v)$, E by $E/(1-v^2)$ and α by $(1+v)\alpha$.**

i.e. $\qquad\qquad\qquad\qquad$ **$E\alpha$ becomes $E\alpha(1-v)$**

In the absence of rotation the equations simplify to

$$\sigma_r = A - \dfrac{B}{r^2} - \dfrac{Ea}{r^2}\int Tr\,dr \qquad (4.29)$$

$$\sigma_H = A + \dfrac{B}{r^2} + \dfrac{Ea}{r^2}\int Tr\,dr - EaT \qquad (4.30)$$

With a *linear variation of temperature* from $T = 0$ at $r = 0$,

i.e. with $\qquad\qquad\qquad\qquad T = Kr$

$$\sigma_r = A - \dfrac{B}{r^2} - \dfrac{EaKr}{3} \qquad (4.31)$$

$$\sigma_H = A + \dfrac{B}{r^2} - 2\dfrac{EaKr}{3} \qquad (4.32)$$

With a *steady heat flow*, for example, in the case of thick cylinders when $E\alpha$ becomes $E\alpha/(1-v)$–see p. 125.

$$\dfrac{r\,dT}{dr} = \text{constant} = b$$

$$\therefore \qquad \frac{dT}{dr} = \frac{b}{r} \quad \text{and} \quad T = a + b \log_e r$$

and the equations become

$$\sigma_r = A - \frac{B}{r^2} - \frac{EaT}{2(1 - v)} \tag{4.33}$$

$$\sigma_H = A + \frac{B}{r^2} - \frac{EaT}{2(1 - v)} - \frac{Eab}{2(1 - v)} \tag{4.34}$$

In practical applications where the temperature is higher on the inside of the disc or thick cylinder than the outside, the thermal stresses are tensile on the outside surface and compressive on the inside. They may thus be considered as favourable in pressurised thick cylinder applications where they will tend to reduce the high tensile stresses on the inside surface due to pressure. However, in the chemical industry, where endothermic reactions may be contained within the walls of a thick cylinder, the reverse situation applies and the two stress systems add to provide a potentially more severe stress condition.

Examples

Example 4.1

A steel ring of outer diameter 300 mm and internal diameter 200 mm is shrunk onto a solid steel shaft. The interference is arranged such that the radial pressure between the mating surfaces will not fall below 30 MN/m^2 whilst the assembly rotates in service. If the maximum circumferential stress on the inside surface of the ring is limited to 240 MN/m^2, determine the maximum speed at which the assembly can be rotated. It may be assumed that no relative slip occurs between the shaft and the ring.

For steel, $\rho = 7470$ kg/m^3, $v = 0.3$, $E = 208$ GN/m^2.

Solution

From eqn. (4.7)

$$\sigma_r = A - \frac{B}{r^2} - \frac{(3 + v)}{8} \rho \omega^2 r^2 \tag{1}$$

Now when $r = 0.15$, $\qquad \sigma_r = 0$

$$\therefore \qquad 0 = A - \frac{B}{0.15^2} - \frac{3.3}{8} \rho \omega^2 (0.15)^2 \tag{2}$$

Also, when $r = 0.1$, $\qquad \sigma_r = -30$ MN/m^2

$$\therefore \qquad -30 \times 10^6 = A - \frac{B}{0.1^2} - \frac{3.3}{8} \rho \omega^2 (0.1)^2 \tag{3}$$

(2)–(3), $\qquad 30 \times 10^6 = B(100 - 44.4) - \frac{3.3}{8} \rho \omega^2 (0.0225 - 0.01)$

$$\therefore \qquad B = \frac{30 \times 10^6}{55.6} + 3.3 \times \frac{0.0125 \times 7470}{8 \times 55.6} \omega^2$$

$$B = 0.54 \times 10^6 + 0.693 \omega^2$$

and from (3),

$$A = 100(0.54 \times 10^6 + 0.693\omega^2) + \frac{3.3 \times 7470 \times 0.01\omega^2}{8} - 30 \times 10^6$$

$$= 54 \times 10^6 + 69.3\omega^2 + 30.8\omega^2 - 30 \times 10^6$$

$$= 24 \times 10^6 + 100.1\omega^2$$

But since the maximum hoop stress at the inside radius is limited to 240 MN/m², from eqn. (4.8)

$$\sigma_H = A + \frac{B}{r^2} - \frac{(1 + 3v)}{8}\rho\omega^2 r^2$$

i.e.

$$240 \times 10^6 = (24 \times 10^6 + 100.1\omega^2) + \frac{(0.54 \times 10^6 + 0.693\omega^2)}{0.1^2} - \frac{1.9}{8} \times 7470 \times 0.01\omega^2$$

$$240 \times 10^6 = 78 \times 10^6 + 169.3\omega^2 - 17.7\omega^2$$

$$\therefore \quad 151.7\omega^2 = 162 \times 10^6$$

$$\omega^2 = \frac{162 \times 10^6}{151.7} = 1.067 \times 10^6$$

$$\omega = 1033 \text{ rad/s} = \mathbf{9860 \text{ rev/min}}$$

Example 4.2

A steel rotor disc which is part of a turbine assembly has a uniform thickness of 40 mm. The disc has an outer diameter of 600 mm and a central hole of 100 mm diameter. If there are 200 blades each of mass 0.153 kg pitched evenly around the periphery of the disc at an effective radius of 320 mm, determine the rotational speed at which yielding of the disc first occurs according to the maximum shear stress criterion of elastic failure.

For steel, $E = 200$ GN/m², $v = 0.3$, $\rho = 7470$ kg/m³ and the yield stress σ_y in simple tension $= 500$ MN/m².

Solution

$$\text{Total mass of blades} = 200 \times 0.153 = 30.6 \text{ kg}$$

$$\text{Effective radius} = 320 \text{ mm}$$

Therefore   centrifugal force on the blades $= m\omega^2 r = 30.6 \times \omega^2 \times 0.32$

Now   the area of the disc rim $= \pi dt = \pi \times 0.6 \times 0.004 = 0.024\pi\text{m}^2$

The centrifugal force acting on this area thus produces an effective radial stress acting on the outside surface of the disc since the blades can be assumed to produce a uniform loading around the periphery.

Therefore radial stress at outside surface

$$= \frac{30.6 \times \omega^2 \times 0.32}{0.024\pi} = 130\omega^2 \text{ N/m}^2 \quad \text{(tensile)}$$

Now eqns. (4.7) and (4.8) give the general form of the expressions for hoop and radial stresses set up owing to rotation,

i.e.
$$\sigma_r = A - \frac{B}{r^2} - \frac{(3+v)}{8}\rho\omega^2 r^2 \qquad (1)$$

$$\sigma_H = A + \frac{B}{r^2} - \frac{(1+3v)}{8}\rho\omega^2 r^2 \qquad (2)$$

When $r = 0.05$, $\sigma_r = 0$

$\therefore$
$$0 = A - 400B - \frac{3.3}{8}\rho\omega^2(0.05)^2 \qquad (3)$$

When $r = 0.3$, $\sigma_r = +130\omega^2$

$\therefore$
$$130\omega^2 = A - 11.1B - \frac{3.3}{8}\rho\omega^2(0.3)^2 \qquad (4)$$

$(4)-(3)$,
$$130\omega^2 = 388.9B - \frac{3.3}{8}\rho\omega^2(9-0.25)10^{-2}$$

$$130\omega^2 = 388.9B - 270\omega^2$$

$$B = \frac{(130+270)}{388.9}\omega^2 = 1.03\omega^2$$

Substituting in (3),

$$A = 412\omega^2 + \frac{3.3}{8} \times 7470(0.05)^2\omega^2$$

$$= 419.7\omega^2 = 420\omega^2$$

Therefore substituting in (2) and (1), the stress conditions at the inside surface are

$$\sigma_H = 420\omega^2 + 412\omega^2 - 4.43\omega^2 = 827\omega^2$$

with $\sigma_r = 0$

and at the outside $\sigma_H = 420\omega^2 + 11.42\omega^2 - 159\omega^2 = 272\omega^2$

with $\sigma_r = 130\omega^2$

The most severe stress conditions therefore occur at the inside radius where the maximum shear stress is greatest

i.e.
$$\tau_{max} = \frac{\sigma_1 - \sigma_3}{2} = \frac{827\omega^2 - 0}{2}$$

Now the maximum shear stress theory of elastic failure states that failure is assumed to occur when this stress equals the value of τ_{max} at the yield point in simple tension,

i.e.
$$\tau_{max} = \frac{\sigma_1 - \sigma_3}{2} = \frac{\sigma_y - 0}{2} = \frac{\sigma_y}{2}$$

Thus, for failure according to this theory,

$$\frac{\sigma_y}{2} = \frac{827\omega^2}{2}$$

i.e. $$827\omega^2 = \sigma_y = 500 \times 10^6$$

∴ $$\omega^2 = \frac{500}{827} \times 10^6 = 0.604 \times 10^6$$

$$\omega = 780 \text{ rad/s} = \mathbf{7450 \ rev/min}$$

Example 4.3

The cross-section of a turbine rotor disc is designed for uniform strength under rotational conditions. The disc is keyed to a 60 mm diameter shaft at which point its thickness is a maximum. It then tapers to a minimum thickness of 10 mm at the outer radius of 250 mm where the blades are attached. If the design stress of the shaft is 250 MN/m² at the design speed of 12 000 rev/min, what is the required maximum thickness? For steel $\rho = 7470$ kg/m³.

Solution

From eqn. (4.22) the thickness of a uniform strength disc is given by

$$t = t_0 e^{(-\rho\omega^2 r^2)/(2\sigma)} \tag{1}$$

where t_0 is the thickness at $r = 0$.

Now at $r = 0.25$,

$$\frac{\rho\omega^2 r^2}{2\sigma} = \frac{7470}{2 \times 250 \times 10^6} \left(12\,000 \times \frac{2\pi}{60} \right)^2 \times 0.25^2 = 1.47$$

and at $r = 0.03$,

$$\frac{\rho\omega^2 r^2}{2\sigma} = \frac{7470}{2 \times 250 \times 10^6} \left(12\,000 \times \frac{2\pi}{60} \right)^2 \times 0.03^2$$

$$= 1.47 \times \frac{9 \times 10^{-4}}{625 \times 10^{-4}} = 0.0212$$

But at $r = 0.25$, $t = 10$ mm

Therefore substituting in (1),

$$0.01 = t_0 e^{-1.47} = 0.2299 \, t_0$$

$$t_0 = \frac{0.01}{0.2299} = 0.0435\text{m} = 43.5 \text{ mm}$$

Therefore at $r = 0.03$

$$t = 0.0435 e^{-0.0212} = 0.0435 \times 0.98$$

$$= 0.0426 \text{ m} = \mathbf{42.6 \ mm}$$

Example 4.4

(a) Derive expressions for the hoop and radial stresses developed in a solid disc of radius R when subjected to a thermal gradient of the form $T = Kr$. Hence determine the position and magnitude of the maximum stresses set up in a steel disc of 150 mm diameter when the temperature rise is 150°C. For steel, $\alpha = 12 \times 10^{-6}$ per °C and $E = 206.8$ GN/m^2.

(b) How would the values be changed if the temperature at the centre of the disc was increased to 30°C, the temperature rise across the disc maintained at 150°C and the thermal gradient now taking the form $T = a + br$?

Solution

(a) The hoop and radial stresses are given by eqns. (4.29) and (4.30) as follows:

$$\sigma_r = A - \frac{B}{r^2} - \frac{\alpha E}{r^2} \int Tr \, dr \tag{1}$$

$$\sigma_H = A + \frac{B}{r^2} + \frac{\alpha E}{r^2} \int Tr \, dr - \alpha E T \tag{2}$$

In this case

$$\int Tr \, dr = K \int r^2 dr = \frac{Kr^3}{3}$$

the constant of integration being incorporated into the general constant A.

$$\therefore \qquad \sigma_r = A - \frac{B}{r^2} - \frac{\alpha EKr}{3} \tag{3}$$

$$\sigma_H = A + \frac{B}{r^2} + \frac{\alpha EKr}{3} - \alpha EKr \tag{4}$$

Now in order that the stresses at the centre of the disc, where $r = 0$, shall not be infinite, B must be zero and hence B/r^2 is zero. Also $\sigma_r = 0$ at $r = R$.

Therefore substituting in (3),

$$0 = A - \frac{\alpha EKR}{3} \quad \text{and} \quad A = \frac{\alpha EKR}{3}$$

Substituting in (3) and (4) and rearranging,

$$\sigma_r = \frac{\alpha EK}{3}(R - r)$$

$$\sigma_H = \frac{\alpha EK}{3}(R - 2r)$$

The variation of both stresses with radius is linear and they will both have maximum values at the centre where $r = 0$.

$$\therefore \qquad \sigma_{r_{max}} = \sigma_{H_{max}} = \frac{\alpha EKR}{3}$$

$$= \frac{12 \times 10^{-6} \times 206.8 \times 10^9 \times K \times 0.075}{3}$$

Now $T = Kr$ and T must therefore be zero at the centre of the disc where r is zero. Thus, with a known temperature rise of 150°C, it follows that the temperature at the outside radius must be 150°C.

$$\therefore \qquad 150 = K \times 0.075$$

$$\therefore \qquad K = 2000°/\text{m}$$

i.e. $\qquad \sigma_{r_{max}} = \sigma_{H_{max}} = \dfrac{12 \times 10^{-6} \times 206.8 \times 10^9 \times 2000 \times 0.075}{3}$

$$= \mathbf{124\ MN/m^2}$$

(b) With the modified form of temperature gradient,

$$\int Tr\,dr = \int (a + br)r\,dr = \int (ar + br^2)\,dr$$

$$= \frac{ar^2}{2} + \frac{br^3}{3}$$

Substituting in (1) and (2),

$$\sigma_r = A - \frac{B}{r^2} - \frac{\alpha E}{r^2}\left[\frac{ar^2}{2} + \frac{br^3}{3}\right] \tag{5}$$

$$\sigma_H = A + \frac{B}{r^2} + \frac{\alpha E}{r^2}\left[\frac{ar^2}{2} + \frac{br^3}{3}\right] - \alpha ET \tag{6}$$

Now $\qquad\qquad T = a + br$

Therefore at the inside of the disc where $r = 0$ and $T = 30°C$,

$$30 = a + b(0) \tag{7}$$

and $\qquad\qquad a = 30$

At the outside of the disc where $T = 180°C$,

$$180 = a + b(0.075) \tag{8}$$

(8) − (7) $\qquad\qquad 150 = 0.075b \quad \therefore\ b = 2000$

Substituting in (5) and (6) and simplifying,

$$\sigma_r = A - \frac{B}{r^2} - \alpha E(15 + 667r) \tag{9}$$

$$\sigma_H = A + \frac{B}{r^2} + \alpha E(15 + 667r) - \alpha ET \tag{10}$$

Now for finite stresses at the centre,

$$B = 0$$

Also, at $r = 0.075$, $\qquad\qquad \sigma_r = 0$ and $T = 180°C$

Therefore substituting in (9),

$$0 = A - 12 \times 10^{-6} \times 206.8 \times 10^9(15 + 667 \times 0.075)$$

$$0 = A - 12 \times 206.8 \times 10^3 \times 65$$

$\therefore \qquad\qquad A = 161.5 \times 10^6$

From (9) and (10) the maximum stresses will again be at the centre where $r = 0$,

i.e. $\qquad\qquad \sigma_{r_{max}} = \sigma_{H_{max}} = A - \alpha ET = \mathbf{124\ MN/m^2}$, as before.

N.B. The same answers would be obtained for any linear gradient with a temperature difference of 150°C. Thus a solution could be obtained with the procedure of part (a) using the form of distribution $T = Kr$ with the value of T at the outside taken to be 150°C (the value at $r = 0$ being automatically zero).

Example 4.5

An initially unstressed short steel cylinder, internal radius 0.2 m and external radius 0.3 m, is subjected to a temperature distribution of the form $T = a + b \log_e r$ to ensure constant heat flow through the cylinder walls. With this form of distribution the radial and circumferential stresses at any radius r, where the temperature is T, are given by

$$\sigma_r = A - \frac{B}{r^2} - \frac{\alpha ET}{2(1 - v)}$$

$$\sigma_H = A + \frac{B}{r^2} - \frac{\alpha ET}{2(1 - v)} - \frac{E\alpha b}{2(1 - v)}$$

If the temperatures at the inside and outside surfaces are maintained at 200°C and 100°C respectively, determine the maximum circumferential stress set up in the cylinder walls. For steel, $E = 207$ GN/m^2, $v = 0.3$ and $\alpha = 11 \times 10^{-6}$ per °C.

Solution

$$T = a + b \log_e r$$

$\therefore$ $$200 = a + b \log_e 0.2 = a + b(0.6931 - 2.3026)$$

$$200 = a - 1.6095 \, b \tag{1}$$

also $$100 = a + b \log_e 0.3 = a + b(1.0986 - 2.3026)$$

$$100 = a - 1.204 \, b \tag{2}$$

$(2) - (1)$, $$100 = -0.4055 \, b$$

$$b = -246.5 = -247$$

Also $$\frac{E\alpha}{2(1 - v)} = \frac{207 \times 10^9 \times 11 \times 10^{-6}}{2(1 - 0.29)}$$

$$= 1.6 \times 10^6$$

Therefore substituting in the given expression for radial stress,

$$\sigma_r = A - \frac{B}{r^2} - 1.6 \times 10^6 T$$

At $r = 0.3$, $\sigma_r = 0$ and $T = 100$

$$0 = A - \frac{B}{0.09} - 1.6 \times 10^6 \times 100 \tag{3}$$

At $r = 0.2$, $\sigma_r = 0$ and $T = 200$

$$0 = A - \frac{B}{0.04} - 1.6 \times 10^6 \times 200 \tag{4}$$

(4) − (3), $0 = B(11.1 - 25) - 1.6 \times 10^8$

$$B = -11.5 \times 10^6$$

and from (4),

$$A = 25B + 3.2 \times 10^8$$
$$= (-2.88 + 3.2)10^8 = 0.32 \times 10^8$$

substituting in the given expression for hoop stress,

$$\sigma_H = 0.32 \times 10^8 - \frac{11.5 \times 10^6}{r^2} - 1.6 \times 10^6 T + 1.6 \times 10^6 \times 247$$

At $r = 0.2$, $\sigma_H = (0.32 - 2.88 - 3.2 + 3.96)10^8 = \mathbf{-180 \ MN/m^2}$

At $r = 0.3$, $\sigma_H = (0.32 - 1.28 - 1.6 + 3.96)10^8 = \mathbf{+140 \ MN/m^2}$

The maximum tensile circumferential stress therefore occurs at the outside radius and has a value of 140 MN/m^2. The maximum compressive stress is 180 MN/m^2 at the inside radius.

Problems

Unless otherwise stated take the following material properties for steel:

$$\rho = 7470 \ \text{kg/m}^3; \quad \nu = 0.3; \quad E = 207 \ \text{GN/m}^2$$

4.1 (B). Determine equations for the hoop and radial stresses set up in a solid rotating disc of radius R commencing with the following relationships:

$$\sigma_r = A - \frac{B}{r^2} - (3 + \nu)\frac{\rho\omega^2 r^2}{8}$$

$$\sigma_H = A + \frac{B}{r^2} - (1 + 3\nu)\frac{\rho\omega^2 r^2}{8}$$

Hence determine the maximum stress and the stress at the outside of a 250 mm diameter disc which rotates at 12 000 rev/min.   [76, 32.3 MN/m^2.]

4.2 (B). Determine from first principles the hoop stress at the inside and outside radius of a thin steel disc of 300 mm diameter having a central hole of 100 mm diameter, if the disc is made to rotate at 5000 rev/min. What will be the position and magnitude of the maximum radial stress?

[38.9, 12.3 MN/m^2; 87 mm rad; 8.4 MN/m^2.]

4.3 (B). Show that the tensile hoop stress set up in a thin rotating ring or cylinder is given by

$$\sigma_H = \rho\omega^2 r^2$$

Hence determine the maximum angular velocity at which the disc can be rotated if the hoop stress is limited to 20 MN/m^2. The ring has a mean diameter of 260 mm.   [3800 rev/min.]

4.4 (B). A solid steel disc 300 mm diameter and of small constant thickness has a steel ring of outer diameter 450 mm and the same thickness shrunk onto it. If the interference pressure is reduced to zero at a rotational speed of 3000 rev/min, calculate

(a) the radial pressure at the interface when stationary;
(b) the difference in diameters of the mating surfaces of the disc and ring before assembly.

The radial and circumferential stresses at radius r in a ring or disc rotating at ω rad/s are obtained from the following relationships:

$$\sigma_r = A - \frac{B}{r^2} - (3 + \nu)\frac{\rho\omega^2 r^2}{8}$$

$$\sigma_H = A + \frac{B}{r^2} - (1+3v)\frac{\rho\omega^2 r^2}{8} \qquad\qquad [8.55 \text{ MN/m}^2, 0.045 \text{ mm.}]$$

4.5 (B). A steel rotor disc of uniform thickness 50 mm has an outer rim of diameter 800 mm and a central hole of diameter 150 mm. There are 200 blades each of weight 2 N at an effective radius of 420 mm pitched evenly around the periphery. Determine the rotational speed at which yielding first occurs according to the maximum shear stress criterion.

Yield stress in simple tension $= 750 \text{ MN/m}^2$.

The basic equations for radial and hoop stresses given in Example 4.4 may be used without proof.
[7300 rev/min.]

4.6 (B). A rod of constant cross-section and of length $2a$ rotates about its centre in its own plane so that each end of the rod describes a circle of radius a. Find the maximum stress in the rod as a function of the peripheral speed V. $[\frac{1}{2}(\rho\omega^2 a^2).]$

4.7 (B). A turbine blade is to be designed for constant tensile stress σ under the action of centrifugal force by varying the area A of the blade section. Consider the equilibrium of an element and show that the condition is

$$A = A_h e^{[-\rho\omega^2(r^2-r_h^2)]/(2\sigma)}$$

where A_h and r_h are the cross-sectional area and radius at the hub (i.e. base of the blade).

4.8 (B). A steel turbine rotor of 800 mm outside diameter and 200 mm inside diameter is 50 mm thick. The rotor carries 100 blades each 200 mm long and of mass 0.5 kg. The rotor runs at 3000 rev/min. Assuming the shaft to be rigid, calculate the expansion of the *inner* bore of the disc due to rotation and hence the initial shrinkage allowance necessary. [0.14 mm.]

4.9 (B). A steel disc of 750 mm diameter is shrunk onto a steel shaft of 80 mm diameter. The interference on the diameter is 0.05 mm.
(a) Find the maximum tangential stress in the disc at standstill.
(b) Find the speed in rev/min at which the contact pressure is zero.
(c) What is the maximum tangential stress at the speed found in (b)? [65 MN/m²; 3725; 65 MN/m².]

4.10 (B). A flat steel turbine disc of 600 mm outside diameter and 120 mm inside diameter rotates at 3000 rev/min at which speed the blades and shrouding cause a tensile rim loading of 5 MN/m². The maximum stress at this speed is to be 120 MN/m². Find the maximum shrinkage allowance on the diameter when the disc is put on the shaft. [0.097 mm.]

4.11 (B). Find the maximum permissible speed of rotation for a steel disc of outer and inner radii 150 mm and 70 mm respectively if the outer radius is not to increase in magnitude due to centrifugal force by more than 0.03 mm. [7900 rev/min.]

4.12 (B). The radial and hoop stresses at any radius r for a disc of uniform thickness rotating at an angular speed ω rad/s are given respectively by

$$\sigma_r = A - \frac{B}{r^2} - (3+v)\frac{\rho\omega^2 r^2}{8}$$

$$\sigma_H = A + \frac{B}{r^2} - (1+3v)\frac{\rho\omega^2 r^2}{8}$$

where A and B are constants, v is Poisson's ratio and ρ is the density of the material. Determine the greatest values of the radial and hoop stresses for a disc in which the outer and inner radii are 300 mm and 150 mm respectively.

Take $\omega = 150$ rad/s, $v = 0.304$ and $\rho = 7470$ kg/m³. [U.L.] [1.56, 13.2 MN/m².]

4.13 (B). Derive an expression for the tangential stress set up when a thin hoop, made from material of density ρ kg/m³, rotates about its polar axis with a tangential velocity of v m/s.

What will be the greatest value of the mean radius of such a hoop, made from flat mild-steel bar, if the maximum allowable tensile stress is 45 MN/m² and the hoop rotates at 300 rev/min.

Density of steel $= 7470$ kg/m³. [2.47 m.]

4.14 (C). Determine the hoop stresses at the inside and outside surfaces of a long thick cylinder inside radius $= 75$ mm, outside radius $= 225$ mm, which is rotated at 4000 rev/min.

Take $v = 0.3$ and $\rho = 7470$ kg/m³. [57.9, 11.9 MN/m².]

4.15 (C). Calculate the maximum principal stress and maximum shear stress set up in a thin disc when rotating at 12 000 rev/min. The disc is of 300 mm outside diameter and 75 mm inside diameter.

Take $v = 0.3$ and $\rho = 7470$ kg/m³. [221, 110.5 MN/m².]

4.16 (B). A thin-walled cylindrical shell made of material of density ρ has a mean radius r and rotates at a constant angular velocity of ω rad/s. Assuming the formula for centrifugal force, establish a formula for the circumferential (hoop) stress induced in the cylindrical shell due to rotation about the longitudinal axis of the cylinder and, if necessary, adjust the derived expression to give the stress in MN/m^2.

A drum rotor is to be used for a speed of 3000 rev/min. The material is steel with an elastic limit stress of 248 MN/m^2 and a density of 7.8 Mg/m^3. Determine the mean diameter allowable if a factor of safety of 2.5 on the elastic limit stress is desired. Calculate also the expansion of this diameter (in millimetres) when the shell is rotating.

For steel, $E = 207$ GN/m^2.   [I.Mech.E.] [0.718 m; 0.344 mm.]

4.17 (B). A forged steel drum, 0.524 m outside diameter and 19 mm wall thickness, has to be mounted in a machine and spun about its longitudinal axis. The centrifugal (hoop) stress induced in the cylindrical shell is not to exceed 83 MN/m^2. Determine the maximum speed (in rev/min) at which the drum can be rotated.

For steel, the density = 7.8 Mg/m^3.   [3630.]

4.18 (B). A cylinder, which can be considered as a thin-walled shell, is made of steel plate 16 mm thick and is 2.14 m internal diameter. The cylinder is subjected to an internal fluid pressure of 0.55 MN/m^2 gauge and, at the same time, rotated about its longitudinal axis at 3000 rev/min. Determine:

(a) the hoop stress induced in the wall of the cylinder due to rotation;
(b) the hoop stress induced in the wall of the cylinder due to the internal pressure;
(c) the factor of safety based on an ultimate stress of the material in simple tension of 456 MN/m^2.

Steel has a density of 7.8 Mg/m^3.   [89.5, 36.8 MN/m^2; 3.6]

4.19 (B). The "bursting" speed of a cast-iron flywheel rim, 3 m mean diameter, is 850 rev/min. Neglecting the effects of the spokes and boss, and assuming that the flywheel rim can be considered as a thin rotating hoop, determine the ultimate tensile strength of the cast iron. Cast iron has a density of 7.3 Mg/m^3.

A flywheel rim is to be made of the-same material and is required to rotate at 400 rev/min. Determine the maximum permissible mean diameter using a factor of safety of 8.   [U.L.C.I.] [2.25 mm]

4.20 (B). An internal combustion engine has a cast-iron flywheel that can be considered to be a uniform thickness disc of 230 mm outside diameter and 50 mm inside diameter. Given that the ultimate tensile stress and density of cast iron are 200 N/mm^2 and 7180 kg/m^3 respectively, calculate the speed at which the flywheel would burst. Ignore any stress concentration effects and assume Poisson's ratio for cast iron to be 0.25.

[C.E.I.] [254.6 rev/s.]

4.21 (B). A thin steel circular disc of uniform thickness and having a central hole rotates at a uniform speed about an axis through its centre and perpendicular to its plane. The outside diameter of the disc is 500 mm and its speed of rotation is 81 rev/s. If the maximum allowable direct stress in the steel is not to exceed 110 MN/m^2 (11.00 h bar), determine the diameter of the central hole.

For steel, density $\rho = 7800$ kg/m^3 and Poisson's ratio $\nu = 0.3$.

Sketch diagrams showing the circumferential and radial stress distribution across the plane of the disc indicating the peak values and state the radius at which the maximum radial stress occurs.   [B.P.] [264 mm.]

4.22 (B). (a) Prove that the differential equation for radial equilibrium in cylindrical coordinates of an element in a uniform thin disc rotating at ω rad/s and subjected to principal direct stresses σ_r and σ_θ is given by the following expression:

$$\sigma_\theta - \sigma_r - r\frac{d\sigma_r}{dr} = \rho\omega^2 r^2$$

(b) A thin solid circular disc of uniform thickness has an outside diameter of 300 mm. Using the maximum shear strain energy per unit volume theory of elastic failure, calculate the rotational speed of the disc to just cause initiation of plastic yielding if the yield stress of the material of the disc is 300 MN/m^2, the density of the material is 7800 kg/m^3 and Poisson's ratio for the material is 0.3.   [B.P.] [324 rev/s.]

Thermal gradients

4.23 (C). Determine expressions for the stresses developed in a hollow disc subjected to a temperature gradient of the form $T = Kr$. What are the maximum stresses for such a case if the internal and external diameters of the cylinder are 80 mm and 160 mm respectively;

$\alpha = 12 \times 10^{-6}$ per $^\circ$C and $E = 206.8$ GN/m^2.

The temperature at the outside radius is -50°C.   [-34.5, 27.6 MN/m^2.]

4.24 (C). Calculate the maximum stress in a solid magnesium alloy disc 60 mm diameter when the temperature rise is linear from 60°C at the centre to 90°C at the outside.

$\alpha = 7 \times 10^{-6}$ per °C and $E = 105$ GN/m^2. [7.4 MN/m^2.]

4.25 (C). Calculate the maximum compressive and tensile stresses in a hollow steel disc, 100 mm outer diameter and 20 mm inner diameter when the temperature rise is linear from 100°C at the inner surface to 50°C at the outer surface.

$\alpha = 10 \times 10^{-6}$ per °C and $E = 206.8$ GN/m^2. [−62.9, +40.3 MN/m^2.]

4.26 (C). Calculate the maximum tensile and compressive stresses in a hollow copper *cylinder* 20 mm outer diameter and 10 mm inner diameter when the temperature rise is linear from 0°C at the inner surface to 100°C at the outer surface.

$\alpha = 16 \times 10^{-6}$ per °C and $E = 104$ GN/m^2. [142, −114 MN/m^2.]

4.27 (C). A hollow steel disc has internal and external diameters of 0.2 m and 0.4 m respectively. Determine the circumferential thermal stresses set up at the inner and outer surfaces when the temperature at the outside surface is 100°C. A temperature distribution through the cylinder walls of the form $T = Kr$ may be assumed, i.e. when $r = $ zero, $T = $ zero.

For steel, $E = 207$ GN/m^2 and $\alpha = 11 \times 10^{-6}$ per °C.

What is the significance of (i) the first two terms of the stress eqns. (4.29) and (4.30), (ii) the remaining terms?

Hence comment on the relative magnitude of the maximum hoop stresses obtained in a high pressure vessel which is used for (iii) a chemical action which is exothermic, i.e. generating heat, (iv) a chemical reaction which is endothermic, i.e. absorbing heat. [63.2, −50.5 MN/m^2.]

4.28 (C). In the previous problem sketch the thermal hoop and radial stress variation diagrams across the wall thickness of the disc inserting the numerical value of the hoop stresses at the inner, mean and outer radii, and also the maximum radial stress, inserting the radius at which it occurs.

[$\sigma_{\text{mean}} = -2.78$ MN/m^2, $\sigma_{r_{\text{max}}} = 9.65$ MN/m^2.]

4.29 (C). A thin uniform steel disc, 254 mm outside diameter with a central hole 50 mm diameter, rotates at 10000 rev/min. The temperature gradient varies linearly such that the *difference* of temperature between the inner and outer (hotter) edges of the plate is 46°C. For the material of the disc, $E = 205$ GN/m^2, Poisson's ratio $= 0.3$ and the coefficient of linear expansion $= 11 \times 10^{-6}$ per °C. The density of the material is 7700 kg/m^3.

Calculate the hoop stresses induced at the inner and outer surfaces. [176−12.1 MN/m^2.]

4.30 (C). An unloaded steel cylinder has internal and external diameters of 204 mm and 304 mm respectively. Determine the circumferential thermal stresses at the inner and outer surfaces where the steady temperatures are 200°C and 100°C respectively.

Take $E = 207$ GN/m^2, $\alpha = 11 \times 10^{-6}$ per °C and Poisson's ratio $= 0.29$.

The temperature distribution through the wall thickness may be regarded as follows:

$$T = a + b \log_e r, \text{ where } a \text{ and } b \text{ are constants}$$

With this form of temperature distribution, the radial and circumferential thermal stresses at radius r where the temperature is T are obtained from

$$\sigma_r = A - \frac{B}{r^2} - \frac{E\alpha T}{2(1-\nu)} \quad \text{and} \quad \sigma_H = A + \frac{B}{r^2} - \frac{E\alpha T}{2(1-\nu)} - \frac{E\alpha b}{2(1-\nu)}$$

[−255, 196 MN/m^2.]

4.31 (C). Determine the hoop stresses at the inside and outside surfaces of a long thick cylinder which is rotated at 4000 rev/min. The cylinder has an internal radius of 80 mm and an external radius of 250 mm and is constructed from steel, the relevant properties of which are given above.

How would these values be modified if, under service conditions, the temperatures of the inside and outside surfaces reached maximum levels of 40°C and 90°C respectively?

A linear thermal gradient may be assumed.

For steel $\alpha = 11 \times 10^{-6}$ per °C.

[71.4, 18.9; 164.5, −46.8 MN/m^2.]

4.32 (C). (a) Determine the wall thickness required for a high pressure cylindrical vessel, 800 mm diameter, in order that yielding shall be prevented according to the Tresca criterion of elastic failure when the vessel is subjected to an internal pressure of 450 bar.

(b) Such a vessel is now required to form part of a chemical plant and to contain exothermic reactions which produce a maximum internal temperature of 120°C at a reaction pressure of 450 bar, the outer surface being cooled to an "ambient" temperature of 20°C. In the knowledge that such a thermal gradient condition will introduce additional stresses to those calculated in part (a) the designer proposes to increase the wall thickness by 20% in order that, once again, yielding shall be prevented according to the Tresca theory. Is this a valid proposal?

You may assume that the thermal gradient is of the form $T = a + br^2$ and that the modifying terms to the Lamé expressions to cover thermal gradient conditions are

for radial stress:
$$-\frac{\alpha E}{r^2} \int Tr\,dr$$

for hoop stress:
$$\frac{\alpha E}{r^2} \int Tr\,dr - \alpha ET.$$

For the material of the vessel, $\sigma_y = 280$ MN/m^2, $\alpha = 12 \times 10^{-6}$ per °C and $E = 208$ GN/m^2.

[52 mm; No-design requires $\sigma_y = 348$ MN/m^2.]

CHAPTER 5

TORSION OF NON-CIRCULAR AND THIN-WALLED SECTIONS

Summary

For torsion of *rectangular sections* the maximum shear stress τ_{max} and angle of twist θ are given by

$$\tau_{max} = \frac{T}{k_1 db^2}$$

$$\frac{\theta}{L} = \frac{T}{k_2 db^3 G}$$

k_1 and k_2 being two constants, their values depending on the ratio d/b and being given in Table 5.1.

For *narrow rectangular sections*, $k_1 = k_2 = \frac{1}{3}$.

Thin-walled open sections may be considered as combinations of narrow rectangular sections so that

$$\tau_{max} = \frac{T}{\Sigma k_1 db^2} = \frac{3T}{\Sigma db^2}$$

$$\frac{\theta}{L} = \frac{T}{\Sigma k_2 db^3 G} = \frac{3T}{G \Sigma db^3}$$

The relevant formulae for other non-rectangular, non-tubular solid shafts are given in Table 5.2.

For *thin-walled closed sections* the stress at any point is given by

$$\tau = \frac{T}{2At}$$

where A is the area enclosed by the median line or mean perimeter and t is the thickness. The maximum stress occurs at the point where t is a minimum.

The angle of twist is then given by

$$\theta = \frac{TL}{4A^2 G} \int \frac{ds}{t}$$

which, for *tubes of constant thickness*, reduces to

$$\frac{\theta}{L} = \frac{Ts}{4A^2 Gt} = \frac{\tau s}{2AG}$$

where s is the length or perimeter of the median line.

Thin-walled cellular sections may be solved using the concept of constant shear flow $q(= \tau t)$, bearing in mind that the angles of twist of all cells or constituent parts are assumed equal.

5.1. Rectangular sections

Detailed analysis of the torsion of non-circular sections which includes the warping of cross-sections is beyond the scope of this text. For *rectangular shafts*, however, with longer side d and shorter side b, it can be shown by experiment that the maximum shearing stress occurs at the centre of the longer side and is given by

$$\tau_{\max} = \frac{T}{k_1 db^2} \tag{5.1}$$

where k_1 is a constant depending on the ratio d/b and given in Table 5.1 below.

Table 5.1. Table of k_1 and k_2 values for rectangular sections in torsion[a].

d/b	1.0	1.5	1.75	2.0	2.5	3.0	4.0	6.0	8.0	10.0	∞
k_1	0.208	0.231	0.239	0.246	0.258	0.267	0.282	0.299	0.307	0.313	0.333
k_2	0.141	0.196	0.214	0.229	0.249	0.263	0.281	0.299	0.307	0.313	0.333

[a] S. Timoshenko, *Strength of Materials*, Part I, *Elementary Theory and Problems*, Van Nostrand, New York.

The essential difference between the shear stress distributions in circular and rectangular members is illustrated in Fig. 5.1, where the shear stress distribution along the major and minor axes of a rectangular section together with that along a "radial" line to the corner of the section are indicated. The maximum shear stress is shown at the centre of the longer side, as noted above, and the stress at the corner is zero.

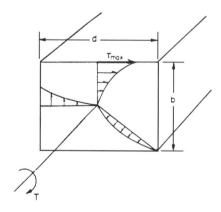

Fig. 5.1. Shear stress distribution in a solid rectangular shaft.

The angle of twist per unit length is given by

$$\frac{\theta}{L} = \frac{T}{k_2 db^3 G} \tag{5.2}$$

k_2 being another constant depending on the ratio d/b and also given in Table 5.1.

In the absence of Table 5.1, however, it is possible to reduce the above equations to the following *approximate* forms:

$$\tau_{max} = \frac{T}{db^2}\left[3 + 1.8\frac{b}{d}\right] = \frac{T}{db^3}[3d + 1.8b] \tag{5.3}$$

and

$$\theta = \frac{42TLJ}{GA^4} = \frac{42TLJ}{Gd^4b^4} \tag{5.4}$$

where A is the cross-sectional area of the section $(= bd)$ and $J = (bd/12)(b^2 + d^2)$.

5.2. Narrow rectangular sections

From Table 5.1 it is evident that as the ratio d/b increases, i.e. the rectangular section becomes longer and thinner, the values of constants k_1 and k_2 approach 0.333. Thus, for narrow rectangular sections in which $d/b > 10$ both k_1 and k_2 are assumed to be 1/3 and eqns. (5.1) and (5.2) reduce to

$$\tau_{max} = \frac{3T}{db^2} \tag{5.5}$$

$$\frac{\theta}{L} = \frac{3T}{db^3G} \tag{5.6}$$

5.3. Thin-walled open sections

There are many cases, particularly in civil engineering applications, where rolled steel or extruded alloy sections are used where some element of torsion is involved. In most cases the sections consist of a combination of rectangles, and the relationships given in eqns. (5.1) and (5.2) can be adapted with reasonable accuracy provided that:

(a) the sections are "open", i.e. angles, channels. T-sections, etc., as shown in Fig. 5.2;
(b) the sections are thin compared with the other dimensions.

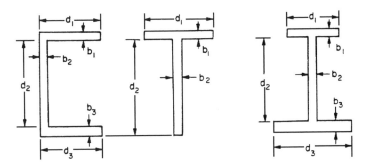

Fig. 5.2. Typical thin-walled open sections.

For such sections eqns. (5.1) and (5.2) may be re-written in the form

$$\tau_{max} = \frac{T}{k_1 db^2} = \frac{T}{Z'} \tag{5.7}$$

and

$$\frac{\theta}{L} = \frac{T}{k_2 db^3 G} = \frac{T}{J_{eq}G} \tag{5.8}$$

where Z' is the torsion section modulus

$$= Z' \text{ web} + Z' \text{ flanges} = k_1 d_1 b_1^2 + k_1 d_2 b_2^2 + \ldots \text{etc.}$$

$$= \Sigma k_1 db^2$$

and J_{eq} is the "effective" polar moment of area or "equivalent J" (see §5.7)

$$= J_{eq} \text{ web} + J_{eq} \text{ flanges} = k_2 d_1 b_1^3 + k_2 d_2 b_2^3 + \cdots \text{etc.}$$

$$= \Sigma k_2 db^3$$

i.e.

$$\tau_{max} = \frac{T}{\sum k_1 db^2} \tag{5.9}$$

and

$$\frac{\theta}{L} = \frac{T}{G \sum k_2 db^3} \tag{5.10}$$

and for d/b ratios in excess of 10, $k_1 = k_2 = \frac{1}{3}$, so that

$$\tau_{max} = \frac{3T}{\sum db^2} \tag{5.11}$$

$$\frac{\theta}{L} = \frac{3T}{G \sum db^3} \tag{5.12}$$

To take account of the stress concentrations at the fillets of such sections, however, Timoshenko and Young[†] suggest that the maximum shear stress as calculated above is multiplied by the factor

$$\left[1 + \frac{b}{4a}\right]$$

(Figure 5.3). This has been shown to be fairly reliable over the range $0 < a/b < 0.5$. In the event of sections containing limbs of different thicknesses the largest value of b should be used.

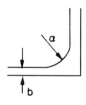

Fig. 5.3.

[†] S. Timoshenko and A.D. Young, *Strength of Materials*, Van Nostrand, New York, 1968 edition.

5.4. Thin-walled split tube

The thin-walled split tube shown in Fig. 5.4 is considered to be a special case of the thin-walled open type of section considered in §5.3. It is therefore treated as an equivalent rectangle with a longer side d equal to the circumference (less the gap), and a width b equal to the thickness.

Then
$$\tau_{max} = \frac{T}{k_1 d b^2}$$

and
$$\frac{\theta}{L} = \frac{T}{k_2 d b^3 G}$$

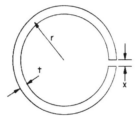

d = mean circumference = $2\pi r$

Fig. 5.4. Thin tube with longitudinal split.

where k_1 and k_2 for thin-walled tubes are usually equal to $\frac{1}{3}$.

It should be noted here that the presence of even a very small cut or gap in a thin-walled tube produces a torsional stiffness (torque per unit angle of twist) very much smaller than that for a complete tube of the same dimensions.

5.5. Other solid (non-tubular) shafts

Table 5.2 (see p. 146) indicates the relevant formulae for maximum shear stress and angle of twist of other standard non-circular sections which may be encountered in practice.

Approximate angles of twist for other solid cross-sections may be obtained by the substitution of an elliptical cross-section of the same area A and the same polar second moment of area J. The relevant equation for the elliptical section in Table 5.2 may then be applied.

Alternatively, a very powerful procedure which applies for all solid sections, however irregular in shape, utilises a so-called "inscribed circle" procedure described in detail by Roark[†]. The procedure is equally applicable to thick-walled standard T, I and channel sections and is outlined briefly below:

Inscribed circle procedure

Roark shows that the maximum shear stress which is set up when any solid section is subjected to torque occurs at, or very near to, one of the points where the largest circle which

[†] R.J. Roark and W.C. Young, *Formulas for Stress & Strain*, 5th edn. McGraw-Hill, Kogakusha.

Table 5.2[a].

Cross-section	Maximum shear stress	Angle of twist per unit length
Elliptic 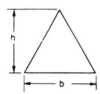	$$\frac{16T}{\pi b^2 h}$$ at end of minor axis XX where $J = \dfrac{\pi}{64}[bh^3 + hb^3]$ and A is the area of cross-section $= \pi bh/4$	$$\frac{4\pi^2 TJ}{A^4 G}$$
Equilateral triangle 	$$\frac{20T}{b^3}$$ at the middle of each side	$$\frac{46.2T}{b^4 G}$$
Regular hexagon	$$\frac{T}{0.217\ Ad}$$ where d is the diameter of inscribed circle and A is the cross-sectional area	$$\frac{T}{0.133\ Ad^2 G}$$

[a] From S. Timoshenko, *Strength of Materials*, Part II, *Advanced Theory and Problems*, Van Nostrand, New York, p. 235. Approximate angles of twist for other solid cross-sections may be obtained by the substitution of an equivalent elliptical cross-section of the same area A and the same polar second moment of area J. The relevant equation for the elliptical section in Table 5.2 may then be applied.

can be constructed within the cross-section touches the section boundary – see Fig. 5.5. Normally it occurs at the point where the curvature of the boundary is algebraically the least, convex curvatures being taken as positive and concave or re-entrant curvatures negative.

The maximum shear stress is then obtained from either:

$$\tau_{\max} = \left(\frac{G\theta}{L}\right) C \quad \text{or} \quad \tau_{\max} = \left(\frac{\tau}{K}\right) C$$

where, *for positive curvatures* (i.e. straight or convex boundaries),

$$C = \frac{D}{1 + \dfrac{\pi^2 D^4}{16A^2}} \left[1 + 0.15 \left(\frac{\pi^2 D^4}{16A^2} - \frac{D}{2r}\right)\right]$$

with D = diameter of the largest inscribed circle,

r = radius of curvature of boundary at selected position (positive),

A = cross-sectional area of section,

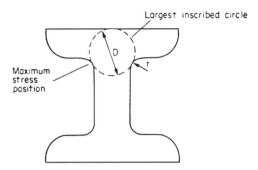

Fig. 5.5. Inscribed circle stress evaluation procedure.

or, *for negative curvatures* (concave or re-entrant boundaries):

$$C = \frac{D}{1 + \dfrac{\pi^2 D^4}{16 A^2}} \left[1 + \left\{ 0.118 \log_e \left(1 - \frac{D}{2r} \right) - 0.238 \frac{D}{2r} \right\} \tanh \frac{2\phi}{\pi} \right]$$

with $\phi =$ angle through which a tangent to the boundary rotates in travelling around the re-entrant position (radians) and r being taken as negative.

For standard thick-walled open sections such as T, I, Z, angle and channel sections Roark also introduces formulae for angles of twist based upon the same inscribed circle procedure parameters.

5.6. Thin-walled closed tubes of non-circular section (Bredt–Batho theory)

Consider the thin-walled closed tube shown in Fig. 5.6 subjected to a torque T about the Z axis, i.e. in a transverse plane. Both the cross-section and the wall thickness around the periphery may be irregular as shown, but for the purposes of this simplified treatment it must be assumed that the thickness does not vary along the length of the tube. Then, if τ is the shear stress at B and τ' is the shear stress at C (where the thickness has increased to t') then, from the equilibrium of the complementary shears on the sides AB and CD of the element shown, it follows that

$$\tau t \, dz = \tau' t' \, dz$$

$$\tau t = \tau' t'$$

i.e. *the product of the shear stress and the thickness is constant* at all points on the periphery of the tube. This constant is termed the *shear flow* and denoted by the symbol q (shear force per unit length).

Thus $$q = \tau t = \textbf{constant} \tag{5.13}$$

The quantity q is termed the shear flow because if one imagines the inner and outer boundaries of the tube section to be those of a channel carrying a flow of water, then, provided that the total quantity of water in the system remains constant, the quantity flowing past any given point is also constant.

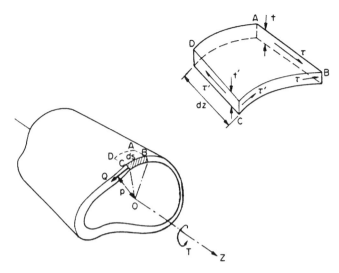

Fig. 5.6. Thin-walled closed section subjected to axial torque.

At any point, then, the shear force Q on an element of length ds is $Q = \tau t\, ds = q\, ds$ and the shear stress is q/t.

Consider now, therefore, the element BC subjected to the shear force $Q = q\, ds = \tau t\, ds$. The moment of this force about O

$$= dT = Qp$$

where p is the perpendicular distance from O to the force Q.

$\therefore$ $$dT = q\, ds\, p$$

Therefore the moment, or torque, for the whole section

$$= \int qp\, ds = q \int p\, ds$$

But the area $COB = \frac{1}{2}$ base $\times$ height $= \frac{1}{2} p\, ds$

i.e. $$dA = \tfrac{1}{2} p\, ds \quad \text{or} \quad 2dA = p\, ds$$

$\therefore$ $$\text{torque } T = 2q \int dA$$

$$T = 2qA \tag{5.14}$$

where A is the area enclosed within the median line of the wall thickness.

Now, since

$$q = \tau t$$

it follows that $$T = 2\tau t A$$

or $$\tau = \frac{T}{2At} \tag{5.15}$$

where t is the thickness at the point in question.

It is evident, therefore, that *the maximum shear stress* in such cases *occurs at the point of minimum thickness*.

Consider now an axial strip of the tube, of length L, along which the thickness and hence the shear stress is constant. The shear strain energy *per unit volume* is given by

$$U = \int \frac{\tau^2}{2G}$$

Thus, with thickness t, width ds and hence $V = tL\,ds$

$$U = \int \frac{\tau^2}{2G} tL\,ds$$

$$= \int \left(\frac{T}{2At}\right)^2 \frac{tL}{2G}\,ds$$

$$= \frac{T^2L}{8A^2G} \int \frac{ds}{t}$$

But the energy stored equals the work done $= \frac{1}{2}T\theta$.

$$\therefore \qquad \frac{1}{2}T\theta = \frac{T^2L}{8A^2G} \int \frac{ds}{t}$$

The angle of twist of the tube is therefore given by

$$\theta = \frac{TL}{4A^2G} \int \frac{ds}{t}$$

For *tubes of constant thickness* this reduces to

$$\theta = \frac{TLs}{4A^2Gt} = \frac{\tau Ls}{2AG} \qquad (5.16)$$

where s is the perimeter of the median line.

The above equations must be used with care and do not apply to cases where there are abrupt changes in thickness or re-entrant corners.

For closed sections which have constant thickness over specified lengths but varying from one part of the perimeter to another:

$$\frac{\theta}{L} = \frac{T}{4A^2G} \left[\frac{s_1}{t_1} + \frac{s_2}{t_2} + \frac{s_3}{t_3} + \cdots \text{etc.}\right]$$

5.7. Use of "equivalent J" for torsion of non-circular sections

The simple torsion theory for circular sections can be written in the form:

$$\frac{\theta}{L} = \frac{T}{GJ}$$

and, as stated on page 143, it is often convenient to express the twist of non-circular sections in similar form:

i.e.
$$\frac{\theta}{L} = \frac{T}{GJ_{eq}}$$

where J_{eq} is the *"equivalent J'* or *"effective polar moment of area"* for the section in question.

Thus, *for open sections:*
$$\frac{\theta}{L} = \frac{T}{\Sigma k_2 db^3 G} = \frac{T}{GJ_{eq}}$$

with $J_{eq} = \Sigma k_2 db^3$ $(= \frac{1}{3}\Sigma db^3$ for $d/b > 10)$.

Similarly, *for square tubes of closed section:*
$$\frac{\theta}{L} = \frac{TLs}{4A^2 Gt} = \frac{T}{G[4A^2 t/s]} = \frac{T}{GJ_{eq}}$$

and $J_{eq} = 4A^2 t/s$.

The *torsional stiffness* of any section, i.e. the ratio of torque divided by angle of twist per unit length, is then directly given by the value of GJ or GJ_{eq} i.e.

$$\textbf{Stiffness} = \frac{T}{\theta/L} = \textbf{GJ (or GJ}_{\textbf{eq}}\textbf{)}.$$

5.8. Thin-walled cellular sections

The Bredt–Batho theory developed in the previous section may be applied to the solution of problems involving cellular sections of the type shown in Fig. 5.7.

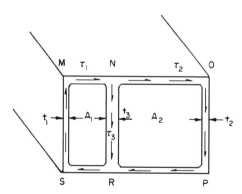

Fig. 5.7. Thin-walled cellular section.

Assume the length $RSMN$ is of constant thickness t_1 and subjected therefore to a constant shear stress τ_1. Similarly, $NOPR$ is of thickness t_2 and stress τ_2 with NR of thickness t_3 and stress τ_3.

Considering the equilibrium of complementary shear stresses on a longitudinal section at N, it follows that

$$\tau_1 t_1 = \tau_2 t_2 + \tau_3 t_3 \tag{5.17}$$

Alternatively, this equation may be obtained considering the arrows shown to be directions of shear flow $q(= \tau t)$. At N the flow q_1 along MN divides into q_2 along NO and q_3 along NR,

i.e.
$$q_1 = q_2 + q_3$$

or
$$\tau_1 t_1 = \tau_2 t_2 + \tau_3 t_3 \quad \text{(as before)}$$

The total torque for the section is then found as the sum of the torques on the two cells by application of eqn. (5.14) to the two cells and adding the result,

i.e.
$$T = 2q_1 A_1 + 2q_2 A_2$$
$$T = 2(\tau_1 t_1 A_1 + \tau_2 t_2 A_2) \tag{5.18}$$

Also, since the angle of twist will be common to both cells, applying eqn. (5.16) to each cell gives

$$\theta = \frac{L}{2G}\left(\frac{\tau_1 s_1 + \tau_3 s_3}{A_1}\right) = \frac{L}{2G}\left(\frac{\tau_2 s_2 - \tau_3 s_3}{A_2}\right)$$

where s_1, s_2 and s_3 are the median line perimeters $RSMN$, $NOPR$ and NR respectively.

The negative sign appears in the final term because the shear flow along NR for this cell opposes that in the remainder of the perimeter.

$\therefore$
$$\frac{2G\theta}{L} = \frac{1}{A_1}(\tau_1 s_1 + \tau_3 s_3) = \frac{1}{A_2}(\tau_2 s_2 - \tau_3 s_3) \tag{5.19}$$

5.9. Torsion of thin-walled stiffened sections

The stiffness of any section has been shown above to be given by its value of GJ or GJ_{eq}.

Consider, therefore, the rectangular polymer extrusion of simple symmetrical cellular constructions shown in Fig. 5.8(a). The shear flow in each cell is indicated.

At A
$$q_1 = q_2 + q_3.$$

But because of symmetry q_1 must equal q_3 $\therefore q_2 = 0$;
i.e., for a symmetrical cellular thin-walled member there is no shear carried by the central web and therefore as far as stiffness of the section is concerned the web can be ignored.

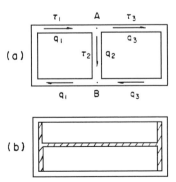

Fig. 5.8(a). Polymer cellular section with symmetrical cells. (b) Polymer cell with central web removed but reinforced by steel I section.

∴ Stiffness of complete section, from eqn. (5.16)

$$= GJ_E = \frac{4A^2t}{s}G$$

where A and s are the area and perimeter of the complete section.

Now since G of the polymer is likely to be small, the stiffness of the section, and its resistance to applied torque, will be low. It can be reinforced by metallic insertions such as that of the I section shown in Fig. 5.8(b).

For the I section, from eqn. (5.8)

$$GJ_E = G\Sigma k_2 db^3$$

and the value represents the increase in stiffness presented by the compound section.

Stress conditions for limiting twist per unit lengths are then given by:
For the tube

$$T = GJ_E(\theta/L) = 2At\tau$$

$$\therefore (\theta/L)_{max} = \frac{2At}{GJ_E} \cdot \tau_{max}$$

and for the I section

$$T = GJ_E(\theta/L) = (\Sigma k_2 db^3 G)\theta/L$$

or
$$T = (\Sigma k_1 db^2)\tau$$

∴
$$(\theta/L)_{max} = \frac{\Sigma k_1\, d}{Gb\Sigma k_2\, d} \cdot \tau_{max}$$

Usually (but not always) this would be considerably greater than that for the polymer tube, making the tube the controlling design factor.

5.10. Membrane analogy

It has been stated earlier that the mathematical solution for the torsion of certain solid and thin-walled sections is complex and beyond the scope of this text. In such cases it is extremely fortunate that an analogy exists known as the *membrane analogy*, which provides a very convenient mental picture of the way in which stresses build up in such components and allows experimental determination of their values.

It can be shown that the mathematical solution for elastic torsion problems involving partial differential equations is identical in form to that for a thin membrane lightly stretched over a hole. The membrane normally used for visualisation is a soap film. Provided that the hole used is the same shape as the cross-section of the shaft in question and that air pressure is maintained on one side of the membrane, the following relationships exist:

(a) the torque carried by the section is equal to twice the volume enclosed by the membrane;

(b) the shear stress at any point in the section is proportional to the slope of the membrane at that point (Fig. 5.9);

(c) the direction of the shear stress at any point in the section is always at right angles to the slope of the membrane at the same point.

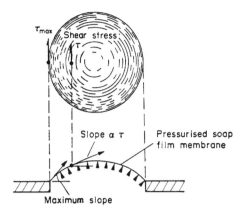

Fig. 5.9. Membrane analog.

Application of the above rules to the open sections of Fig. 5.2 shows that each section will carry approximately the same torque at the same maximum shear stress since the volumes enclosed by the membranes and the maximum slopes of the membranes are approximately equal in each case.

The membrane analogy is particularly powerful in the study of the comparative torsional properties of different sections without the need for detailed calculations. For example, it should be evident from the volume relationship (a) above that if two cross-sections have the same area, that which is nearer to circular will be the stronger in torsion since it will produce the greatest enclosed volume.

The analogy also helps to support the theory used for thin-walled open sections in §5.3 when thin rectangular sections are taken to have the same torsional stiffness be they left as a single rectangle or bent into open tubes, angle sections, channel sections, etc.

From the slope relationship (b) the greatest shear stresses usually occur at the boundary of the thickest parts of the section. They are usually high at positions where the boundary is sharply concave but low at the ends of outstanding flanges.

5.11. Effect of warping of open sections

In the preceding paragraphs it has been assumed that the torque is applied at the ends of the member and that all sections are free to warp. In practice, however, there are often cases where one or more sections of a member are constrained in some way so that cross-sections remain plane, i.e. warping is prevented. Whilst this has little effect on the angle of twist of certain solid cross-sections, e.g. rectangular or elliptical sections where the length is significantly greater than the section dimensions, it may have a considerable effect on the twist of open sections. In the latter case the constraint of warping is often accompanied by considerable bending of the flanges. Detailed treatment of warping is beyond the scope of this text[†] and it is sufficient to note here that when warping is restrained, angles of twist are generally reduced and hence torsional stiffnesses increased.

[†] S. Timoshenko and J.N. Goodier, *Theory of Elasticity*, McGraw-Hill, New York.

Examples

Example 5.1

A rectangular steel bar 25 mm wide and 38 mm deep is subjected to a torque of 450 Nm. Estimate the maximum shear stress set up in the material of the bar and the angle of twist, using the experimentally derived formulate stated in §5.1.

What percentage error would be involved in each case if the approximate equations are used?

For steel, take $G = 80$ GN/m^2.

Solution

The maximum shear stress is given by eqn. (5.1):

$$\tau_{max} = \frac{T}{k_1 db^2}$$

In this case $d = 38$ mm, $b = 25$ mm, i.e. $d/b = 1.52$ and k_1 for d/b of $1.5 = 0.231$.

$$\therefore \qquad \tau_{max} = \frac{450}{0.231 \times 38 \times 10^{-3} \times (25 \times 10^{-3})^2} = \textbf{82 MN/m}^2$$

The angle of twist per unit length is given by eqn. (5.2):

$$\frac{\theta}{L} = \frac{T}{k_2 db^3 G}$$

and from the tables, for $d/b = 1.5$, k_2 is 0.196.

$$\therefore \qquad \theta = \frac{450}{0.196 \times 38 \times 10^{-3} \times (25 \times 10^{-3})^3 \times 80 \times 10^9}$$

$$= 0.0483 \text{ rad/m}$$

$$= \textbf{2.77 degrees/m}$$

Approximately

$$\tau_{max} = \frac{T}{db^2}(3 + 1.8b/d)$$

$$= \frac{450}{38 \times 10^{-3} \times (25 \times 10^{-3})^2}\left(3 + 1.8 \times \frac{25}{38}\right)$$

$$= \frac{450}{2.375 \times 10^{-5}}(3 + 1.184) = \textbf{79.3 MN/m}^2$$

Therefore percentage error

$$= \left(\frac{79.3 - 82.02}{82.02}\right)100 = \textbf{-3.3\%}$$

Again, approximately,

$$\theta = \frac{42TJ}{GA^4} \text{ per metre}$$

Now
$$J = I_{xx} + I_{yy} = \frac{bd^3}{12} + \frac{db^3}{12} = \frac{bd}{12}(d^2 + b^2)$$

$$= \frac{25 \times 38(25^2 + 38^2)}{12 \times 10^{12}} = 0.1638 \times 10^{-6} \text{ m}^4$$

$\therefore$
$$\theta = \frac{42 \times 450 \times 0.164 \times 10^{-6}}{80 \times 10^9 \times (25 \times 38 \times 10^{-6})^4} = 0.0476 \text{ rad/m}$$

$$= 2.73 \text{ degrees/m}$$

$$\text{Percentage error} = \left(\frac{2.73 - 2.77}{2.77}\right) 100 = -1.44\%$$

Example 5.2

Compare the torsional stiffness of the following cross-sections which can be assumed to be of unit length. Compare also the maximum shear stresses set up in each case:

(a) a hollow tube 40 mm mean diameter and 2 mm wall thickness;

(b) the same tube with a 2 mm wide saw-cut along its length;

(c) a rectangular solid bar, side ratio 4 to 1, having the same cross-sectional area as that enclosed by the mean diameter of the hollow tube;

(d) an equal-leg angle section having the same perimeter and thickness as the tube;

(e) a square box section having the same perimeter and thickness as the tube.

Solution

(a) In the case of the *closed hollow tube* we can apply the standard torsion equation

$$\frac{T}{J} = \frac{G\theta}{L} = \frac{\tau}{r}$$

together with the simplified formula for the polar moment of area J of thin tubes,

$$J = 2\pi r^3 t$$

$\therefore$
$$\text{torsional stiffness} = \frac{T}{\theta} = \frac{GJ}{L} = \frac{2\pi \times (20 \times 10^{-3})^3 \times 2 \times 10^{-3} G}{1}$$

$$= 100.5 \times 10^{-9} G$$

$$\text{maximum shear stress} = \frac{TR}{J} = \frac{20 \times 10^{-3} \times T}{2\pi \times (20 \times 10^{-3})^3 \times 2 \times 10^{-3}}$$

$$= \mathbf{0.198 \times 10^6 T}$$

(b) *Tube with split*
From the work of §5.4,

$$\text{angle of twist/unit length} = \frac{\theta}{L} = \frac{T}{k_2 \, db^3 G} = \frac{T}{k_2(2\pi r - x)t^3 G}$$

$$\therefore \quad \text{torsional stiffness} = \frac{T}{\theta} = \frac{k_2(2\pi r - x)t^3 G}{L}$$

$$= \frac{0.333[2\pi \times 20 \times 10^{-3} - 2 \times 10^{-3}](2 \times 10^{-3})^3 G}{1}$$

$$= 0.333(125.8 - 2)8 \times 10^{-12} G$$

$$= \mathbf{329.8 \times 10^{-12} G}$$

$$\text{Maximum shear stress} = \frac{T}{k_1 d b^2}$$

$$= \frac{T}{0.333 \times 123.8 \times 10^{-3} \times (2 \times 10^{-8})^2}$$

$$= \mathbf{6.06 \times 10^6 T}$$

i.e. splitting the tube along its length has reduced the stiffness by a factor of approximately **300**, the maximum stress increasing by approximately **30** times.

(c) *Rectangular bar*

$$\text{Area of hollow tube} = \text{area of bar}$$

$$= \pi \times (20 \times 10^{-3})^2$$

$$\therefore \quad 4b^2 = 8\pi \times 10^{-4}$$

$$b^2 = 2\pi \times 10^{-4}$$

$$\therefore \quad b = 2.5 \times 10^{-2} \text{ m} = 25 \text{ mm}$$

$$\therefore \quad d = 4b = 100 \text{ mm}$$

d/b ratio $= 4$

$$\therefore \quad k_1 = 0.282 \quad \text{and} \quad k_2 = 0.281$$

Therefore from eqn. (5.2),

$$\frac{\theta}{L} = \frac{T}{k_2 d b^3 G}$$

$$\therefore \quad \frac{T}{\theta} = 0.281 \times 10 \times 10^{-2} \times (2.5 \times 10^{-2})^3 G$$

$$= 43.9 \times 10^{-8} G$$

$$= \mathbf{439 \times 10^{-9} G}$$

From eqn. (5.1),

$$\tau_{\text{max}} = \frac{T}{k_1 d b^2} = \frac{T}{0.282 \times 10 \times 10^{-2} \times (2.5 \times 10^{-2})^2}$$

$$= \mathbf{0.057 \times 10^6 T}$$

(d) *Equal-leg angle section*

$$\text{Perimeter of angle} = \text{perimeter of tube}$$

$$= 2\pi \times 20 \times 10^{-3} \text{ m}$$

$$\therefore \quad \text{Length of side } d = 20\pi \times 10^{-3} \text{ m}$$

Therefore applying eqn. (5.12),

$$\frac{\theta}{L} = \frac{3T}{G\Sigma db^3}$$

$$= \frac{3T}{2G \times 20\pi \times 10^{-3} \times (2 \times 10^{-3})^3}$$

$$\therefore \quad \frac{T}{\theta} = (2G \times 20\pi \times 8 \times 10^{-12})/3$$

$$= \mathbf{0.335 \times 10^{-9} G}$$

And from eqn. (5.11)

$$\tau_{\text{max}} = \frac{3T}{\Sigma db^2}$$

$$= \frac{3T}{2 \times 20\pi \times 10^{-3} \times (2 \times 10^{-3})^2}$$

$$= \mathbf{5.97 \times 10^6 \ T}$$

(e) *Square box section (closed)*

$$\text{Perimeter } s = \text{tube perimeter} = 2\pi \times 20 \times 10^{-3} \text{ m}$$

$$\therefore \quad \text{side length} = \frac{2\pi \times 20 \times 10^{-3}}{4} = \pi \times 10^{-2} \text{ m}$$

Therefore area enclosed by median line

$$= A = (\pi \times 10^{-2})^2$$

From eqn. (5.16),

$$\theta = \frac{TLs}{4A^2 Gt}$$

$$\therefore \quad \frac{T}{\theta} = \frac{4 \times (\pi \times 10^{-2})^4 G \times 2 \times 10^{-3}}{1 \times 2\pi \times 20 \times 10^{-3}}$$

$$= \mathbf{62 \times 10^{-9} G}$$

From eqn. (5.15)

$$\tau_{\text{max}} = \frac{T}{2At} = \frac{T}{2 \times (\pi \times 10^{-2})^2 \times 2 \times 10^{-3}}$$

$$= \mathbf{0.253 \times 10^6 T}$$

Example 5.3

A thin-walled member 1.2 m long has the cross-section shown in Fig. 5.10. Determine the maximum torque which can be carried by the section if the angle of twist is limited to 10°. What will be the maximum shear stress when this maximum torque is applied? For the material of the member $G = 80 \text{ GN/m}^2$.

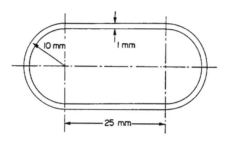

Fig. 5.10.

Solution

This problem is of the type considered in §5.6, a solution depending upon the length of, and the area enclosed by, the median line.

Now,   perimeter of median line $= s = (2 \times 25 + 2\pi \times 10)$ mm

$$= 112.8 \text{ mm}$$

area enclosed by median $= A = (20 \times 25 + \pi \times 10^2)$ mm^2

$$= 814.2 \text{ mm}^2$$

From eqn (5.16), $\theta = \dfrac{TLs}{4A^2\, Gt}$

$\therefore$ $\dfrac{10 \times 2\pi}{360} = \dfrac{T \times 1.2 \times 112.8 \times 10^{-3}}{4(814.2 \times 10^{-6})^2 \times 80 \times 10^9 \times 1 \times 10^{-3}}$

i.e. maximum torque possible,

$$T = \frac{20\pi \times 4 \times 814.2^2 \times 80 \times 10^{-6}}{360 \times 1.2 \times 112.8 \times 10^{-3}}$$

$$= \textbf{273 Nm}$$

From eqn. (5.15), $\tau_{max} = \dfrac{T}{2At}$

$$= \frac{273}{2 \times 814.2 \times 10^{-6} \times 1 \times 10^{-3}}$$

$$= 168 \times 10^6 = \textbf{168 MN/m}^2$$

The maximum stress produced is 168 MN/m^2.

Example 5.4

The median dimensions of the two cells shown in the cellular section of Fig. 5.6 are $A_1 = 20 \text{ mm} \times 40 \text{ mm}$ and $A_2 = 50 \text{ mm} \times 40 \text{ mm}$ with wall thicknesses $t_1 = 2$ mm, $t_2 = 1.5$ mm and $t_3 = 3$ mm. If the section is subjected to a torque of 320 Nm, determine the angle of twist per unit length and the maximum shear stress set up. The section is constructed from a light alloy with a modulus of rigidity $G = 30$ GN/m^2.

Solution

From eqn. (5.18), $320 = 2(\tau_1 \times 2 \times 20 \times 40 + \tau_2 \times 1.5 \times 50 \times 40)10^{-9}$ (1)

From eqn. (5.19),

$$2 \times 30 \times 10^9 \times \theta = \frac{1}{20 \times 40 \times 10^{-6}}[\tau_1(40 + 2 \times 20)10^{-3} + \tau_3 \times 40 \times 10^{-3}] \quad (2)$$

and $2 \times 30 \times 10^9 \times \theta = \dfrac{1}{50 \times 40 \times 10^{-6}}[\tau_2(40 + 2 \times 50)10^{-3} - \tau_3 \times 40 \times 10^{-3}]$ (3)

Equating (2) and (3),

From eqn. (5.17), $2\tau_1 = 1.5\tau_2 + 3\tau_3$ (4)

$$\tfrac{1}{8}[80\tau_1 + 40\tau_3] = \tfrac{1}{20}[140\tau_2 - 40\tau_3]$$

Multiply through by 40,

$$400\tau_1 + 200\tau_3 = 280\tau_2 - 80\tau_3$$

$$40\tau_1 = 28\tau_2 - 28\tau_3 \quad (5)$$

(5) $\times$ 60/28 $85.7\tau_1 = 60\tau_2 - 60\tau_3$ (6)

But, from (4), multiplied by 20,

$$40\tau_1 = 30\tau_2 + 60\tau_3 \quad (7)$$

(6) + (7), $125.7\tau_1 = 90\tau_2$ (8)

and from (1), $320 = (3.2\tau_1 + 6\tau_2)10^{-6}$

$$320 \times 10^6 = 3.2\tau_1 + 6\tau_2 \quad (9)$$

substituting for τ_2 from (8),

$$320 \times 10^6 = 3.2\tau_1 + 6 \times \frac{125.7}{90}\tau_1$$

$$= 3.2\tau_1 + 8.4\tau_1$$

$\therefore$ $\tau_1 = \dfrac{320 \times 10^6}{11.6} = 27.6 \times 10^6 = 27.6$ MN/m^2

From (8),

$$\tau_2 = \frac{125.7}{90} \times 27.6 = 38.6 \text{ MN/m}^2$$

From (4),

$$\tau_3 = \tfrac{1}{3}(2 \times 27.6 - 1.5 \times 38.6)$$

$$= \tfrac{1}{3}(55.2 - 57.9) = \tfrac{1}{3} \times (-2.7) = -0.9 \text{ MN/m}^2$$

The negative sign indicates that the direction of shear flow in the wall of thickness t_3 is reversed from that shown in Fig. 5.6.

The maximum shear stress present in the section is thus **38.6 MN/m²** in the 1.5 mm wall thickness.

From (3),

$$2 \times 30 \times 10^9 \times \theta = \frac{(140t_2 - 40t_3)}{50 \times 40 \times 10^{-3}}$$

$$= \frac{140 \times 38.6 \times 10^6 - 40(-0.9 \times 10^6)}{50 \times 40 \times 10^{-3} \times 2 \times 30 \times 10^9}$$

$$= \frac{(5.40 + 0.036)}{120} \text{ radian}$$

$$= \frac{5.440}{120} \times \frac{360}{2\pi} = \mathbf{2.6°}$$

The angle of twist of the section is 2.6°.

Problems

5.1 (A). A 40 mm × 20 mm rectangular steel shaft is subjected to a torque of 1 kNm. What will be the approximate position and magnitude of the maximum shear stress set up in the shaft? Determine also the corresponding angle of twist per metre length of the shaft.

For the bar material $G = 80$ GN/m². [254 MN/m²; 9.78°/m.]

5.2 (B). An extruded light alloy angle section has dimensions 80 mm × 60 mm × 4 mm and is subjected to a torque of 20 Nm. If $G = 30$ GN/m² determine the maximum shear stress and the angle of twist per unit length. How would the former answer change if one considered the stress concentration effect at the fillet owing to a fillet radius of 10 mm? [27.6 MN/m²; 13.2°/m; 30.4 MN/m².]

5.3 (B). Compare the torsional rigidities of the following sections:

(a) a hollow tube 30 mm outside diameter and 1.5 mm thick; [2.7 × 10⁻⁸ *G*.]
(b) the same tube split along its length with a 1 mm gap; [0.0996 × 10⁻⁹ *G*.]
(c) an equal leg angle section having the same perimeter and thickness as (b); [0.0996 × 10⁻⁹ *G*.]
(d) a square box section with side length 30 mm and 1.5 mm wall thickness; [3.48 × 10⁻⁸ *G*.]
(e) a rectangular solid bar, side ratio 2.5 to 1, having the same metal cross-sectional area as the hollow tube. [1.79 × 10⁻⁹ *G*.]

Compare also the maximum stresses arising in each case.

[0.522 ×10⁶*T*; 15 × 10⁶*T*; 15 × 10⁶*T*; 0.41 × 10⁶*T*; 4.05 × 10⁶*T*.]

5.4 (B). The spring return of an interlocking device for a cold room door is to be made of a rectangular strip of spring steel loaded in torsion. The width of the strip cannot be greater than 10 mm and the effective length 100 mm. Calculate the thickness of the strip if the torque is to be 15 Nm at an angle of 10° and if the torsion yield stress of 420 MN/m² is not to be exceeded at this angle. Take G as 83 GN/m².

Assume $k_1 = k_2 = \frac{1}{3}$. [3.27 mm.]

5.5 (B). A thin-walled member of 2 m long has the section shown in Fig. 5.11. Determine the torque that can be applied and the angle of twist achieved if the maximum shear stress is limited to 30 MN/m². $G = 250$ GN/m². [42.85 Nm; 0.99°.]

5.6 (B). A steel sheet, 400 mm wide by 2 mm thick, is to be formed into a hollow section by bending through 360° and butt-welding the long edges together. The shape may be (a) circular, (b) square, (c) a rectangle 140 mm × 60 mm. Assume a median length of 400 mm in each case (i.e. no stretching) and square corners for non-circular sections. The allowable shearing stress is 90 MN/m². For each of the shapes listed determine the magnitude of the maximum permissible torque and the angles of twist per metre length if $G = 80$ GN/m². [4.58, 3.6, 3.01 kNm; 1°, 1°17′, 1°31′.]

5.7 (B). Figure 5.12 represents the cross-section of an aircraft fuselage made of aluminium alloy. The sheet thicknesses are: 1 mm from A to B and C to D; 0.8 mm from B to C and 0.7 mm from D to A. For a maximum torque of 5000 Nm determine the magnitude of the maximum shear stress and the angle of twist/metre length. $G = 30$ GN/m². [50 MN/m²; 0.0097 rad.]

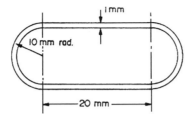

Fig. 5.11.

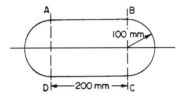

Fig. 5.12.

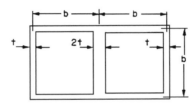

Fig. 5.13.

5.8 (B/C). Show that for the symmetrical section shown in Fig. 5.13 there is no stress in the central web. Show also that the shear stress in the remainder of the section has a value of $T/4tb^2$.

5.9 (C). A washing machine agitator of the cross-section shown in Fig. 5.14 acts as a torsional member subjected to a torque T. The central tube is 100 mm internal diameter and 12 mm thick; the rectangular bars are 50 mm × 18 mm section. Assuming that the total torque carried by the member is given by

$$T = T_{\text{tube}} + 4T_{\text{bar}}$$

determine the maximum value of T which the shaft can carry if the maximum stress is limited to 80 MN/m².
    (Hint: equate angles of twist of tube and bar.)   [19.1 kNm.]

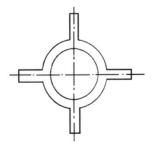

Fig. 5.14.

5.10 (C). The cross-section of an aeroplane elevator is shown in Fig. 5.15. If the elevator is 2 m long and constructed from aluminium alloy with $G = 30$ GN/m^2, calculate the total angle of twist of the section and the magnitude of the shear stress in each part for an applied torque of 40 Nm.

$$[0.0169°; 3.43, 2.58, 1.15 \times 10^5 \text{ N/m}^2.]$$

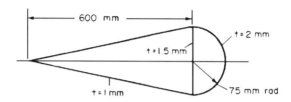

Fig. 5.15.

5.11 (B/C). Develop a relationship between torque and angle of twist for a closed uniform tube of thin-walled non-circular section and use this to derive the twist per unit length for a strip of thin rectangular cross-section.

Use the above relationship to show that, for the same torque, the ratio of angular twist per until length for a closed square-section tube to that for the same section but opened by a longitudinal slit and free to warp is approximately $4t^2/3b^2$, where t, the material thickness, is much less than the mean width b of the cross-section.

[C.E.I.]

5.12 (C). A torsional member used for stirring a chemical process is made of a circular tube to which is welded four rectangular strips as shown in Fig. 5.16. The tube has inner and outer diameters of 94 mm and 100 mm respectively, each strip is 50 mm × 18 mm, and the stirrer is 3 m in length.

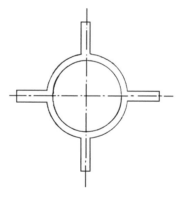

Fig. 5.16.

If the maximum shearing stress in any part of the cross-section is limited to 56 MN/m^2, neglecting any stress concentration, calculate the maximum torque which can be carried by the stirrer and the resulting angle of twist over the full length.

For torsion of rectangular sections the torque T is related to the maximum shearing stress, τ_{max}, and angle of twist, θ, in radians per unit length, as follows:

$$T = k_1 b d^2 \tau_{max} = k_2 b d^3 G\theta$$

where b is the longer and d the shorter side of the rectangle and in this case, $k_1 = 0.264$, $k_2 = 0.258$ and $G = 83$ GN/m^2.

[C.E.I.] [2.83 kNm, 2.4°.]

5.13 (C). A long tube is subjected to a torque of 200 Nm. The tube has the double-cell, thin-walled, effective cross-section illustrated in Fig. 5.17. Assuming that no buckling occurs and that the twist per unit length of the tube is constant, determine the maximum shear stresses in each wall of the tube.

[C.E.I.] [0.76, 1.01, 0.19 MN/m^2.]

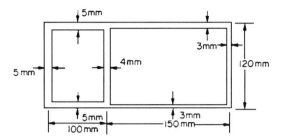

Fig. 5.17.

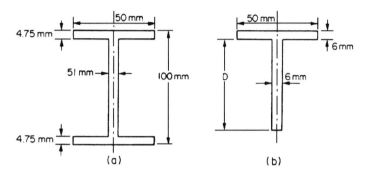

Fig. 5.18.

5.14 (B/C). An I-section has the dimensions shown in Fig. 5.18(a), and is subjected to an axial torque. Find the maximum value of the torque if the shear stress in the material is limited to 56 MN/m^2 and the twist per metre length is limited to 9°. Assume the modulus of rigidity G for the material is 82 GN/m^2.

If the I-section is replaced by a T-section made of the same material and transmits the same torque, what will be the limb length, D, of the T-section and the angle of twist per metre length? Assume the T-section is subjected to the same limiting conditioning as the I-section and that it has the dimensions shown in Fig. 5.18(b). For narrow rectangular sections assume k values of $\frac{1}{3}$ in the formulae for torque and angle of twist.

[B.P.] [0.081 m; 6.5°/m.]

5.15 (B/C). (a) An aluminium sheet, 600 mm wide and 4 mm thick, is to be formed into a hollow section tube by bending through 360° and butt-welding the long edges together. The cross-section shape may be either circular or square.

Assuming a median length of 600 mm in each case, i.e. assuming no stretching occurs, determine the maximum torque that can be carried and the resulting angle of twist per metre length in each case.

Maximum allowable shearing stress = 65 MN/m^2, shear modulus $G = 40$ GN/m^2.

(b) What would be the effect on the stiffness per metre length of each type of section of a narrow saw-cut through the tube wall along the length of the tube? In the case of the square section assume that the cut is taken along the centre of one face.

[B.P.] [14.9 kNm, 0.975°; 11.7k Nm, 1.24°; reduction 1690 times, reduction 1050 times.]

5.16 (B). The two sections shown in Fig. 5.19 are under consideration for an engineering application which includes both bending and applied torque. Make a critical comparison of the strengths of the two sections under the two modes of loading and make a recommendation as to the section which should be adopted. The material to be used is to be the same for both sections.

The rectangular section torsion constants k_1 and k_2 may be found in terms of the section d/b ratio from Table 3.1.

[Tubular]

5.17 (B). Compare the angles of twist of the following sections when each is subjected to the same torque of 3 kNm;

(a) circular tube, 80 mm outside diameter, 6 mm thick;

(b) square tube, 52 mm side length (median dimension), 6 mm thick;

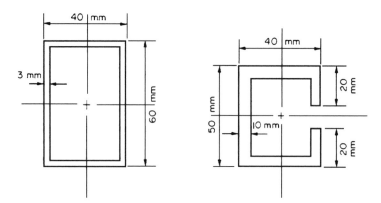

Fig. 5.19.

(c) circular tube as (a) but with additional four rectangular fins 80 mm long by 15 mm wide symmetrically placed around the tube periphery.

All sections have the same length of 2 m and $G = 80$ GN/m^2 [0.039 rad; 0.088 rad; 0.038 rad]
To what maximum torque can sections (a), (b) and (c) be subjected if the maximum shear stress is limited to 100 MN/m^2?   [4.8 kNm; 3.24 kNm; 5.7 kNm]
    What maximum angle of twist can be accepted by tube (c) for the same limiting shear stress?   [0.0625 rad]

5.18 (B). Figure 5.20 shows part of the stirring mechanism for a chemical process, consisting of a circular stainless-steel tube of length 2 m, outside diameter 75 mm and wall thickness 6 mm, welded onto a square mild-steel tube of length 1.5 m. Four blades of rectangular section stainless-steel, 100 mm × 15 mm, are welded along the full length of the stainless-steel tube as shown.

(a) Select a suitable section for the square tube from the available stock list below so that when the maximum allowable shear stress of 58 MN/m^2 is reached in the stainless-steel, the shear stress in the mild steel of the square tube does not exceed 130 MN/m^2.

Section	Dimension	Wall thickness	Torsion constant (J equiv)
1	50 × 50 mm	5 mm	476 000 mm^4
2	60 × 60 mm	4 mm	724 000 mm^4
3	70 × 70 mm	3.6 mm	1 080 000 mm^4.

(b) Having selected an appropriate mild steel tube, determine how much the entire mechanism will twist during operation at a constant torque of 3 kNm.

The shear modulus of stainless steel is 78 GN/m^2 and of mild steel is 83 GN/m^2. Neglect the effect of any stress concentration.   [50 mm × 50 mm; 0.152 rad]

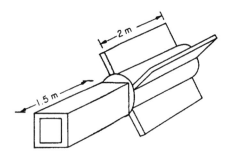

Fig. 5.20.

5.19 (B). Figure 5.21 shows the cross-section of a thin-walled fabricated service conduit used for the protection of long runs of electrical wiring in a production plant. The lower plate AB may be removed for inspection and re-cabling purposes.

Owing to the method by which the conduit is supported and the weight of pipes/wires that it carries, the section is subjected to a torque of 130 Nm. With plate AB assumed in position, determine the maximum shear stress set up in the walls of the conduit. What will be the angle of twist per unit length?

By consideration of maximum stress levels and angles of twist, establish whether the section design is appropriate for the removal of plate AB for maintenance purposes assuming that the same torque remains applied. If modifications are deemed to be necessary suggest suitable measures.

For the conduit material $G = 80$ GN/m^2 and maximum allowable shear stress $= 180$ MN/m^2.

[167 MN/m^2; 39°/m]

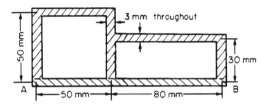

3 mm throughout

50 mm

30 mm

A   50 mm   80 mm   B

Fig. 5.21.

5.20 (B). (a) Figure 5.22 shows the cross-section of a thin-walled duct which forms part of a fluid transfer system. The wire mesh, FC, through which sediment is allowed to pass, may be assumed to contribute no strength

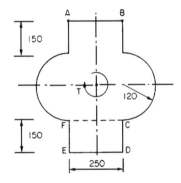

Fig. 5.22. All dimensions (mm) may be taken as median dimensions.

to the section. Owing to the method of support, the weight of the fluid and duct introduces a torque to the section which may be assumed uniform.

If the maximum shear stress in the duct material is limited to 150 MN/m^2; determine the maximum torque which can be tolerated and the angle of twist per metre length when this maximum torque is applied. For the duct material $G = 85$ GN/m^2. [432.6 kNm; 0.516°/m]

(b) In order to facilitate cleaning and inspection of the duct, plates AB and ED are removable. What would be the effect on the results of part (a) if plate AB were inadvertently left off over part of the duct length after inspection? [5.12 kNm; 12.6°/m]

5.21 (C). Figure 5.8 shows a polymer extrusion of wall thickness 4 mm. The section is to be stiffened by the insertion of an aluminium I section as shown, the centre web of the polymer extrusion having been removed. The I section wall thickness is also 4 mm.

If $G = 3.3$ GN/m^2 for the polymer and 70 GN/m^2 for the aluminium, what increase in stiffness is achieved? What increase in torque is allowable, if the design is governed by maximum allowable stresses of 5 MN/m^2 and 100 MN/m^2 in the polymer and aluminium respectively? [258%, 7.4%]

CHAPTER 6

EXPERIMENTAL STRESS ANALYSIS

Introduction

We live today in a complex world of manmade structures and machines. We work in buildings which may be many storeys high and travel in cars and ships, trains and planes; we build huge bridges and concrete dams and send mammoth rockets into space. Such is our confidence in the modern engineer that we take these manmade structures for granted. We assume that the bridge will not collapse under the weight of the car and that the wings will not fall away from the aircraft. We are confident that the engineer has assessed the stresses within these structures and has built in sufficient strength to meet all eventualities.

This attitude of mind is a tribute to the competence and reliability of the modern engineer. However, the commonly held belief that the engineer has been able to calculate mathematically the stresses within the complex structures is generally ill-founded. When he is dealing with familiar design problems and following conventional practice, the engineer draws on past experience in assessing the strength that must be built into a structure. A competent civil engineer, for example, has little difficulty in selecting the size of steel girder that he needs to support a wall. When he departs from conventional practice, however, and is called upon to design unfamiliar structures or to use new materials or techniques, the engineer can no longer depend upon past experience. The mathematical analysis of the stresses in complex components may not, in some cases, be a practical proposition owing to the high cost of computer time involved. If the engineer has no other way of assessing stresses except by recourse to the nearest standard shape and hence analytical solution available, he builds in greater strength than he judges to be necessary (i.e. he incorporates a factor of safety) in the hope of ensuring that the component will not fail in practice. Inevitably, this means unnecessary weight, size and cost, not only in the component itself but also in the other members of the structure which are associated with it.

To overcome this situation the modern engineer makes use of experimental techniques of stress measurement and analysis. Some of these consist of "reassurance" testing of completed structures which have been designed and built on the basis of existing analytical knowledge and past experience: others make use of scale models to predict the stresses, often before final designs have been completed.

Over the past few years these *experimental stress analysis* or *strain measurement* techniques have served an increasingly important role in aiding designers to produce not only efficient but economic designs. In some cases substantial reductions in weight and easier manufacturing processes have been achieved.

A large number of problems where experimental stress analysis techniques have been of particular value are those involving fatigue loading. Under such conditions failure usually starts when a fatigue crack develops at some position of high localised stress and propagates until final rupture occurs. As this often requires several thousand repeated cycles of load under service conditions, full-scale production is normally well under way when failure occurs. Delays at this stage can be very expensive, and the time saved by stress analysis techniques in locating the source of the trouble can far outweigh the initial cost of the equipment involved.

The main techniques of experimental stress analysis which are in use today are:

(1) brittle lacquers
(2) strain gauges
(3) photoelasticity
(4) photoelastic coatings

The aim of this chapter is to introduce the fundamental principles of these techniques, together with limited details of the principles of application, in order that the reader can appreciate (a) the role of the experimental techniques as against the theoretical procedures described in the other chapters, (b) the relative merits of each technique, and (c) the more specialised literature which is available on the techniques, to which reference will be made.

6.1. Brittle lacquers

The brittle-lacquer technique of experimental stress analysis relies on the failure by cracking of a layer of a brittle coating which has been applied to the surface under investigation. The coating is normally sprayed onto the surface and allowed to air- or heat-cure to attain its brittle properties. When the component is loaded, this coating will crack as its so-called *threshold strain* or *strain sensitivity* is exceeded. A typical crack pattern obtained on an engineering component is shown in Fig. 6.1. Cracking occurs where the strain is

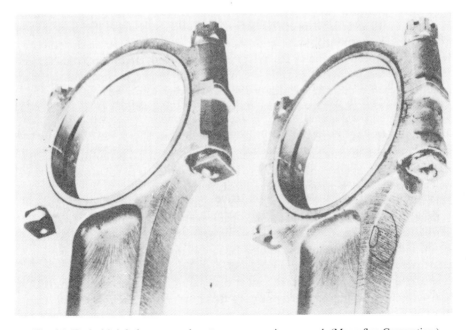

Fig. 6.1. Typical brittle-lacquer crack pattern on an engine con-rod. (Magnaflux Corporation.)

greatest, so that an immediate indication is given of the presence of stress concentrations. The cracks also indicate the directions of maximum strain at these points since they are always aligned at right angles to the direction of the maximum principal tensile strain. The method is thus of great value in determining the optimum positions in which to place strain gauges (see §6.2) in order to record accurately the measurements of strain in these directions.

The brittle-coating technique was first used successfully in 1932 by Dietrich and Lehr in Germany despite the fact that references relating to observation of the phenomenon can be traced back to Clarke's investigations of tubular bridges in 1850. The most important advance in brittle-lacquer technology, however, came in the United States in 1937–41 when Ellis, De Forrest and Stern produced a series of lacquers known as "Stresscoat" which, in a modified form, remain widely used in the world today.

There are many every-day examples of brittle coatings which can be readily observed by the reader to exhibit cracks indicating local yielding when the strain is sufficiently large, e.g. cellulose, vitreous or enamel finishes. Cellulose paints, in fact, are used by some engineering companies as a brittle lacquer on rubber models where the strains are quite large.

As an interesting experiment, try spraying a comb with several thin coats of hair-spray lacquer, giving each layer an opportunity to dry before application of the next coat. Finally, allow the whole coating several hours to fully cure; cracks should then become visible when the comb is bent between your fingers.

In engineering applications a little more care is necessary in the preparation of the component and application of the lacquer, but the technique remains a relatively simple and hence attractive one. The surface of the component should be relatively smooth and clean, standard solvents being used to remove all traces of grease and dirt. The lacquer can then be applied, the actual application procedure depending on the type of lacquer used. Most lacquers may be sprayed or painted onto the surface, spraying being generally more favoured since this produces a more uniform thickness of coating and allows a greater control of the thickness. Other lacquers, for example, are in wax or powder form and require pre-heating of the component surface in order that the lacquer will melt and run over the surface. Optimum coating thicknesses depend on the lacquer used but are generally of the order of 1 mm.

In order to determine the strain sensitivity of the lacquer, and hence to achieve an approximate idea of the strains existing in the component, it is necessary to coat calibration bars at the same time and in exactly the same manner as the specimen itself. These bars are normally simple rectangular bars which fit into the calibration jig shown in Fig. 6.2 to form a simple cantilever with an offset cam at the end producing a known strain distribution along the cantilever length. When the lacquer on the bar is fully cured, the lever on the cam is moved forward to depress the end of the bar by a known amount, and the position at which the cracking of the lacquer begins gives the strain sensitivity when compared with the marked strain scale. This enables quantitative measurements of strain levels to be made on the components under test since if, for example, the calibration sensitivity is shown to be 800 microstrain (strain $\times 10^{-6}$), then the strain at the point on the component at which cracks first appear is also 800 microstrain.

This type of quantitative measurement is generally accurate to no better than 10–20%, and brittle-lacquer techniques are normally used to locate the *positions* of stress maxima, the actual values then being determined by subsequent strain-gauge testing.

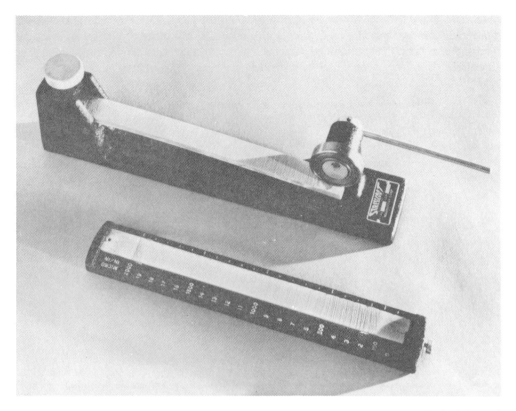

Fig. 6.2. (*Top*) Brittle-lacquer calibration bar in a calibration jig with the cam depressed to apply load. (*Bottom*) Calibration of approximately 100 microstrain. (Magnaflux Corporation.)

Loading is normally applied to the component in increments, held for a few minutes and released to zero prior to application of the next increment; the time interval between increments should be several times greater than that of the loading cycle. With this procedure *creep* effects in the lacquer, where strain in the lacquer changes at constant load, are completely overcome. After each load application, cracks should be sought and, when located, encircled and identified with the load at that stage using a chinagraph pencil. This enables an accurate record of the development of strain throughout the component to be built up.

There are a number of methods which can be used to aid crack detection including (a) pre-coating the component with an aluminium undercoat to provide a background of uniform colour and intensity, (b) use of a portable torch which, when held close to the surface, highlights the cracks by reflection on the crack faces, (c) use of dye-etchants or special electrified particle inspection techniques, details of which may be found in standard reference texts.[3]

Given good conditions, however, and a uniform base colour, cracks are often visible without any artificial aid, viewing the surface from various angles generally proving sufficient.

Figures 6.3 and 6.4 show further examples of brittle-lacquer crack patterns on typical engineering components. The procedure is simple, quick and relatively inexpensive; it can be carried out by relatively untrained personnel, and immediate qualitative information, such as positions of stress concentration, is provided on the most complicated shapes.

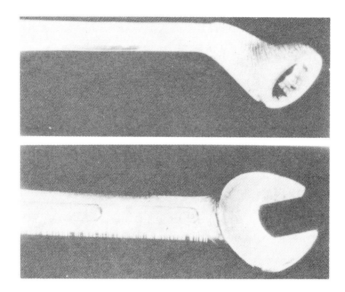

Fig. 6.3. Brittle-lacquer crack patterns on an open-ended spanner and a ring spanner. In the former the cracks appear at right angles to the maximum bending stress in the edge of the spanner whilst in the ring spanner the presence of torsion produces an inclination of the principal stress and hence of the cracks in the lacquer.

Fig. 6.4. Brittle-lacquer crack pattern highlighting the positions of stress concentration on a motor vehicle component. (Magnaflux Corporation.)

Various types of lacquer are available, including a special ceramic lacquer which is particularly useful for investigation under adverse environmental conditions such as in the presence of water, oil or heavy vibration.

Refinements to the general technique allow the study of residual stresses, compressive stress fields, dynamic situations, plastic yielding and miniature components with little increased difficulty. For a full treatment of these and other applications, the reader is referred to ref. 3.

6.2. Strain gauges

The accurate assessment of stresses, strains and loads in components under working conditions is an essential requirement of successful engineering design. In particular, the location of peak stress values and stress concentrations, and subsequently their reduction or removal by suitable design, has applications in every field of engineering. The most widely used experimental stress-analysis technique in industry today, particularly under working conditions, is that of strain gauges.

Whilst a number of different types of strain gauge are commercially available, this section will deal almost exclusively with the electrical resistance type of gauge introduced in 1939 by Ruge and Simmons in the United States.

The *electrical resistance strain gauge* is simply a length of wire or foil formed into the shape of a continuous grid, as shown in Fig. 6.5, cemented to a non-conductive backing. The gauge is then bonded securely to the surface of the component under investigation so that any strain in the surface will be experienced by the gauge itself. Since the fundamental equation for the electrical resistance R of a length of wire is

$$R = \frac{\rho L}{A} \qquad (6.1)$$

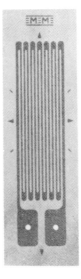

Fig. 6.5. Electric resistance strain gauge. (Welwyn Strain Measurement Ltd.)

where L is the length, A is the cross-sectional area and ρ is the *specific resistance* or *resistivity*, it follows that any change in length, and hence sectional area, will result in a change of resistance. Thus measurement of this resistance change with suitably calibrated equipment enables a direct reading of linear strain to be obtained. This is made possible by the relationship which exists for a number of alloys over a considerable strain range between change of resistance and strain which may be expressed as follows:

$$\frac{\Delta R}{R} = K \times \frac{\Delta L}{L} \tag{6.2}$$

where ΔR and ΔL are the changes in resistance and length respectively and K is termed the *gauge factor*.

Thus

$$\textbf{gauge factor } K = \frac{\Delta R/R}{\Delta L/L} = \frac{\Delta R/R}{\varepsilon} \tag{6.3}$$

where ε is the strain. The value of the gauge factor is always supplied by the manufacturer and can be checked using simple calibration procedures if required. Typical values of K for most conventional gauges lie in the region of 2 to 2.2, and most modern strain-gauge instruments allow the value of K to be set accordingly, thus enabling strain values to be recorded directly.

The changes in resistance produced by normal strain levels experienced in engineering components are very small, and sensitive instrumentation is required. Strain-gauge instruments are basically *Wheatstone bridge* networks as shown in Fig. 6.6, the condition of balance for this network being (i.e. the galvanometer reading zero when)

$$R_1 \times R_3 = R_2 \times R_4 \tag{6.4}$$

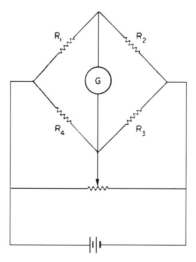

Fig. 6.6. Wheatstone bridge circuit.

In the simplest half-bridge wiring system, gauge 1 is the *active* gauge, i.e. that actually being strained. Gauge 2 is so-called *dummy* gauge which is bonded to an unstrained piece of metal similar to that being strained, its purpose being to cancel out any resistance change in R_1 that occurs due to temperature fluctuations in the vicinity of the gauges. Gauges 1 and 2 then represent the working half of the network – hence the name "half-bridge" system – and gauges 3 and 4 are standard resistors built into the instrument. Alternative wiring systems utilise one (*quarter-bridge*) or all four (*full-bridge*) of the bridge resistance arms.

6.3. Unbalanced bridge circuit

With the Wheatstone bridge initially balanced to zero any strain on gauge R_1 will cause the galvanometer needle to deflect. This deflection can be calibrated to read strain, as noted above, by including in the circuit an arrangement whereby gauge-factor adjustment can be achieved. Strain readings are therefore taken with the pointer off the zero position and the bridge is thus *unbalanced*.

6.4. Null balance or balanced bridge circuit

An alternative measurement procedure makes use of a variable resistance in one arm of the bridge to cancel any deflection of the galvanometer needle. This adjustment can then be calibrated directly as strain and readings are therefore taken with the pointer on zero, i.e. in the *balanced* position.

6.5. Gauge construction

The basic forms of wire and foil gauges are shown in Fig. 6.7. Foil gauges are produced by a printed-circuit process from selected melt alloys which have been rolled to a thin film, and these have largely superseded the previously popular wire gauge. Because of the increased area of metal in the gauge at the ends, the foil gauge is not so sensitive to strains at right angles to the direction in which the major axis of the gauge is aligned, i.e. it has a low transverse or cross-sensitivity – one of the reasons for its adoption in preference to the wire gauge. There are many other advantages of foil gauges over wire gauges, including better strain transmission from the substrate to the grid and better heat transmission from the grid

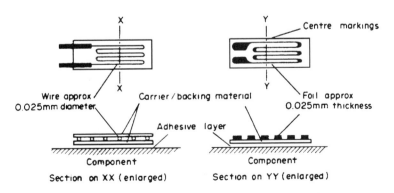

Fig. 6.7. Basic format of wire and foil gauges. (Merrow)

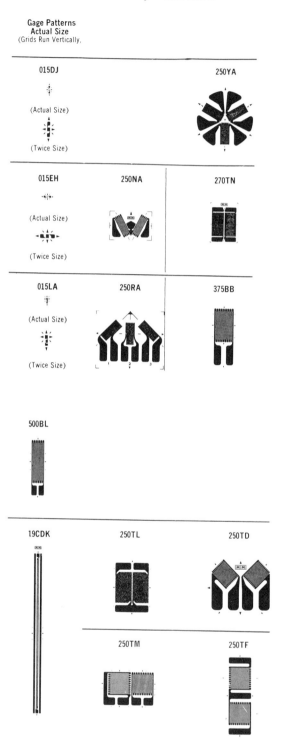

Fig. 6.8. Typical gauge sizes and formats. (Welwyn Strain Measurement Division)

to the substrate; as a result of which they are usually more stable. Additionally, the grids of foil gauges can be made much smaller and there is almost unlimited freedom of grid configuration, solder tab arrangement, multiple grid configuration, etc.

6.6. Gauge selection

Figure 6.8 shows but a few of the many types and size of gauge which are available. So vast is the available range that it is difficult to foresee any situation for which there is no gauge suitable. Most manufacturers' catalogues[13] give full information on gauge selection, and any detailed treatment would be out of context in this section. Essentially, the choice of a suitable gauge incorporates consideration of physical size and form, resistance and sensitivity, operating temperature, temperature compensation, strain limits, flexibility of the gauge backing (and hence relative stiffness) and cost.

6.7. Temperature compensation

Unfortunately, in addition to strain, other factors affect the resistance of a strain gauge, the major one being temperature change. It can be shown that temperature change of only a few degrees completely dwarfs any readings due to the typical strains encountered in engineering applications. Thus it is vitally important that any temperature effects should be cancelled out, leaving only the mechanical strain required. This is achieved either by using the conventional dummy gauge, *half-bridge*, system noted earlier, or, alternatively, by the use of *self-temperature-compensated gauges*. These are gauges constructed from material which has been subjected to particular metallurgical processes and which produce very small (and calibrated) thermal output over a specified range of temperature when bonded onto the material for which the gauge has been specifically designed (see Fig. 6.9).

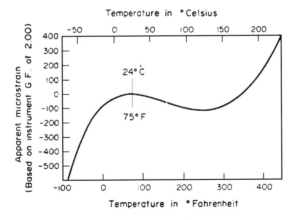

Fig. 6.9. Typical output from self-temperature-compensated gauge (Vishay)

In addition to the gauges, the lead-wire system must also be compensated, and it is normal practice to use the three-lead-wire system shown in Fig. 6.10. In this technique, two of the leads are in opposite arms of the bridge so that their resistance cancels, and the third lead, being in series with the power supply, does not influence the bridge balance. All leads must be of equal length and wound tightly together so that they experience the same temperature conditions.

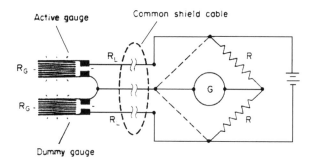

Fig. 6.10. Three-lead wire system for half-bridge (dummy-active) operation.

In applications where a single self-temperature-compensated gauge is used in a quarter-bridge arrangement the three-wire circuit becomes that shown in Fig. 6.11. Again, only one of the current-carrying lead-wires is in series with the active strain gauge, the other is in series with the bridge completion resistor (occasionally still referred to as a "dummy") in the adjacent arm of the bridge. The third wire, connected to the lower solder tab of the active gauge, carries essentially no current but acts simply as a voltage-sensing lead. Provided the two lead-wires (resistance R_L) are of the same size and length and maintained at the same temperature (i.e. kept physically close to each other) then any resistance changes due to temperature will cancel.

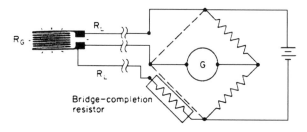

Fig. 6.11. Three-lead-wire system for quarter-bridge operation with single self-temperature-compensated gauge.

6.8. Installation procedure

The quality and success of any strain-gauge installation is influenced greatly by the care and precision of the installation procedure and correct choice of the adhesive. The apparently mundane procedure of actually cementing the gauge in place is a critical step in the operation. Every precaution must be taken to ensure a chemically clean surface if perfect adhesion is to be attained. Full details of typical procedures and equipment necessary are given in refs 6 and 13, as are the methods which may be used to test the validity of the installation prior to recording measurements. Techniques for protection of gauge installations are also covered. Typical strain-gauge installations are shown in Figs. 6.12 and 6.13.

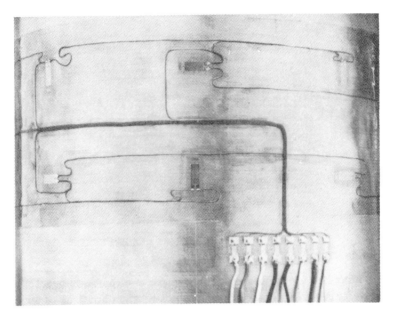

Fig. 6.12. Typical strain-gauge installation showing six of eight linear gauges bonded to the surface of a cylinder to record longitudinal and hoop strains. (Crown copyright.)

6.9. Basic measurement systems

(a) For direct strain

The standard procedure for the measurement of tensile or compressive direct strains utilises the *full-bridge* circuit of Fig. 6.14 in which not only are the effects of any bending eliminated but the sensitivity is increased by a factor of 2.6 over that which would be achieved using a single linear gauge.

Bearing in mind the balance requirement of the Wheatstone bridge, i.e. $R_1, R_3 = R_2 R_4$, each pair of gauges on either side of the equation will have an additive effect if their signs are similar or will cancel if opposite. Thus the opposite signs produced by bending cancel on both pairs whilst the similar signs of the direct strains are additive. The value 2.6 arises from twice the applied axial strain (R_1 and R_3) plus twice the Poisson's ratio strain (R_2 and R_4), assuming $\nu = 0.3$. The latter is compressive, i.e. negative, on the opposite side of the bridge from R_1 and R_3, and hence is an added signal to that of R_1 and R_3.

(b) Bending

Figure 5.15(a) shows the arrangement used to record bending strains independently of direct strains. It is normal to bond linear gauges on opposite surfaces of the component and to use the *half-bridge* system shown in Fig. 6.6; this gives a sensitivity of twice that which would be

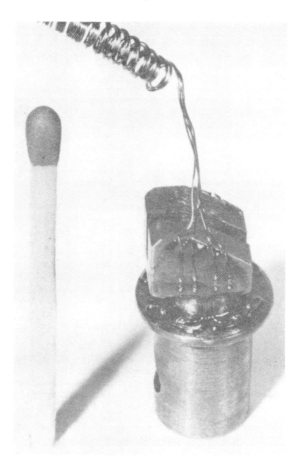

Fig. 6.13. Miniature strain-gauge installation. (Welwyn.)

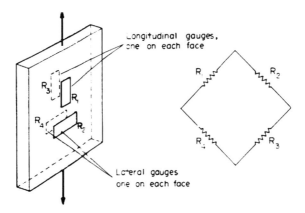

Fig. 6.14. "Full bridge" circuit arranged to eliminate any bending strains produced by unintentional eccentricities of loading in a nominal axial load application. The arrangement also produces a sensitivity 26 times that of a single active gauge. (Merrow.)

achieved with a single-linear gauge. Alternatively, it is possible to utilise again the Poisson strains as in §6.9(a) by bonding additional lateral gauges (i.e. perpendicular to the other gauges) on each surface and using a full-bridge circuit to achieve a sensitivity of 2.6. In this case, however, gauges R_2 and R_4 would be interchanged from the arrangement shown in Fig. 6.14 and would appear as in Fig. 6.15(b).

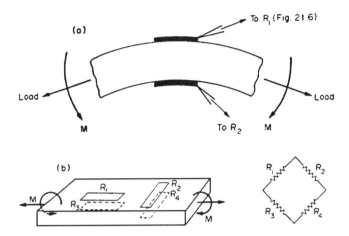

Fig. 6.15. (a) Determination of bending strains independent of end loads: "half-bridge" method. Sensitivity twice that of a single active gauge. (b) Determination of bending strains independent of end loads: "full-bridge" procedure. Sensitivity 2.6.

(c) Torsion

It has been shown that pure torsion produces direct stresses on planes at 45° to the shaft axis – one set tensile, the other compressive. Measurements of torque or shear stress using strain-gauge techniques therefore utilise gauges bonded at 45° to the axis in order to record the direct strains. Again, it is convenient to use a wiring system which automatically cancels unwanted signals, i.e. in this case the signals arising due to unwanted direct or bending strains which may be present. Once again, a full-bridge system is used and a sensitivity of four times that of a single gauge is achieved (Fig. 6.16).

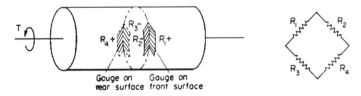

Fig. 6.16. Torque measurement using full-bridge circuit-sensitivity four times that of a single active gauge.

6.10. D.C. and A.C. systems

The basic Wheatstone bridge circuit shown in all preceding diagrams is capable of using either a direct current (d.c.) or an alternating current (a.c.) source; Fig. 6.6, for example, shows the circuit excited by means of a standard battery (d.c.) source. Figure 6.17, however, shows a typical arrangement for a so-called a.c. *carrier frequency* system, the main advantage of this being that all unwanted signals such as noise are eliminated and a stable signal of gauge output is produced. The relative merits and disadvantages of the two types of system are outside the scope of this section but may be found in any standard reference text (refs. 4, 6, 7 and 13).

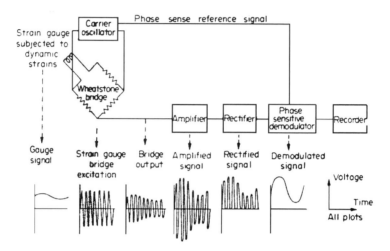

Fig. 6.17. Schematic arrangement of a typical carrier frequency system. (Merrow.)

6.11. Other types of strain gauge

The previous discussion has related entirely to the electrical resistance type of strain gauge and, indeed, this is by far the most extensively used type of gauge in industry today. It should be noted, however, that many other forms of strain gauge are available. They include:

(a) **mechanical gauges** or **extensometers** using optical or mechanical lever systems;

(b) **pneumatic gauges** using changes in pressure;

(c) **acoustic gauges** using the change in frequency of a vibrating wire;

(d) **semiconductor** or **piezo-resistive gauges** using the piezo-resistive effect in silicon to produce resistance changes;

(e) **inductance gauges** using changes in inductance of, e.g., differential transformer systems;

(f) **capacitance gauges** using changes in capacitance between two parallel or near-parallel plates.

Each type of gauge has a particular field of application in which it can compete on equal, or even favourable, terms with the electrical resistance form of gauge. None, however, are as versatile and generally applicable as the resistance gauge. For further information on each type of gauge the reader is referred to the references listed at the end of this chapter.

6.12. Photoelasticity

In recent years, photoelastic stress analysis has become a technique of outstanding importance to engineers. When polarised light is passed through a stressed transparent model, interference patterns or *fringes* are formed. These patterns provide immediate qualitative information about the general distribution of stress, positions of stress concentrations and of areas of low stress. On the basis of these results, designs may be modified to reduce or disperse concentrations of stress or to remove excess material from areas of low stress, thereby achieving reductions in weight and material costs. As photoelastic analysis may be carried out at the design stage, stress conditions are taken into account before production has commenced; component failures during production, necessitating expensive design modifications and re-tooling, may thus be avoided. Even when service failures do occur, photoelastic analysis provides an effective method of failure investigation and often produces valuable information leading to successful re-design, typical photoelastic fringe patterns are shown in Fig. 6.18.

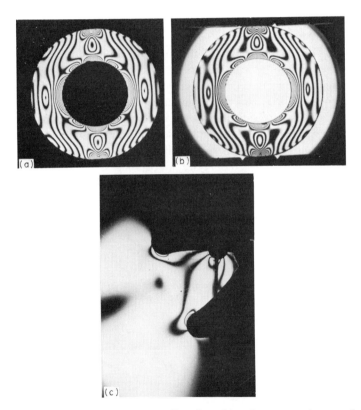

Fig. 6.18. Typical photoelastic fringe patterns. (a) Hollow disc subjected to compression on a diameter (dark field background). (b) As (a) but with a light field background. (c) Stress concentrations as the roots of a gear tooth.

Conventional or *transmission photoelasticity* has for many years been a powerful tool in the hands of trained stress analysts. However, untrained personnel interested in the technique have often been dissuaded from attempting it by the large volume of advanced mathematical and optical theory contained in reference texts on the subject. Whilst this theory is, no doubt, essential for a complete understanding of the phenomena involved and of some of the more advanced techniques, it is important to accept that a wealth of valuable information can be obtained by those who are not fully conversant with all the complex detail. A major feature of the technique is that it allows one to effectively "look into" the component and pin-point flaws or weaknesses in design which are otherwise difficult or impossible to detect. Stress concentrations are immediately visible, stress values around the edge or boundary of the model are easily obtained and, with a little more effort, the separate principal stresses within the model can also be determined.

6.13. Plane-polarised light – basic polariscope arrangements

Before proceeding with the details of the photoelastic technique it is necessary to introduce the meaning and significance of *plane-polarised light* and its use in the equipment termed *polariscopes* used for photoelastic stress analysis. If light from an ordinary light bulb is passed through a polarising sheet or *polariser*, the sheet will act like a series of vertical slots so that the emergent beam will consist of light vibrating in one plane only: the plane of the slots. The light is then said to be *plane polarised*.

When directed onto an unstressed photoelastic model, the plane-polarised light passes through unaltered and may be completely extinguished by a second polarising sheet, termed an *analyser*, whose axis is perpendicular to that of the polariser: This is then the simplest form of polariscope arrangement which can be used for photoelastic stress analysis and it is termed a "*crossed*" *set-up* (see Fig. 6.19). Alternatively, a "*parallel*" *set-up*" may be used in which the axes of the polariser and analyser are parallel, as in Fig. 6.20. With the model unstressed, the plane-polarised light will then pass through both the model and analyser unaltered and maximum illumination will be achieved. When the model is stressed in the parallel set-up, the resulting fringe pattern will be seen against a light background or "field", whilst with the crossed arrangement there will be a completely black or "dark field".

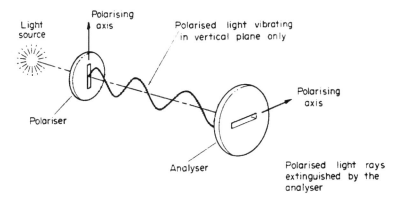

Fig. 6.19. "Crossed" set-up. Polariser and analyser arranged with their polarising axes at right angles; plane polarised light from the polariser is completely extinguished by the analyser. (Merrow.)

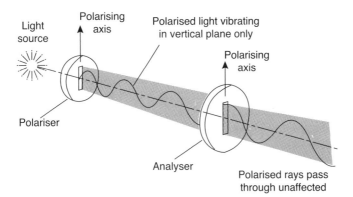

Fig. 6.20. "Parallel" set-up. Polariser and analyser axes parallel; plane-polarised light from the polariser passes through the analyser unaffected, producing a so-called "light field" arrangement. (Merrow.)

6.14. Temporary birefringence

Photoelastic models are constructed from a special class of transparent materials which exhibit a property known as *birefringence*, i.e. they have the ability to split an incident plane-polarised ray into two component rays; they are *double refracting*. This property is only exhibited when the material is under stress, hence the qualified term "*temporary bire-fringence*", and the direction of the component rays always coincides with the directions of the principal stresses (Fig. 6.21). Further, the speeds of the rays are proportional to the magnitudes of the respective stresses in each direction, so that the rays emerging from the model are out of phase and hence produce interference patterns when combined.

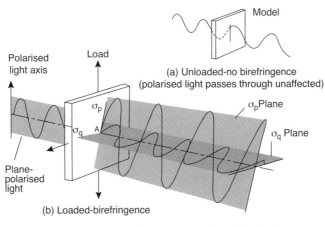

Fig. 6.21. Temporary birefringence. (a) Plane-polarised light directed onto an unstressed model passes through unaltered. (b) When the model is stressed the incident plane-polarised light is split into two component rays. The directions of the rays coincide with the directions of the principal stresses, and the speeds of the rays are proportional to the magnitudes of the respective stresses in their directions. The emerging rays are out of phase, and produce an interference pattern of fringes. (Merrow.)

6.15. Production of fringe patterns

When a model of an engineering component constructed from a birefringent material is stressed, it has been shown above that the incident plane-polarised light will be split into two component rays, the directions of which at any point coincide with the directions of the principal stresses at the point. The rays pass through the model at speeds proportional to the principal stresses in their directions and emerge out of phase. When they reach the analyser, shown in the crossed position in Fig. 6.22, only their horizontal components are transmitted and these will combine to produce interference fringes as shown in Fig. 6.23.

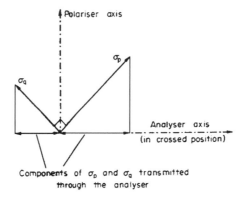

Fig. 6.22. Transmission through the analyser. (Merrow.)

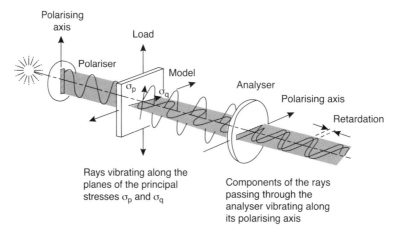

Fig. 6.23. Loaded model viewed in a plane polariscope arrangement with a "crossed set-up".

The difference in speeds of the rays, and hence the amount of interference produced, is proportional to the difference in the principal stress values $(\sigma_p - \sigma_q)$ at the point in question. Since the maximum shear stress in any two-dimensional stress system is given by

$$\tau_{\max} = \tfrac{1}{2}(\sigma_p - \sigma_q)$$

it follows that the interference or fringe pattern produced by the photoelastic technique will give an immediate indication of the variation of shear stress throughout the model. Only at a free, unloaded boundary of a model, where one of the principal stresses is zero, will the fringe pattern yield a direct indication of the principal direct stress (in this case the tangential boundary stress). However, since the majority of engineering failures are caused by fatigue cracks commencing at the point of maximum tensile stress at the boundary, this is not a severe limitation. Further discussion of the interpretation of fringe patterns is referred to the following section.

If the original light source is *monochromatic*, e.g. mercury green or sodium yellow, the fringe pattern will appear as a series of distinct black lines on a uniform green or yellow background. These black lines or fringes correspond to points where the two rays are exactly 180° out of phase and therefore cancel. If white light is used, however, each composite wavelength of the white light is cancelled in turn and a multicoloured pattern of fringes termed *isochromatics* is obtained.

Monochromatic sources are preferred for accurate quantitative photoelastic measurements since a high number of fringes can be clearly discerned at, e.g., stress concentration positions. With a white light source the isochromatics become very pale at high stress regions and clear fringe boundaries are no longer obtained. White light sources are therefore normally reserved for general qualitative assessment of models, for isolation of zero fringe order positions (i.e. zero shear stress) which appear black on the multicoloured background, and for the investigation of stress directions using *isoclinics*. These are defined in detail in §6.19.

6.16. Interpretation of fringe patterns

It has been stated above that the pattern of fringes achieved by the photoelastic technique yields:

(a) *A complete indication of the variation of shear stress throughout the entire model*. Since ductile materials will generally fail in shear rather than by direct stress, this is an important feature of the technique. At points where the fringes are most numerous and closely spaced, the stress is highest; at points where they are widely spaced or absent, the stress is low. With a white-light source such areas appear black, indicating zero shear stress, but it cannot be emphasised too strongly that this does not necessarily mean zero stress since if the values of σ_p and σ_q (however large) are equal, then $(\sigma_p - \sigma_q)$ will be zero and a black area will be produced. Extreme care must therefore be taken in the interpretation of fringe patterns. Generally, however, fringe patterns may be compared with contour lines on a map, where close spacing relates to steep slopes and wide spacing to gentle inclines. Peaks and valleys are immediately evident, and actual heights are readily determined by counting the contours and converting to height by the known scale factor. In an exactly similar way, photoelastic fringes are counted from the known zero (black) positions and the resulting number or order of fringe at the point in question is converted to stress by a calibration constant known as the *material fringe value*. Details of the calibration procedure will be given later.

(b) *Individual values of the principal stresses at free unloaded boundaries, one of these always being zero*. The particular relevance of this result to fatigue failures has been mentioned, and the use of photoelasticity to produce modifications to boundary profiles in order to reduce boundary stress concentrations and hence the likelihood of fatigue failures has been a major use of the technique. In addition to the immediate indication of high stress locations, the photoelastic model will show regions of low stress from which material can be conveniently **removed** without weakening the component to effect a reduction in weight and material cost. Surprisingly, perhaps, a reduction in material at or near a high stress concentration can also produce a significant reduction in maximum stress. Re-design can be carried out on a "file-it-and-see" basis, models being modified or re-shaped within minutes in order to achieve the required distribution of stress.

Whilst considerable valuable qualitative information can be readily obtained from photo-elastic models without any calculations at all, there are obviously occasions where the precise values of the stresses are required. These are obtained using the following basic equation of photoelasticity,

$$\sigma_p - \sigma_q = \frac{nf}{t} \tag{6.5}$$

where σ_p and σ_q are the values of the maximum and minimum principal stresses at the point under consideration, n is the fringe number or *fringe order* at the point, f is the *material fringe value* or *coefficient*, and t is the model thickness.

Thus with a knowledge of the material fringe value obtained by calibration as described below, the required value of $(\sigma_p - \sigma_q)$ at any point can be obtained readily by simply counting the fringes from zero to achieve the value n at the point in question and substitution in the above relatively simple expression.

Maximum shear stress or boundary stress values are then easily obtained and the application of one of the so-called *stress-separation* procedures will yield the separate value of the principal stress at other points in the model with just a little more effort. These may be of particular interest in the design of components using brittle materials which are known to be relatively weak under the action of direct stresses.

6.17. Calibration

The value of f, which, it will be remembered, is analogous to the height scale for contours on a survey map, is determined by a simple calibration experiment in which the known stress at some point in a convenient model is plotted against the fringe value at that point under various loads. One of the most popular loading systems is diametral compression of a disc, when the relevant equation for the stress at the centre is

$$\sigma_p - \sigma_q = \frac{8P}{\pi Dt} \tag{6.6}$$

where P is the applied load, D is the disc diameter and t is the thickness.

Thus, comparing with the photoelastic eqn. (6.1),

$$\frac{nf}{t} = \frac{8P}{\pi Dt}$$

The slope of the load versus fringe order graph is given by

$$\frac{P}{n} = f \times \frac{\pi D}{8} \tag{6.7}$$

Hence f can be evaluated.

6.18. Fractional fringe order determination – compensation techniques

The accuracy of the photoelastic technique is limited, among other things, to the accuracy with which the fringe order at the point under investigation can be evaluated. It is not sufficiently accurate to count to the nearest whole number of fringes, and precise determination of fractions of fringe order at points lying between fringes is required. Conventional methods for determining these fractions of fringe order are termed *compensation techniques* and allow estimation of fringe orders to an accuracy of one-fiftieth of a fringe. The two methods most often used are the Tardy and Senarmont techniques. Before either technique can be adopted, the directions of the polariser and analyser must be aligned with the directions of the principal stresses at the point. This is achieved by rotating both units together in the plane polariscope arrangement until an *isoclinic* (§6.19) crosses the point. In most modern polariscopes facilities exist to couple the polariser and analyser together in order to facilitate synchronous rotation. The procedure for the two techniques then varies slightly.

(a) Tardy method

Quarter-wave plates are inserted at 45° to the polariser and analyser as the dark field circular polariscope set-up of Fig. 6.24. Normal fringe patterns will then be visible in the absence of isoclinics.

(b) Senarmont method

The polariser and analyser are rotated through a further 45° retaining the dark field, thus moving the polarising axes at 45° to the principal stress directions at the point. Only one quarter-wave plate is then inserted between the model and the analyser and rotated to again achieve a dark field. The normal fringe pattern is then visible as with the Tardy method.

Thus, having identified the integral value n of the fringe order at the point, i.e. between 1 and 2, or 2 and 3, for instance, the fractional part can now be established for both methods in the same way.

The analyser is rotated on its own to produce movement of the fringes. In particular, the nearest *lower order* of fringe is moved to the point of interest and the angle θ moved by the analyser recorded.

The fringe order at the chosen point is then $n + \dfrac{\theta°}{180°}$.

N.B.–Rotation of the analyser in the opposite direction $\phi°$ would move the nearest *highest order* fringe $(n + 1)$ back to the point. In this case the fringe order at the point would be

$$(n + 1) - \frac{\phi}{180}$$

It can be shown easily by trial that the sum of the two angles θ and ϕ is always 180°.

There is little to choose between the two methods in terms of accuracy; some workers prefer to use Tardy, others to use Senarmont.

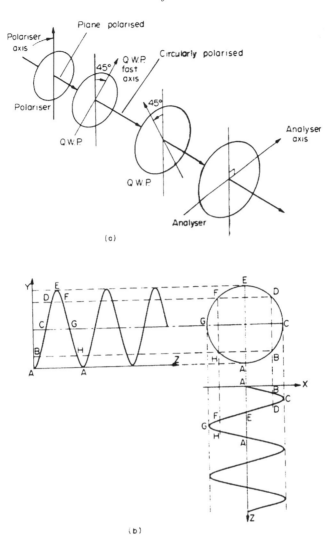

Fig. 6.24. (a) Circular polariscope arrangement. Isoclinics are removed optically by inserting quarter-wave plates (Q.W.P.) with optical axes at 45° to those of the polariser and analyser. Circularly polarised light is produced. (Merrow.) (b) Graphical construction for the addition of two rays at right angles a quarter-wavelength out of phase, producing resultant circular envelope, i.e. circularly polarised light.

6.19. Isoclinics – circular polarisation

If plane-polarised light is used for photoelastic studies as suggested in the preceding text, the fringes or isochromatics will be partially obscured by a set of black lines known as isoclinics (Fig. 6.25). With the coloured isochromatics of a white light source, these are easily identified, but with a monochromatic source confusion can easily arise between the black fringes and the black isoclinics.

It is therefore convenient to use a different optical system which eliminates the isoclinics but retains the basic fringe pattern. The procedure adopted is outlined below.

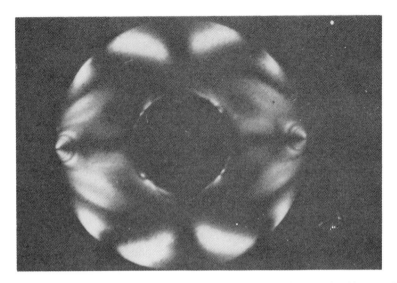

Fig. 6.25. Hollow disc subjected to diametral compression as in Fig. 6.18(a) but in this case showing the isoclinics superimposed.

An *isoclinic* line is a locus of points at which the principal stresses have the same inclination; the 20° isoclinic, for example, passes-through all points at which the principal stresses are inclined at 20° to the vertical and horizontal (Fig. 6.26). Thus isoclinics are not peculiar to photoelastic studies; it is simply that they have a particular relevance in this case and they are readily visualised. For the purpose of this introduction it is sufficient to note that they are used as the basis for construction of *stress trajectories* which show the directions of the principal stresses at all points in the model, and hence in the component. Further details may be found in the relevant standard texts.

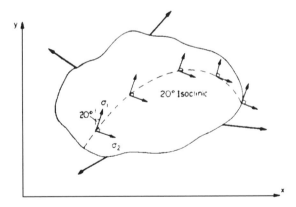

Fig. 6.26. The 20° isoclinic in a body subjected to a general stress system. The isoclinic is given by the locus of all points at which the principal stresses are inclined at 20° to the reference x and y axes.

To prevent the isoclinics interfering with the analysis of stress magnitudes represented by the basic fringe pattern, they are removed optically by inserting quarter-wave plates with their axes at 45° to those of the polariser and analyser as shown in Fig. 6.24. These eliminate all unidirectional properties of the light by converting it into *circularly polarised* light. The amount of interference between the component rays emerging from the model, and hence the fringe patterns, remains unchanged and is now clearly visible in the absence of the isoclinics.

6.20. Stress separation procedures

The photoelastic technique has been shown to provide principal stress difference and hence maximum shear stresses at all points in the model, boundary stress values and stress directions. It has also been noted that there are occasions where the separate values of the principal stresses are required at points other than at the boundary, e.g. in the design of components using brittle materials. In this case it is necessary to employ one of the many *stress separation* procedures which are available. It is beyond the scope of this section to introduce these in detail, and full information can be obtained if desired from standard texts.[8,9,11] The principal techniques which find most application are (a) the oblique incidence method, and (b) the shear slope or "shear difference" method.

6.21. Three-dimensional photoelasticity

In the preceding text, reference has been made to models of uniform thickness, i.e. two-dimensional models. Most engineering problems, however, arise in the design of components which are three-dimensional. In such cases the stresses vary not only as a function of the shape in any one plane but also throughout the "thickness" or third dimension. Often a proportion of the more simple three-dimensional model or loading cases can be represented by equivalent two-dimensional systems, particularly if the models are symmetrical, but there remains a greater proportion which cannot be handled by the two-dimensional approach. These, however, can also be studied using the photoelastic method by means of the so-called *stress-freezing* technique.

Three-dimensional photoelastic models constructed from the same birefringent material introduced previously are loaded, heated to a critical temperature and cooled very slowly back to room temperature. It is then found that a fringe pattern associated with the elastic stress distribution in the component has been locked or "frozen" into the model. It is then possible to cut the model into thin slices of uniform thickness, each slice then being examined as if it were a two-dimensional model. Special procedures for model manufacture, slicing of the model and fringe interpretation are required, but these are readily obtained with practice.

6.22. Reflective coating technique[12]

A special adaptation of the photoelastic technique utilises a thin sheet of photoelastic material which is bonded onto the surface of a metal component using a special adhesive containing an aluminium pigment which produces a reflective layer. Polarised light is directed onto the photoelastic coating and viewed through an analyser after reflection off the metal surface using a *reflection polariscope* as shown in Fig. 6.27.

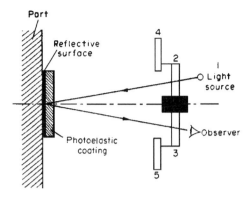

Fig. 6.27. Reflection polariscope principle and equipment.

A fringe pattern is observed which relates to the strain in the metal component. The technique is thus no longer a model technique and allows the evaluation of strains under loading conditions. Static and dynamic loading conditions can be observed, the latter with the aid of a stroboscope or high-speed camera, and the technique gives a full field view of the strain distribution in the surface of the component. Unlike the transmission technique, however, it gives no information as to the stresses *within* the material.

Standard photoelastic sheet can be used for bonding to flat components, but special casting techniques are available which enable the photoelastic material to be obtained in a partially polymerised, very flexible, stage, and hence allows it to be contoured or moulded around complex shapes without undue thickness changes. After a period has been allowed for complete polymerisation to occur in the moulded position, the sheet is removed and bonded firmly back into place with the reflective adhesive.

The reflective technique is particularly useful for the observation of service loading conditions over wide areas of structure and is often used to highlight the stress concentration positions which can subsequently become the subject of detailed strain-gauge investigations.

6.23. Other methods of strain measurement

In addition to the widely used methods of experimental stress analysis or strain measurement covered above, there are a number of lesser-used techniques which have particular advantages in certain specialised conditions. These techniques can be referred to under the general title of grid methods, although in some cases a more explicit title would be "interference methods".

The standard **grid technique** consists of marking a grid, either mechanically or chemically, on the surface of the material under investigation and measuring the distortions of this grid under strain. A direct modification of this procedure, known as the **"replica" technique**, involves the firing of special pellets from a gun at the grid both before and during load. The surface of the pellets are coated with "Woods metal" which is heated in the gun prior to firing. Replicas of the undeformed and deformed grids are then obtained in the soft metal on contact with the grid-marked surface. These are viewed in a vernier comparison microscope to obtain strain readings.

A further modification of the grid procedure, known as the **moiré technique**. superimposes the deformed grid on an undeformed master (or vice versa). An interference pattern, known as **moiré fringes**, similar to those obtained when two layers of net curtain are superimposed, is produced and can be analysed to yield strain values.

X-rays can be used to obtain surface strain values from measurements of crystal lattice deformation. **Acoustoelasticity**, based on a principle similar to photoelasticity but using polarised ultrasonic sound waves, has been proposed but is not universally accepted to date. **Holography**, using the laser as a source of coherent light, and again relying on the interference obtained between holograms of deformed and undeformed components, has recently created considerable interest, but none of these techniques appear at the moment to represent a formidable challenge to the major techniques listed earlier.

Bibliography

1. A.J. Durelli, E.A. Phillips and C.H. Tsao, *Analysis of Stress and Strain*, McGraw-Hill, New York, 1958.
2. Magnaflux Corporation, *Principles of Stresscoat*.
3. E.J. Hearn, *Brittle Lacquers for Strain Measurement*, Merrow Publishing Co., Watford, England, 1971.
4. C.C. Perry and H.P. Lissner, *Strain Gauge Primer*, McGraw-Hill, New York.
5. T. Potma, *Strain Gauges*, Iliffe, London, 1967.
6. E.J. Hearn, *Strain Gauges*, Merrow Publishing Co., Watford, England, 1971.
7. R. Murray and P. Stein, *Strain Gauge Techniques*, M.I.T. Press, Cambridge, Mass., 1956.
8. E.J. Hearn, *Photoelasticity*, Merrow Publishing Co., Watford, England, 1971.
9. M.M. Frocht, *Photoelasticity*, vols. I and II, Wiley, 1961.
10. H.T. Jessop and F.C. Harris, *Photoelasticity*; Cleaver-Hume, 1949.
11. E.G. Coker and L.N.G. Filon, *Photoelasticity*, Cambridge University Press, 1957.
12. F. Zandman, S. Redner, J.W. Dally, *Photoelastic Coatings* Iowa State/S.E.S.A. 1977.
13. J. Pople, B.S.S.M. *Strain Measurement Reference Book*, B.S.S.M. Newcastle, England.

CHAPTER 7

CIRCULAR PLATES AND DIAPHRAGMS

Summary

The slope and deflection of circular plates under various loading and support conditions are given by the fundamental deflection equation

$$\frac{d}{dr}\left[\frac{1}{r}\frac{d}{dr}\left(r\frac{dy}{dr}\right)\right] = -\frac{Q}{D}$$

where y is the deflection at radius r; dy/dr is the slope θ at radius r; Q is the applied load or shear force per unit length, usually given as a function of r; D is a constant termed the "flexural stiffness" or "flexural rigidity" $= Et^3/[12(1-v^2)]$ and t is the plate thickness.

For applied uniformly distributed load (i.e. pressure q) the equation becomes

$$\frac{d}{dr}\left[\frac{1}{r}\frac{d}{dr}\left(r\frac{dy}{dr}\right)\right] = -\frac{qr}{2D}$$

For central concentrated load F

$$Q = \frac{F}{2\pi r} \quad \text{and the right-hand-side becomes} \quad -\frac{F}{2\pi rD}$$

For axisymmetric non-uniform pressure (e.g. impacting gas or water jet)

$$q = K/r \quad \text{and the right-hand-side becomes} \quad -K/2D$$

The *bending moments per unit length* at any point in the plate are:

$$M_r = M_{xy} = D\left[\frac{d\theta}{dr} + v\frac{\theta}{r}\right]$$

$$M_z = M_{yz} = D\left[v\frac{d\theta}{dr} + \frac{\theta}{r}\right]$$

Similarly, the *radial and tangential stresses* at any radius r are given by:

$$\text{radial stress } \sigma_r = \frac{Eu}{(1-v^2)}\left[\frac{d\theta}{dr} + v\frac{\theta}{r}\right]$$

$$\text{tangential stress } \sigma_z = \frac{Eu}{(1-v^2)}\left[v\frac{d\theta}{dr} + \frac{\theta}{r}\right]$$

Alternatively,
$$\sigma_r = \frac{12u}{t^3}M_r \quad \text{and} \quad \sigma_z = \frac{12u}{t^3}M_z$$

For a **circular plate**, radius R, *freely supported* at its edge and subjected to a *load F distributed around a circle radius R_1*

$$y_{max} = \frac{F}{8\pi D}\left[\frac{(3+v)}{2(1+v)}(R^2 - R_1^2) - R_1^2 \log_e \frac{R}{R_1}\right]$$

and

$$\sigma_{r_{max}} = \frac{3F}{4\pi t^2}\left[2(1+v)\log_e \frac{R}{R_1} + (1-v)\frac{(R^2 - R_1^2)}{R^2}\right]$$

$$= \sigma_{z_{max}}$$

Table 7.1. Summary of maximum deflections and stresses.

Loading condition	Maximum deflection	Maximum stresses	
	(y_{max})	$\sigma_{r_{max}}$	$\sigma_{z_{max}}$
Uniformly loaded, edges clamped	$\dfrac{3qR^4}{16Et^3}(1-v^2)$	$\dfrac{3qR^2}{4t^2}$	$\dfrac{3qR^2}{8t^2}(1+v)$
Uniformly loaded, edges freely supported	$\dfrac{3qR^4}{16Et^3}(5+v)(1-v)$	$\dfrac{3qR^2}{8t^2}(3+v)$	$\dfrac{3qR^2}{8t^2}(3+v)$
Central load F, edges clamped	$\dfrac{3FR^2}{4\pi Et^3}(1-v^2)$	$\dfrac{3F}{2\pi t^2}$	$\dfrac{3vF}{2\pi t^2}$
Central load F, edges freely supported	$\dfrac{3FR^2}{4\pi Et^3}(3+v)(1-v)$	From $\dfrac{3F}{2\pi t^2}(1+v)\log_e \dfrac{R}{r}$	From $\dfrac{3F}{2\pi t^2}\left[(1+v)\log_e \dfrac{R}{r} + (1-v)\right]$

For an **annular ring**, *freely supported* at its outside edge, with total *load F applied around the inside radius R_1*, the maximum stress is tangential at the inside radius,

i.e.

$$\sigma_{z_{max}} = \frac{3F(1+v)}{\pi t^2}\left[\frac{R^2}{(R-R_1)}\log_e \frac{R}{R_1}\right]$$

If the outside edge is *clamped* the maximum stress becomes

$$\sigma_{max} = \frac{3F}{2\pi t^2}\left[\frac{(R^2 - R_1^2)}{R^2}\right]$$

For **thin membranes** subjected to *uniform pressure q* the maximum deflection is given by

$$y_{max} = 0.662\, R\left[\frac{qR}{Et}\right]^{1/3}$$

For **rectangular plates** subjected to *uniform loads* the maximum deflection and bending moments are given by equations of the form

$$y_{max} = \alpha\frac{qb^4}{Et^3}$$

$$M = \beta q b^2$$

the constants α and β depending on the method of support and plate dimensions. Typical values are listed later in Tables 7.3 and 7.4.

A. CIRCULAR PLATES

7.1. Stresses

Consider the portion of a thin plate or diaphragm shown in Fig. 7.1 bent to a radius R_{XY} in the XY plane and R_{YZ} in the YZ plane. The relationship between stresses and strains in a three-dimensional strain system is given by eqn. (7.2),[†]

$$\varepsilon_x = \frac{1}{E}[\sigma_x - \nu\sigma_y - \nu\sigma_z]$$

$$\varepsilon_z = \frac{1}{E}[\sigma_z - \nu\sigma_x - \nu\sigma_y]$$

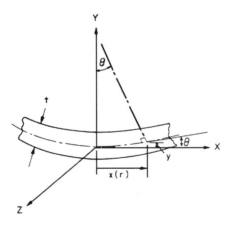

Fig. 7.1.

Now for thin plates, provided deflections are restricted to no greater than half the plate thickness,[‡] the direct stress in the Y direction may be assumed to be zero and the above equations give

$$\sigma_x = \frac{E}{(1 - \nu^2)}[\varepsilon_x + \nu\varepsilon_z] \tag{7.1}$$

$$\sigma_z = \frac{E}{(1 - \nu^2)}[\varepsilon_z + \nu\varepsilon_x] \tag{7.2}$$

[†] E.J. Hearn, *Mechanics of Materials 1*, Butterworth-Heinemann, 1997.
[‡] S. Timoshenko, *Theory of Plates and Shells*, 2nd edn., McGraw-Hill, 1959.

If u is the distance of any fibre from the neutral axis, then, for pure bending in the XY and YZ planes,

$$\frac{M}{I} = \frac{\sigma}{y} = \frac{E}{R} \quad \text{and} \quad \frac{\sigma}{E} = \frac{u}{R} = \varepsilon$$

$$\therefore \qquad \varepsilon_x = \frac{u}{R_{XY}} \quad \text{and} \quad \varepsilon_z = \frac{u}{R_{YZ}}$$

Now $\dfrac{1}{R} = \dfrac{d^2 y}{dx^2}$ and, for small deflections, $\dfrac{du}{dx} = \tan\theta = \theta$ (radians).

$$\therefore \qquad \frac{1}{R_{XY}} = \frac{d^2 y}{dx^2} = \frac{d\theta}{dx}$$

and

$$\varepsilon_x = u\frac{d\theta}{dx} \quad (= \text{radial strain}) \tag{7.3}$$

Consider now the diagram Fig. 7.2 in which the radii of the concentric circles through C_1 and D_1 on the unloaded plate increase to $[(x + dx) + (\theta + d\theta)u]$ and $[x + u\theta]$, respectively, when the plate is loaded.

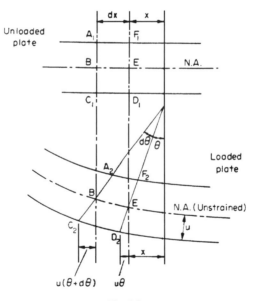

Fig. 7.2.

Circumferential strain at D_2

$$= \varepsilon_z = \frac{2\pi(x + u\theta) - 2\pi x}{2\pi x}$$

$$= \frac{u\theta}{x} \quad (= \text{circumferential strain}) \tag{7.4}$$

Substituting eqns. (7.3) and (7.4) in eqns. (7.1) and (7.2) yields

$$\sigma_x = \frac{E}{(1-v^2)}\left[u\frac{d\theta}{dx}+v\frac{u\theta}{x}\right]$$

i.e.

$$\sigma_x = \frac{Eu}{(1-v^2)}\left[\frac{d\theta}{dx}+v\frac{\theta}{x}\right] \tag{7.5}$$

Similarly,

$$\sigma_z = \frac{Eu}{(1-v^2)}\left[\frac{\theta}{x}+v\frac{d\theta}{dx}\right] \tag{7.6}$$

Thus we have equations for the stresses in terms of the slope θ and rate of change of slope $d\theta/dx$. We shall now proceed to evaluate the bending moments in the two planes in similar form and hence to the procedure for determination of θ and $d\theta/dx$ from a knowledge of the applied loading.

7.2. Bending moments

Consider the small section of plate shown in Fig. 7.3, which is of unit length. Defining the moments M as *moments per unit length* and applying the simple bending theory,

$$M = \frac{\sigma I}{y} = \frac{\sigma}{u}\left[\frac{1\times t^3}{12}\right] = \frac{\sigma t^3}{12u}$$

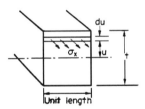

Fig. 7.3.

Substituting eqns. (7.5) and (7.6),

$$M_{XY} = \frac{Et^3}{12(1-v^2)}\left[\frac{d\theta}{dx}+v\frac{\theta}{x}\right] \tag{7.7}$$

Now $\dfrac{Et^3}{12(1-v^2)} = D$ is a constant and termed the *flexural stiffness*

so that

$$M_{XY} = D\left[\frac{d\theta}{dx}+v\frac{\theta}{x}\right] \tag{7.8}$$

and, similarly,

$$M_{YZ} = D\left[\frac{\theta}{x}+v\frac{d\theta}{dx}\right] \tag{7.9}$$

It is now possible to write the stress equations in terms of the applied moments,

i.e.
$$\sigma_x = M_{XY} \frac{12u}{t^3} \tag{7.10}$$

$$\sigma_z = M_{YZ} \frac{12u}{t^3} \tag{7.11}$$

7.3. General equation for slope and deflection

Consider now Fig. 7.4 which shows the forces and moments *per unit length* acting on a small element of the plate subtending an angle $\delta\phi$ at the centre. Thus M_{XY} and M_{YZ} are the moments per unit length in the two planes as described above and Q is the shearing force per unit length in the direction OY.

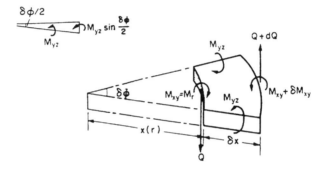

Fig. 7.4. Small element of circular plate showing applied moments and forces per unit length.

For equilibrium of moments in the radial XY plane, taking moments about the outside edge,

$$(M_{XY} + \delta M_{XY})(x + \delta x)\delta\phi - M_{XY}x\delta\phi - 2M_{YZ}\delta x \sin \tfrac{1}{2}\delta\phi + Qx\delta\phi\delta x = 0$$

which, neglecting squares of small quantities, reduces to

$$M_{XY}\delta x + \delta M_{XY}x - M_{YZ}\delta x + Qx\delta x = 0$$

In the limit, therefore,

$$M_{XY} + x\frac{dM_{XY}}{dx} - M_{YZ} = -Qx$$

Substituting eqns. (7.8) and (7.9), and simplifying

$$\frac{d^2\theta}{dx^2} + \frac{1}{x}\frac{d\theta}{dx} - \frac{\theta}{x^2} = -\frac{Q}{D}$$

This may be re-written in the form

$$\frac{d}{dx}\left[\frac{1}{x}\frac{d(x\theta)}{dx}\right] = -\frac{Q}{D} \tag{7.12}$$

This is then the general equation for slopes and deflections of circular plates or diaphragms. Provided that the applied loading Q is known as a function of x the expression can be treated in a similar manner to the equation

$$M = EI\frac{d^2 y}{dx^2}$$

used in the Macaulay beam method, i.e. it may be successively integrated to determine θ, and hence y, in terms of constants of integration, and these can then be evaluated from known end conditions of the plate.

It will be noted that the expressions have been derived using cartesian coordinates (X, Y and Z). For circular plates, however, it is convenient to replace the variable x with the general radius r when the equations derived above may be re-written as follows:

$$\frac{d}{dr}\left[\frac{1}{r}\frac{d}{dr}\left(r\frac{dy}{dr}\right)\right] = -\frac{Q}{D} \tag{7.13}$$

radial stress

$$\sigma_r = \frac{Eu}{(1 - v^2)}\left[\frac{d\theta}{dr} + v\frac{\theta}{r}\right] \tag{7.14}$$

tangential stress

$$\sigma_z = \frac{Eu}{(1 - v^2)}\left[v\frac{d\theta}{dr} + \frac{\theta}{r}\right] \tag{7.15}$$

moments

$$M_r = D\left[\frac{d\theta}{dr} + v\frac{\theta}{r}\right] \tag{7.16}$$

$$M_z = D\left[v\frac{d\theta}{dr} + \frac{\theta}{r}\right] \tag{7.17}$$

In the case of applied uniformly distributed loads, i.e. pressures q, the effective shear load Q per unit length for use in eqn. (7.13) is found as follows.

At any radius r, for equilibrium,

$$Q \times 2\pi r = q \times \pi r^2$$

i.e.
$$Q = \frac{qr}{2}$$

Thus *for applied pressures* eqn. (7.13) may be re-written

$$\frac{d}{dr}\left[\frac{1}{r}\frac{d}{dr}\left(r\frac{dy}{dr}\right)\right] = -\frac{qr}{2D} \tag{7.18}$$

7.4. General case of a circular plate or diaphragm subjected to combined uniformly distributed load q (pressure) and central concentrated load F

For this general case the equivalent shear Q per unit length is given by

$$Q \times 2\pi r = q \times \pi r^2 + F$$

$$\therefore \qquad Q = \frac{qr}{2} + \frac{F}{2\pi r}$$

Substituting in eqn. (7.18)

$$\frac{d}{dr}\left[\frac{1}{r}\frac{d}{dr}\left(r\frac{dy}{dr}\right)\right] = \left[-\frac{qr}{2} - \frac{F}{2\pi r}\right]\frac{1}{D}$$

Integrating,

$$\frac{1}{r}\frac{d}{dr}\left(r\frac{dy}{dr}\right) = -\frac{1}{D}\int\left[\frac{qr}{2} + \frac{F}{2\pi r}\right]dr$$

$$= -\frac{1}{D}\left[\frac{qr^2}{4} + \frac{F}{2\pi}\log_e r\right] + C_1$$

$$\therefore \qquad \frac{d}{dr}\left(r\frac{dy}{dr}\right) = -\frac{1}{D}\left[\frac{qr^3}{4} + \frac{Fr}{2\pi}\log_e r\right] + C_1 r$$

Integrating,

$$r\frac{dy}{dr} = -\frac{1}{D}\left[\frac{qr^4}{16} + \frac{F}{2\pi}\left\{\frac{r^2}{2}\log_e r - \frac{r^2}{4}\right\}\right] + \frac{C_1 r^2}{2} + C_2$$

$$\therefore \qquad \text{slope } \theta = \frac{dy}{dr} = -\frac{qr^3}{16D} - \frac{Fr}{8\pi D}[2\log_e r - 1] + C_1\frac{r}{2} + \frac{C_2}{r} \qquad (7.19)$$

Integrating again and simplifying,

$$\text{deflection } y = -\frac{qr^4}{64D} - \frac{Fr^2}{8\pi D}[\log_e r - 1] + C_1\frac{r^2}{4} + C_2\log_e r + C_3 \qquad (7.20)$$

The values of the constants of integration will be determined from known end conditions of the plate; slopes and deflections at any radius can then be evaluated. As an example of the procedure used it is now convenient to consider a number of standard loading cases and to determine the maximum deflections and stresses for each.

7.5. Uniformly loaded circular plate with edges clamped

The relevant fundamental equation for this loading condition has been shown to be

$$\frac{d}{dr}\left[\frac{1}{r}\frac{d}{dr}\left(r\frac{dy}{dr}\right)\right] = -\frac{qr}{2D}$$

Integrating,

$$\frac{1}{r}\frac{d}{dr}\left(r\frac{dy}{dr}\right) = -\frac{qr^2}{4D} + C_1$$

$$\frac{d}{dr}\left(r\frac{dy}{dr}\right) = -\frac{qr^3}{4D} + C_1 r$$

Integrating,

$$r\frac{dy}{dr} = -\frac{qr^4}{16D} + C_1\frac{r^2}{2} + C_2$$

$$\therefore \qquad \text{slope } \theta = \frac{dy}{dr} = -\frac{qr^3}{16D} + C_1\frac{r}{2} + \frac{C_2}{r} \qquad (7.21)$$

Integrating,

$$\text{deflection } y = \frac{-qr^4}{64D} + \frac{C_1 r^2}{4} + C_2 \log_e r + C_3 \tag{7.22}$$

Now if the slope θ is not to be infinite at the centre of the plate, $C_2 = 0$.
Taking the origin at the centre of the deflected plate, $y = 0$ when $r = 0$.
Therefore, from eqn. (7.22), $C_3 = 0$.
At the outside, clamped edge where $r = R$, $\theta = dy/dr = 0$.
Therefore substituting in the slope eqn. (6.21),

$$-\frac{qR^3}{16D} + \frac{C_1 R}{2} = 0$$

$$\therefore \qquad C_1 = \frac{qR^2}{8D}$$

The maximum deflection of the plate will be at the centre, but since this has been used as the origin the deflection equation will yield $y = 0$ at $r = 0$; indeed, this was one of the conditions used to evaluate the constants. We must therefore determine the equivalent amount by which the end supports are assumed to move up relative to the "fixed" centre.
Substituting $r = R$ in the deflection eqn. (7.22) yields

$$\text{maximum deflection} = -\frac{qR^4}{64D} + \frac{qR^4}{32D} = \frac{qR^4}{64D}$$

The positive value indicates, as usual, upwards deflection of the ends relative to the centre, i.e. along the positive y direction. The central deflection of the plate is thus, as expected, in the same direction as the loading, along the negative y direction (downwards).
Substituting for D,

$$y_{\text{max}} = \frac{qR^4}{64} \left[\frac{12(1 - v^2)}{Et^3} \right]$$

$$= \frac{3qR^4}{16Et^3}(1 - v^2) \tag{7.23}$$

Similarly, from eqn. (7.21),

$$\text{slope } \theta = -\frac{qr^3}{16D} + \frac{qR^2 r}{16D} = -\frac{qr}{16D}[r^2 - R^2]$$

$$\therefore \qquad \frac{d\theta}{dr} = -\frac{3qr^2}{16D} + \frac{qR^2}{16D} = -\frac{q}{16D}[3r^2 - R^2]$$

Now, from eqn. (7.14)

$$\sigma_r = \frac{Eu}{(1 - v^2)} \left[\frac{d\theta}{dr} + v\frac{\theta}{r} \right]$$

$$= \frac{Eu}{(1 - v^2)} \left[-\frac{qr^2}{16D}(3 + v) + \frac{qR^2}{16D}(1 + v) \right]$$

The maximum stress for the clamped edge condition will thus be obtained at the edge where $r = R$ and at the surface of the plate where $u = t/2$,

i.e.
$$\sigma_{r_{max}} = \frac{E}{(1 - v^2)} \frac{t}{2} \frac{2qR^2}{16D} = \frac{3qR^2}{4t^2}$$
(7.24)

N.B.–It is not possible to determine the maximum stress by equating $d\sigma_r/dr$ to zero since this only gives the point where the slope of the σ_r curve is zero (see Fig. 7.7). The value of the stress at this point is not as great as the value at the edge.

Similarly,

$$\sigma_z = \frac{Eu}{(1 - v^2)} \left[\frac{\theta}{r} + v \frac{d\theta}{dr} \right]$$

$$= \frac{Eu}{(1 - v^2)} \left[-\frac{qr^2}{16D}(3v + 1) + \frac{qR^2}{16D}(1 + v) \right]$$

Unlike σ_r, this has a maximum value when $r = 0$, i.e. at the centre.

$$\sigma_{z_{max}} = \frac{E}{(1 - v^2)} \frac{t}{2} \frac{qR^2}{16D}(1 + v)$$

$$= \frac{3qR^2}{8t^2}(1 + v)$$
(7.25)

7.6. Uniformly loaded circular plate with edges freely supported

Since the loading, and hence fundamental equation, is the same as for §7.4, the slope and deflection equations will be of the same form, i.e. eqns (7.21) and (7.22) will apply. Further, the constants C_2 and C_3 will again be zero for the same reasons as before and only one new condition to solve for the constant C_1 is required.

Here we must make use of the fact that the bending moment is always zero at any free support,

i.e. at $r = R$. $M_r = 0$

Therefore from eqn. (7.16),

$$D \left[\frac{d\theta}{dr} + v\frac{\theta}{r} \right] = 0$$

∴
$$\frac{d\theta}{dr} = -v\frac{\theta}{r}$$

Substituting from eqn. (7.21) with $r = R$ and $C_2 = 0$,

$$-\frac{3qR^2}{16D} + \frac{C_1}{2} = -v \left[-\frac{qR^2}{16D} + \frac{C_1}{2} \right]$$

∴
$$C_1 = \frac{qR^2}{8D} \left[\frac{(3 + v)}{(1 + v)} \right]$$

The maximum deflection is at the centre and again equal to the deflection of the supports relative to the centre.

Substituting for the constants with $r = R$ in eqn. (7.22),

$$\text{maximum deflection} = -\frac{qR^4}{64D} + \frac{qR^2}{8D}\frac{(3+v)}{(1+v)}\frac{R^2}{4}$$

$$= \frac{qR^4}{64D}\left[\frac{(5+v)}{(1+v)}\right]$$

i.e. substituting for D,

$$y_{max} = \frac{3qR^4}{16Et^3}(5+v)(1-v) \tag{7.26}$$

With $v = 0.3$ this value is approximately *four times* that for the clamped edge condition.

As before, the stresses are obtained from eqns. (7.14) and (7.15) by substituting for $d\theta/dr$ and θ/r from eqn. (7.21),

$$\sigma_r = \frac{Eu}{(1-v^2)}\left[-\frac{qr^2}{16D}(3+v) + \frac{qR^2}{16D}(3+v)\right]$$

This gives a maximum stress at the centre where $r = 0$

$$\sigma_{r_{max}} = \frac{E}{(1-v^2)}\frac{t}{2}\frac{qR^2}{16D}(3+v)$$

$$= \frac{3qR^2}{8t^2}(3+v)$$

Similarly, $\sigma_{z_{max}} = \dfrac{3qR^2}{8t^2}(3+v)$ also at the centre

i.e. for a uniformly loaded circular plate with edges freely supported,

$$\sigma_{r_{max}} = \sigma_{z_{max}} = \frac{3qR^2}{8t^2}(3+v) \tag{7.27}$$

7.7. Circular plate with central concentrated load F and edges clamped

For a central concentrated load,

$$Q \times 2\pi r = F$$

$$\therefore \qquad Q = \frac{F}{2\pi r}$$

The fundamental equation for slope and deflection is, therefore,

$$\frac{d}{dr}\left[\frac{1}{r}\frac{d}{dr}\left(r\frac{dy}{dr}\right)\right] = -\frac{F}{2\pi r D}$$

Integrating, $\dfrac{1}{r}\dfrac{d}{dr}\left(r\dfrac{dy}{dr}\right) = -\dfrac{F}{2\pi D}\log_e r + C_1$

$$\frac{d}{dr}\left(r\frac{dy}{dr}\right) = -\frac{Fr}{2\pi D}\log_e r + C_1 r$$

Integrating,

$$r\frac{dy}{dr} = -\frac{F}{2\pi D}\left[\frac{r^2}{2}\log_e r - \frac{r^2}{4}\right] + \frac{C_1 r^2}{2} + C_2$$

$$\therefore \qquad \theta = \frac{dy}{dr} = -\frac{F}{2\pi D}\left[\frac{r}{2}\log_e r - \frac{r}{4}\right] + C_1\frac{r}{2} + \frac{C_2}{r} \qquad (7.28)$$

Integrating,

$$y = -\frac{Fr^2}{8\pi D}\left[\log_e r - 1\right] + \frac{C_1 r^2}{4} + C_2\log_e r + C_3 \qquad (7.29)$$

Again, taking the origin at the centre of the deflected plate as shown in Fig. 7.5, the following conditions apply:

For a non-infinite slope at the centre $C_2 = 0$ and at $r = 0$, $y = 0$, $\therefore$ $C_3 = 0$.

Also, at $r = R$, slope $\theta = dy/dr = 0$.

Therefore from eqn. (7.28),

$$\frac{F}{2\pi D}\left[\frac{R}{2}\log_e R - \frac{R}{4}\right] = \frac{C_1 R}{2}$$

$$\therefore \qquad \frac{F}{\pi D}\left[\frac{\log_e R}{2} - \frac{1}{4}\right] = C_1$$

The maximum deflection will be at the centre and again equivalent to that obtained when $r = R$, i.e. from eqn. (7.29),

$$\text{maximum deflection} = -\frac{FR^2}{8\pi D}[\log_e R - 1] + \frac{FR^2}{4\pi D}\left[\frac{\log_e R}{2} - \frac{1}{4}\right]$$

$$= \frac{FR^2}{16\pi D}[-2\log_e R + 2 + 2\log_e R - 1]$$

$$= \frac{FR^2}{16\pi D}$$

Substituting for D,

$$y_{max} = \frac{FR^2}{16\pi}\frac{12(1 - v^2)}{Et^3}$$

$$= \frac{3FR^2}{4\pi Et^3}(1 - v^2) \qquad (7.30)$$

Again substituting for $d\theta/dr$ and θ/r from eqn. (7.28) into eqns (7.14) and (7.15) yields

$$\sigma_{r_{max}} = \frac{3F}{2\pi t^2} \qquad (7.31)$$

$$\sigma_{z_{max}} = \frac{3vF}{2\pi t^2} \qquad (7.32)$$

7.8. Circular plate with central concentrated load F and edges freely supported

The fundamental equation and hence the slope and deflection expressions will be as for the previous section (§7.7),

i.e.
$$\theta = \frac{dy}{dr} = -\frac{F}{2\pi D}\left[\frac{r}{2}\log_e r - \frac{r}{4}\right] + \frac{C_1 r}{2} \tag{7.33}$$

$$y = -\frac{Fr^2}{8\pi D}[\log_e r - 1] + \frac{C_1 r^2}{4} \tag{7.34}$$

constants C_2 and C_3 being zero as before.

As for the uniformly loaded plate with freely supported edges, the constant C_1 is determined from the knowledge that the bending moment M_r is zero at the free support,

i.e. at $r = R$,
$$M_r = 0$$

Therefore from eqn. (7.16),

$$D\left[\frac{d\theta}{dr} + v\frac{\theta}{r}\right] = 0 \quad \text{and} \quad \frac{d\theta}{dr} = -v\frac{\theta}{r}$$

and, substituting from eqn. (7.33) with $r = R$,

$$-\frac{F}{8\pi D}[2\log_e R - 1] + \frac{C_1}{2} = -\frac{vF}{8\pi D}[2\log_e R - 1] - \frac{vC_1}{2}$$

$$\therefore \qquad \frac{C_1}{2}(1 + v) = \frac{F}{8\pi D}[2(1 + v)\log_e R - (1 - v)]$$

$$C_1 = \frac{F}{4\pi D}\left[2\log_e R + \frac{(1 - v)}{(1 + v)}\right]$$

As before, the maximum deflection is at the centre and equivalent to that obtained with $r = R$. Substituting in eqn. (7.34),

$$\text{maximum deflection} = \frac{FR^2}{8\pi D}[\log_e R - 1] + \frac{FR^2}{16\pi D}\left[2\log_e R + \frac{(1 - v)}{(1 + v)}\right]$$

$$= \frac{FR^2}{16\pi D}\frac{(3 + v)}{(1 + v)}$$

Substituting for D

$$y_{max} = \frac{3FR^2}{4\pi Et^3}(3 + v)(1 - v) \tag{7.35}$$

For $v = 0.3$ this is approximately 2.5 times that for the clamped edge condition.

From eqn. (7.14),

$$\sigma_r = \frac{Eu}{(1 - v^2)}\left[\frac{d\theta}{dr} + v\frac{\theta}{r}\right]$$

Substituting for $d\theta/dr$ and θ/r as above,

$$\sigma_r = \frac{Eu}{(1 - v^2)}\left[\frac{F}{4\pi D}(1 + v)\log_e \frac{R}{r}\right]$$

$$= \frac{3F}{2\pi t^2}(1 + v)\log_e \frac{R}{r} \tag{7.36}$$

Thus the radial stress σ_r will be zero at the edge and will rise to a maximum value (theoretically infinite) at the centre. However, in practice, load cannot be applied strictly at a point but must contact over a finite area. Provided this area is known the maximum stress can be calculated.

Similarly, from eqn. (7.15)

$$\sigma_z = \frac{Eu}{(1 - v^2)}\left[v\frac{d\theta}{dr} + \frac{\theta}{r}\right]$$

and, again substituting for $d\theta/dr$ and θ/r,

$$\sigma_z = \frac{3F}{2\pi t^2}\left[(1 + v)\log_e \frac{R}{r} + (1 - v)\right] \tag{7.37}$$

7.9. Circular plate subjected to a load F distributed round a circle

Consider the circular plate of Fig. 7.5 subjected to a total load F distributed round a circle of radius R_1. A solution is obtained to this problem by considering the plate as consisting of two parts $r < R_1$ and $r > R_1$, bearing in mind that the values of θ, y and M_r must be the same for both parts at the common radius $r = R_1$.

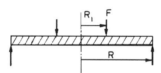

Fig. 7.5. Solid circular plate subjected to total load F distributed around a circle of radius R_1.

Thus, for $r < R_1$, we have a plate with zero distributed load and zero central concentrated load,

i.e.
$$q = F = 0$$

Therefore from eqn. (7.20),

$$y = \frac{C_1 r^2}{4} + C_2 \log_e r + C_3$$

and from eqn. (7.19)

$$\theta = \frac{dy}{dr} = \frac{C_1 r}{2} + \frac{C_2}{r}$$

For non-infinite slope at the centre, $C_2 = 0$ and with the axis for deflections at the centre of the plate, $y = 0$ when $r = 0$, $\therefore$ $C_3 = 0$.

Therefore for the inner portion of the plate

$$y = \frac{C_1 r^2}{4} \quad \text{and} \quad \theta = \frac{dy}{dr} = \frac{C_1 r}{2}$$

For the outer portion of the plate $r > R_1$ and eqn. (22.20) reduces to

$$y = -\frac{Fr^2}{8\pi D}[\log_e r - 1] + \frac{C_1' r^2}{4} + C_2' \log_e r + C_3' \tag{7.38}$$

and from eqn. (7.19)

$$\theta = \frac{dy}{dr} = -\frac{Fr}{8\pi D}[2\log_e r - 1] + \frac{C_1' r}{2} + \frac{C_2'}{r} \tag{7.39}$$

Equating these values with those obtained for the inner portions,

$$\frac{C_1 R_1^2}{4} = -\frac{FR_1^2}{8\pi D}[\log_e R_1 - 1] + \frac{C_1' R_1^2}{4} + C_2' \log_e R_1 + C_3'$$

and

$$\frac{C_1 R_1}{2} = -\frac{FR_1}{8\pi D}[2\log_e R_1 - 1] + \frac{C_1' R_1}{2} + \frac{C_2'}{R_1}$$

Similarly, from (7.16), equating the values of M_r at the common radius R_1 yields

$$-\frac{F}{8\pi D}[2(1 + v)\log_e R_1 + (1 - v)] + \frac{C_1'}{2}(1 + v) - \frac{C_2'(1 - v)}{R_1^2} = \frac{C_1}{2}(1 + v)$$

Further, with $M_r = 0$ at $r = R$, the outside edge, from eqn. (7.16)

$$-\frac{F}{8\pi D}[2(1 + v)\log_e R + (1 - v)] + \frac{C_1'}{2}(1 + v) - \frac{C_2'}{R^2}(1 - v) = 0 \tag{7.40}$$

There are thus four equations with four unknowns C_1, C_1', C_2' and C_3' and a solution using standard simultaneous equation procedures is possible. Such a solution yields the following values:

$$C_1' = \frac{F}{4\pi D}\left[2\log_e R + \frac{(1 - v)(R^2 - R_1^2)}{(1 + v)R^2}\right]$$

$$C_2' = -\frac{FR_1^2}{8\pi D}$$

$$C_3' = \frac{FR_1^2}{8\pi D}[\log_e R_1 - 1]$$

The central deflection is found, as before, from the deflection of the edge, $r = R$, relative to the centre.

Substituting in eqn. (7.38) yields

$$y_{max} = \frac{F}{8\pi D}\left[\frac{(3 + v)}{2(1 + v)}(R^2 - R_1^2) - R_1^2) - R_1^2 \log_e \frac{R}{R_1}\right] \tag{7.41}$$

The maximum radial bending moment and hence radial stress occurs at $r = R_1$, giving

$$\sigma_{r_{max}} = \frac{3F}{4\pi t^2}\left[2(1 + v)\log_e \frac{R}{R_1} + (1 - v)\frac{(R^2 - R_1^2)}{R^2}\right] \tag{7.42}$$

It can also be shown similarly that the maximum tangential stress is of equal value to the maximum radial stress.

7.10. Application to the loading of annular rings

The general eqns. (7.38) and (7.39) derived above apply also for annular rings with a total load F applied around the inner edge of radius R_1 as shown in Fig. 7.6.

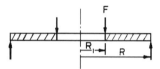

Fig. 7.6. Annular ring with total load F distributed around inner radius.

Here, however, the radial bending M_r is zero at both $r = R_1$ and $r = R$. Thus, applying the condition of eqn. (7.40) for both these radii yields

$$-\frac{F}{8\pi D}[2(1+v)\log_e R + (1-v)] + \frac{C_1}{2}(1+v) - \frac{C_2}{R^2}(1-v) = 0$$

and

$$-\frac{F}{8\pi D}[2(1+v)\log_e R_1 + (1-v)] + \frac{C_1}{2}(1+v) - \frac{C_2}{R_1^2}(1-v) = 0$$

Subtracting to eliminate C_1 gives

$$C_2 = \frac{F}{4\pi D}\left[\frac{(1+v)}{(1-v)}\frac{R^2 R_1^2}{(R^2 - R_1^2)}\log_e \frac{R}{R_1}\right]$$

and hence

$$C_1 = \frac{F}{4\pi D}\left[\frac{2(R^2 \log_e R - R_1^2 \log_e R_1)}{(R^2 - R_1^2)} + \frac{(1-v)}{(1+v)}\right]$$

It can then be shown that the maximum stress set up is the tangential stress at $r = R_1$ of value

$$\sigma_{z_{\max}} = \frac{3F(1+v)}{\pi t^2}\left[\frac{R^2}{(R^2 - R_1^2)}\right]\log_e \frac{R}{R_1} \tag{7.43}$$

If the **outside edge of the plate is clamped** instead of freely supported the maximum stress becomes

$$\sigma_{\max} = \frac{3F}{2\pi t^2}\left[\frac{(R^2 - R_1^2)}{R^2}\right]$$

7.11. Summary of end conditions

Axes can be selected to move with the plate as shown in Fig. 7.7(a) or stay at the initial, undeflected position Fig. 7.7(b).

For the former case, i.e. axes origin at the centre of the deflected plate, the end conditions which should be used for solution of the constants of integration are:

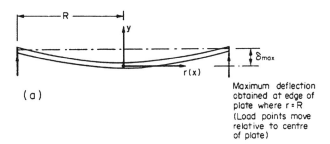

Maximum deflection
obtained at edge of
plate where r = R
(Load points move
relative to centre
of plate)

(a)

Fig. 7.7(a). Origin of reference axes taken to move with the plate.

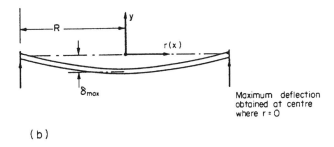

Maximum deflection
obtained at centre
where r = O

(b)

Fig. 7.7(b). Origin of reference axes remaining in the undeflected plate position.

Edges freely supported:
  (i) Slope θ and deflection y non-infinite at the centre. $\therefore C_2 = 0$.
 (ii) At $x = 0$, $y = 0$ giving $C_3 = 0$.
(iii) At $x = R$, $M_{xy} = 0$; hence C_1.

The maximum deflection is then that given at $x = R$.

Edges clamped:
  (i) Slope θ and deflection y non-infinite at the centre. $\therefore C_2 = 0$.
 (ii) At $x = 0$, $y = 0$ $\therefore C_3 = 0$.
(iii) At $x = R$, $\dfrac{dy}{dx} = 0$; hence C_1.

Again the maximum deflection is that given at $x = R$.

7.12. Stress distributions in circular plates and diaphragms subjected to lateral pressures

It is now convenient to consider the stress distribution in plates subjected to lateral, uniformly distributed loads or pressures in more detail since this represents the loading condition encountered most often in practice.

Figures 7.8(a) and 7.8(b) show the radial and tangential stress distributions on the lower surface of a thin plate subjected to uniform pressure as given by the equations obtained in §§7.5 and 7.6.

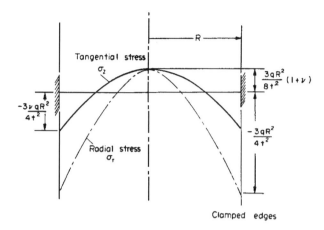

Fig. 7.8(a). Radial and tangential stress distributions in circular plates with clamped edges.

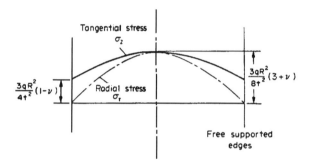

Fig. 7.8(b). Radial and tangential stress distributions in circular plates with freely supported edges.

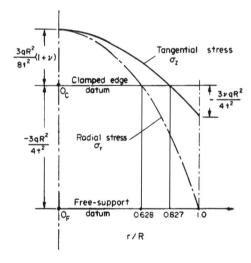

Fig. 7.9. Comparison of bending stresses in circular plates for clamped and freely supported edge conditions.

The two figures may be combined on to common axes as in Fig. 7.9 to facilitate comparison of the stress distributions for freely supported and clamped-edge conditions. Then if ordinates are measured from the horizontal axis through origin O_c, the curves give the values of radial and tangential stress for clamped-edge conditions.

Alternatively, measuring the ordinates from the horizontal axis passing through origin O_F in Fig. 7.9, i.e. adding to the clamped-edges stresses the constant value $\frac{3}{4}qR^2/t^2$, we obtain the stresses for a simply supported edge condition. The combined diagram clearly illustrates that a more favourable stress distribution is obtained when the edges of a plate are clamped.

7.13. Discussion of results – limitations of theory

The results of the preceding paragraphs are summarised in Table 7.1 at the start of the chapter. From this table the following approximate relationships are seen to apply:

(1) The maximum deflection of a uniformly loaded circular plate with freely supported edges is approximately four times that for the clamped-edge condition.

(2) Similarly, for a central concentrated load, the maximum deflection in the freely supported edge condition is 2.5 times that for clamped edges.

(3) With clamped edges the maximum deflection for a central concentrated load is four times that for the equivalent u.d.l. (i.e. $F = q \times \pi R^2$) and the maximum stresses are doubled.

(4) With freely supported edges, the maximum deflection for a central concentrated load is 2.5 times that for the equivalent u.d.l.

It must be remembered that the theory developed in this chapter has been based upon the assumption that deflections are small in comparison with the thickness of the plate. If deflections exceed half the plate thickness, then stretching of the middle surface of the plate must be considered. Under these conditions deflections are no longer proportional to the loads applied, e.g. for circular plates with clamped edges deflections δ can be determined from the equation

$$\delta + 0.58\frac{\delta^3}{t^2} = \frac{qR^4}{64D} \tag{7.44}$$

For very thin diaphragms or membranes subjected to uniform pressure, stresses due to stretching of the middle surface may far exceed those due to bending and under these conditions the central deflection is given by

$$y_{\max} = 0.0662\, R \left[\frac{qR}{Et}\right]^{1/3} \tag{7.45}$$

In the design of circular plates subjected to central concentrated loading, the maximum tensile stress on the lower surface of the plate is of prime interest since the often higher compressive stresses in the upper surface are generally much more localised. Local yielding of ductile materials in these regions will not generally affect the overall deformation of the plate provided that the lower surface tensile stresses are kept within safe limits. The situation is similar for plates constructed from brittle materials since their compressive strengths far exceed their strength in tension so that a limit on the latter is normally a safe design procedure. The theory covered in this text has involved certain simplifying assumptions; a full treatment of the problem shows that the limiting tensile stress is more accurately given by the equation

$$\sigma_{r_{max}} = \frac{F}{t^2}(1 + v)(0.485 \log_e R/t + 0.52)$$ (7.46)

7.14. Other loading cases of practical importance

In addition to the standard cases covered in the previous sections there are a number of other loading cases which are often encountered in practice; these are illustrated in Fig. 7.10[†]. The method of solution for such cases is introduced briefly below.[‡]

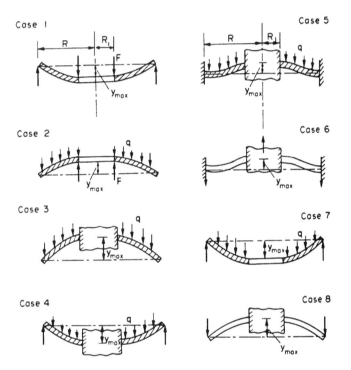

Fig. 7.10. Circular plates and diaphragms: various loading cases encountered in practice.

In all the cases illustrated the **maximum stress** is obtained from the following standard form of equations:
For *uniformly distributed loads*

$$\sigma_{max} = k_1 \frac{qR^2}{t^2}$$ (7.47)

For *loads concentrated around the edge of the central hole,*

$$\sigma_{max} = \frac{k_1 F}{t^2}$$ (7.48)

[†] S. Timoshenko, *Strength of Materials*, Part II, *Advanced Theory and Problems*, Van Nostrand.
[‡] A.M. Wahl and G. Lobo, *Trans. ASME* 52 (1929).

Similarly, the **maximum deflections** in each case are given by the following equations:
For *uniformly distributed loads*,

$$y_{\max} = k_2 \frac{qR^4}{Et^3} \tag{7.49}$$

For *loads concentrated around the central hole*,

$$y_{\max} = k_2 \frac{FR^2}{Et^3} \tag{7.50}$$

The values of the factors k_1 and k_2 for the loading cases of Fig. 7.10 are given in Table 7.2, assuming a Poisson's ratio ν of 0.3.

Table 7.2. Coefficients k_1 and k_2 for the eight cases shown in Fig. 7.10[a].

$\dfrac{R}{R_1}$	1.25		1.5		2		3		4		5	
Case	k_1	k_2	k_1	k_2	k_1	k_2	k_1	k_2	k_1	k_2	k_1	k_2
1	1.10	0.341	1.26	0.519	1.48	0.672	1.88	0.734	2.17	0.724	2.34	0.704
2	0.66	0.202	1.19	0.491	2.04	0.902	3.34	1.220	4.30	1.300	5.10	1.310
3	0.135	0.00231	0.410	0.0183	1.04	0.0938	2.15	0.293	2.99	0.448	3.69	0.564
4	0.122	0.00343	0.336	0.0313	0.74	0.1250	1.21	0.291	1.45	0.417	1.59	0.492
5	0.090	0.00077	0.273	0.0062	0.71	0.0329	1.54	0.110	2.23	0.179	2.80	0.234
6	0.115	0.00129	0.220	0.0064	0.405	0.0237	0.703	0.062	0.933	0.092	1.13	0.114
7	0.592	0.184	0.976	0.414	1.440	0.664	1.880	0.824	2.08	0.830	2.19	0.813
8	0.227	0.00510	0.428	0.0249	0.753	0.0877	1.205	0.209	1.514	0.293	1.745	0.350

[a] S. Timoshenko, *Strength of Materials*, Part II, *Advanced Theory and Problems*, Van Nostrand, p. 113.

B. BENDING OF RECTANGULAR PLATES

The theory of bending of rectangular plates is beyond the scope of this text and will not be introduced here. The standard formulae obtained from the theory,[†] however, may be presented in simple form and are relatively easy to apply. The results for the two most frequently used loading conditions are therefore summarised below.

7.15. Rectangular plates with simply supported edges carrying uniformly distributed loads

For a rectangular plate length d, shorter side b and thickness t, the *maximum deflection* is found to occur at the centre of the plate and given by

$$y_{\max} = \alpha \frac{qb^4}{Et^3} \tag{7.51}$$

the value of the factor α depending on the ratio d/b and given in Table 7.3.

[†] S. Timoshenko, *Theory of Plates and Shells*, 2nd edn., McGraw-Hill, New York, 1959.

Table 7.3. Constants for uniformly loaded rectangular plates with simply supported edges[a].

d/b	1.0	1.1	1.2	1.3	1.4	1.5	1.6	1.7
α	0.0443	0.0530	0.0616	0.0697	0.0770	0.0843	0.0906	0.0964
β_1	0.0479	0.0553	0.0626	0.0693	0.0753	0.0812	0.0862	0.0908
β_2	0.0479	0.0494	0.0501	0.0504	0.0506	0.0500	0.0493	0.0486
d/b	1.8	1.9	2.0	3.0	4.0	5.0	∞	
α	0.1017	0.1064	0.1106	0.1336	0.1400	0.1416	0.1422	
β_1	0.0948	0.0985	0.1017	0.1189	0.1235	0.1246	0.1250	
β_2	0.0479	0.0471	0.0464	0.0404	0.0384	0.0375	0.0375	

[a] S. Timoshenko, *Theory of Plates and Shells*, 2nd edn., McGraw-Hill, New York, 1959.

The *maximum bending moments*, per unit length, also occur at the centre of the plate and are given by

$$M_{XY_{max}} = \beta_1 q b^2 \qquad (7.52)$$

$$M_{YZ_{max}} = \beta_2 q b^2 \qquad (7.53)$$

the factors β_1 and β_2 being given in Table 7.4 for an assumed value of Poisson's ratio v equal to 0.3.

It will be observed that for length ratios d/b in excess of 3 the values of the factors α, β_1, and β_2 remain practically constant as also will the corresponding maximum deflections and bending moments.

7.16. Rectangular plates with clamped edges carrying uniformly distributed loads

Here again the *maximum deflection* takes place at the centre of the plate, the value being given by an equation of similar form to eqn. (7.51) for the simply-supported edge case but with different values of α,

i.e.
$$y_{max} = \alpha \frac{q b^4}{E t^3}$$

The *bending moment* equations are also similar in form, the *numerical maximum occurring at the middle of the longer side* and given by

$$M_{max} = \beta q b^2$$

Typical values for α and β are given in Table 7.4. In this case values are practically constant for $d/b > 2$.

Table 7.4. Constants for uniformly loaded rectangular plates with clamped edges.[a]

d/b	1.00	1.25	1.50	1.75	2.00	∞
α	0.0138	0.0199	0.0240	0.0264	0.0277	0.0284
β	0.0513	0.0665	0.0757	0.0806	0.0829	0.0833

[a] S. Timoshenko, *Theory of Plates and Shells*, 2nd edn., McGraw-Hill, New York, 1959.

It will be observed, by comparison of the values of the factors in Tables 7.3 and 7.4, that when the edges of a plate are clamped the maximum deflection is considerably reduced from the freely supported condition but the maximum bending moments, and hence maximum stresses, are not greatly affected.

Examples

Example 7.1

A circular flat plate of diameter 120 mm and thickness 10 mm is constructed from steel with $E = 208$ GN/m^2 and $v = 0.3$. The plate is subjected to a uniform pressure of 5 MN/m^2 on one side only. If the plate is clamped at the edges determine:

(a) the maximum deflection;
(b) the position and magnitude of the maximum radial stress.

What percentage change in the results will be obtained if the edge conditions are changed such that the plate can be assumed to be freely supported?

Solution

(a) From eqn. (7.23) the maximum deflection with clamped edges is given by

$$y_{max} = \frac{3qR^4}{16Et^3}(1 - v^2)$$

$$= \frac{3 \times 5 \times 10^6 \times (60 \times 10^{-3})^4(1 - 0.3^2)}{16 \times 208 \times 10^9 \times (10 \times 10^{-3})^3}$$

$$= 0.053 \times 10^{-3} = \mathbf{0.053 \ mm}$$

(b) From eqn. (7.24) the maximum radial stress occurs at the outside edge and is given by

$$\sigma_{r_{max}} = \frac{3qR^2}{4t^2}$$

$$= \frac{3 \times 5 \times 10^6 \times (60 \times 10^{-3})^2}{4 \times (10 \times 10^{-3})^2}$$

$$= 135 \times 10^6 = \mathbf{135 \ MN/m^2}$$

When the edges are freely supported, eqn. (7.26) gives

$$y'_{max} = \frac{3qR^4}{16Et^3}(5 + v)(1 - v)$$

$$= \frac{(5 + v)(1 - v)}{(1 - v^2)}y_{max}$$

$$= \frac{(5.3 \times 0.7)}{0.91} \times 0.053 = \mathbf{0.216 \ mm}$$

and eqn. (7.27) gives

$$\sigma'_{r_{max}} = \frac{3qR^2}{8t^2}(3+v)$$

$$\sigma'_{r_{max}} = \frac{(3+v)}{2}\sigma_{r_{max}}$$

$$= \frac{3.3}{2} \times 135 = \mathbf{223 \ MN/m^2}$$

Thus the percentage increase in maximum deflection

$$= \frac{(0.216 - 0.053)}{0.053}100 = \mathbf{308\%}$$

and the percentage increase in maximum radial stress

$$= \frac{(223 - 135)}{135}100 = \mathbf{65\%}$$

Example 7.2

A circular disc 150 mm diameter and 12 mm thickness is clamped around the periphery and built into a piston of diameter 60 mm at the centre. Assuming that the piston remains rigid, determine the maximum deflection of the disc when the piston carries a load of 5 kN. For the material of the disc $E = 208$ GN/m^2 and $v = 0.3$.

Solution

From eqn. (7.29) the deflection of the disc is given by

$$y = \frac{-Fr^2}{8\pi D}[\log_e r - 1] + \frac{C_1 r^2}{4} + C_2 \log_e r + C_3 \qquad (1)$$

and from eqn. (7.28)

$$\text{slope } \theta = \frac{-Fr}{8\pi D}[2\log_e r - 1] + \frac{C_1 r}{2} + \frac{C_2}{r} \qquad (2)$$

Now slope $= 0$ at $r = 0.03$ m.
Therefore from eqn. (2)

$$0 = \frac{-5000 \times 0.03}{8\pi D}[2\log_e 0.03 - 1] + 0.015C_1 + 33.3C_2$$

But
$$D = \frac{Et^3}{12(1-v^2)} = \frac{208 \times 10^9 \times (12 \times 10^{-3})^3}{12(1-0.09)}$$

$$= \frac{208 \times 1728}{12 \times 0.91} = 32900$$

$\therefore$
$$0 = \frac{-5000 \times 0.03}{8\pi \times 32900}[2(-3.5066) - 1] + 0.015C_1 + 33.3C_2$$

$\therefore$
$$-1.45 \times 10^{-3} = 0.015C_1 + 33.3C_2 \qquad (3)$$

Also the slope $= 0$ at $r = 0.075$.
Therefore from eqn. (2) again,

$$0 = \frac{-5000 \times 0.075}{8\pi \times 32900}[2\log_e 0.075 - 1] + 0.0375C_1 + \frac{C_2}{0.075}$$

$$= -4.54 \times 10^{-4}[2(-2.5903) - 1] + 0.0375C_1 + 13.33C_2$$

$$- 2.8 \times 10^{-3} = 0.0375C_1 + 13.33C_2 \qquad (4)$$

$(3) \times \dfrac{0.0375}{0.015}$,

$$-3.625 \times 10^{-3} = 0.0375C_1 + 83.25C_2 \qquad (5)$$

$(5) - (4)$,

$$-0.825 \times 10^{-3} = 69.92C_2$$

$$\therefore \qquad C_2 = -11.8 \times 10^{-6}$$

Substituting in (5),

$$-3.625 \times 10^{-3} = 0.0375C_1 - 9.82 \times 10^{-4}$$

$$C_1 = -\frac{(3.625 - 0.982)}{0.0375}10^{-3}$$

$$= -7.048 \times 10^{-2}$$

Now taking $y = 0$ at $r = 0.075$, from eqn. (1)

$$0 = \frac{-5000 \times (0.075)^2}{8\pi \times 32900}[\log_e 0.075 - 1] - \frac{7.048 \times 10^{-2}}{4}(0.075)^2$$

$$- 11.8 \times 10^{-6}\log_e 0.075 + C_3$$

$$= -3.4 \times 10^{-5}(-3.5903) - 99.1 \times 10^{-6} + 30.6 \times 10^{-6} + C_3$$

$$= 10^{-6}(122 - 99.1 + 30.6) + C_3$$

$$\therefore \qquad C_3 = -53.5 \times 10^{-6}$$

Therefore deflection at $r = 0.03$ is given by eqn. (1),

$$\delta_{max} = \frac{-5000 \times (0.03)^2}{8\pi \times 32900}[\log_e 0.03 - 1] - \frac{7.048 \times 10^{-2}}{4}(0.03)^2$$

$$- 11.8 \times 10^{-6}\log_e 0.03 - 53.5 \times 10^{-6}$$

$$= 10^{-6}[+24.5 - 15.9 + 41.4 - 53.5] = \mathbf{-3.5 \times 10^{-6}\ m}$$

Problems

In the following examples assume that

$$\frac{d}{dr}\left[\frac{1}{r}\frac{d}{dr}\left(r\frac{dy}{dr}\right)\right] = -\frac{Q}{D} \quad \text{or} \quad = -\frac{qr}{2D}$$

with conventional notations.

Unless otherwise stated, $E = 207$ GN/m^2 and $\nu = 0.3$.

7.1 (B/C). A circular flat plate of 120 mm diameter and 6.35 mm thickness is clamped at the edges and subjected to a uniform lateral pressure of 345 kN/m^2. Evaluate (a) the central deflection, (b) the position and magnitude of the maximum radial stress. [1.45×10^{-5} m, 23.1 MN/m^2; $r = 60$ mm.]

7.2 (B/C). The plate of Problem 7.1 is subjected to the same load but is simply supported round the edges. Calculate the central deflection. [58×10^{-6} m.]

7.3 (B/C). An aluminium plate diaphragm is 500 mm diameter and 6 mm thick. It is clamped around its periphery and subjected to a uniform pressure q of 70 kN/m^2. Calculate the values of maximum bending stress and deflection.

Take $Q = qR/2$, $E = 70$ GN/m^2 and $\nu = 0.3$. [91, 59.1 MN/m^2; 3.1 mm.]

7.4 (B/C). A circular disc of uniform thickness 1.5 mm and diameter 150 mm is clamped around the periphery and built into a piston, diameter 50 mm, at the centre. The piston may be assumed rigid and carries a central load of 450 N. Determine the maximum deflection. [0.21 mm.]

7.5 (C). A circular steel plate 5 mm thick, outside diameter 120 mm, inside diameter 30 mm, is clamped at its outer edge and loaded by a ring of edge moments $M_r = 8$ kN/m of circumference at its inner edge. Calculate the deflection at the inside edge. [4.68 mm.]

7.6 (C). A solid circular steel plate 5 mm thick, 120 mm outside diameter, is clamped at its outer edge and loaded by a ring of loads at $r = 20$ mm. The total load on the plate is 10 kN. Calculate the central deflection of the plate. [0.195 mm.]

7.7 (C). A pressure vessel is fitted with a circular manhole 600 mm diameter, the cover of which is 25 mm thick. If the edges are clamped, determine the maximum allowable pressure, given that the maximum principal strain in the cover plate must not exceed that produced by a simple direct stress of 140 MN/m^2. [1.19 MN/m^2.]

7.8 (B/C). The crown of a gas engine piston may be treated as a cast-iron diaphragm 300 mm diameter and 10 mm thick, clamped at its edges. If the gas pressure is 3 MN/m^2, determine the maximum principal stresses and the central deflection.

$\nu = 0.3$ and $E = 100$ GN/m^2. [506, 329 MN/m^2; 2.59 mm.]

7.9 (B/C). How would the values for Problem 7.8 change if the edges are released from clamping and freely supported? [835,835 MN/m^2; 10.6 mm.]

7.10 (B/C). A circular flat plate of diameter 305 mm and thickness 6.35 mm is clamped at the edges and subjected to a uniform lateral pressure of 345 kN/m^2.

Evaluate: (a) the central deflection, (b) the position and magnitude of the maximum radial stress. [6.1×10^{-4} m; 149.2 MN/m^2.]

7.11 (B/C). The plate in Problem 7.10 is subjected to the same load, but simply supported round the edges. Evaluate the central deflection. [24.7×10^{-4} m.]

7.12 (B/C). The flat end-plate of a 2 m diameter container can be regarded as clamped around its edge. Under operating conditions the plate will be subjected to a uniformly distributed pressure of 0.02 MN/m^2. Calculate from first principles the required thickness of the end plate if the bending stress in the plate should not exceed 150 MN/m^2. For the plate material $E = 200$ GN/m^2 and $\nu = 0.3$. [C.E.I.] [10 mm.]

7.13 (C). A cylinder head valve of diameter 38 mm is subjected to a gas pressure of 1.4 MN/m^2. It may be regarded as a uniform thin circular plate simply supported around the periphery. Assuming that the valve stem applies a concentrated force at the centre of the plate, calculate the movement of the stem necessary to lift the valve from its seat. The flexural rigidity of the valve is 260 Nm and Poisson's ratio for the material is 0.3. [C.E.I.] [0.067 mm.]

7.14 (C). A diaphragm of light alloy is 200 mm diameter, 2 mm thick and firmly clamped around its periphery before and after loading. Calculate the maximum deflection of the diaphragm due to the application of a uniform pressure of 20 kN/m^2 normal to the surface of the plate.

Determine also the value of the maximum radial stress set up in the material of the diaphragm.

Assume $E = 70$ GN/m^2 and Poisson's ratio $v = 0.3$. [B.P.] [0.61 mm; 37.5 MN/m^2.]

7.15 (C). A thin plate of light alloy and 200 mm diameter is firmly clamped around its periphery. Under service conditions the plate is to be subjected to a uniform pressure p of 20 kN/m^2 acting normally over its whole surface area.

Determine the required minimum thickness t of the plate if the following design criteria apply;

(a) the maximum deflection is not to exceed 6 mm;

(b) the maximum radial stress is not to exceed 50 MN/m^2.

Take $E = 70$ GN/m^2 and $v = 0.3$. [B.P.] [1.732 mm.]

7.16 (C). Determine equations for the maximum deflection and maximum radial stress for a circular plate, radius R, subjected to a distributed pressure of the form $q = K/r$. Assume simply supported edge conditions:

$$\left[\delta_{max} = \frac{-KR^3(4+v)}{36D(1+v)}, \sigma_{max} = \frac{EtRK(2+v)}{12D(1-v^2)} \right]$$

7.17 (C). The cover of the access hole for a large steel pressure vessel may be considered as a circular plate of 500 mm diameter which is firmly clamped around its periphery. Under service conditions the vessel operates with an internal pressure of 0.65 MN/m^2.

Determine the minimum thickness of plate required in order to achieve the following design criteria:

(a) the maximum deflection is limited to 5 mm;

(b) the maximum radial stress is limited to 200 MN/m^2.

For the steel, $E = 208$ GN/m^2 and $v = 0.3$.

You may commence your solution on the assumption that the deflection y at radius r for a uniform circular plate under the action of a uniform pressure q is given by:

$$\frac{d}{dr}\left[\frac{1}{r} \cdot \frac{d}{dr}\left(r \cdot \frac{dy}{dr} \right) \right] = -\frac{qr}{2D}$$

where D is the "flexural stiffness" of the plate. [9.95 mm.]

7.18 (C). A circular plate, 300 mm diameter and 5 mm thick, is built-in at its periphery. In order to strengthen the plate against a concentrated central axial load P the plate is stiffened by radial ribs and a prototype is found to have a stiffness of 11300 N per mm central deflection.

(a) Check that the equation:

$$y = \frac{Pr^2}{8\pi D}\left[l_n\left(\frac{r}{R_2} \right) - \frac{1}{2} \right] + \frac{PR_2^2}{16\pi D}$$

satisfies the boundary conditions for the *unstiffened* plate.

(b) Hence determine the stiffness of the plate without the ribs in terms of central deflection and calculate the relative stiffening effect of the ribs.

(c) What additional thickness would be required for an unstiffened plate to produce the same effect? For the plate material $E = 200$ GN/m^2 and $v = 0.28$. [5050 N/mm; 124%; 1.54 mm.]

CHAPTER 8

INTRODUCTION TO ADVANCED ELASTICITY THEORY

8.1. Type of stress

Any element of material may be subjected to three independent types of stress. Two of these have been considered in detai previously, namely *direct stresses* and *shear stresses*, and need not be considered further here. The third type, however, has not been specifically mentioned previously although it has in fact been present in some of the loading cases considered in earlier chapters; these are the so-called *body-force stresses*. These body forces arise by virtue of the bulk of the material, typical examples being:

(a) gravitational force due to a component's own weight: this has particular significance in civil engineering applications, e.g. dam and chimney design;

(b) centrifugal force, depending on radius and speed of rotation, with particular significance in high-speed engine or turbine design;

(c) magnetic field forces.

In many practical engineering applications the only body force present is the gravitational one, and in the majority of cases its effect is minimal compared with the other applied forces due to mechanical loading. In such cases it is therefore normally neglected. In high-speed dynamic loading situations such as the instances quoted in (b) above, however, the centrifugal forces far exceed any other form of loading and are therefore the primary factor for consideration.

Unlike direct and shear stresses, body force stresses are defined as **force per unit volume**, and particular note must be taken of this definition in relation to the proofs of formulae which follow.

8.2. The cartesian stress components: notation and sign convention

Consider an element of material subjected to a complex stress system in three dimensions. Whatever the type of applied loading the resulting stresses can always be reduced to the nine components, i.e. three direct and six shear, shown in Fig. 8.1.

It will be observed that in this case a modified notation is used for the stresses. This is termed the double-suffix notation and it is particularly useful in the detailed study of stress problems since it indicates both the direction of the stress *and* the plane on which it acts.

The *first* suffix gives the *direction* of the stress.

The *second* suffix gives the *direction of the normal of the plane* on which the stress acts.

Thus, for example,

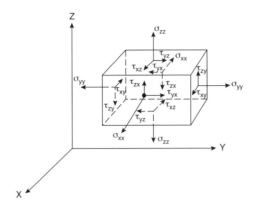

Fig. 8.1. The cartesian stress components.

σ_{xx} is the stress in the X direction on the X facing face (i.e. a *direct* stress). Common suffices therefore always indicate that the stress is a direct stress. Similarly, σ_{xy} is the stress in the X direction on the Y facing face (i.e. a *shear* stress). Mixed suffices always indicate the presence of shear stresses and thus allow the alternative symbols σ_{xy} or τ_{xy}. Indeed, the alternative symbol τ is not strictly necessary now since the suffices indicate whether the stress σ is a direct one or a shear.

8.2.1. Sign conventions

(a) *Direct stresses*. As always, direct stresses are assumed positive when tensile and negative when compressive.

(b) *Shear stresses*. Shear stresses are taken to be positive if they act in a positive cartesian (X, Y or Z) direction whilst acting on a plane whose outer normal points also in a positive cartesian direction.

Thus positive shear is assumed with $+$ direction and $+$ facing face.

Alternatively', positive shear is also given with $-$ direction *and* $-$ facing face (a double negative making a positive, as usual).

A careful study of Fig. 8.1 will now reveal that all stresses shown are positive in nature.

The *cartesian stress components* considered here relate to the three mutually perpendicular axes X, Y and Z. In certain loading cases, notably those involving axial symmetry, this system of components is inconvenient and an alternative set known as *cylindrical components* is used. These involve the variables, radius r, angle θ and axial distance z, and will be considered in detail later.

8.3. The state of stress at a point

Consider any point Q within a stressed material, the nine cartesian stress components at Q being known. It is now possible to determine the normal, direct and resultant stresses which act on any plane through Q whatever its inclination relative to the cartesian axes. Suppose one such plane ABC has a normal n which makes angles nx, ny and nz with the YZ, XZ and XY planes respectively as shown in Figs. 8.2 and 8.3. (Angles between planes

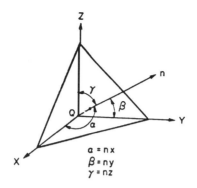

Fig. 8.2.

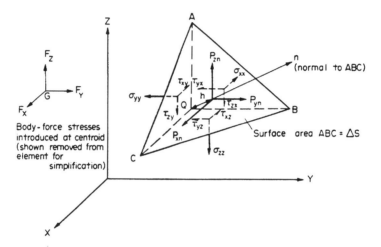

Fig. 8.3. The state of stress on an inclined plane through any given point in a three-dimensional cartesian stress system.

ABC and *YZ* are given by the angle between the normals to both planes *n* and *x*, etc.) For convenience, let the plane *ABC* initially be some perpendicular distance *h* from *Q* so that the cartesian stress components actually acting at *Q* can be shown on the sides of the tetrahdedron element *ABCQ* so formed (Fig. 8.3). In the derivation below the value of *h* will be reduced to zero so that the equations obtained will relate to the condition when *ABC* passes through *Q*.

In addition to the cartesian components, the unknown components of the stress on the plane *ABC*, i.e. p_{xn}, p_{yn} and p_{zn}, are also indicated, as are the body-force field stress components which act at the centre of gravity of the tetrahedron. (To improve clarity of the diagram they are shown displaced from the element.)

Since body-force stresses are defined as forces/unit volume, the components in the *X*, *Y* and *Z* directions are of the form

$$F \times \Delta S \frac{h}{3}$$

where $\Delta S h/3$ is the volume of the tetrahedron. If the area of the surface *ABC*, i.e. ΔS, is assumed small then all stresses can be taken to be uniform and the component of force in the X direction due to σ_{xx} is given by

$$\sigma_{xx}\Delta S \cos nx$$

Stress components in the other axial directions will be similar in form.
Thus, for equilibrium of forces in the X direction,

$$p_{xn}\Delta S + F_x\Delta S\frac{h}{3} = \sigma_{xx}\Delta S \cos nx + \tau_{xy}\Delta S \cos ny + \tau_{xz}\Delta S \cos nz$$

As $h \to 0$ (i.e. plane ABC passes through Q), the second term above becomes very small
and can be neglected. The above equation then reduces to

$$p_{xn} = \sigma_{xx} \cos nx + \tau_{xy} \cos ny + \tau_{xz} \cos nz \tag{8.1}$$

Similarly, for equilibrium of forces in the y and z directions,

$$p_{yn} = \sigma_{yy} \cos ny + \tau_{yx} \cos nx + \tau_{yz} \cos nz \tag{8.2}$$

$$p_{zn} = \sigma_{zz} \cos nz + \tau_{zx} \cos nx + \tau_{zy} \cos ny \tag{8.3}$$

The resultant stress p_n on the plane ABC is then given by

$$p_n = \sqrt{(p_{xn}^2 + p_{yn}^2 + p_{zn}^2)} \tag{8.4}$$

The normal stress σ_n is given by resolution perpendicular to the face ABC,

i.e. $$\sigma_n = p_{xn} \cos nx + p_{yn} \cos ny + p_{zn} \cos nz \tag{8.5}$$

and, by Pythagoras' theorem (Fig. 8.4), the shear stress τ_n is given by

$$\tau_n = \sqrt{(p_n^2 - \sigma_n^2)} \tag{8.6}$$

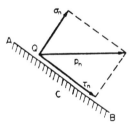

Fig. 8.4. Normal, shear and resultant stresses on the plane ABC.

It is often convenient and quicker to define the line of action of the resultant stress p_n by
the *direction cosines*

$$l' = \cos(p_n x) = p_{xn}/p_n \tag{8.7}$$

$$m' = \cos(p_n y) = p_{yn}/p_n \tag{8.8}$$

$$n' = \cos(p_n z) = p_{zn}/p_n \tag{8.9}$$

The direction of the plane *ABC* being given by other direction cosines

$$l = \cos nx, \quad m = \cos ny, \quad n = \cos ny$$

It can be shown by simple geometry that

$$l^2 + m^2 + n^2 = 1 \quad \text{and} \quad (l')^2 + (m')^2 + (n')^2 = 1$$

Equations (8.1), (8.2) and (8.3) may now be written in two alternative ways.

(a) *Using the common symbol* σ for stress and relying on the double suffix notation to discriminate between shear and direct stresses:

$$p_{xn} = \sigma_{xx} \cos nx + \sigma_{xy} \cos ny + \sigma_{xz} \cos nz \tag{8.10}$$

$$p_{yn} = \sigma_{yx} \cos nx + \sigma_{yy} \cos ny + \sigma_{yz} \cos nz \tag{8.11}$$

$$p_{zn} = \sigma_{zx} \cos nx + \sigma_{zy} \cos ny + \sigma_{zz} \cos nz \tag{8.12}$$

In each of the above equations the first suffix is common throughout, the second suffix on the right-hand-side terms are in the order x, y, z throughout, and in each case the cosine term relates to the second suffix. These points should aid memorisation of the equations.

(b) *Using the direction cosine form*:

$$p_{xn} = \sigma_{xx}l + \sigma_{xy}m + \sigma_{xz}n \tag{8.13}$$

$$p_{yn} = \sigma_{yx}l + \sigma_{yy}m + \sigma_{yz}n \tag{8.14}$$

$$p_{zn} = \sigma_{zx}l + \sigma_{zy}m + \sigma_{zz}n \tag{8.15}$$

Memory is again aided by the notes above, but in this case it is the direction cosines, l, m and n which relate to the appropriate second suffices x, y and z.

Thus, provided that the direction cosines of a plane are known, together with the cartesian stress components at some point Q on the plane, the direct, normal and shear stresses on the plane at Q may be determined using, firstly, eqns. (8.13–15) and, subsequently, eqns. (8.4–6).

Alternatively the procedure may be carried out graphically as will be shown in §8.9.

8.4. Direct, shear and resultant stresses on an oblique plane

Consider again the oblique plane *ABC* having direction cosines l, m and n, i.e. these are the cosines of the angle between the normal to plane and the x, y, z directions.

In general, the resultant stress on the plane p_n will not be normal to the plane and it can therefore be resolved into two alternative sets of components.

(a) In the co-ordinate directions giving components p_{xn}, p_{yn} and p_{zn}, as shown in Fig. 8.5, with values given by eqns. (8.13), (8.14) and (8.15).

(b) Normal and tangential to the plane as shown in Fig. 8.6, giving components, of σ_n (normal or direct stress) and τ_n (shear stress) with values given by eqns. (8.5) and (8.6).

The value of the resultant stress can thus be obtained from either of the following equations:

$$p_n^2 = \sigma_n^2 + \tau_n^2 \tag{8.16}$$

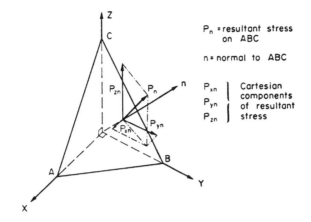

Fig. 8.5. Cartesian components of resultant stress on an inclined plane.

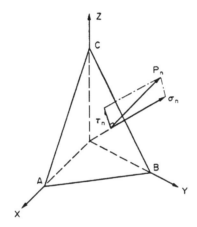

Fig. 8.6. Normal and tangential components of resultant stress on an inclined plane.

or
$$p_n^2 = p_{xn}^2 + p_{yn}^2 + p_{zn}^2 \tag{8.17}$$

these being alternative forms of eqns. (8.6) and (8.4) respectively.

From eqn. (8.5) the normal stress on the plane is given by:

$$\sigma_n = p_{xn} \cdot l + p_{yn} \cdot m + p_{zn} \cdot n$$

But from eqns. (8.13), (8.14) and (8.15)

$$p_{xn} = \sigma_{xx} \cdot l + \sigma_{xy} \cdot m + \sigma_{xz} \cdot n$$

$$p_{yn} = \sigma_{yx} \cdot l + \sigma_{yy} \cdot m + \sigma_{yz} \cdot n$$

$$p_{zn} = \sigma_{zx} \cdot l + \sigma_{zy} \cdot m + \sigma_{zz} \cdot n$$

$\therefore$ Substituting into eqn (8.5) and using the relationships $\sigma_{xy} = \sigma_{yx}$; $\sigma_{xz} = \sigma_{zx}$ and $\sigma_{yz} = \sigma_{zy}$ which will be proved in §8.12

$$\sigma_n = \sigma_{xx} \cdot l^2 + \sigma_{yy} \cdot m^2 + \sigma_{zz} \cdot n^2 + 2\sigma_{xy} \cdot lm + 2\sigma_{yz} \cdot mn + 2\sigma_{xz} \cdot ln. \tag{8.18}$$

and from eqn. (8.6) the shear stress on the plane will be given by

$$\tau_n^2 = p_{xn}^2 + p_{yn}^2 + p_{zn}^2 - \sigma_n^2 \tag{8.19}$$

In the particular case where plane *ABC* is a principal plane (i.e. no shear stress):

$$\sigma_{xy} = \sigma_{xz} = \sigma_{yz} = 0$$

and

$$\sigma_{xx} = \sigma_1, \quad \sigma_{yy} = \sigma_2 \quad \text{and} \quad \sigma_{zz} = \sigma_3$$

the above equations reduce to:

$$\sigma_n = \sigma_1 \cdot l^2 + \sigma_2 \cdot m^2 + \sigma_3 \cdot n^2 \tag{8.20}$$

and since

$$p_{xn} = \sigma_1 \cdot l \quad p_{yn} = \sigma_2 \cdot m \quad \text{and} \quad p_{zn} = \sigma_3 \cdot n$$

$$\tau_n^2 = \sigma_1^2 \cdot l^2 + \sigma_2^2 \cdot m^2 + \sigma_3^2 \cdot n^2 - \sigma_n^2 \tag{8.21}$$

8.4.1. Line of action of resultant stress

As stated above, the resultant stress p_n is generally not normal to the plane *ABC* but inclined to the *x*, *y* and *z* axes at angles θ_x, θ_y and θ_z – see Fig. 8.7.

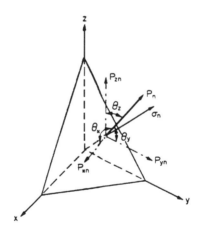

Fig. 8.7. Line of action of resultant stress.

The components of p_n in the *x*, *y* and *z* directions are then

$$\left.\begin{array}{l} p_{xn} = p_n . \cos \theta_x \\ p_{yn} = p_n . \cos \theta_y \\ p_{zn} = p_n . \cos \theta_z \end{array}\right\} \tag{8.22}$$

and the direction cosines which define the line of actions of the resultant stress are

$$\left.\begin{array}{l} l' = \cos \theta_x = p_{xn}/p_n \\ m' = \cos \theta_y = p_{yn}/p_n \\ n' = \cos \theta_z = p_{zn}/p_n \end{array}\right\} \tag{8.23}$$

8.4.2. Line of action of normal stress

By definition the normal stress is that which acts normal to the plane, i.e. the line of action of the normal stress has the same direction cosines as the normal to plane viz: l, m and n.

8.4.3. Line of action of shear stress

As shown in §8.4 the resultant stress p_n can be considered to have two components; one normal to the plane (σ_n) and one along the plane (the shear stress τ_n) – see Fig. 8.6.

Let the direction cosines of the line of action of this shear stress be l_s, m_s and n_s.

The alternative components of the resultant stress, p_{xn}, p_{yn} and p_{zn}, can then either be obtained from eqn (8.22) or by resolution of the normal and shear components along the x, y and z directions as follows:

$$\left.\begin{array}{l} p_{xn} = \sigma_n \cdot l + \tau_n \cdot l_s \\ p_{yn} = \sigma_n \cdot m + \tau_n \cdot m_s \\ p_{zn} = \sigma_n \cdot n + \tau_n \cdot n_s \end{array}\right\} \tag{8.24}$$

Thus the direction cosines of the line of action of the shear stress τ_n are:

$$\left.\begin{array}{l} l_s = \dfrac{p_{xn} - l \cdot \sigma_n}{\tau_n} \\[2mm] m_s = \dfrac{p_{yn} - m \cdot \sigma_n}{\tau_n} \\[2mm] n_s = \dfrac{p_{zn} - n \cdot \sigma_n}{\tau_n} \end{array}\right\} \tag{8.25}$$

8.4.4. Shear stress in any other direction on the plane

Let ϕ be the angle between the direction of the shear stress τ_n and the required direction. Then, since the angle between any two lines in space is given by,

$$\cos \phi = l_s \cdot l_\phi + m_s \cdot m_\phi + n_s \cdot n_\phi \tag{8.26}$$

where l_ϕ, m_ϕ, n_ϕ are the direction cosines of the new shear stress direction, it follows that the required magnitude of the shear stress on the "ϕ" plane will be given by

$$\tau_\phi = \tau_n \cdot \cos \phi \tag{8.27}$$

Alternatively, resolving the components of the resultant stress (p_{xn}, p_{yn} and p_{zn}) along the new direction we have:

$$\tau_\phi = p_{xn} \cdot l_\phi + p_{yn} \cdot m_\phi + p_{zn} \cdot n_\phi \tag{8.28}$$

and substituting eqns. (8.13), (8.14) and (8.15)

$$\tau_\phi = \sigma_{xx} \cdot ll_\phi + \sigma_{yy} \cdot mm_\phi + \sigma_{zz} \cdot nn_\phi + \sigma_{xy}(lm_\phi + l_\phi \cdot m)$$
$$+ \sigma_{xz}(ln_\phi + nl_\phi) + \sigma_{yz}(mn_\phi + nm_\phi) \tag{8.29}$$

Whilst eqn. (8.28) has been derived for the shear stress τ_ϕ it will, in fact, apply equally for any type of stress (i.e. shear or normal) which acts on the plane *ABC* in the ϕ direction.

In the case of the shear stress, however, its line of action must always be perpendicular to the normal to the plane so that

$$ll_\phi + mm_\phi + nn_\phi = 0.$$

In the case of a normal stress the relationship between the direction cosines is simply

$$l = l_\phi, m = m_\phi \text{ and } n = n_\phi$$

since the stress and the normal to the plane are in the same direction. Eqn. (8.29) then reduces to that found previously, viz. eqn. (8.18).

8.5. Principal stresses and strains in three dimensions – Mohr's circle representation

The procedure used for constructing Mohr's circle representation for a three-dimensional principal *stress* system has previously been introduced in §13.7[†]. For convenience of reference the resulting diagram is repeated here as Fig. 8.8. A similar representation for a three-dimensional principal *strain* system is shown in Fig. 8.9.

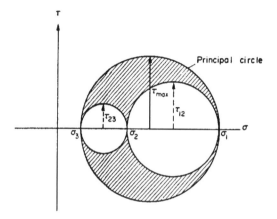

Fig. 8.8. Mohr circle representation of three-dimensional stress state showing the principal circle, the radius of which is equal to the greatest shear stress present in the system.

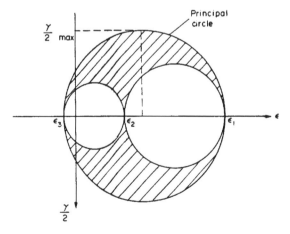

Fig. 8.9. Mohr representation for a three-dimensional principal strain system.

[†] E.J. Hearn, *Mechanics of Materials 1*, Butterworth-Heinemann, 1977.

In both cases the **principal circle** is indicated, the radius of which gives the maximum shear stress and *half* the maximum shear strain, respectively, in the three-dimensional system.

This form of representation utilises different diagrams for the stress and strain systems. An alternative procedure uses a single *combined diagram* for both cases and this is described in detail §§8.6 and 8.7.

8.6. Graphical determination of the direction of the shear stress τ_n on an inclined plane in a three-dimensional principal stress system

As before, let the inclined plane have direction cosines l, m and n. A true representation of this plane is given by constructing a so-called "true shape triangle" the ratio of the lengths of its sides being the ratio of the direction cosines–Fig. 8.10.

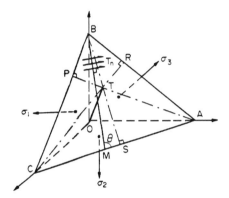

Fig. 8.10. Graphical determination of direction of shear stress on an inclined plane.

If lines are drawn perpendicular to each side from the opposite vertex, meeting the sides at points P, R and S, they will intersect at point T the "orthocentre". This is also the point through which the normal to the plane from O passes.

If σ_1, σ_2 and σ_3 are the three principal stresses then point M is positioned on AC such that

$$\frac{CM}{CA} = \frac{(\sigma_2 - \sigma_3)}{(\sigma_1 - \sigma_2)}$$

The required direction of the shear stress is then perpendicular to the line BD.

The equivalent procedure on the Mohr circle construction is as follows (see Fig. 8.11).

Construct the three stress circles corresponding to the three principal stresses σ_1, σ_2 and σ_3.
Set off line AB at an angle $\alpha = \cos^{-1} l$ to the left of the vertical through A.
Set off line CB at an angle $\gamma = \cos^{-1} n$ to the right of the vertical through C to meet AB at B.
Mark the points where these lines cut the principal circle R and P respectively.
Join AP and CR to cut at point T.
Join BT and extend to cut horizontal axis AC at S.
With point M the σ_2 position, join BM.

The required shear stress direction is then perpendicular to the line BM.

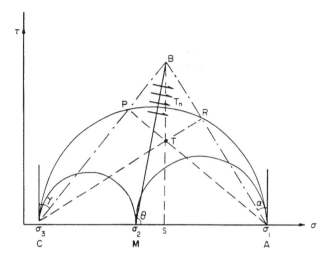

Fig. 8.11. Mohr circle equivalent procedure to that of Fig. 8.10.

8.7. The combined Mohr diagram for three-dimensional stress and strain systems

Consider any three-dimensional stress system with principal stresses σ_1, σ_2 and σ_3 (all assumed tensile). Principal strains are then related to the principal stresses as follows:

$$\varepsilon_1 = \frac{1}{E}(\sigma_1 - v\sigma_2 - v\sigma_3), \text{ etc.}$$

$$E\varepsilon_1 = \sigma_1 - v(\sigma_2 + \sigma_3)$$

$$= \sigma_1 - v(\sigma_1 + \sigma_2 + \sigma_3) + v\sigma_1 \tag{1}$$

Now the *hydrostatic, volumetric* or *mean stress* $\bar{\sigma}$ is defined as

$$\bar{\sigma} = \tfrac{1}{3}(\sigma_1 + \sigma_2 + \sigma_3)$$

Therefore substituting in (1),

$$E\varepsilon_1 = \sigma_1(1 + v) - 3v\bar{\sigma} \tag{2}$$

But the volumetric stress $\bar{\sigma}$ may also be written in terms of the bulk modulus,

i.e. $$\text{bulk modulus } K = \frac{\text{volumetric stress}}{\text{volumetric strain}}$$

and

$$\text{volumetric strain} = \text{sum of the three linear strains}$$

$$= \varepsilon_1 + \varepsilon_2 + \varepsilon_3 = \Delta$$

$$\therefore \qquad K = \frac{\bar{\sigma}}{\Delta}$$

but $$E = 3K(1 - 2v)$$

$$\therefore \qquad \bar{\sigma} = \Delta K = \Delta \frac{E}{3(1 - 2v)}$$

Substituting in (2),

$$E\varepsilon_1 = \sigma_1(1+v) - \frac{3v\Delta E}{3(1-2v)}$$

and, since $E = 2G(1+v)$,

$$2G(1+v)\varepsilon_1 = \sigma_1(1+v) - \frac{v\Delta 2G(1+v)}{(1-2v)}$$

$$\therefore \qquad \sigma_1 = 2G\left[\varepsilon_1 + \frac{v\Delta}{(1-2v)}\right]$$

But, mean strain

$$\bar{\varepsilon} = \tfrac{1}{3}(\varepsilon_1 + \varepsilon_2 + \varepsilon_3) = \tfrac{1}{3}\Delta$$

$$\therefore \qquad \sigma_1 = 2G\left[\varepsilon_1 + \frac{3v\bar{\varepsilon}}{(1-2v)}\right] \qquad (8.30)$$

Alternatively, re-writing eqn. (8.16) in terms of ε_1,

$$\varepsilon_1 = \frac{\sigma_1}{2G} - \frac{3v}{(1-2v)}\bar{\varepsilon}$$

But

$$\bar{\varepsilon} = \frac{\Delta}{3} = \frac{\bar{\sigma}}{3K}$$

But

$$E = 2G(1+v) = 3K(1-2v)$$

i.e.

$$3K = 2G\frac{(1+v)}{(1-2v)}$$

$$\therefore \qquad \bar{\varepsilon} = \frac{\bar{\sigma}(1-2v)}{2G(1+v)}$$

$$\therefore \qquad \varepsilon_1 = \frac{\sigma_1}{2G} + \frac{\bar{\sigma}}{2G}\frac{(1-2v)}{(1+v)}$$

i.e.

$$\varepsilon_1 = \frac{1}{2G}\left[\sigma_1 - \frac{3v\bar{\sigma}}{(1+v)}\right] \qquad (8.31)$$

In the above derivation the cartesian stresses σ_{xx}, σ_{yy} and σ_{zz} could have been used in place of the principal stresses σ_1, σ_2 and σ_3 to yield more general expressions but of identical form. It therefore follows that the stress and associated strain in *any* given direction within a complex three-dimensional stress system is given by eqns. (8.30) and (8.31) which must satisfy the three-dimensional Mohr's circle construction.

Comparison of eqns. (8.30) and (8.31) indicates that

$$2G\varepsilon_1 = \sigma_1 - \frac{3v}{(1+v)}\bar{\sigma}$$

Thus, having constructed the three-dimensional Mohr's *stress* circle representations, the equivalent *strain* values may be obtained simply by reference to a new axis displaced a distance $(3v/(1+v))\bar{\sigma}$ as shown in Fig. 8.12 bringing the new axis origin to O'.

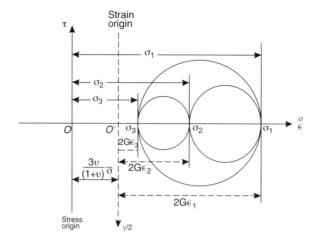

Fig. 8.12. The "combined Mohr diagram" for three-dimensional stress and strain systems.

Distances from the new axis to any principal stress value, e.g. σ_1, will then be $2G$ *times* the corresponding ε_1 principal strain value,

i.e. $$O'\sigma_1 \div 2G = \varepsilon_1$$

Thus the same circle construction will apply for both stresses and strains provided that:

(a) the shear strain axis is offset a distance $\dfrac{3v}{(1+v)}\overline{\sigma}$ to the right of the shear stress axis;

(b) a scale factor of $2G$, $[= E/(1+v)]$, is applied to measurements from the new axis.

8.8. Application of the combined circle to two-dimensional stress systems

The procedure of §14.13[†] uses a common set of axes and a common centre for Mohr's stress and strain circles, each having an appropriate radius and scale factor. An alternative procedure utilises the combined circle approach introduced above where a single circle can be used in association with two different origins to obtain both stress and strain values.

As in the above section the relationship between the stress and strain scales is

$$\frac{\text{stress scale}}{\text{strain scale}} = \frac{E}{(1+v)} = 2G$$

This is in fact the condition for both the stress and strain circles to have the same radius[‡] and should not be confused with the condition required in §14.13[†] of the alternative approach for the two circles to be concentric, when the ratio of scales is $E/(1-v)$.

[†] E.J. Hearn, *Mechanics of Materials 1*, Butterworth-Heinemann, 1997.

[‡] For equal radii of both the stress and strain circles

$$\frac{(\sigma_1 - \sigma_2)}{2 \times \text{stress scale}} = \frac{(\varepsilon_1 - \varepsilon_2)}{2 \times \text{strain scale}}$$

$$\therefore \qquad \frac{\text{stress scale}}{\text{strain scale}} = \frac{(\sigma_1 - \sigma_2)}{(\varepsilon_1 - \varepsilon_2)} = \frac{(\sigma_1 - \sigma_2)}{(\sigma_1 - \sigma_2)}\frac{E}{(1+v)} = \frac{E}{(1+v)}$$

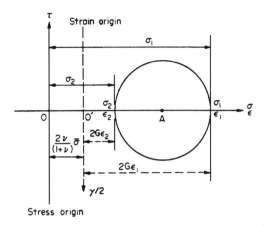

Fig. 8.13. Combined Mohr diagram for two-dimensional stress and strain systems.

With reference to Fig. 8.13 the two origins must then be positioned such that

$$OA = \frac{(\sigma_1 + \sigma_2)}{2 \times \text{stress scale}}$$

$$O'A = \frac{(\varepsilon_1 + \varepsilon_2)}{2 \times \text{strain scale}}$$

$$\therefore \qquad \frac{OA}{O'A} = \frac{(\sigma_1 + \sigma_2)}{(\varepsilon_1 + \varepsilon_2)} \times \frac{\text{strain scale}}{\text{stress scale}}$$

$$= \frac{(\sigma_1 + \sigma_2)}{(\varepsilon_1 + \varepsilon_2)} \times \frac{(1 + \nu)}{E}$$

But
$$\varepsilon_1 = \frac{1}{E}(\sigma_1 - \nu\sigma_2)$$

$$\varepsilon_2 = \frac{1}{E}(\sigma_2 - \nu\sigma_1)$$

$$\therefore \qquad \varepsilon_1 + \varepsilon_2 = \frac{1}{E}(\sigma_1 + \sigma_2)(1 - \nu)$$

$$\therefore \qquad \frac{OA}{O'A} = \frac{(\sigma_1 + \sigma_2)}{(\sigma_1 + \sigma_2)} \frac{E}{(1 - \nu)} \frac{(1 + \nu)}{E} = \frac{(1 + \nu)}{(1 - \nu)}$$

Thus the distance between the two origins is given by

$$OO' = OA - O'A = OA - \frac{(1 - \nu)}{(1 + \nu)}OA$$

$$= \frac{(\sigma_1 + \sigma_2)}{2}\left[1 - \frac{(1 - \nu)}{(1 + \nu)}\right]$$

$$= \frac{(\sigma_1 + \sigma_2)(2\nu)}{2(1 + \nu)} = \frac{\nu}{(1 + \nu)}(\sigma_1 + \sigma_2)$$

$$= \frac{2\nu}{(1 + \nu)}\bar{\sigma} \qquad\qquad (8.32)$$

where $\bar{\sigma}$ is the mean stress in the two-dimensional stress system $= \frac{1}{2}(\sigma_1 + \sigma_2) =$ position of centre of stress circle.

The relationship is thus identical in form to the three-dimensional equivalent with 2 replacing 3 for the *two*-dimensional system.

Again, therefore, the *single-circle construction applies for both stresses and strain provided that the axes are offset by the appropriate amount and a scale factor for strains of 2G is applied.*

8.9. Graphical construction for the state of stress at a point

The following procedure enables the determination of the direct (σ_n) and shear (τ_n) stresses at any point on a plane whose direction cosines are known and, in particular, on the *octahedral planes* (see §8.19).

The construction procedure for Mohr's circle representation of three-dimensional stress systems has been introduced in §8.4. Thus, for a given state of stress producing principal stress σ_1, σ_2 and σ_3, Mohr's circles are as shown in Fig. 8.8.

For a given plane S characterised by direction cosines l, m and n the remainder of the required construction proceeds as follows (Fig. 8.14). (Only half the complete Mohr's circle representation is shown since this is sufficient for the execution of the construction procedure.)

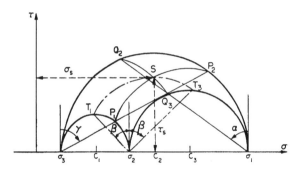

Fig. 8.14. Graphical construction for the state of stress on a general stress plane.

(1) Set off angle $\alpha = \cos^{-1} l$ from the vertical at σ_1 to cut the circles in Q_2 and Q_3.

(2) With centre C_1 (centre of σ_2, σ_3 circle) draw arc Q_2Q_3.

(3) Set off angle $\gamma = \cos^{-1} n$ from the vertical at σ_3 to cut the circles at P_1 and P_2.

(4) With centre C_3 (centre of σ_1, σ_2 circle) draw arc P_1P_2.

(5) The position S representing the required plane is then given by the point where the two arcs Q_2Q_3 and P_1P_2 intersect. *The stresses on this plane are then σ_s and τ_s as shown.* Careful study of the above construction procedure shows that the suffices of points considered in each step always complete the grouping 1, 2, 3. This should aid memorisation of the procedure.

(6) As a check on the accuracy of the drawing, set off angles $\beta = \cos^{-1} m$ on either side of the vertical through σ_2 to cut the $\sigma_2\sigma_3$ circle in T_1 and the $\sigma_1\sigma_2$ circle in T_3.

(7) With centre C_2 (centre of the $\sigma_1\sigma_3$ circle) draw arc T_1T_3 which should then pass through S if all steps have been carried out correctly and the diagram is accurate. The construction is very much easier to follow if all steps connected with points P, Q and T are carried out in different colours.

8.10. Construction for the state of strain on a general strain plane

The construction detailed above for determination of the state of *stress* on a general stress plane applies equally to the determination of *strains* when the symbols σ_1, σ_2 and σ_3 are replaced by the principal *strain* values ε_1, ε_2 and ε_3.

Thus, having constructed the three-dimensional Mohr representation of the principal strains as described in §8.4, the general plane is located as described above and illustrated in Fig. 8.15.

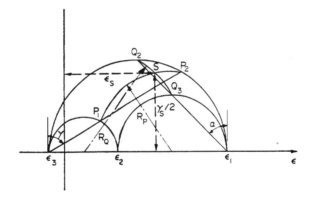

Fig. 8.15. Graphical construction for the state of strain on a general strain plane.

8.11. State of stress—tensor notation

The state of stress equations for any three-dimensional system of cartesian stress components have been obtained in §8.3 as:

$$p_{xn} = \sigma_{xx} \cdot l + \sigma_{xy} \cdot m + \sigma_{xz} \cdot n$$

$$p_{yn} = \sigma_{yx} \cdot l + \sigma_{yy} \cdot m + \sigma_{yz} \cdot n$$

$$p_{zn} = \sigma_{zx} \cdot l + \sigma_{zy} \cdot m + \sigma_{zz} \cdot n$$

The cartesian stress components within this equation can then be remembered conveniently in *tensor notation* as:

$$\begin{bmatrix} \sigma_{xx} & \sigma_{xy} & \sigma_{xz} \\ \sigma_{yx} & \sigma_{yy} & \sigma_{yz} \\ \sigma_{zx} & \sigma_{zy} & \sigma_{zz} \end{bmatrix} \; (\textit{general stress tensor}) \tag{8.33}$$

For a *principal stress system*, i.e. no shear, this reduces to:

$$\begin{bmatrix} \sigma_1 & 0 & 0 \\ 0 & \sigma_2 & 0 \\ 0 & 0 & \sigma_3 \end{bmatrix} (principal\ stress\ tensor) \tag{8.34}$$

and a special case of this is the so-called "*hydrostatic*" stress system with equal principal stresses in all three directions, i.e. $\sigma_1 = \sigma_2 = \sigma_3 = \bar{\sigma}$, and the tensor becomes:

$$\begin{bmatrix} \bar{\sigma} & 0 & 0 \\ 0 & \bar{\sigma} & 0 \\ 0 & 0 & \bar{\sigma} \end{bmatrix} (hydrostatic\ stress\ tensor) \tag{8.35}$$

As shown in §23.16 it is often convenient to divide a general stress into two parts, one due to a hydrostatic stress $\bar{\sigma} = \frac{1}{3}(\sigma_1 + \sigma_2 + \sigma_3)$, the other due to shearing deformations.

Another convenient tensor notation is therefore that for pure shear, ie $\sigma_{xx} = \sigma_{yy} = \sigma_{zz} = 0$ giving the tensor:

$$\begin{bmatrix} 0 & \sigma_{xy} & \sigma_{xz} \\ \sigma_{yx} & 0 & \sigma_{yz} \\ \sigma_{zx} & \sigma_{zy} & 0 \end{bmatrix} (pure\ shear\ tensor) \tag{8.36}$$

The general stress tensor (8.33) is then the combination of the hydrostatic stress tensor and the pure shear tensor.

i.e. *General three-dimensional stress state = hydrostatic stress state + pure shear state*.

This approach is utilised in other sections of this text, notably: §8.16, §8.19 and §8.20.

It therefore follows that an alternative method of presentation of a *pure shear state of stress* is, in tensor form:

$$\begin{bmatrix} (\sigma_1 - \bar{\sigma}) & 0 & 0 \\ 0 & (\sigma_2 - \bar{\sigma}) & 0 \\ 0 & 0 & (\sigma_3 - \bar{\sigma}) \end{bmatrix} \tag{8.37}$$

N.B.: It can be shown that the condition for a state of stress to be one of pure shear is that the first stress invariant is zero.

i.e. $\qquad\qquad I_1 = \sigma_{xx} + \sigma_{yy} + \sigma_{zz} = 0 \qquad$ (see 8.15)

8.12. The stress equations of equilibrium

(a) In cartesian components

In all the previous work on complex stress systems it has been assumed that the stresses acting on the sides of any element are constant. In many cases, however, a general system of direct, shear and body forces, as encountered in practical engineering applications, will produce stresses of variable magnitude throughout a component. Despite this, however, the distribution of these stresses must always be such that overall equilibrium both of the component, and of any element of material within the component, is maintained, and it is a consideration of the conditions necessary to produce this equilibrium which produces the so-called *stress equations of equilibrium*.

Consider, therefore, a body subjected to such a general system of forces resulting in the cartesian stress components described in §8.2 together with the body-force stresses F_x,

F_y and F_z. The element shown in Fig. 8.16 then displays, for simplicity, only the stress components in the X direction together with the body-force stress components. It must be realised, however, that similar components act in the Y and Z directions and these must be considered when deriving equations for equilibrium in these directions: they, of course, have no effect on equilibrium in the X direction.

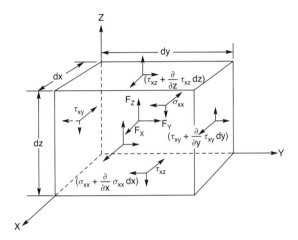

Fig. 8.16. Small element showing body force stresses and other stresses in the X direction only.

It will be observed that on each pair of opposite faces the stress changes in magnitude in the following manner,

e.g.   stress on one face $= \sigma_{xx}$

stress on opposite face $= \sigma_{xx} +$ change in stress

$= \sigma_{xx} +$ rate of change $\times$ distance between faces

Now the rate of change of σ_{xx} with x is given by $\partial\sigma_{xx}/\partial x$, partial differentials being used since σ_{xx} may well be a function of y and z as well as of x.
    Therefore

$$\text{stress on opposite face} = \sigma_{xx} + \frac{\partial\sigma_{xx}}{\partial x}dx$$

Multiplying by the area $dy\,dz$ of the face on which this stress acts produces the force in the X direction.
    Thus, for equilibrium of *forces* in the X direction,

$$\left[\sigma_{xx} + \frac{\partial}{\partial x}\sigma_{xx}\,dx - \sigma_{xx}\right]dy\,dz + \left[\tau_{xy} + \frac{\partial}{\partial y}\tau_{xy}\,dy - \tau_{xy}\right]dx\,dz$$

$$+ \left[\tau_{xz} + \frac{\partial}{\partial z}\tau_{xz}\,dz - \tau_{xz}\right]dx\,dy + F_x\,dx\,dy\,dz = 0$$

(The body-force term being defined as a stress per unit volume is multiplied by the volume $(dx\,dy\,dz)$ to obtain the corresponding force.)

Dividing through by $dx\,dy\,dz$ and simplifying,

$$\frac{\partial\sigma_{xx}}{\partial x} + \frac{\partial\tau_{xy}}{\partial y} + \frac{\partial\tau_{xz}}{\partial z} + F_x = 0$$

Similarly, for equilibrium in the Y direction,

$$\frac{\partial\tau_{yx}}{\partial x} + \frac{\partial\sigma_{yy}}{\partial y} + \frac{\partial\tau_{yz}}{\partial z} + F_Y = 0 \qquad (8.38)$$

and in the Z direction,

$$\frac{\partial\tau_{zx}}{\partial x} + \frac{\partial\tau_{zy}}{\partial y} + \frac{\partial\sigma_{zz}}{\partial z} + F_z = 0$$

these equations being termed the *general stress equations of equilibrium*.

Bearing in mind the comments of §8.2, the symbol τ in the above equations may be replaced by σ, the mixed suffix denoting the fact that it is a shear stress, and the above equations can be remembered quite easily using a similar procedure to that used in §8.2 based on the suffices, i.e. first suffices and body-force terms are constant for each horizontal row and in the normal order x, y and z.

	X	Y	Z	
X	$\frac{xx}{\partial x}$	$\frac{xy}{\partial y}$	$\frac{xz}{\partial z}$	$+F_X = 0$
Y	$\frac{yx}{\partial x}$	$\frac{yy}{\partial y}$	$\frac{yz}{\partial z}$	$+F_Y = 0$
Z	$\frac{zx}{\partial x}$	$\frac{zy}{\partial y}$	$\frac{zz}{\partial z}$	$+F_Z = 0$

The above equations have been derived by consideration of equilibrium of *forces* only, and this does not represent a complete check on the equilibrium of the system. This can only be achieved by an additional consideration of the *moments of the forces* which must also be in balance.

Consider, therefore, the element shown in Fig. 8.17 which, again for simplicity, shows only the stresses which produce moments about the Y axis. For convenience the origin of the cartesian coordinates has in this case been chosen to coincide with the centroid of the element. In this way the direct stress and body-force stress terms will be eliminated since the forces produced by these will have no moment about axes through the centroid.

It has been assumed that shear stresses τ_{xy}, τ_{yz} and τ_{xz} act on the coordinate planes passing through G so that they will each increase and decrease on either side of these planes as described above.

Thus, for equilibrium of moments about the Y axis,

$$\left[\tau_{xz} + \frac{\partial}{\partial z}(\tau_{xz})\frac{dz}{2}\right]dx\,dy\frac{dz}{2} + \left[\tau_{xz} - \frac{\partial}{\partial z}(\tau_{xz})\frac{dz}{2}\right]dx\,dy\frac{dz}{2}$$

$$- \left[\tau_{zx} + \frac{\partial}{\partial x}(\tau_{zx})\frac{dx}{2}\right]dy\,dz\frac{dx}{2} - \left[\tau_{zx} - \frac{\partial}{\partial x}(\tau_{zx})\frac{dx}{2}\right]dy\,dz\frac{dx}{2} = 0$$

Dividing through by $(dx\,dy\,dz)$ and simplifying, this reduces to

$$\tau_{xz} = \tau_{zx}$$

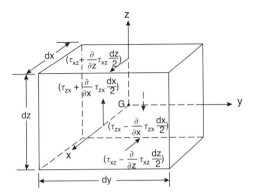

Fig. 8.17. Element showing only stresses which contribute to a moment about the Y axis.

Similarly, by consideration of the equilibrium of moments about the X and Z axes,

$$\tau_{zy} = \tau_{yz}$$

$$\tau_{xy} = \tau_{yx}$$

Thus the shears and complementary shears on adjacent faces are equal as in the simple two-dimensional case. The nine cartesian stress components thus reduce to *six independent* values,

i.e.
$$\begin{pmatrix} \sigma_{xx} & \sigma_{xy} & \sigma_{xz} \\ \sigma_{yx} & \sigma_{yy} & \sigma_{yz} \\ \sigma_{zx} & \sigma_{zy} & \sigma_{zz} \end{pmatrix} \quad \text{or} \quad \begin{pmatrix} \sigma_{xx} & \tau_{xy} & \tau_{xz} \\ \tau_{yx} & \sigma_{yy} & \tau_{yz} \\ \tau_{zx} & \tau_{zy} & \sigma_{zz} \end{pmatrix}$$

(b) In cylindrical coordinates

The equations of equilibrium derived above in cartesian components are very useful for components and stress systems which can easily be referred to a set of three mutually perpendicular axes. There are many cases, however, e.g. those components with *axial symmetry*, where other coordinate axes prove far more convenient. One such set of axes is the *cylindrical coordinate* system with variables r, θ and z as shown in Fig. 8.18.

Consider, therefore, the equilibrium in a *radial* direction of the element shown in Fig. 8.19(a). Again, for simplicity, only those stresses which produce force components in this direction are indicated. It must be observed, however, that in this case the $\sigma_{\theta\theta}$ terms will also produce components in the radial direction as shown by Fig. 8.19(b). The body-force stress components are denoted by F_R, F_Z and F_θ.

Therefore, resolving forces radially,

$$\left[\sigma_{rr} + \frac{\partial}{\partial r}(\sigma_{rr})\,dr\right](r+dr)\,d\theta\,dz - \sigma_{rr}\,rd\theta\,dz + \left[\sigma_{r\theta} + \frac{\partial}{\partial\theta}(\sigma_{r\theta})\,d\theta\right]dr\,dz\cos\frac{d\theta}{2}$$

$$- \sigma_{r\theta}\,dr\,dz\cos\frac{d\theta}{2} + \left[\left(\sigma_{rz} + \frac{\partial(\sigma_{rz})}{\partial z}dz\right) - \sigma_{rz}\right]\left(r+\frac{dr}{2}\right)d\theta\,dr$$

$$- \sigma_{\theta\theta}\,dr\,dz\sin\frac{d\theta}{2} - \left[\sigma_{\theta\theta} + \frac{\partial}{\partial\theta}(\sigma_{\theta\theta})\,d\theta\right]dr\,dz\sin\frac{d\theta}{2} + F_R\,rdr\,d\theta\,dz = 0$$

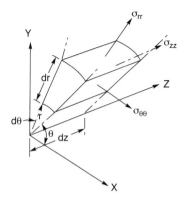

Fig. 8.18. Cylindrical coordinates.

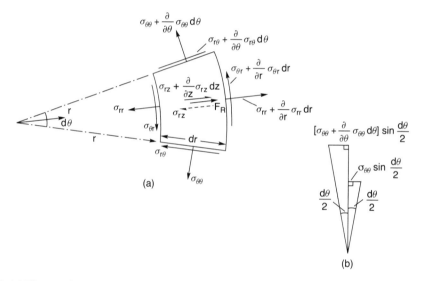

Fig. 8.19. (a) Element showing stresses which contribute to equilibrium in the radial and circumferential directions. (b) Radial components of hoop stresses.

With $\cos \dfrac{d\theta}{2} \simeq 1$ and $\sin \dfrac{d\theta}{2} \simeq \dfrac{d\theta}{2}$, this equation reduces to

$$\frac{\partial}{\partial r}(\sigma_{rr}) + \frac{1}{r}\frac{\partial}{\partial \theta}(\sigma_{r\theta}) + \frac{\partial}{\partial z}(\sigma_{rz}) + \frac{(\sigma_{rr} - \sigma_{\theta\theta})}{r} + F_R = 0$$

Similarly, in the θ direction, the relevant equilibrium equation reduces to

$$\frac{\partial}{\partial r}(\sigma_{r\theta}) + \frac{1}{r}\frac{\partial(\sigma_{\theta\theta})}{\partial \theta} + \frac{\partial}{\partial z}(\sigma_{\theta z}) + \frac{2\sigma_{r\theta}}{r} + F_\theta = 0 \qquad (8.39)$$

and in the Z direction (Fig. 8.20)

$$\frac{\partial(\sigma_{rz})}{\partial r} + \frac{1}{r}\frac{\partial(\sigma_{\theta z})}{\partial \theta} + \frac{\partial}{\partial z}(\sigma_{zz}) + \frac{\sigma_{rz}}{r} + F_z = 0$$

These are, then, the *stress equations of equilibrium in cylindrical coordinates* and in their most general form. Clearly these are difficult to memorise and, fortunately, very few problems arise in which the equations in this form are required. In many cases *axial symmetry* exists and circular sections remain concentric and circular throughout loading, i.e. $\sigma_{r\theta} = 0$.

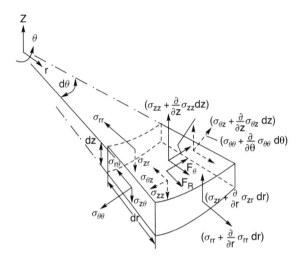

Fig. 8.20. Element indicating additional stresses which contribute to equilibrium in the axial (z) direction.

Thus **for axial symmetry** the equations reduce to

$$\left.\begin{aligned}
\frac{\partial}{\partial r}(\sigma_{rr}) + \frac{\partial}{\partial z}(\sigma_{rz}) + \frac{(\sigma_{rr} - \sigma_{\theta\theta})}{r} + F_R &= 0 \\[2mm]
\frac{1}{r}\frac{\partial(\sigma_{\theta\theta})}{\partial\theta} + \frac{\partial(\sigma_{\theta z})}{\partial z} + F_\theta &= 0 \\[2mm]
\frac{\partial}{\partial r}(\sigma_{rz}) + \frac{1}{r}\frac{\partial\sigma_{\theta z}}{\partial\theta} + \frac{\partial(\sigma_{zz})}{\partial z} + \frac{\sigma_{rz}}{r} + F_z &= 0
\end{aligned}\right\} \qquad (8.40)$$

Further simplification applies in cases where the **coordinate axes** can be selected to **coincide with principal stress directions** as in the case of thick cylinders subjected to uniform pressure or thermal gradients. In such cases there will be no shear, and in the absence of body forces the equations reduce to the relatively simple forms

$$\left.\begin{aligned}
\frac{\partial}{\partial r}(\sigma_{rr}) + \frac{(\sigma_{rr} - \sigma_{\theta\theta})}{r} &= 0 \\[2mm]
\frac{\partial(\sigma_{\theta\theta})}{\partial\theta} &= 0 \\[2mm]
\frac{\partial(\sigma_{zz})}{\partial z} &= 0
\end{aligned}\right\} \qquad (8.41)$$

8.13. Principal stresses in a three-dimensional cartesian stress system

As an alternative to the graphical Mohr's circle procedures the principal stresses in three-dimensional complex stress systems can be determined analytically as follows.

The equations for the state of stress at a point derived in §8.3 may be combined to give the equation

$$\sigma_n^3 - (\sigma_{xx} + \sigma_{yy} + \sigma_{zz})\sigma_n^2 + (\sigma_{xx}\sigma_{yy} + \sigma_{yy}\sigma_{zz} + \sigma_{xx}\sigma_{zz} - \tau_{xy}^2 - \tau_{yz}^2 - \tau_{zx}^2)\sigma_n$$

$$- (\sigma_{xx}\sigma_{yy}\sigma_{zz} - \sigma_{xx}\tau_{yz}^2 - \sigma_{yy}\tau_{zx}^2 - \sigma_{zz}\tau_{xy}^2 + 2\tau_{xy}\tau_{yz}\tau_{xz}) = 0 \qquad (8.42)$$

With a knowledge of the cartesian stress components this cubic equation can be solved for σ_n to produce the three principal stress values required. A general procedure for the solution of cubic equations is given below.

8.13.1. Solution of cubic equations

Consider the cubic equation

$$x^3 + ax^2 + bx + c = 0 \qquad (1)$$

Substituting,
$$x = y - a/3 \qquad (2)$$

with
$$p = b - a^2/3 \qquad (3)$$

and
$$q = c - \frac{ab}{3} + \frac{2a^3}{27} \qquad (4)$$

we obtain the modified equation

$$y^3 + py + q = 0 \qquad (5)$$

Substituting,
$$y = rz \qquad (6)$$

$$z^3 + \frac{pz}{r^2} + \frac{q}{r^3} = 0 \qquad (7)$$

Now consider the standard trigonometric identity

$$\cos 3\theta = 4\cos^3\theta - 3\cos\theta \qquad (8)$$

Rearranging and substituting $z = \cos\theta$, $\qquad (9)$

$$z^3 - \frac{3z}{4} - \frac{1}{4}\cos 3\theta = 0 \qquad (10)$$

(7) and (10) are of similar form and will be identical provided that

$$r = \sqrt{-\frac{4p}{3}} \qquad (11)$$

and
$$\cos 3\theta = -\frac{4q}{r^3} \qquad (12)$$

Three values of θ may be obtained to satisfy (12),

i.e.
$$\theta, \theta + 120° \quad \text{and} \quad \theta + 240°$$

Then, from (9), three corresponding values of z are obtained, namely

$$z_1 = \cos \theta°$$

$$z_2 = \cos(\theta + 120°)$$

$$z_3 = \cos(\theta + 240°)$$

(6) then yields appropriate values of y and hence the required values of x via (2).

8.14. Stress invariants; Eigen values and Eigen vectors

Consider the special case of the "stress at a point" tetrahedron Fig. 8.3 where plane ABC is a principal plane subjected to a principal stress σ_p and, by definition, zero shear stress. The normal stress is thus coincident with the resultant stress and both equal to σ_p.

If the direction cosines of σ_p (and hence of the principal plane) are l_p, m_p, n_p then:

$$p_{xn} = \sigma_p \cdot l_p$$

$$p_{yn} = \sigma_p \cdot m_p$$

$$p_{zn} = \sigma_p \cdot n_p$$

i.e. substituting in eqns. (8.13), (8.14) and (8.15) we have:

$$\sigma_p \cdot l_p = \sigma_{xx} \cdot l_p + \sigma_{xy} \cdot m_p + \sigma_{xz} \cdot n_p$$

$$\sigma_p \cdot m_p = \sigma_{yx} \cdot l_p + \sigma_{yy} \cdot m_p + \sigma_{yz} \cdot n_p$$

$$\sigma_p \cdot n_p = \sigma_{zx} \cdot l_p + \sigma_{zy} \cdot m_p + \sigma_{zz} \cdot n_p$$

or

$$\left.\begin{array}{l} 0 = (\sigma_{xx} - \sigma_p)l_p + \sigma_{xy} \cdot m_p + \sigma_{xz} \cdot n_p \\ 0 = \sigma_{yx} l_p + (\sigma_{yy} - \sigma_p)m_p + \sigma_{yz} \cdot n_p \\ 0 = \sigma_{zx} l_p + \sigma_{zy} \cdot m_p + (\sigma_{zz} - \sigma_p)n_p \end{array}\right\} \tag{8.43}$$

Considering eqn. (8.43) as a set of three homogeneous linear equations in unknowns l_p, m_p and n_p, the direction cosines of the principal plane, one possible solution, viz. $l_p = m_p = n_p = 0$, can be dismissed since $l^2 + m^2 + n^2 = 1$ must always be maintained. The only other solution which gives real values for the direction cosines is that obtained by equating the determinant of the R.H.S. to zero:

i.e.

$$\begin{vmatrix} (\sigma_{xz} - \sigma_p) & \sigma_{xy} & \sigma_{xz} \\ \sigma_{yx} & (\sigma_{yy} - \sigma_p) & \sigma_{yz} \\ \sigma_{zx} & \sigma_{zy} & (\sigma_{zz} - \sigma_p) \end{vmatrix} = 0$$

Evaluating the determinant yields the so-called "*characteristic equation*"

$$\sigma_p^3 - (\sigma_{xx} + \sigma_{yy} + \sigma_{zz})\,\sigma_p^2 + [(\sigma_{xx}\sigma_{yy} + \sigma_{yy}\sigma_{zz} + \sigma_{zz}\sigma_{xx}) - (\sigma_{xy}^2 + \sigma_{yz}^2 + \sigma_{zx}^2)]\,\sigma_p$$

$$- [\sigma_{xx}\sigma_{yy}\sigma_{zz} + 2\sigma_{xy}\sigma_{yz}\sigma_{zx} - (\sigma_{xx}\sigma_{yz}^2 + \sigma_{yy}\sigma_{zx}^2 + \sigma_{zz}\sigma_{xy}^2)] = 0 \tag{8.44}$$

Thus, for any given set of cartesian stress components in three dimensions a solution of this cubic equation is required before principal stress value can be determined; a graphical solution is not possible.

Eigen values

The solutions for the principal stresses σ_1, σ_2 and σ_3 from the characteristic equation are known as the **Eigen values** whilst the associated direction cosines l_p, m_p and n_p are termed the **Eigen vectors.**

One procedure for solution of the cubic characteristic equation is given in §8.10.

8.15. Stress invariants

If, for the same applied stress system, the stress components had been given relative to some other set of cartesian co-ordinates x', y' and z', the above equation would still apply (with x' replacing x, y' replacing y and z' replacing z) and would still produce the same principal stress values. It follows, therefore, that whatever axis system is chosen the coefficients of the various terms of the characteristics equation must have the same values, i.e. they are "non-varying quantities" or "*invariant*".

The equation can thus be re-written in the form:

$$\sigma_p^3 - I_1\sigma_p^2 - I_2\sigma_p - I_3 = 0 \tag{8.45}$$

with

$$\left. \begin{array}{l} I_1 = \sigma_{xx} + \sigma_{yy} + \sigma_{zz} \\[2mm] I_2 = (\sigma_{xy}^2 + \sigma_{yz}^2 + \sigma_{zx}^2) - (\sigma_{xx} \cdot \sigma_{yy} + \sigma_{yy} \cdot \sigma_{zz} + \sigma_{zz} \cdot \sigma_{xx}) \\[2mm] I_3 = \sigma_{xx} \cdot \sigma_{yy} \cdot \sigma_{zz} + 2\sigma_{xy} \cdot \sigma_{yz}\sigma_{zx} - \sigma_{xx}\sigma_{yz}^2 - \sigma_{yy}\sigma_{zx}^2 - \sigma_{zz}\sigma_{xy}^2 \end{array} \right\} \tag{8.46}$$

the three quantities I_1, I_2 and I_3 being termed the **stress invariants.**

If the reference axes selected are the principal stress axes in the system then all shear components reduce to zero and the equations (8.46) reduce to:

$$\left. \begin{array}{ll} I_1 = \sigma_1 + \sigma_2 + \sigma_3 & = \Sigma\sigma_p \\[2mm] I_2 = -(\sigma_1\sigma_2 + \sigma_2\sigma_3 + \sigma_3\sigma_1) & = \Sigma\sigma_p^2 \\[2mm] I_3 = \sigma_1\sigma_2\sigma_3 & = \Sigma\sigma_p^3 \end{array} \right\} \tag{8.47}$$

The first and second invariants are particularly important in development of the theory of plasticity since it is assumed that:

(a) I_1 has no influence on initial yielding

(b) $I_2 =$ constant can be taken as an important criterion of yielding.

For biaxial stress conditions, i.e. $\sigma_3 = 0$, the third stress invariant vanishes and the others reduce to

$$\left. \begin{array}{l} I_1 = \sigma_1 + \sigma_2 \\[2mm] I_2 = \sigma_1\sigma_2 \end{array} \right\} \tag{8.48}$$

or, in the xy plane, from eqn. (8.46)

$$\left. \begin{array}{l} I_1 = \sigma_{xx} + \sigma_{yy} \\[2mm] I_2 = \sigma_{xx}\sigma_{yy} - \sigma_{xy}^2 \end{array} \right\} \tag{8.49}$$

Now from eqn. $(13.11)^{\dagger}$ the principal stresses in a two-dimensional stress system are given by:

$$\sigma_{1,2} = \tfrac{1}{2}(\sigma_{xx} + \sigma_{yy}) \pm \tfrac{1}{2}[(\sigma_{xx} - \sigma_{yy})^2 + 4\sigma_{xy}^2]^{\frac{1}{2}}$$

$$= \tfrac{1}{2}(\sigma_{xx} + \sigma_{yy}) \pm \tfrac{1}{2}[(\sigma_{xx} + \sigma_{yy})^2 - 4\sigma_{xx}\sigma_{yy} - \sigma_{xy}^2]^{\frac{1}{2}}$$

which is the general solution of the following quadratic equation:

$$\sigma_p^2 = (\sigma_{xx} + \sigma_{yy})\sigma_p + (\sigma_{xx}\sigma_{yy} - \sigma_{xy}^2) = 0$$

i.e. $$\sigma_p^2 - I_1\sigma_p + I_2 = 0 \tag{8.50}$$

The graphical solution of this equation is as follows (see Fig. 8.21):

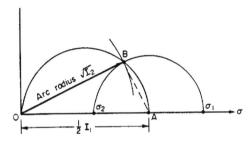

Fig. 8.21. Graphical determination of principal stresses in a two-dimensional stress system from known stress invariant I values (solution for positive I_2 value)

(i) On a horizontal (direct stress) axis mark off a length $OA = \tfrac{1}{2}I_1$.

(ii) Draw semi-circle on OA as diameter.

(iii) With centre O draw arc OB, radius $\sqrt{I_2}$, to cut the semi-circle at B.

(iv) With centre A and radius AB draw semi-circle to cut stress axis at σ_1 and σ_2 the required principal stress values.

N.B. If I_2 is negative (see §8.46), algebraically $\sqrt{I_2} > \tfrac{1}{2}I_1$ and the line OB cannot cut the semi-circle on OA as diameter and no solution can be obtained. In this case an alternative construction is required – see Fig. 8.22.

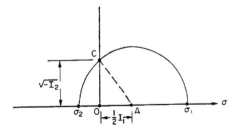

Fig. 8.22. As Fig. 8.21 but for negative I_2 value.

(i) Again mark off length $OA = \tfrac{1}{2}I_1$.

(ii) Erect perpendicular at O of length $OC = \sqrt{-I_2}$.

† E.J. Hearn, *Mechanics of Materials 1*, Butterworth-Heinemann, 1977.

(iii) With centre A and radius AC draw a circle to cut OA (produced as necessary) at σ_1 and σ_2 the required principal stress values.

Returning to a three-dimensional principal stress system a further interesting graphical relationship is obtained from the 3D Mohr circle construction – see Fig. 8.23.[*]

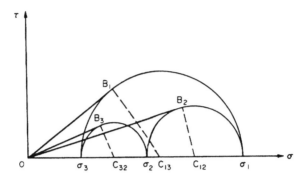

Fig. 8.23. Stress invariants for a three-dimensional stress system in terms of tangents to the Mohr stress circles
$$I_1 = \sigma_1 + \sigma_2 + \sigma_3, I_2 = OB_1^2 + OB_2^2 + OB_3^2, I_3 = OB_1 \cdot OB_2 \cdot OB_3.$$

The three stress invariants are given in Fig. 8.23 in terms of the tangents to the three circles from the origin 0 as:

$$I_1 = \sigma_1 + \sigma_2 + \sigma_3$$

$$I_2 = \sigma_1\sigma_2 + \sigma_2\sigma_3 + \sigma_3\sigma_1 = OB_1^2 + OB_2^2 + OB_3^2$$

$$I_3 = \sigma_1\sigma_2\sigma_3 = OB_1 \times OB_2 \times OB_3$$

8.16. Reduced stresses

An alternative form of the cubic characteristic equation is obtained if a "*hydrostatic stress*" of $I_1/3$ is substracted from the original stress system to produce "*reduced stresses*" $\sigma' = \sigma - I_1/3$.

Thus, replacing σ_p by $(\sigma' + I_1/3)$ in eqn. (8.45) we have:

$$\sigma'^3 - \left(\frac{I_1^2 + 3I_2}{3}\right)\sigma' - \left(\frac{2I_1^3 + 9I_1I_2 + 27I_3}{27}\right) = 0$$

or

$$\sigma'^3 - J_1\sigma'^2 - J_2\sigma' - J_3 = 0 \qquad (8.51)$$

with

$$J_1 = 0$$

$$J_2 = \frac{1}{3}[I_1^2 + 3I_2]$$

$$J_3 = \frac{1}{27}[2I_1^3 + 9I_1I_2 + 27I_3]$$

[*] M.G. Derrington and W. Johnson, *The Defect of Mohr's Circle for Three-Dimensional Stress States.*

The terms J_1, J_2 and J_3 are termed the *invariants of reduced stress* and, again, have special significance in the consideration of yielding of metals and associated plastic theory.

It will be shown in §8.20 that the hydrostatic stress component does not affect the yield of metals and

$$\text{hydrostatic stress} = \tfrac{1}{3}(\sigma_1 + \sigma_2 + \sigma_3) = \tfrac{1}{3}I_1$$

It therefore follows that first stress invariant I_1 also has no significance on yielding and since the principal stress system can be written, as above, in terms of reduced stresses $\sigma' = (\sigma - 1/3\ I_1)$ it also follows that it must be the reduced stress components which influence yielding.

(N.B.: "Reduced stresses" are synonymous with the deviatoric stresses introduced in §8.20.)

Other useful relationships which can be derived from the above equations are:

$$(\sigma_1 - \sigma_2)^2 + (\sigma_2 - \sigma_3)^2 + (\sigma_3 - \sigma_1)^2 = 6J_2 \tag{8.52}$$

and

$$(\sigma_{xx} - \sigma_{yy})^2 + (\sigma_{yy} - \sigma_{zz})^2 + (\sigma_{zz} - \sigma_{xx})^2 + 6(\tau_{xy}^2 + \tau_{yz}^2 + \tau_{zx}^2) = 2I_1^2 + 6I_2 \tag{8.53}$$

The left-hand sides of both equations are thus, in themselves, invariant and are useful in further considerations of strain energy, yielding and failure.

For example, the shear strain energy theory of elastic failure uses the criterion:

$$(\sigma_1 - \sigma_2)^2 + (\sigma_2 - \sigma_3)^2 + (\sigma_3 - \sigma_1)^2 = 2\sigma_y^2 = \text{constant}$$

which, from eqn. (8.52), can be simply re-written as

$$J_2 = \text{constant}.$$

N.B.: It should be remembered that eqns. (8.52) and (8.53) are merely different ways of presenting the same information since:

$$6J_2 = 2I_1^2 + 6I_2.$$

8.17. Strain invariants

It has been shown in §14.10[†] that the basic transformation equations for stress and strain have identical form provided that ε is used in place of σ and $\gamma/2$ in place of τ. The equations derived above for the stress invariants will therefore apply equally for strain conditions provided that the same rules are followed.

8.18. Alternative procedure for determination of principal stresses (eigen values)

An alternative solution to the characteristic cubic equation expressed in stress invariant format, viz. eqn. (8.45), is as follows:

Given the basic equation:

$$\sigma_p^3 - I_1\sigma_p^2 - I_2\sigma_p - I_3 = 0 \tag{8.45}bis$$

[†] E.J. Hearn, *Mechanics of Materials 1*, Butterworth-Heinemann, 1997.

the stress invariants may be calculated from:

$$I_1 = \sigma_{xx} + \sigma_{yy} + \sigma_{zz}$$

$$I_2 = -(\sigma_{xx}\sigma_{yy} + \sigma_{yy}\sigma_{zz} + \sigma_{zz}\sigma_{xx}) + \tau_{xy}^2 + \tau_{yz}^2 + \tau_{zx}^2 \qquad (8.46)\text{bis}$$

$$I_3 = \sigma_{xx}\sigma_{yy}\sigma_{zz} + 2\tau_{xy}\tau_{yz}\tau_{zx} - \sigma_{xx}\tau_{yz}^2 - \sigma_{yy}\tau_{zx}^2 - \sigma_{zz}\tau_{xy}^2$$

and the required principal stresses obtained from[‡]:

$$\sigma_{p_1} = 2S\cos{(a/3)} + I_1/3$$

$$\sigma_{p_2} = 2S\cos{[(a/3) + 120°]} + I_1/3 \qquad (8.54)$$

$$\sigma_{p_3} = 2S\cos{[(a/3 + 240°]} + I_1/3$$

with
$$S = (R/3)^{1/2} \quad \text{and} \quad \alpha = \cos^{-1}(-Q/2T)$$

and
$$R = \frac{1}{3}I_1^2 - I_2$$

$$Q = \frac{1}{3}I_1 I_2 - I_3 - \frac{2}{27}I_1^3$$

$$T = \left(\frac{1}{27}R^3\right)^{1/2}$$

After calculation of the three principal stress values, they can be placed in their normal conventional order of magnitude, viz. σ_1, σ_2 and σ_3.

The procedure is, in effect, the same as that of §8.13 but carried out in terms of the stress invariants.

8.18.1. Evaluation of direction cosines for principal stresses (eigen vectors)

Having determined the three principal stress values for a given three-dimensional complex stress state using the procedures of §8.13.1 or §8.18, above, a complete solution of the problem generally requires a determination of the directions in which these stresses act−as given by their respective direction cosines or eigen vector values.

The relationship between a particular principal stress σ_p and the cartesian stress components is given by eqn (8.43)

i.e.
$$(\sigma_{xx} - \sigma_p)l + \tau_{xy} \cdot m + \tau_{xz} \cdot n = 0$$

$$\tau_{xy} \cdot l + (\sigma_{yy} - \sigma_p)m + \tau_{yz}n = 0$$

$$\tau_{xz} \cdot l + \tau_{yz} \cdot m + (\sigma_{zz} - \sigma_p)n = 0$$

If one of the known principal stress values, say σ_1, is substituted in the above equations together with the given cartesian stress components, three equations result in the three unknown direction cosines for that principal stress i.e. l_1, m_1 and n_1.

However, only two of these are independent equations and the additional identity $l_1^2 + m_1^2 + n_1^2 = 1$ is required in order to evaluate $l_1 m_1$ and n_1.

[‡] E.E. Messal, "Finding true maximum shear stress", *Machine Design*, Dec. 1978.

The procedure can then be repeated substituting the other principal stress values σ_2 and σ_3, in turn, to produce eigen vectors for these stresses but it is tedious and an alternative matrix approach is recommended as follows:

Equation (8.43) above can be expressed in matrix form, thus:

$$\begin{bmatrix} (\sigma_{xx} - \sigma_p) & \tau_{xy} & \tau_{xz} \\ \tau_{xy} & (\sigma_{yy} - \sigma_p) & \tau_{yz} \\ \tau_{xz} & \tau_{yz} & (\sigma_{zz} - \sigma_p) \end{bmatrix} \begin{Bmatrix} l \\ m \\ n \end{Bmatrix} = 0$$

with cofactors of the determinant on the elements of the first row of:

$$a = \begin{vmatrix} (\sigma_{yy} - \sigma_p) & \tau_{yz} \\ \tau_{yz} & (\sigma_{zz} - \sigma_p) \end{vmatrix}$$

$$b = - \begin{vmatrix} \tau_{xy} & \tau_{yz} \\ \tau_{xz} & (\sigma_{zz} - \sigma_p) \end{vmatrix}$$

$$c = \begin{vmatrix} \tau_{xy} & (\sigma_{yy} - \sigma_p) \\ \tau_{xz} & \tau_{yz} \end{vmatrix}$$

with the direction cosines or eigen vectors of the principal stresses given by:

$$l_p = ak \qquad m_p = bk \qquad n_p = ck$$

with
$$k = \frac{1}{\sqrt{a^2 + b^2 + c^2}}$$

thus satisfying the identity $l_p^2 + m_p^2 + n_p^2 = 1$.

Substitution of any principal stress value, again say σ_1, into the above equations together with the given cartesian stress components allows solution of the determinants and yields values for a_1, b_1 and c_1, hence k_1 and hence l_1, m_1 and n_1, the desired eigen vectors. The process can then be repeated for the other principal stress values $\sigma_2 + \sigma_3$.

8.19. Octahedral planes and stresses

Any complex three-dimensional stress system produces three mutually perpendicular principal stresses σ_1, σ_2, and σ_3. Associated with this stress state are so-called *octahedral planes* each of which cuts across the corners of a principal element such as that shown in Fig. 8.24 to produce the octahedron (8-sided figure) shown in Fig. 8.25. The stresses acting on the octahedral planes have particular significance.

The normal stresses acting on each of the octahedral planes are equal in value and tend to compress or enlarge the octahedron without distorting its shape. They are thus said to be *hydrostatic* stresses and have values given by

$$\sigma_{\text{oct}} = \tfrac{1}{3}[\sigma_1 + \sigma_2 + \sigma_3] = \bar{\sigma} \tag{8.55}$$

Similarly, the shear stresses acting on each of the octahedral planes are also identical and tend to distort the octahedron without changing its volume.

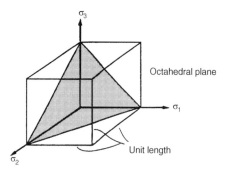

Fig. 8.24. Cubical principal stress element showing one of the octahedral planes.

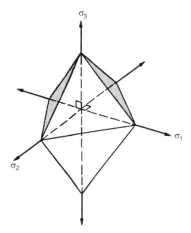

Fig. 8.25. Principal stress system showing the eight octahedral planes forming an octahedron.

The value of the *octahedral shear stresses*[†] is given by

$$\tau_{oct} = \tfrac{1}{3} - [(\sigma_1 - \sigma_2)^2 + (\sigma_2 - \sigma_3)^2 + (\sigma_3 - \sigma_1)^2]^{1/2} \qquad (8.56)$$

$$= \tfrac{2}{3} - [\tau_{12}^2 + \tau_{23}^2 + \tau_{13}^2]^{1/2} \qquad (8.57)$$

τ_{12}, τ_{23} and τ_{13} being the maximum shear stresses in the 1−2, 2−3 and 1−3 planes respectively.

Thus the general state of stress may be represented on octahedral planes as shown in Fig. 8.26, the *direction cosines* of the octahedral planes being given by

$$l = m = n = \pm 1/\sqrt{1^2 + 1^2 + 1^2} = \pm 1/\sqrt{3} \qquad (8.58)$$

The values of the octahedral shear and direct stresses may also be obtained by the graphical construction of §8.9 since they are represented by a point in the shaded area of the three-dimensional Mohr's circle construction of Figs. 8.8 and 8.9.

[†] A.J. Durelli, E.A. Phillips and C.H. Tsao. *Analysis of Stress and Strain*, chap. 3. p. 26, McGraw-Hill, New York, 1958.

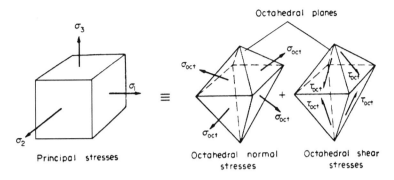

Fig. 8.26. Representation of a general state of stress on the octahedral planes.

The octahedral shear stress has a particular significance in relation to the elastic failure of materials. Whilst its value is always smaller than the greatest numerical (principal) shear stress, it nevertheless has a value which is influenced by all three principal stress values and has been shown to be a reliable criterion for predicting yielding under complex loading conditions.

The **maximum octahedral shear stress theory of elastic failure** thus assumes that yield or failure under complex stress conditions will occur when the octahedral shear stress has a value equal to that obtained in the simple tensile test at yield.

Now for uniaxial tension, $\sigma_2 = \sigma_3 = 0$ and $\sigma_1 = \sigma_y$ and from eqn. (8.56)

$$\tau_{\text{oct}} = \frac{\sqrt{2}}{3}\sigma_y$$

Therefore the criterion of failure becomes

$$\frac{\sqrt{2}}{3}\sigma_y = \frac{1}{3}[(\sigma_1 - \sigma_2)^2 + (\sigma_2 - \sigma_3)^2 + (\sigma_3 - \sigma_1)^2]^{1/2}$$

i.e.
$$2\sigma_y^2 = (\sigma_1 - \sigma_2)^2 + (\sigma_2 - \sigma_3)^2 + (\sigma_3 - \sigma_1)^2 \qquad (8.59)$$

This is clearly the same criterion as that referred to earlier as the Maxwell/von Mises *distortion* or *shear strain energy* theory.

8.20. Deviatoric stresses

It is sometimes convenient to consider stresses with reference to some false zero, i.e. to measure their values above or below some selected datum stress value, and not their absolute values. This is particularly useful in advanced analysis using the theory of plasticity.

The selected datum stress $\bar{\sigma}$ or "false zero" is taken to be that stress which produces only a change in volume. This is the stress which acts equally in all directions and is referred to earlier (page 251) as the *hydrostatic* or *dilatational* stress. This is defined in terms of the principal stresses or the cartesian stresses as follows:

$$\bar{\sigma} = \tfrac{1}{3}(\sigma_1 + \sigma_2 + \sigma_3) = \tfrac{1}{3}(\sigma_{xx} + \sigma_{yy} + \sigma_{zz}) \qquad (8.60)$$

i.e.
$$\bar{\sigma} = \text{mean of the three principal stress values.}$$

The principal stresses in any three-dimensional complex stress system may now be written in the form

$$\sigma_1 = \text{mean stress} + \text{deviation from the mean}$$

$$= hydrostatic \text{ stress} + deviatoric \text{ stress}$$

Thus the additional terms required to make up any stress value from the datum to the absolute value are termed the *deviatoric stresses* and written with a prime superscript,

i.e. $$\sigma_1 = \bar{\sigma} + \sigma', \quad \text{etc.}$$

Cartesian stresses σ_{xx}, σ_{yy} and σ_{zz} can now be referred to the new datum as follows:

$$\left.\begin{aligned}
\sigma'_{xx} &= \sigma_{xx} - \bar{\sigma} = \tfrac{1}{3}(2\sigma_{xx} - \sigma_{yy} - \sigma_{zz}) \\
\sigma'_{yy} &= \sigma_{yy} - \bar{\sigma} = \tfrac{1}{3}(2\sigma_{yy} - \sigma_{xx} - \sigma_{zz}) \\
\sigma'_{zz} &= \sigma_{zz} - \bar{\sigma} = \tfrac{1}{3}(2\sigma_{zz} - \sigma_{xx} - \sigma_{yy})
\end{aligned}\right\} \qquad (8.61)$$

All the above values then represent deviatoric stresses.

It may be observed that the system used for representing stresses in terms of the datum stress and the deviation from the datum is, in effect, a consideration of the normal and shear stresses respectively, on the octahedral planes, since the octahedral and deviatoric planes are equally inclined to all three axes ($l = m = n = \pm 1/\sqrt{3}$) and the selected datum stress

$$\bar{\sigma} = \tfrac{1}{3}(\sigma_1 + \sigma_2 + \sigma_3)$$

is also the octahedral normal stress value.

As stated earlier when discussing octahedral stresses, this has a particular relevance to the yield behaviour of materials.

Whilst any detailed study of the theory of plasticity is beyond the scope of this text, the fundamental requirements of the theory should be understood. These are:

(a) the volume of material remains constant under plastic deformation;
(b) the hydrostatic stress component $\bar{\sigma}$ does not cause yielding of the material;
(c) the hydrostatic stress component $\bar{\sigma}$ does not influence the point at which yielding occurs.

From these points it is clear that it is therefore **the deviatoric or octahedral shear stresses which must govern the yield behaviour of materials**. This is supported by the accuracy of the octahedral shear stress (distortion energy) theory and, to a lesser extent, the maximum shear stress theory, in predicting the elastic failure of *ductile* materials. Both theories involve stress differences, i.e. shear stresses, and are therefore independent of the hydrostatic stress as indicated by (b) above.

The representation of a principal stress system in terms of the octahedral and deviatoric stresses may thus be shown as in Fig. 8.27.

It should now be clear that the terms *hydrostatic, volumetric, mean, dilational and octahedral normal stresses* all indicate the same quantity.

The standard elastic stress–strain relationships of eqn. (8.71)

$$\varepsilon_{xx} = \tfrac{1}{E}[\sigma_{xx} - \nu\sigma_{yy} - \nu\sigma_{zz}]$$

$$\varepsilon_{yy} = \tfrac{1}{E}[\sigma_{yy} - \nu\sigma_{xx} - \nu\sigma_{zz}]$$

$$\varepsilon_{zz} = \tfrac{1}{E}[\sigma_{zz} - \nu\sigma_{xx} - \nu\sigma_{yy}]$$

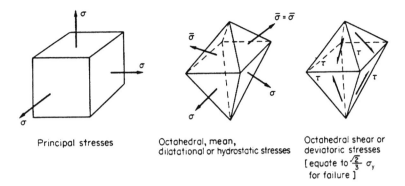

Principal stresses   Octahedral, mean,   Octahedral shear or
                      dilatational or hydrostatic stresses   deviatoric stresses
                                                   [equate to $\sqrt{\frac{2}{3}}\,\sigma_y$
                                                       for failure]

Fig. 8.27. Representation of a principal stress system in terms of octahedral and deviatoric stresses.

may be re-written in a form which distinguishes between those parts which contribute only to a change in volume and those producing a change of shape.

Thus, for a hydrostatic or mean stress $\sigma_m = \frac{1}{3}(\sigma_{xx} + \sigma_{yy} + \sigma_{zz})$ and remembering the relationship between the elastic constants $E = 2G(1 + v)$

$$\left.\begin{aligned}
\varepsilon_{xx} &= \frac{1}{E}(1 - 2v)\sigma_m + \frac{1}{2G}(\sigma_{xx} - \sigma_m) \\[4pt]
\varepsilon_{yy} &= \frac{1}{E}(1 - 2v)\sigma_m + \frac{1}{2G}(\sigma_{yy} - \sigma_m) \\[4pt]
\varepsilon_{zz} &= \frac{1}{E}(1 - 2v)\sigma_m + \frac{1}{2G}(\sigma_{zz} - \sigma_m)
\end{aligned}\right\} \tag{8.62}$$

with $\gamma_{xy} = \tau_{xy}/2G$; $\gamma_{yz} = \tau_{yz}/2G$; $\gamma_{zx} = \tau_{zx}/2G$.

The terms $(\sigma_{xx} - \sigma_m)$, $(\sigma_{yy} - \sigma_m)$ and $(\sigma_{zz} - \sigma_m)$ are the *deviatoric* components of stress.

The volumetric strain ε_m associated with the hydrostatic or mean stress σ_m is then:

$$\varepsilon_m = \frac{\sigma_m}{K} = \varepsilon_{xx} + \varepsilon_{yy} + \varepsilon_{zz}$$

where K is the bulk modulus.

8.21. Deviatoric strains

As for the deviatoric stresses the *deviatoric strains* are also defined with reference to some selected "false zero" or datum value,

$$\bar{\varepsilon} = \tfrac{1}{3}(\varepsilon_1 + \varepsilon_2 + \varepsilon_3) \tag{8.63}$$

$$= \text{mean of the three principal strain values.}$$

Thus, referred to the new datum, the principal strain values become

$$\varepsilon_1' = \varepsilon_1 - \bar{\varepsilon} = \varepsilon_1 - \tfrac{1}{3}(\varepsilon_1 + \varepsilon_2 + \varepsilon_3)$$

$\therefore$

$$\varepsilon'_1 = \tfrac{1}{3}(2\varepsilon_1 - \varepsilon_2 - \varepsilon_3)$$

Similarly,

$$\varepsilon'_2 = \tfrac{1}{3}(2\varepsilon_2 - \varepsilon_1 - \varepsilon_3) \qquad (8.64)$$

$$\varepsilon'_3 = \tfrac{1}{3}(2\varepsilon_3 - \varepsilon_1 - \varepsilon_2)$$

and these are the so-called *deviatoric strains*. It may now be observed that the following relationship applies:

$$\varepsilon'_1 + \varepsilon'_2 + \varepsilon'_3 = 0 \qquad (8.65)$$

It can also be shown that the deviatoric strains are related to the principal strains as follows:

$$(\varepsilon'_1)^2 + (\varepsilon'_2)^2 + (\varepsilon'_3)^2 = \tfrac{1}{3}[(\varepsilon_1 - \varepsilon_2)^2 + (\varepsilon_2 - \varepsilon_3)^2 + (\varepsilon_3 - \varepsilon_1)^2] \qquad (8.66)$$

8.22. Plane stress and plane strain

If a body consists of two parallel planes a constant thickness apart and bounded by any closed surface as shown in Fig. 8.28, it is said to be a *plane body*. Associated with this type of body there is a particular class of problems within the general theory of elasticity which are termed *plane elastic* problems, and these allow a number of simplifying assumptions in their treatment.

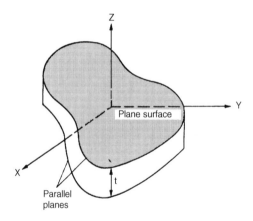

Fig. 8.28. A plane element.

In order to qualify for these simplifications, however, there are a number of restrictions which must be placed on the applied load system:

(1) no loads may be applied to the top and bottom plane surfaces (in practice there is often a uniform stress in the Z direction on the planes but this can always be reduced to zero by superimposing a suitable stress σ_{zz} of opposite sign);

(2) the loads on the lateral boundaries (and the surface shears) must be in the plane of the body and must be uniformly distributed across the thickness;

(3) similarly, body forces in the X and Y directions directions must be uniform across the thickness and the body force in the Z direction must be zero.

There is no limitation on the thickness of the plane body and, indeed, the thickness serves as a means of classification within the general type of problem. Normally a *plane stress* approach is applied to members which are relatively thin in relation to their other dimensions, whereas *plane strain* methods are employed for relatively thick members. The terms plane stress and plane strain are defined in detail below.

The plane elastic type of problem may thus be defined as one in which stresses and strains do not vary in the Z direction. Additionally, lines parallel to the Z axis remain straight and parallel to the axis throughout loading.

i.e.
$$\gamma_{zx} = \gamma_{zy} = 0$$

(The problem of torsion provides an exception to this rule.)

8.22.1. Plane stress

A plane stress problem is taken to be one in which σ_{zz} is zero. As stated above, cases where a uniform stress is applied to the plane surfaces can easily be reduced to this condition by application of a suitable σ_{zz} stress of opposite sign. Shear components in the Z direction must also be zero.

i.e.
$$\tau_{zx} = \tau_{zy} = 0 \qquad\qquad (8.67)$$

Under these conditions the stress equations of equilibrium in cartesian coordinates reduce to

$$\left.\begin{array}{l} \dfrac{\partial \sigma_{xx}}{\partial x} + \dfrac{\partial \tau_{xy}}{\partial y} + X = 0 \\[3mm] \dfrac{\partial \tau_{xy}}{\partial x} + \dfrac{\partial \sigma_{yy}}{\partial y} + Y = 0 \end{array}\right\} \qquad\qquad (8.68)$$

The following stress and strain relationships then apply:

$$\varepsilon_{xx} = \frac{1}{E}(\sigma_{xx} - v\sigma_{yy}) \qquad \varepsilon_{yy} = \frac{1}{E}(\sigma_{yy} - v\sigma_{xx})$$

$$\sigma_{xx} = \frac{E}{(1 - v^2)}[\varepsilon_{xx} + v\varepsilon_{yy}] \qquad \sigma_{yy} = \frac{E}{(1 - v^2)}[\varepsilon_{yy} + v\varepsilon_{xx}]$$

$$\tau_{xy} = G\gamma_{xy} = \frac{E}{2(1 + v)}\gamma_{xy}$$

Plane stress systems are often referred to as *two-dimensional* or *bi-axial* stress systems, a typical example of which is the case of thin plates loaded at their edges with forces applied in the plane of the plate.

8.22.2. Plane strain

Plane strain problems are normally defined as those in which the strains in the Z direction are zero. Again, problems with a uniform strain in the Z direction at all points on the plane surface can be reduced to the above case by the addition of a suitable uniform stress σ_{zz}, the additional lateral strains and displacements so introduced being easily calculated.
Thus
$$\varepsilon_{zz} = \gamma_{yz} = \gamma_{zx} = 0 \qquad\qquad (8.69)$$

Also, from the basic assumptions of plane elastic problems,

$$\tau_{zy} = \tau_{zx} = 0$$

The equations of stress equilibrium in this case reduce to

$$\left. \begin{array}{c} \dfrac{\partial \sigma_{xx}}{\partial x} + \dfrac{\partial \tau_{xy}}{\partial y} + X = 0 \\[3mm] \dfrac{\partial \tau_{xy}}{\partial x} + \dfrac{\partial \sigma_{yy}}{\partial y} + Y = 0 \end{array} \right\} \tag{8.70}$$

The stress–strain relations are then as follows:

$$\varepsilon_{xx} = \frac{(1 - v^2)}{E}\left[\sigma_{xx} - \frac{v}{(1 - v)}\sigma_{yy}\right] \quad \text{and} \quad \varepsilon_{yy} = \frac{(1 - v^2)}{E}\left[\sigma_{yy} - \frac{v}{(1 - v)}\sigma_{xx}\right]$$

$$\sigma_{xx} = \frac{E(1 - v)}{(1 + v)(1 - 2v)}\left[\varepsilon_{xx} + \frac{v}{(1 - v)}\varepsilon_{yy}\right]$$

$$\sigma_{yy} = \frac{E(1 - v)}{(1 + v)(1 - 2v)}\left[\varepsilon_{yy} + \frac{v}{(1 - v)}\varepsilon_{xx}\right]$$

Also
$$\tau_{xy} = G\gamma_{xy}$$

It should be noted that the plane strain equations can be derived simply from the plane stress equations by replacing

$$v \text{ by } \frac{v}{(1 - v)} \quad \text{and} \quad E \text{ by } \frac{E}{(1 - v^2)}$$

A typical example of plane strain is the pressurisation of long cylinders where the above equations given accurate results, particularly in the middle portion of the cylinder, whether the end conditions are free, partially fixed or rigidly fixed.

An example of the transfer of a plane stress to a corresponding plane strain solution is given when the relevant equations for the hoop and radial stresses present in rotating thick cylinders are readily obtained from those of rotating thin discs by use of the substitution $v/(1 - v)$ in place of v (see §4.4).

8.23. The stress–strain relations

The following formulae form a useful summary of the relationships which exist between the stresses and strains in a general three-dimensional stress system.

(a) Strains in terms of stress

$$\left. \begin{array}{ll} \varepsilon_{xx} = \dfrac{1}{E}[\sigma_{xx} - v(\sigma_{yy} + \sigma_{zz})] & \gamma_{xy} = \dfrac{2(1 + v)}{E}\tau_{xy} = \dfrac{\tau_{xy}}{G} \\[3mm] \varepsilon_{yy} = \dfrac{1}{E}[\sigma_{yy} - v(\sigma_{xx} + \sigma_{zz})] & \gamma_{yz} = \dfrac{2(1 + v)}{E}\tau_{yz} = \dfrac{\tau_{yz}}{G} \\[3mm] \varepsilon_{zz} = \dfrac{1}{E}[\sigma_{zz} - v(\sigma_{xx} + \sigma_{yy})] & \gamma_{zx} = \dfrac{2(1 + v)}{E}\tau_{zx} = \dfrac{\tau_{zx}}{G} \end{array} \right\} \tag{8.71}$$

(b) Stresses in terms of strains

$$\left.\begin{array}{ll}
\sigma_{xx} = \dfrac{E}{(1+v)(1-2v)}[\varepsilon_{xx} + v(\varepsilon_{yy} + \varepsilon_{zz} - \varepsilon_{xx}) & \tau_{xy} = \dfrac{E}{2(1+v)}\gamma_{xy} = G\gamma_{xy} \\[3mm]
\sigma_{yy} = \dfrac{E}{(1+v)(1-2v)}[\varepsilon_{yy} + v(\varepsilon_{xx} + \varepsilon_{zz} - \varepsilon_{yy}) & \tau_{yz} = \dfrac{E}{2(1+v)}\gamma_{yz} = G\gamma_{yz} \\[3mm]
\sigma_{zz} = \dfrac{E}{(1+v)(1-2v)}[\varepsilon_{zz} + v(\varepsilon_{xx} + \varepsilon_{yy} - \varepsilon_{zz}) & \tau_{zx} = \dfrac{E}{2(1+v)}\gamma_{zx} = G\gamma_{zx}
\end{array}\right\} \quad (8.72)$$

with $\qquad\qquad E = 2G(1+v) \quad\text{and}\quad E = 3K(1-2v)$

hence $\qquad\qquad K = \dfrac{2G(1+v)}{3(1-2v)}$

(c) For biaxial stress conditions:

(a) Strains in terms of stresses

$$\varepsilon_{xx} = \frac{1}{E}[\sigma_{xx} - v\sigma_{yy}]$$

$$\varepsilon_{yy} = \frac{1}{E}[\sigma_{yy} - v\sigma_{xx}] \quad\text{and}\quad \gamma_{xy} = \frac{2(1+v)}{E}\tau_{xy}$$

$$\varepsilon_{zz} = -\frac{v}{E}[\sigma_{xx} + \sigma_{yy}]$$

(b) Stresses in terms of strains

$$\sigma_{xx} = \frac{E}{(1-v^2)}[\varepsilon_{xx} + v\varepsilon_{yy}] \quad\text{and}\quad \tau_{xy} = \frac{E}{2(1+v)}\gamma_{xy}$$

$$\sigma_{yy} = \frac{E}{(1-v^2)}[\varepsilon_{yy} + v\varepsilon_{xx}]$$

Equivalent expressions apply for polar coordinates with r, θ and z replacing x, y and z respectively.

8.24. The strain–displacement relationships

Consider the deformation of a cubic element of material as load is applied. Any corner of the element, e.g. P, will then move to some position P', the movement having components u, v and w in the X, Y and Z directions respectively as shown in Fig. 8.29. Other points in the cube will also be displaced but generally by different amounts.

The movement in the X direction will be given by

$$u = \left(\frac{\partial u}{\partial x}\right)\delta x$$

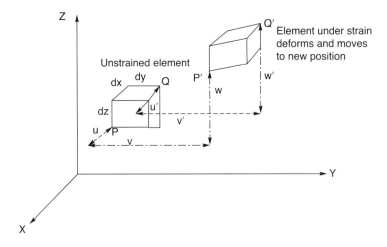

Fig. 8.29. Deformation of a cubical element under load.

The *strain* in the X direction will then be

$$\varepsilon_{xx} = \frac{\text{change in length}}{\text{original length}} = \frac{\left(\dfrac{\partial u}{\partial x}\right)\delta x}{\delta x}$$

i.e.
$$\left. \begin{array}{l} \varepsilon_{xx} = \dfrac{\partial u}{\partial x} \\[2mm] \text{Similarly,} \quad \varepsilon_{yy} = \dfrac{\partial v}{\partial y} \\[2mm] \varepsilon_{zz} = \dfrac{\partial w}{\partial z} \end{array} \right\} \tag{8.73}$$

Consider now Fig. 8.30 which shows the deformations in the XY plane enlarged.

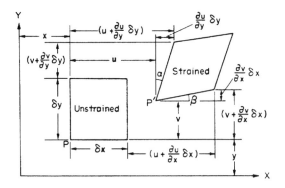

Fig. 8.30. Deformations under load in the XY plane.

Shear strains are defined as angles of deformation or changes in angles between two perpendicular segments. Thus γ_{xy} is the change in angle between two perpendicular segments in the *XY* plane as load is applied,

i.e. $$\gamma_{xy} = \alpha + \beta = \frac{\left(\dfrac{\partial v}{\partial x}\right)\delta x}{\delta x} + \frac{\left(\dfrac{\partial u}{\partial y}\right)\delta y}{\delta y}$$

$\therefore$

$$\left.\begin{aligned} \gamma_{xy} &= \frac{\partial v}{\partial x} + \frac{\partial u}{\partial y} \\[2mm] \gamma_{yz} &= \frac{\partial w}{\partial y} + \frac{\partial v}{\partial z} \\[2mm] \gamma_{zx} &= \frac{\partial u}{\partial z} + \frac{\partial w}{\partial x} \end{aligned}\right\} \tag{8.74}$$

Similarly,

and

Summary of the strain–displacement equations

(a) *In cartesian coordinates* with displacements u, v and w along x, y and z respectively.

$$\varepsilon_{xx} = \frac{\partial u}{\partial x} \qquad \gamma_{xy} = \frac{\partial v}{\partial x} + \frac{\partial u}{\partial y}$$

$$\varepsilon_{yy} = \frac{\partial v}{\partial y} \qquad \gamma_{yz} = \frac{\partial w}{\partial y} + \frac{\partial v}{\partial z}$$

$$\varepsilon_{zz} = \frac{\partial w}{\partial z} \qquad \gamma_{zx} = \frac{\partial u}{\partial z} + \frac{\partial w}{\partial x}$$

(b) *In polar coordinates* with displacements u_r, u_θ and u_z along r, θ and z respectively: these equations become:

$$\varepsilon_{rr} = \frac{\partial u_r}{\partial r}$$

$$\varepsilon_{\theta\theta} = \frac{u_r}{r} + \frac{1}{r}\cdot\frac{\partial u_\theta}{\partial \theta}$$

$$\varepsilon_{zz} = \frac{\partial u_z}{\partial z}$$

with

$$\gamma_{r\theta} = \frac{1}{r}\cdot\frac{\partial u_r}{\partial \theta} + \frac{\partial u_\theta}{\partial r} - \frac{u_\theta}{r}$$

$$\gamma_{\theta z} = \frac{1}{r}\cdot\frac{\partial u_z}{\partial \theta} + \frac{\partial u_\theta}{\partial z}$$

$$\gamma_{zr} = \frac{\partial u_r}{\partial z} + \frac{\partial u_z}{\partial r}$$

8.25. The strain equations of transformation

Using the experimental or theoretical procedures described in earlier sections it is possible to derive the values of the direct and shear stresses acting at a point on a body. These are normally obtained with reference to some convenient set of X, Y coordinates which, for example, may be parallel to the edges of the component considered. Sometimes, however, it may be more convenient to refer the values obtained to some other set of axes $X'Y'$ at an angle θ to the original axes.

In this case the two-dimensional versions of eqns. (8.73) and (8.74) apply equally well to the new axes (Fig. 8.31),

i.e.
$$\varepsilon_{x'x'} = \frac{\partial u'}{\partial x'} \quad \varepsilon_{y'y'} = \frac{\partial v'}{\partial y'} \quad \text{and} \quad \gamma_{x'y'} = \frac{\partial v'}{\partial x'} + \frac{\partial u'}{\partial y'}$$

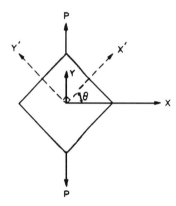

Fig. 8.31. Alternative coordinates to which strains may be referred.

Now, using the partial differentiation chain rule,

$$\frac{\partial u'}{\partial x'} = \left[\frac{\partial}{\partial x}\frac{\partial x}{\partial x'} + \frac{\partial}{\partial y}\frac{\partial y}{\partial x'} \right] u'$$

$$= \left[\cos\theta \frac{\partial}{\partial x} + \sin\theta \frac{\partial}{\partial y} \right] (u\cos\theta + v\sin\theta)$$

$$= \cos^2\theta \frac{\partial u}{\partial x} + \sin^2\theta \frac{\partial v}{\partial y} + \left(\frac{\partial v}{\partial x} + \frac{\partial u}{\partial y} \right) \sin\theta \cos\theta$$

$$\therefore \qquad \varepsilon_{x'x'} = \varepsilon_{xx}\cos^2\theta + \varepsilon_{yy}\sin^2\theta + \gamma_{xy}\sin\theta\cos\theta$$

Or, in terms of the double angle 2θ,

$$\varepsilon_{x'x'} = \tfrac{1}{2}(\varepsilon_{xx} + \varepsilon_{yy}) + \tfrac{1}{2}(\varepsilon_{xx} - \varepsilon_{yy})\cos 2\theta + \tfrac{1}{2}\gamma_{xy}\sin 2\theta \qquad (8.75)$$

This is the same as eqn. (14.14) obtained in §14.10[†] for the normal strain on any plane in terms of the coordinate strains. Indeed, the above represents an alternative proof for what are really similar requirements.

[†] E.J. Hearn, *Mechanics of Materials 1*, Butterworth-Heinemann, 1997.

8.26. Compatibility

Equations (8.73) and (8.74) relate the six components of strain (three direct and three shear) to the equivalent displacements under a three-dimensional stress system. If, however, the situation arises where the six strain components are known, as they could well be following some theoretical or experimental strain analysis, then the above equations represent three in excess of that required for solution of the three unknown displacements (three unknowns require only three equations for solution). Thus, unless the solution obtained from any three equations satisfies the other three equations, then the values cannot be accepted as a valid solution. Certain specific relations must therefore be satisfied before a valid solution is obtained and these are termed *the compatibility relations*.

The problem can be considered physically as follows: consider a body divided into a large number of small cubic elements. When load is applied the elements deform and simple measurements of length and angle changes will yield the direct and shear strains in each element. These can be summated to produce the overall component strains if required. If, however, the deformed elements are separated and provided in their deformed shapes as a jigsaw puzzle, the puzzle can only be completed, i.e. the elements fully assembled without voids or discontinuities, if each element is correctly strained or deformed. The procedure used to check this condition then represents the compatibility equations. The compatibility relationships in terms of strain are derived as follows:

$$\varepsilon_{xx} = \frac{\partial u}{\partial x} \quad \therefore \quad \frac{\partial^2 \varepsilon_{xx}}{\partial y^2} = \frac{\partial^3 u}{\partial x \partial y^2}$$

$$\varepsilon_{yy} = \frac{\partial v}{\partial y} \quad \therefore \quad \frac{\partial^2 \varepsilon_{yy}}{\partial x^2} = \frac{\partial^2 v}{\partial x^2 \partial y}$$

But

$$\gamma_{xy} = \frac{\partial u}{\partial y} + \frac{\partial v}{\partial x}$$

Therefore differentiating once with respect to x and once with respect to y,

$$\frac{\partial^2 \gamma_{xy}}{\partial x \partial y} = \frac{\partial^3 u}{\partial x \partial y^2} + \frac{\partial^3 v}{\partial x^2 \partial y}$$

i.e.

$$\left. \begin{aligned} \frac{\partial^2 \gamma_{xy}}{\partial x \partial y} &= \frac{\partial^2 \varepsilon_{xx}}{\partial y^2} + \frac{\partial^2 \varepsilon_{yy}}{\partial x^2} \\ \frac{\partial^2 \gamma_{yz}}{\partial y \partial z} &= \frac{\partial^2 \varepsilon_{yy}}{\partial z^2} + \frac{\partial^2 \varepsilon_{zz}}{\partial y^2} \\ \frac{\partial^2 \gamma_{zx}}{\partial z \partial x} &= \frac{\partial^2 \varepsilon_{zz}}{\partial x^2} + \frac{\partial^2 \varepsilon_{xx}}{\partial z^2} \end{aligned} \right\}$$

Similarly,

and

(8.76)

These are three of the compatibility equations.

It can also be shown† that a further three compatibility relationships apply, namely

† A.E.H. Love, *Treatise on the Mathematical Theory of Elasticity*, 4th edn., Dover Press, New York, 1944.

$$2\frac{\partial^2 \varepsilon_{xx}}{\partial y\,\partial z} = \frac{\partial}{\partial x}\left[\frac{\partial \gamma_{xy}}{\partial z} + \frac{\partial \gamma_{zx}}{\partial y} - \frac{\partial \gamma_{yz}}{\partial x}\right] \left.\begin{array}{c} \\ \\ \\ \end{array}\right\}$$

$$2\frac{\partial^2 \varepsilon_{yy}}{\partial z\,\partial x} = \frac{\partial}{\partial y}\left[\frac{\partial \gamma_{xy}}{\partial z} + \frac{\partial \gamma_{zx}}{\partial y} - \frac{\partial \gamma_{yz}}{\partial x}\right] \left.\begin{array}{c} \\ \\ \\ \end{array}\right\} \quad (8.77)$$

$$2\frac{\partial^2 \varepsilon_{zz}}{\partial x\,\partial y} = \frac{\partial}{\partial z}\left[-\frac{\partial \gamma_{xy}}{\partial z} + \frac{\partial \gamma_{zx}}{\partial y} + \frac{\partial \gamma_{yz}}{\partial x}\right]$$

The compatibility equations can also be written in terms of stress as follows:
Consider the first of the strain compatibility relationships given in eqn. (8.41).

i.e.
$$\frac{\partial^2 \varepsilon_{xx}}{\partial y^2} + \frac{\partial^2 \varepsilon_{yy}}{\partial x^2} = \frac{\partial^2 \gamma_{xy}}{\partial x\partial y}$$

For plane strain conditions (and a similar derivation shows that the equation derived is equally appropriate for plane stress) we have:

$$\varepsilon_{xx} = \frac{1}{E}(\sigma_{xx} - v\sigma_{yy})$$

$$\varepsilon_{yy} = \frac{1}{E}(\sigma_{yy} - v\sigma_{xx})$$

and
$$\gamma_{xy} = \frac{2(1+v)}{E}\tau_{xy}$$

Substituting:

$$\frac{1}{E}\frac{\partial^2 \sigma_{xx}}{\partial y^2} - \frac{v}{E}\frac{\partial^2 \sigma_{yy}}{\partial y^2} + \frac{1}{E}\frac{\partial^2 \sigma_{yy}}{\partial x^2} - \frac{v}{E}\frac{\partial^2 \sigma_{xx}}{\partial x^2} = \frac{2(1+v)}{E}\frac{\partial^2 \tau_{xy}}{\partial x\partial y} \quad (1)$$

Now from the equilibrium equations assuming plane stress and zero body force stresses we have:

$$\frac{\partial \sigma_{xx}}{\partial x} + \frac{\partial \tau_{xy}}{\partial y} = 0 \quad (2)$$

$$\frac{\partial \tau_{xy}}{\partial x} + \frac{\partial \sigma_{yy}}{\partial y} = 0 \quad (3)$$

Differentiating (2) with respect to x and (3) with respect to y and adding we have:

$$\frac{\partial^2 \sigma_{xx}}{\partial x^2} + \frac{\partial^2 \sigma_{yy}}{\partial y^2} = -2\frac{\partial^2 \tau_{xy}}{\partial x\partial y} \quad (4)$$

Eliminating τ_{xy} between eqns (4) and (1) we obtain:

$$\frac{1}{E}\frac{\partial^2 \sigma_{xx}}{\partial y^2} - \frac{v}{E}\frac{\partial^2 \sigma_{yy}}{\partial y^2} + \frac{1}{E}\frac{\partial^2 \sigma_{yy}}{\partial x^2} - \frac{v}{E}\frac{\partial^2 \sigma_{xx}}{\partial x^2} = -\frac{(1+v)}{E}\frac{\partial^2 \sigma_{xx}}{\partial x^2} + \frac{\partial^2 \sigma_{yy}}{\partial y^2}$$

i.e.
$$\frac{\partial^2 \sigma_{xx}}{\partial x^2} + \frac{\partial^2 \sigma_{yy}}{\partial x^2} + \frac{\partial^2 \sigma_{xx}}{\partial y^2} + \frac{\partial^2 \sigma_{yy}}{\partial y^2} = 0$$

or
$$\left[\frac{\partial^2}{\partial x^2} + \frac{\partial^2}{\partial y^2}\right](\sigma_x + \sigma_y) = 0 \quad (8.78)$$

A similar development for cylindrical coordinates yields the stress equation of compability

$$\left[\frac{\partial^2}{\partial r^2} + \frac{1}{r} \cdot \frac{\partial}{\partial r} \right] (\sigma_{rr} + \sigma_{\theta\theta}) = 0 \tag{8.79}$$

which in the case of axial symmetry (where stresses are independent of θ) reduces to:

$$\left[\frac{\partial^2}{\partial r^2} + \frac{1}{r} \cdot \frac{\partial}{\partial r} + \frac{1}{r^2} \frac{\partial^2}{\partial \theta^2} \right] (\sigma_{rr} + \sigma_{\theta\theta}) = 0. \tag{8.80}$$

8.27. The stress function concept

From the earlier work of this chapter it should now be evident that in elastic stress analysis there are generally fifteen unknown quantities to be determined; six stresses, six strains and three displacements. These are functions of the independent variables x, y and z (in cartesian coordinates) or r, θ and z (in cylindrical polar coordinates). A quick look at the governing equations presented earlier in the chapter will convince the reader that the equations are difficult to solve for these unknowns, except for a number of relatively simple problems.

In order to extend the range of useful solutions several techniques are available. In the first instance one may make certain assumptions about the physical problem in an effort to simplify the equations. For example, are the loading and boundary conditions such that:

 (i) the plane stress assumption is adequate – as in a thin-walled pressure vessel? or,
(ii) does plane strain exist – as in the case of a pressurised thick cylinder?

If we can convince ourselves that these assumptions are valid we reduce the three-dimensional problem to the two-dimensional case.

Having simplified the governing differential equations one must then devise techniques to solve, or further reduce, their complexity. One such concept was that proposed by Sir George B. Airy.[†]. His approach was to assume that the stresses in the two-dimensional problem σ_{xx}, σ_{yy} and τ_{xy} could be described by a single function of x and y. This function ϕ is referred to as a "*stress function*" (later the "*Airy stress function*") and it appears to be the first time that such a concept was used. Airy's approach was later generalised for the three-dimensional case by Clerk Maxwell.[‡]

Airy proposed that the stresses be derived from a particular function ϕ such that:

$$\left. \begin{array}{l} \sigma_{xx} = \dfrac{\partial^2 \phi}{\partial y^2} \\[2mm] \sigma_{yy} = \dfrac{\partial^2 \phi}{\partial x^2} \\[2mm] \tau_{xy} = -\dfrac{\partial^2 \phi}{\partial x \partial y} \end{array} \right\} \tag{8.81}$$

[†] G.B. Airy, *Brit. Assoc. Advancement of Sci. Rep.* 1862; *Phil. Trans. Roy. Soc.* **153** (1863), 49–80
[‡] J.C. Maxwell *Edinburgh Roy. Soc. Trans.*, **26** (1872), 1–40.

It should be noted that these equations satisfy the two-dimensional versions of equilibrium equations (8.38):

i.e.
$$\left.\begin{array}{c} \dfrac{\partial \sigma_{xx}}{\partial x} + \dfrac{\partial \tau_{xy}}{\partial y} = 0 \\[4mm] \dfrac{\partial \tau_{xy}}{\partial x} + \dfrac{\partial \sigma_{yy}}{\partial y} = 0 \end{array}\right\} \qquad (8.82)$$

It is also necessary that the stress function ϕ must not only satisfy the equilibrium conditions of the problem but must also satisfy the compatibility relationships, i.e. eqn. 8.76. For the two-dimensional case these reduce to:

$$\frac{\partial^2 \gamma_{xy}}{\partial x \partial y} = \frac{\partial^2 \varepsilon_{xx}}{\partial y^2} + \frac{\partial^2 \varepsilon_{yy}}{\partial x^2} \qquad (8.76)\text{bis.}$$

This equation can be written in terms of stress using the appropriate constitutive (stress–strain) relations. To illustrate the procedure the *plane strain* case will be considered. In this the relevant equations are:

$$\left.\begin{array}{c} \varepsilon_{xx} = \dfrac{(1 - v^2)}{E}\left[\sigma_{xx} - \dfrac{v}{(1 - v)}\sigma_{yy}\right] \\[4mm] \varepsilon_{yy} = \dfrac{(1 - v^2)}{E}\left[\sigma_{yy} - \dfrac{v}{(1 - v)}\sigma_{xx}\right] \\[4mm] \gamma_{xy} = \dfrac{\tau_{xy}}{G} = \dfrac{2(1 + v)}{E}\tau_{xy} \end{array}\right\} \qquad (8.83)$$

By substituting these into the compatibility equation (8.76) the following is obtained:

$$2\frac{\partial^2 \tau_{xy}}{\partial x \partial y} = \frac{\partial^2}{\partial y^2}[(1 - v)\sigma_{xx} - v\sigma_{yy}] + \frac{\partial^2}{\partial x^2}[(1 - v)\sigma_{yy} - v\sigma_{xx}]$$

From the equilibrium eqn. (8.70) we get:

$$2\frac{\partial^2 \tau_{xy}}{\partial x \partial y} = -\left(\frac{\partial^2 \sigma_{xx}}{\partial x^2} + \frac{\partial^2 \sigma_{yy}}{\partial y^2}\right) - \left(\frac{\partial X}{\partial x} + \frac{\partial Y}{\partial y}\right)$$

Combining these equations to eliminate the shear stress τ_{xy}, gives:

$$\left(\frac{\partial^2}{\partial x^2} + \frac{\partial^2}{\partial y^2}\right)(\sigma_{xx} + \sigma_{yy}) = -\frac{1}{(1 - v)}\left(\frac{\partial X}{\partial x} + \frac{\partial Y}{\partial y}\right) \qquad (8.84)$$

A similar equation can be obtained for the *plane stress* case, namely:

$$\left(\frac{\partial^2}{\partial x^2} + \frac{\partial^2}{\partial y^2}\right)(\sigma_{xx} + \sigma_{yy}) = -(1 + v)\left(\frac{\partial X}{\partial x} + \frac{\partial Y}{\partial y}\right) \qquad (8.85)$$

If the body forces X and Y have constant values the same equation holds for both plane stress and plane strain, namely:

$$\left(\frac{\partial^2}{\partial x^2} + \frac{\partial^2}{\partial y^2}\right)(\sigma_{xx} + \sigma_{yy}) = \nabla^2(\sigma_{xx} + \sigma_{yy}) = 0 \qquad (8.86)$$

This equation is known as the "*Laplace differential equation*" or the "*harmonic differential equation*." The function $(\sigma_{xx} + \sigma_{yy})$ is referred to as a "harmonic" function. It is interesting to note that the Laplace equation, which of course incorporates all the previous equations, does not contain the elastic constants of the material. **This is an important conclusion for the experimentalist since, providing there exists geometric similarity, material isotropy and linearity and similar applied loading of both model and prototype, then the stress distribution per unit load will be identical in each**. The stress function, previously defined, must satisfy the 'Laplace equation' (8.86). Thus:

$$\left(\frac{\partial^2}{\partial x^2} + \frac{\partial^2}{\partial y^2} \right) \left(\frac{\partial^2}{\partial x^2} + \frac{\partial^2}{\partial y^2} \right) \phi = 0$$

or,
$$\frac{\partial^4 \phi}{\partial x^4} + \frac{2 \partial^4 \phi}{\partial x^2 \partial y^2} + \frac{\partial^4 \phi}{\partial y^4} = 0 \tag{8.87}$$

Alternatively, this can be re-written in the form

$$\left(\frac{\partial^2}{\partial x^2} + \frac{\partial^2}{\partial y^2} \right)^2 \phi = 0$$

or abbreviated to
$$\nabla^4 \phi = 0 \tag{8.88}$$

indicating that the stress function must be a biharmonic function. Equation (8.87) is often referred to as the "*biharmonic equation*" with ϕ known as the "*Airy stress function*".

It is worth noting, at this point in the development, that the problem of plane strain, or plane stress, has been reduced to seeking a solution of the biharmonic equation (8.87) such that the stress components satisfy the boundary conditions of the problem.

Thus, provided that a suitable polynomial expression in x and y (or r and θ) is used for the stress function ϕ then both equilibrium and compatibility are automatically assured. Consideration of the boundary conditions associated with any particular stress system will then yield the appropriate coefficients of the various terms of the polynomial and a complete solution is obtained.

8.27.1. Forms of Airy stress function in Cartesian coordinates

The stress function concept was developed by Airy initially to investigate the bending theory of straight rectangular beams. It was thus natural that a rectangular cartesian coordinate system be used. As an introduction to this topic, therefore, forms of stress function in cartesian coordinates will be explored and applied to a number of fairly simple beam problems. It is hoped that the reader will gain confidence in using the approach and be able to tackle a range of more interesting problems where cylindrical polars (r, θ) is an appropriate alternative coordinate system.

(a) The eqns. (8.81) which define the stress function imply that the most simple function of ϕ to produce a stress is $\phi = Ax^2$, since the lower orders when differentiated twice give a zero result. Substituting this into eqns. (8.81) gives:

$$\sigma_{xx} = 0, \quad \sigma_{yy} = 2A \quad \text{and} \quad \tau_{xy} = 0$$

Thus a stress function of the form $\phi = Ax^2$ can be used to describe a condition of constant stress $2A$ in the y direction over the entire region of a component, e.g. uniform tension or compression testing

(b) $$\phi = By^3.$$

For this stress function

$$\sigma_{xx} = \frac{\partial^2 \phi}{\partial y^2} = 6By$$

with σ_{yy} and τ_{xy} zero.

Thus σ_{xx} is a linear function of vertical dimension y, a situation typical of beam bending.

(c) $$\phi = Ax^2 + Bxy + Cy^2.$$

In this case

$$\sigma_{xx} = \frac{\partial^2 \phi}{\partial y^2} = 2C \quad \text{(a constant)}$$

$$\sigma_{yy} = \frac{\partial^2 \phi}{\partial x^2} = 2A \quad \text{(a constant)}$$

$$\tau_{xy} = \frac{\partial^2 \phi}{\partial x \partial y} = -B \quad \text{(a constant)}$$

and the stress function is suitable for any uniform plane stress state.

(d) $$\phi = Ax^3 + Bx^2y + Cxy^2 + Dy^3.$$

Then

$$\sigma_{xx} = \frac{\partial^2 \phi}{\partial y^2} = 2Cx + 6Dy$$

$$\sigma_{yy} = \frac{\partial^2 \phi}{\partial x^2} = 6Ax + 2By$$

$$\tau_{xy} = -\frac{\partial^2 \phi}{\partial x \partial y} = -2Bx - 2Cy$$

and all stresses may be seen to vary linearly with x and y.

For the particular case where $A = B = C = 0$ the situation resolves itself into that of case (b) i.e. suitable for pure bending.

For many problems an extension of the above function to a comprehensive polynomial expression is found to be rather useful. An appropriate technique is to postulate a general form which will adequately represent the applied loading and boundary conditions. The form of this could be:

$$\phi = Ax^2 + Bxy + Cy^2 + Dx^3 + Ex^2y + Fxy^2 + Gy^3$$
$$+ Hx^4 + Jx^3y + Kx^2y^2 + Lxy^3 + My^4 + Nx^5 + Px^4y$$
$$+ Qx^3y^2 + Rx^2y^3 + Sxy^4 + Ty^5 + \cdots \qquad (8.89)$$

Any term containing x or y up to the third power will automatically satisfy the biharmonic equation $\nabla^4(\phi) = 0$. However, terms containing x^4 or y^4, or higher powers, will appear in

the biharmonic equation. Relations of the associated coefficients can thereby be found which
will satisfy $\nabla^4(\phi) = 0$.

Although beyond the scope of the present text, it is worth noting that the polynomial
approach has severe limitations when applied to cases with discontinuous loads on the
boundary. For such cases, a stress function in the form of a trigonometric series – a Fourier
series for example – should be used.

8.27.2. *Case 1 – Bending of a simply supported beam by a uniformly distributed loading*

An end-supported beam of length $2L$, depth $2d$ and unit width is loaded with a uniformly
distributed load w/unit length as shown in Fig. 8.32. From the work of Chapter 4† the reader
will be aware of the solution of this problem using the simple bending theory sometimes
known as "engineers bending". Using this simple approach it is possible to obtain values for
the longitudinal stress σ_{xx} and the shear stress τ_{xy}. However, the stress function provides the
stress analyst with information about *all* the two-dimensional stresses and thereby the regions
of applicability where the more straightforward methods can be used with confidence.
The boundary conditions of this problem are:

(i) at $y = +d$; $\sigma_{yy} = 0$ for all values of x,
(ii) at $y = -d$; $\sigma_{yy} = -w$ for all values of x,
(iii) at $y = \pm d$; $\tau_{xy} = 0$ for all values of x.

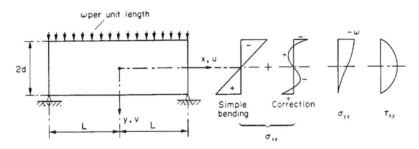

Fig. 8.32. The bending of a simply supported beam by a uniformly distributed load w/unit length.

The overall equilibrium requirements are: –

(iv) $\int_{-d}^{d} \sigma_{xx} y \cdot dy = w(L^2 - x^2)/2$ for the equilibrium of moments at any position x,
(v) $\int_{-d}^{d} \sigma_{xx} dy = 0$ for the equilibrium of forces at any position x.

The biharmonic equation:

(vi) $\nabla^4(\phi) = 0$ must also be satisfied.

To deal with these conditions it is necessary to use the 5th-order polynomial as given in
eqn. (8.89) containing eighteen coefficients A to T.

† E.J. Hearn, *Mechanics of Materials 1*, Butterworth-Heinemann, 1997.

From eqn. (8.81)

$$\left.\begin{aligned}
\sigma_{xx} &= \frac{\partial^2 \phi}{\partial y^2} = 2C + 2Fx + 6Gy + 2Kx^2 + 6Lxy + 12My^2 + 2Qx^3 + 6Rx^2y \\
&\quad + 12Sxy^2 + 20Ty^3 \\
\sigma_{yy} &= \frac{\partial^2 \phi}{\partial x^2} = 2A + 6Dx + 2Ey + 12Hx^2 + 6Jxy + 2Ky^2 + 20Nx^3 + 12Px^2y \\
&\quad + 6Qxy^2 + 2Ry^3 \\
\tau_{xy} &= -\frac{\partial^2 \phi}{\partial x \partial y} = -[B + 2Ex + 2Fy + 3Jx^2 + 4Kxy + 3Ly^2 + 4Px^3 + 6Qx^2y \\
&\quad + 6Rxy^2 + 4Sy^3]
\end{aligned}\right\} \quad (8.90)$$

Using the conditions (i) to (vi) it is possible to set up a series of algebraic equations to determine the values of the eighteen coefficients A to T. Since these conditions must be satisfied for all x values it is appropriate to equate the coefficients of the x terms, for example x^3, x^2, x and the constants, on both sides of the equations. In the case of the biharmonic equation, condition (vi), all x and y values must be satisfied. This procedure gives the following results:

$$
\begin{array}{lll}
A = -w/4 & G = (wL^2/8d^3) - w/20d & N = 0 \\
B = 0 & H = 0 & P = 0 \\
C = 0 & J = 0 & Q = 0 \\
D = 0 & K = 0 & R = -w/8d^3 \\
E = 3w/8d & L = 0 & S = 0 \\
F = 0 & M = 0 & T = w/40d^3
\end{array}
$$

The stress function ϕ can thus be written:

$$\phi = -\frac{w}{4}x^2 + \frac{3w}{8d}x^2y + \left(\frac{wL^2}{8d^3} - \frac{w}{20d}\right)y^3 - \frac{w}{8d^3}x^2y^3 + \frac{5}{40d^3}y^5 \quad (8.91)$$

The values for the stresses follow using eqn. (8.90) with $I = 2d^3/3$

$$\left.\begin{aligned}
\sigma_{xx} &= \frac{w(L^2 - x^2)y}{2I} + \frac{w}{2I}\left(-\frac{2}{5}d^2y + \frac{2}{3}y^3\right) \\
\sigma_{yy} &= -\frac{w}{2I}\left(\frac{2}{3}d^3 - d^2y + \frac{y^3}{3}\right) \\
\tau_{xy} &= -\frac{wx}{2I}(d^2 - y^2)
\end{aligned}\right\} \quad (8.92\text{a–c})$$

These stresses are plotted in Fig. 8.32. The longitudinal stress σ_{xx} consists of two parts. The first term $w(L^2 - x^2)y/2I$ is that given by simple bending theory ($\sigma_{xx} = My/I$). The second term may be considered as a correction term which arises because of the effect of the σ_{yy} compressive stress between the longitudinal fibres. The term is independent of x and therefore constant along the beam. It thus has a value on the ends of the beam given by $x = \pm L$. The expression for σ_{xx} in eqn. (8.92a) is, therefore, only an exact solution if normal forces on the end exist and are distributed in such a manner as to produce the σ_{xx} values given by eqn. (8.92a) at $x = \pm L$. That is as shown by the correction term in Fig. 8.32. However, conditions (iv) and (v) have guaranteed that forces and moments are in equilibrium at the ends $x = \pm L$ and thus, from Saint-Venant's principle, one could conclude that at distance larger than, say, the depth of the beam, the stress distribution given by eqn. (8.92a) is accurate even when the ends are free. Such correction stresses are, however, of small magnitude compared with the simple bending terms when the span of the beam is large in comparison with its depth.

The equation for the shear stress (8.92c) predicts a parabolic distribution of τ_{xy} on every section x. This implies that at the ends $x = \pm L$ the beam must be supported in such a way that these shear stresses are developed. The values predicted by eqn. (8.92c) coincide with the simple solution. The σ_{yy} stress decreases from its maximum on the top surface to zero at the bottom edge. This again is of small magnitude compared to σ_{xx} in a thin beam type component. However, these stresses can be of importance in a deep beam, or a slab arrangement.

Derivation of the displacements in the beam

From the strain displacement relations, the constitutive relations and the derived stresses it is possible to obtain the displacements in the beam. Although this approach is not really part of the stress function concept, it is included for interest at this point in the development. The procedure is as follows:

$$\left.\begin{aligned}
\varepsilon_{xx} &= \frac{\partial v}{\partial x} = \frac{1}{E}(\sigma_{xx} - v\sigma_{yy}) \\
\varepsilon_{yy} &= \frac{\partial v}{\partial y} = \frac{1}{E}(\sigma_{yy} - v\sigma_{xx}) \\
\gamma_{xy} &= \frac{\partial u}{\partial y} + \frac{\partial v}{\partial x} = \tau_{xy}/G
\end{aligned}\right\} \qquad (8.93\text{a-c})$$

Substituting for σ_{xx} and σ_{yy} from eqns (8.92a,b) and integrating (8.93a,b) the following is obtained:

$$u = \frac{w}{2EI}\left[\left(L^2 x - \frac{x^3}{3}\right)y + \left(\frac{2}{3}y^3 - \frac{2}{5}d^2 y\right)x + vx\left(\frac{1}{3}\cdot y^3 - d^3 y + \frac{2}{3}d^3\right)\right] + u_0(y) \tag{8.94}$$

where $u_0(y)$ is a function of y,

$$v = -\frac{w}{2EI}\left[\frac{y^4}{12} - \frac{d^2 y^2}{2} + \frac{2d^3 y}{3} + \frac{v}{2}(L^2 - x^2)y^2 + \frac{v}{6}y^4 - \frac{v}{5}d^2 y\right] + v_0(x) \tag{8.95}$$

where $v_0(x)$ is a function of x.

From eqns (8.92c) and (8.93c)

$$\gamma_{xy} = -\frac{w(1+v)}{EI}(d^2 - y^2)x \tag{8.96}$$

Differentiating u with respect to y and v with respect to x and adding as in eqn. (8.93c) one can equate the result to the right hand side of eqn. 8.96. After simplifying:

$$\frac{w}{2EI}\left[L^2x - \frac{x^3}{3} - \frac{2}{5}xd^2 - vxd^2\right] + \frac{\partial u_0(y)}{\partial y} + \frac{\partial v_0(x)}{\partial x} = -\frac{w(1+v)}{EI}d^2x \qquad (8.97)$$

In eqn. (8.97) some terms are functions of x alone and some are functions of y alone. There is no constant term. Denoting the functions of x and y by $F(x)$ and $G(y)$ respectively, we have:

$$F(x) = \frac{w}{2EI}\left[L^2x - \frac{x^3}{3} - \frac{2}{5}xd^2 - vxd^2\right] + \frac{w(1+v)d^2x}{EI} + \frac{\partial v_0(x)}{\partial x}$$

$$G(y) = \frac{\partial u_0(y)}{\partial y}.$$

Equation (8.97) is thus written

$$F(x) + G(y) = 0$$

If such an equation is to apply for all values of x and y then the functions $F(x)$ and $G(y)$ must themselves be constants and they must be equal in value but opposite in sign. That is in this case, $F(x) = A_1$ and $G(y) = -A_1$.

Thus:$\quad \dfrac{\partial u_0(y)}{\partial y} = -A_1 \quad \therefore \ u_0(y) = -A_1y + B_1 \qquad\qquad\qquad (8.98)$

$$\frac{\partial v_0(x)}{\partial x} = -\frac{w}{2EI}\left[L^2x - \frac{x^3}{3} - \frac{2}{5}xd^2 - vxd^2 + 2(1+v)d^2x\right] + A_1$$

$$\therefore\qquad v_0(x) = -\frac{w}{2EI}\left[L^2\frac{x^2}{2} - \frac{x^4}{12} - \frac{1}{5}x^2d^2 - \frac{v}{2}x^2d^2 + (1+v)d^2x^2\right] + A_1x + C_1 \quad (8.99)$$

Using the boundary conditions of the problem:

at $x = 0$, $y = 0$, $u = 0$: substituting eqn. (8.98) into (8.94) gives $B_1 = 0$,

at $x = 0$, $y = 0$, $v = \delta$: substituting eqn. (8.99) into (8.95) gives $C_1 = \delta$,

at $x = 0$, $y = 0$, $\dfrac{\partial v}{\partial x} = 0$ thus $A_1 = 0$.

Thus:

$$u = \frac{w}{2EI}\left[\left(L^2x - \frac{x^3}{3}\right)y + x\left(\frac{2}{3}y^2 - \frac{2}{5}d^2y\right) + vx\left(\frac{y^3}{3} - d^2y + \frac{2}{3}d^3\right)\right]$$

$$\left.\begin{array}{l} v = -\dfrac{w}{2EI}\left[\dfrac{y^4}{12} - \dfrac{d^2y^2}{2} + \dfrac{2}{3}d^3y + v(L^2 - x^2)\dfrac{y^2}{2} + \dfrac{v}{6}y^4 - \dfrac{v}{5}d^2y\right. \\[3mm] \qquad\qquad \left. + L^2\dfrac{x^2}{2} - \dfrac{x^4}{12} - \dfrac{d^2x^2}{5} + \left(1 + \dfrac{v}{2}\right)d^2x^2\right] + \delta \end{array}\right\} \qquad (8.100)$$

To determine the vertical deflection of the central axis we put $y = 0$ in the above equation, that is:

$$v_{y\,=\,0} = \delta - \frac{w}{2EI}\left[L^2\frac{x^2}{2} - \frac{x^4}{12} - \frac{d^2x^2}{5} + \left(1 + \frac{v}{2}\right)d^2x^2\right]$$

Using the fact that $v = 0$ at $x = \pm L$ we find that the central deflection δ is given by:

$$\delta = \frac{5}{24} \frac{wL^4}{EI} \left[1 + \frac{d^2}{L^2} \frac{12}{5} \left(\frac{4}{5} + \frac{v}{2} \right) \right] \tag{8.101}$$

The first term is the central deflection predicted by the simple bending theory. The second term is the correction to include deflection due to shear. As indicated by the form of eqn. (8.101) the latter is small when the span/depth ratio is large, but is more significant for deep beams. By combining equations (8.100) and (8.101) the displacements u and v can be obtained at any point (x, y) in the beam.

8.27.3. The use of polar coordinates in two dimensions

Many engineering components have a degree of axial symmetry, that is they are either rotationally symmetric about a central axis, as in a circular ring, disc and thick cylinder, or contain circular holes which dominate the stress field, or yet again are made up from parts of hollow discs, like a curved bar. In such cases it is advantageous to use cylindrical polar coordinates (r, θ, z), where r and θ are measured from a fixed origin and axis, respectively and z is in the axial direction. The equilibrium equations for this case are given in eqns. (8.40) and (8.41).

The form of applied loading for these components need not be restricted to the simple rotationally symmetric cases dealt with in earlier chapters. In fact the great value of the stress function concept is that complex loading patterns can be adequately represented by the use of either $\cos n\theta$ and/or $\sin n\theta$, where n is the harmonic order.

A two-dimensional stress field $(\sigma_{rr}, \sigma_{\theta\theta}, \tau_{r\theta})$ is again used for these cases. That is plane stress or plane strain is assumed to provide an adequate approximation of the three-dimensional problem. The next step is to transform the biharmonic eqn. (8.87) to the relevant polar form, namely:

$$\left(\frac{\partial^2}{\partial r^2} + \frac{1}{r} \frac{\partial}{\partial r} + \frac{1}{r^2} \frac{\partial^2}{\partial \theta^2} \right) \left(\frac{\partial^2}{\partial r^2} + \frac{1}{r} \frac{\partial}{\partial r} + \frac{1}{r^2} \frac{\partial^2}{\partial \theta^2} \right) \phi = 0 \tag{8.102}$$

The stresses σ_{rr}, $\sigma_{\theta\theta}$ and $\tau_{r\theta}$ are related to the stress function ϕ in a similar manner to σ_{xx} and σ_{yy}. The resulting values are:

$$\left. \begin{aligned} \sigma_{rr} &= \frac{1}{r} \frac{\partial \phi}{\partial r} + \frac{1}{r^2} \frac{\partial^2 \phi}{\partial \theta^2} \\[2mm] \sigma_{\theta\theta} &= \frac{\partial^2 \phi}{\partial r^2} \\[2mm] \tau_{r\theta} &= \frac{1}{r^2} \frac{\partial \phi}{\partial \theta} - \frac{1}{r} \frac{\partial^2 \phi}{\partial r \partial \theta} = -\frac{\partial}{\partial r} \left(\frac{1}{r} \frac{\partial \phi}{\partial \theta} \right) \end{aligned} \right\} \tag{8.103}$$

The derivation of these from the corresponding cartesian coordinate values is a worthwhile exercise for a winter evening.

8.27.4. Forms of stress function in polar coordinates

In cylindrical polars the stress function is, in general, of the form:

$$\phi = f(r)\cos n\theta \quad \text{or} \quad \phi = f(r)\sin n\theta \tag{8.104}$$

where $f(r)$ is a function of r alone and n is an integer.

In exploring the form of ϕ in polars one can avoid the somewhat tedious polynomial expression used for the cartesian coordinates, by considering the following three cases:

(a) *The axi-symmetric case when $n = 0$* (independent of θ), $\phi = f(r)$. Here the biharmonic eqn. (8.102) reduces to:

$$\left(\frac{d^2}{dr^2} + \frac{1}{r}\frac{d}{dr}\right)^2 \phi = 0 \tag{8.105}$$

and the stresses in eqn. (8.103) to:

$$\sigma_{rr} = \frac{1}{r}\frac{d\phi}{dr}, \quad \sigma_{\theta\theta} = \frac{d^2\phi}{dr^2}, \quad \tau_{r\theta} = 0 \tag{8.106}$$

Equation (8.105) has a general solution:

$$\phi = Ar^2 \ln r + Br^2 + C \ln r + D \tag{8.107}$$

(b) *The asymmetric case $n = 1$*

$$\phi = f_1(r)\sin\theta \quad \text{or} \quad \phi = f_1(r)\cos\theta.$$

Equation (8.102) has the solution for

$$f_1(r) = A_1 r^3 + B_1/r + C_1 r + D_1 r \ln r \tag{8.108}$$

i.e. $$\phi = (A_1 r^3 + B_1/r + C_1 r + D_1 r \ln r)\sin\theta \quad (\text{or} \;\cos\theta)$$

(c) *The asymmetric cases $n \geqslant 2$.*

$$\phi = f_n(r)\sin n\theta \quad \text{or} \quad \phi = f_n(r)\cos n\theta$$

$$f_n(r) = A_n r^n + B_n r^{-n} + C_n r^{n+2} + D_n r^{-n+2} \tag{8.109}$$

i.e. $$\phi = (A_n r^n + B_n r^{-n} + C_n r^{n+2} + D_n r^{-n+2})\sin n\theta \quad (\text{or} \cos n\theta)$$

Other useful solutions are $$\phi = Cr\sin\theta \quad \text{or} \quad \phi = Cr\cos\theta \tag{8.110}$$

In the above A, B, C and D are constants of integration which enable formulation of the various problems.

As in the case of the cartesian coordinate system these stress functions must satisfy the compatibility relation embodied in the biharmonic equation (8.102). Although the reader is assured that they are satisfactory functions, checking them is always a beneficial exercise.

In those cases when it is not possible to adequately represent the form of the applied loading by a single term, say $\cos 2\theta$, then a Fourier series representation using eqn. (8.109) can be used. Details of this are given by Timoshenko and Goodier.[†]

[†] S. Timoshenko and J.N. Goodier, *Theory of Elasticity*, McGraw-Hill, 1951.

In the presentation that follows examples of these cases are given. It will be appreciated that the scope of these are by no means exhaustive but a number of worthwhile solutions are given to problems that would otherwise be intractable. Only the stress values are presented for these cases, although the derivation of the displacements is a natural extension.

8.27.5. Case 2 – Axi-symmetric case: solid shaft and thick cylinder radially loaded with uniform pressure

This obvious case will be briefly discussed since the Lamé equations which govern this problem are so well known and do provide a familiar starting point.

Substituting eqn. (8.107) into the stress equations (8.106) results in

$$\left.\begin{array}{l} \sigma_{rr} = A(1 + 2\ln r) + 2B + C/r^2 \\ \sigma_{\theta\theta} = A(3 + 2\ln r) + 2B - C/r^2 \\ \tau_{r\theta} = 0 \end{array}\right\} \qquad (8.111)$$

When a *solid shaft* is loaded on the external surface, the constants A and C must vanish to avoid the singularity condition at $r = 0$. Hence $\sigma_{rr} = \sigma_{\theta\theta} = 2B$. That is uniform tension, or compression over the cross section.

In the case of the *thick cylinder*, three constants, A, B, and C have to be determined. The constant A is found by examining the form of the tangential displacement v in the cylinder. The expression for this turns out to be a multi-valued expression in θ, thus predicting a different displacement every time θ is increased to $\theta + 2\pi$. That is every time we scan one complete revolution and arrive at the same point again we get a different value for v. To avoid this difficulty we put $A = 0$. Equations (8.111) are thus identical in form to the Lamé eqns. (10.3 and 10.4).[†] The two unknown constants are determined from the applied load conditions at the surface.

8.27.6. Case 3 – The pure bending of a rectangular section curved beam

Consider a circular arc curved beam of narrow rectangular cross-section and unit width, bent in the plane of curvature by end couples M (Fig. 8.33). The beam has a constant cross-section and the bending moment is constant along the beam. In view of this one would expect that the stress distribution will be the same on each radial cross-section, that is, it will be independent of θ. The axi-symmetric form of ϕ, as given in eqn. (8.107), can thus be used:-

i.e. $$\phi = Ar^2 \ln r + Br^2 + C \ln r + D$$

The corresponding stress values are those of eqns (8.111)

$$\sigma_{rr} = A(1 + 2\ln r) + 2B + C/r^2$$

$$\sigma_{\theta\theta} = A(3 + 2\ln r) + 2B - C/r^2$$

$$\tau_{r\theta} = 0$$

[†] E.J. Hearn, *Mechanics of Materials 1*, Butterworth-Heinemann, 1997.

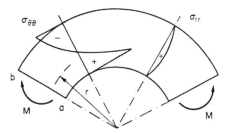

Fig. 8.33. Pure bending of a curved beam.

The boundary conditions for the curved beam case are:

(i) $\sigma_{rr} = 0$ at $r = a$ and $r = b$ (a and b are the inside and outside radii, respectively);
(ii) $\int_a^b \sigma_{\theta\theta} = 0$, for the equilibrium of forces, over any cross-section;
(iii) $\int_a^b \sigma_{\theta\theta} \, r \, dr = -M$, for the equilibrium of moments, over any cross-section;
(iv) $\tau_{r\theta} = 0$, at the boundary $r = a$ and $r = b$.

Using these conditions the constants A, B and C can be determined. The final stress equations are as follows:

$$\left.\begin{aligned}
\sigma_{rr} &= \frac{4M}{Q}\left(\frac{a^2 b^2}{r^2}\ln\frac{b}{a} - a^2\ln\frac{r}{a} - b^2\ln\frac{b}{r}\right) \\[2mm]
\sigma_{\theta\theta} &= \frac{4M}{Q}\left(-\frac{a^2 b^2}{r^2}\ln\frac{b}{a} - a^2\ln\frac{r}{a} - b^2\ln\frac{b}{r} + b^2 - a^2\right) \\[2mm]
\tau_{r\theta} &= 0
\end{aligned}\right\} \qquad (8.112)$$

where $Q = 4a^2 b^2\left(\ln\frac{b}{a}\right)^2 - (b^2 - a^2)^2$

The distributions of these stresses are shown on Fig. 8.33. Of particular note is the nonlinear distribution of the $\sigma_{\theta\theta}$ stress. This predicts a higher inner fibre stress than the simple bending ($\sigma = My/I$) theory.

8.27.7. Case 4. Asymmetric case n = 1. Shear loading of a circular arc cantilever beam

To illustrate this form of stress function the curved beam is again selected; however, in this case the loading is a shear loading as shown in Fig. 8.34.

As previously the beam is of narrow rectangular cross-section and unit width. Under the shear loading P the bending moment at any cross-section is proportional to $\sin\theta$ and, therefore it is reasonable to assume that the circumferential stress $\sigma_{\theta\theta}$ would also be associated with $\sin\theta$. This points to the case $n = 1$ and a stress function given in eqn. (8.108).

i.e. $\phi = (A_1 r^3 + B_1/r + C_1 r + D_1 r \ln r)\sin\theta$ (8.113)

Using eqns. (8.103) the three stresses can be written

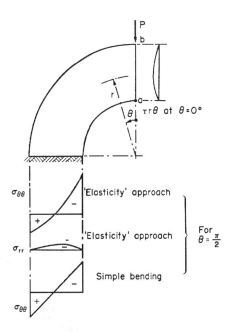

Fig. 8.34. Shear loading of a curved cantilever.

$$\left.\begin{aligned}
\sigma_{rr} &= (2A_1 r - 2B_1/r^3 + D_1/r)\sin\theta \\
\sigma_{\theta\theta} &= (6A_1 r + 2B_1/r^3 + D_1/r)\sin\theta \\
\tau_{r\theta} &= -(2A_1 r - 2B_1/r^3 + D_1/r)\cos\theta
\end{aligned}\right\} \tag{8.114}$$

The boundary conditions are:

(i) $\sigma_{rr} = \tau_{r\theta} = 0$, for $r = a$ and $r = b$.
(ii) $\int_a^b \tau_{r\theta}\, dr = P$, for equilibrium of vertical forces at $\theta = 0$.

Using these conditions the constants A_1, B_1 and D_1 can be determined. The final stress values are:

$$\left.\begin{aligned}
\sigma_{rr} &= \frac{P}{S}\left(r + \frac{a^2 b^2}{r^3} - \frac{a^2 + b^2}{r}\right)\sin\theta \\
\sigma_{\theta\theta} &= \frac{P}{S}\left(3r - \frac{a^2 b^2}{r^3} - \frac{a^2 + b^2}{r}\right)\sin\theta \\
\tau_{r\theta} &= -\frac{P}{S}\left(r + \frac{a^2 b^2}{r^3} - \frac{a^2 + b^2}{r}\right)\cos\theta
\end{aligned}\right\} \tag{8.115}$$

where $s = a^2 - b^2 + (a^2 + b^2)\ln b/a$.
It is noted from these equations that at the load point $\theta = 0$,

$$\left.\begin{aligned}
\sigma_{rr} &= \sigma_{\theta\theta} = 0 \\
\tau_{r\theta} &= -\frac{P}{S}\left(r + \frac{a^2 b^2}{r^3} - \frac{a^2 + b^2}{r}\right)
\end{aligned}\right\} \tag{8.116}$$

As in the previous cases the load P must be applied to the cantilever according to eqn. (8.116) – see Fig. 8.34.

At the fixed end, $\theta = \dfrac{\pi}{2}$;
$$
\left.
\begin{aligned}
\sigma_{rr} &= \frac{P}{S}\left(r + \frac{a^2 b^2}{r^3} - \frac{a^2 + b^2}{r}\right) \\[2mm]
\sigma_{\theta\theta} &= \frac{P}{S}\left(3r - \frac{a^2 b^2}{r^3} - \frac{a^2 + b^2}{r}\right) \\[2mm]
\tau_{r\theta} &= 0
\end{aligned}
\right\}
\tag{8.117}
$$

The distributions of these stresses are shown in Fig. 8.34. They are similar to that for the pure moment application. The simple bending ($\sigma = My/I$) result is also shown. As in the previous case it is noted that the simple approach underestimates the stresses on the inner fibre.

8.27.8. Case 5–The asymmetric cases $n \geqslant 2$–stress concentration at a circular hole in a tension field

The example chosen to illustrate this category concerns the derivation of the stress concentration due to the presence of a circular hole in a tension field. A large number of stress concentrations arise because of geometric discontinuities–such as holes, notches, fillets, etc., and the derivation of the peak stress values, in these cases, is clearly of importance to the stress analyst and the designer.

The distribution of stress round a small circular hole in a flat plate of unit thickness subject to a uniform tension σ_{xx}, in the x direction was first obtained by Prof. G. Kirsch in 1898.[†] The width of the plate is considered large compared with the diameter of the hole as shown in Fig. 8.35. Using the Saint-Venant's[‡] principle the small central hole will not affect the

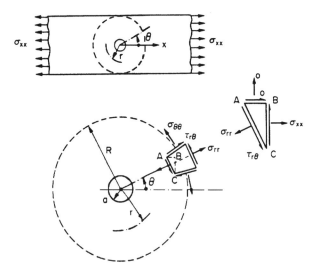

Fig. 8.35. Elements in a stress field some distance from a circular hole.

[†] G. Kirsch Verein Deutsher Ingenieure (V.D.I.) *Zeitschrift*, **42** (1898), 797–807.
[‡] B. de Saint-Venant, *Mem. Acad. Sc. Savants E'trangers*, **14** (1855), 233–250.

stress distribution at distances which are large compared with the diameter of the hole–say the width of the plate. Thus on a circle of large radius R the stress in the x direction, on $\theta = 0$ will be σ_{xx}. Beyond the circle one can expect that the stresses are effectively the same as in the plate without the hole.

Thus at an angle θ, equilibrium of the element ABC, at radius $r = R$, will give

$$\sigma_{rr}.AC = \sigma_{xx}BC\cos\theta, \quad \text{and since, } \cos\theta = BC/AC$$

$$\sigma_{rr} = \sigma_{xx}\cos^2\theta,$$

or $$\sigma_{rr} = \frac{\sigma_{xx}}{2}(1 + \cos 2\theta).$$

Similarly, $$\tau_{r\theta}.AC = -\sigma_{xx}BC\sin\theta$$

$$\therefore \tau_{r\theta} = -\sigma_{xx}\cos\theta\sin\theta = -\frac{\sigma_{xx}}{2}\sin 2\theta.$$

Note the sign of $\tau_{r\theta}$ indicates a direction opposite to that shown on Fig. 8.35.

Kirsch noted that the total stress distribution at $r = R$ can be considered in two parts:

(a) a constant radial stress $\sigma_{xx}/2$

(b) a condition varying with 2θ, that is; $\sigma_{rr} = \dfrac{\sigma_{xx}}{2}\cos 2\theta, \; \tau_{r\theta} = -\dfrac{\sigma_{xx}}{2}\sin 2\theta.$

The final result is obtained by combining the distributions from (a) and (b). *Part (a), shown in Fig. 8.36*, can be treated using the Lamé equations; The boundary conditions are:

$$\text{at } r = a \quad \sigma_{rr} = 0$$

$$r = R \quad \sigma_{rr} = \sigma_{xx}/2$$

Using these in the Lamé equation, $\sigma_{rr} = A + B/r^2$

gives, $$A = \frac{\sigma_{xx}}{2}\left(\frac{R^2}{R^2 - a^2}\right) \quad \text{and} \quad B = -\frac{\sigma_{xx}}{2}\left(\frac{R^2 a^2}{R^2 - a^2}.\right)$$

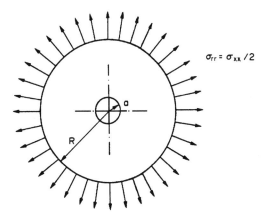

$\sigma_{rr} = \sigma_{xx}/2$

Fig. 8.36. A circular plate loaded at the periphery with a uniform tension.

When $R \gg a$ these can be modified to $A = \dfrac{\sigma_{xx}}{2}$ and $B = -\dfrac{\sigma_{xx}}{2}a^2$

Thus

$$\left. \begin{aligned}
\sigma_{rr} &= \frac{\sigma_{xx}}{2}\left(1 - \frac{a^2}{r^2}\right) \\[4pt]
\sigma_{\theta\theta} &= \frac{\sigma_{xx}}{2}\left(1 + \frac{a^2}{r^2}\right) \\[4pt]
\tau_{r\theta} &= 0
\end{aligned} \right\} \tag{8.118}$$

Part (b), shown in Fig 8.37 is a new case with normal stresses varying with $\cos 2\theta$ and shear stresses with $\sin 2\theta$.

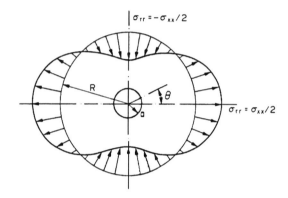

Fig. 8.37. A circular plate loaded at the periphery with a radial stress $= \dfrac{\sigma_{xx}}{2}\cos 2\theta$ (shown above) and a shear stress $= -\dfrac{\sigma_{xx}}{2}\sin 2\theta$.

This fits into the category of $n = 2$ with a stress function eqn. (8.109);

i.e. $$\phi = (A_2 r^2 + B_2/r^2 + C_2 r^4 + D_2)\cos 2\theta \tag{8.119}$$

Using eqns. (8.103) the stresses can be written:

$$\left. \begin{aligned}
\sigma_{rr} &= -(2A_2 + 6B_2/r^4 + 4D_2/r^2)\cos 2\theta \\
\sigma_{\theta\theta} &= (2A_2 + 6B_2/r^4 + 12C_2 r^2)\cos 2\theta \\
\tau_{r\theta} &= (2A_2 - 6B_2/r^4 + 6C_2 r^2 - 2D_2/r^2)\sin 2\theta
\end{aligned} \right\} \tag{8.120}$$

The four constants are found such that σ_{rr} and $\tau_{r\theta}$ satisfy the boundary conditions:

at $r = a$, $\sigma_{rr} = \tau_{r\theta} = 0$

at $r = R \to \infty$, $\sigma_{rr} = \dfrac{\sigma_{xx}}{2}\cos 2\theta$, $\tau_{r\theta} = -\dfrac{\sigma_{xx}}{2}\sin 2\theta$

From these,

$$A_2 = -\sigma_{xx}/4, \quad B_2 = -\sigma_{xx}a^4/4$$
$$C_2 = 0, \quad D_2 = \sigma_{xx}a^2/2$$

Thus:

$$\sigma_{rr} = \frac{\sigma_{xx}}{2}\left(1 - \frac{4a^2}{r^2} + \frac{3a^4}{r^4}\right)\cos 2\theta$$

$$\sigma_{\theta\theta} = -\frac{\sigma_{xx}}{2}\left(1 + \frac{3a^4}{r^4}\right)\cos 2\theta \tag{8.121}$$

$$\tau_{r\theta} = -\frac{\sigma_{xx}}{2}(1 + 2a^2/r^2 - 3a^4/r^4)\sin 2\theta$$

The sum of the stresses given by eqns. (8.120) and (8.121) is that proposed by Kirsch. At the edge of the hole σ_{rr} and $\tau_{r\theta}$ should be zero and this can be verified by substituting $r = a$ into these equations.

The distribution of $\sigma_{\theta\theta}$ round the hole, i.e. $r = a$, is obtained by combining eqns. (8.120) and (8.121):

i.e. $$\sigma_{\theta\theta} = \sigma_{xx}(1 - 2\cos 2\theta) \tag{8.122}$$

and is shown on Fig. 8.38(a).

When $\theta = 0$; $\sigma_{\theta\theta} = -\sigma_{xx}$ and when $\theta = \frac{\pi}{2}$; $\sigma_{\theta\theta} = 3\sigma_{xx}$.

The stress concentration factor (S.C.F) defined as Peak stress/Average stress, gives an S.C.F. = 3 for this case.

The distribution across the plate from point $A\left(\theta = \frac{\pi}{2}\right)$ is:

$$\sigma_{\theta\theta} = \frac{\sigma_{xx}}{2}\left(2 + \frac{a^2}{r^2} + \frac{3a^4}{r^4}\right) \tag{8.123}$$

This is shown in Fig. 8.38(b), which indicates the rapid way in which $\sigma_{\theta\theta}$ approaches σ_{xx} as r increases. Although the solution is based on the fact that $R \gg a$, it can be shown that even when $R = 4a$, that is the width of the plate is four times the diameter of the hole, the error in the S.C.F. is less than 6%.

Using the stress distribution derived for this case it is possible, using superposition, to obtain S.C.F. values for a range of other stress fields where the circular hole is present, see problem No. 8.52 for solution at the end of this chapter.

A similar, though more complicated, analysis can be carried out for an elliptical hole of major diameter $2a$ across the plate and minor diameter $2b$ in the stress direction. In this case the S.C.F. $= 1 + 2a/b$ (see also §8.3). Note that for the circular hole $a = b$, and the S.C.F. $= 3$, as above.

8.27.9. Other useful solutions of the biharmonic equation

(a) Concentrated line load across a plate

The way in which an elastic medium responds to a concentrated line of force is the final illustrative example to be presented in this section. In practice it is neither possible to apply a genuine line load nor possible for the plate to sustain a load without local plastic deformation. However, despite these local perturbations in the immediate region of the load, the rest of the plate behaves in an elastic manner which can be adequately represented by the governing equations obtained earlier. It is thus possible to use the techniques developed above to analyse the concentrated load problem.

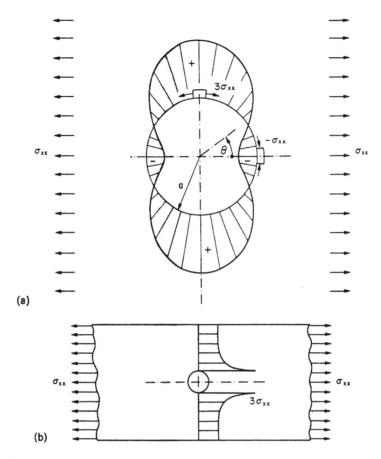

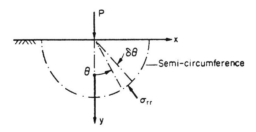

Fig. 8.38. (a) Distribution of circumferential stress $\sigma_{\theta\theta}$ round the hole in a tension field; (b) distribution of circumferential stress $\sigma_{\theta\theta}$ across the plate.

Fig. 8.39. Concentrated load on a semi-infinite plate.

Consider a force *P per unit width* of the plate applied as a line load normal to the surface – see Fig. 8.39. The plate will be considered as equivalent to a semi-infinite solid, that is, one that extends to infinity in the x and y directions below the horizon, $\theta = \pm\frac{\pi}{2}$. The plate is assumed to be of unit width. It is convenient to use cylindrical polars again for this problem.

Using Boussinesq's solutions[†] for a semi-infinite body, Alfred-Aimé Flamant obtained (in 1892)[‡] the stress distribution for the present case. He showed that on any semi-circumference round the load point the stress is entirely radial, that is: $\sigma_{\theta\theta} = \tau_{r\theta} = 0$ and σ_{rr} will be a principal stress. He used a stress function of the type given in eqn. (8.110), namely: $\phi = Cr\theta \sin\theta$ which predicts stresses:

$$\sigma_{rr} = \frac{2C}{r}\cos\theta, \ \sigma_{\theta\theta} = \tau_{r\theta} = 0$$

Applying overall equilibrium to this case it is noted that the resultant vertical force over any semi-circle, of radius r, must equal the applied force P:

$$P = -\int_{-\pi/2}^{\pi/2} (\sigma_{rr} \cdot r\, d\theta) \cos\theta = -\int_{-\pi/2}^{\pi/2} (2C \cos^2\theta)\, d\theta = -C\pi$$

Thus

$$\phi = -\frac{Pr\theta}{\pi}\sin\theta$$

and

$$\sigma_{rr} = -\frac{2P}{\pi}\frac{\cos\theta}{r} \tag{8.124}$$

This can be transformed into x and y coordinates:

$$\left.\begin{array}{l} \sigma_{yy} = \sigma_{rr}\cos^2\theta \\ \sigma_{xx} = \sigma_{rr}\sin^2\theta \\ \tau_{xy} = \sigma_{rr}\sin\theta\cos\theta \end{array}\right\} \tag{8.125}$$

See also §8.3.3 for further transformation of these equations.

This type of solution can be extended to consider the wedge problem, again subject to a line load as shown in Figs. 8.40(a) and (b).

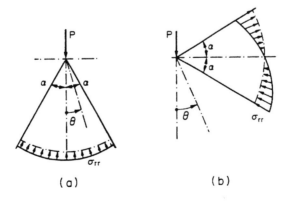

(a)   (b)

Fig. 8.40. Forces on a wedge.

[†] J. Boussinesq, *Application de potentiels a l'étude de l'equilibre!* Paris, 1885; also *Comptes Rendus Acad Sci.*, **114** (1892), 1510–1516.

[‡] Flamant AA *Comptes Rendus Acad. Sci.*, **114** (1892), 1465–1468.

(b) The wedge subject to an axial load – Figure 8.40(a)

For this case,
$$P = -\int_{-\alpha}^{\alpha} (\sigma_{rr} \cdot r \, d\theta) \cos \theta$$

$$P = -\int_{-\alpha}^{\alpha} 2C \cdot \cos^2 \theta \, d\theta$$

$$P = -C(2\alpha + \sin 2\alpha)$$

Thus,
$$\sigma_{rr} = -\frac{2P \cos \theta}{r(2\alpha + \sin 2\alpha)} \tag{8.126}$$

(c) The wedge subject to a normal end load – Figure 8.40(a)

Here,
$$P = -\int_{\frac{\pi}{2}-\alpha}^{\frac{\pi}{2}+\alpha} (\sigma_{rr} \cdot r \, d\theta) \cos \theta$$

$$P = -\int_{\frac{\pi}{2}-\alpha}^{\frac{\pi}{2}+\alpha} 2C \cdot \cos^2 \theta \, d\theta$$

$$P = -C(2\alpha - 2\sin 2\alpha).$$

Thus,
$$\sigma_{rr} = -\frac{2p \cos \theta}{r(2\alpha - \sin 2\alpha)} \tag{8.127}$$

From a combination of these cases any inclination of the load can easily be handled.

(d) Uniformly distributed normal load on part of the surface – Fig. 8.41

The result for σ_{rr} obtained in eqn. (8.124) can be used to examine the case of a uniformly distributed normal load q per unit length over part of a surface-say $\theta = \frac{\pi}{2}$. It is required to find the values of the normal and shear stresses (σ_{xx}, σ_{yy}, τ_{xy}) at the point A situated as indicated in Fig. 8.41. In this case the load is divided into a series of discrete lengths δx over which the load is δP, that is $\delta P = q\delta x$. To make use of eqn. (8.124) we must transform this into polars (r, θ). That is

$$dx = r \, d\theta / \cos \theta. \quad \text{Thus, } dP = q \cdot r \, d\theta / \cos \theta \tag{8.128}$$

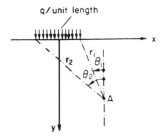

Fig. 8.41. A distributed force on a semi-infinite plate.

Then from eqn. (8.124)

$$d\sigma_{rr} = -\frac{2}{\pi r}dP\cos\theta$$

Substituting eqn. (8.128): $$d\sigma_{rr} = -\frac{2}{\pi r} \cdot q \cdot r d\theta = -\frac{2q}{\pi}d\theta$$

Making use of eqns. (8.125): $$d\sigma_{yy} = -\frac{2q}{\pi}\cos^2\theta\, d\theta$$

$$d\sigma_{xx} = -\frac{2q}{\pi}\sin^2\theta\, d\theta$$

$$d\tau_{xy} = -\frac{2q}{\pi}\sin\theta\cos\theta$$

The total stress values at the point A due to all the discrete loads over θ_1 to θ_2 can then be written,

$$
\left.
\begin{aligned}
\sigma_{yy} &= -\frac{2q}{\pi}\int_{\theta_1}^{\theta_2}\cos^2\theta\, d\theta \\
&= -\frac{q}{2\pi}[2(\theta_2 - \theta_1) + (\sin 2\theta_2 - \sin 2\theta_1)] \\
\sigma_{xx} &= -\frac{q}{2\pi}[2(\theta_2 - \theta_1) - (\sin 2\theta_2 - \sin 2\theta_1)] \\
\tau_{xy} &= -\frac{q}{2\pi}[\cos 2\theta_1 - \cos 2\theta_2]
\end{aligned}
\right\}
\qquad (8.129)
$$

Closure

The stress function concept described above was developed over 100 years ago. Despite this, however, the ideas contained are still of relevance today in providing a series of classical solutions to otherwise intractable problems, particularly in the study of plates and shells.

Examples

Example 8.1

At a point in a material subjected to a three-dimensional stress system the cartesian stress coordinates are:

$$\sigma_{xx} = 100 \text{ MN/m}^2 \quad \sigma_{yy} = 80 \text{ MN/m}^2 \quad \sigma_{zz} = 150 \text{ MN/m}^2$$

$$\sigma_{xy} = 40 \text{ MN/m}^2 \quad \sigma_{yz} = -30 \text{ MN/m}^2 \quad \sigma_{zx} = 50 \text{ MN/m}^2$$

Determine the normal, shear and resultant stresses on a plane whose normal makes angles of 52° with the X axis and 68° with the Y axis.

Solution

The direction cosines for the plane are as follows:

$$l = \cos 52° = 0.6157$$

$$m = \cos 68° = 0.3746$$

and, since $l^2 + m^2 + n^2 = 1$,

$$n^2 = 1 - (0.6157^2 + 0.3746^2)$$

$$= 1 - (0.3791 + 0.1403) = 0.481$$

$$\therefore \qquad n = 0.6935$$

Now from eqns. (8.13–15) the components of the resultant stress on the plane in the X,Y and Z directions are given by

$$p_{xn} = \sigma_{xx}l + \sigma_{xy}m + \sigma_{xz}n$$

$$p_{yn} = \sigma_{yy}m + \sigma_{yx}l + \sigma_{yz}n$$

$$p_{zn} = \sigma_{zz}n + \sigma_{zx}l + \sigma_{zy}m$$

$$\therefore \qquad p_{xn} = (100 \times 0.6157) + (40 \times 0.3746) + (50 \times 0.6935) = 111.2 \text{ MN/m}^2$$

$$p_{yn} = (80 \times 0.3746) + (40 \times 0.6157) + (-30 \times 0.6935) = 33.8 \text{ MN/m}^2$$

$$p_{zn} = (150 \times 0.6935) + (50 \times 0.6157) + (-30 \times 0.3746) = 123.6 \text{ MN/m}^2$$

Therefore from eqn. (8.4) the resultant stress p_n is given by

$$p_n = \left[p_{xn}^2 + p_{yn}^2 + p_{zn}^2 \right]^{1/2} = \left[111.2^2 + 33.8^2 + 123.6^2 \right]^{1/2}$$

$$= \mathbf{169.7 \text{ MN/m}^2}$$

The normal stress σ_n is given by eqn. (8.5),

$$\sigma_n = p_{xn}l + p_{yn}m + p_{zn}n$$

$$= (111.2 \times 0.6157) + (33.8 \times 0.3746) + (123.6 \times 0.6935)$$

$$= \mathbf{166.8 \text{ MN/m}^2}$$

and the shear stress τ_n is found from eqn. (8.6),

$$\tau_n = \sqrt{(p_n^2 - \sigma_n^2)} = (28798 - 27830)^{1/2}$$

$$= \mathbf{31 \text{ MN/m}^2}$$

Example 8.2

Show how the equation of equilibrium in the radial direction of a cylindrical coordinate system can be reduced to the form

$$\frac{\partial \sigma_{rr}}{\partial r} + \frac{(\sigma_{rr} - \sigma_{\theta\theta})}{r} = 0$$

for use in applications involving long cylinders of thin uniform wall thickness.

Hence show that for such a cylinder of internal radius R_0, external radius R and wall thickness T (Fig. 8.42) the radial stress σ_{rr} at any thickness t is given by

$$\sigma_{rr} = -p \frac{R_0}{T} \frac{(T - t)}{(R_0 + t)}$$

where p is the internal pressure, the external pressure being zero.

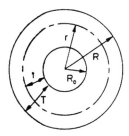

Fig. 8.42.

For thin-walled cylinders the circumferential stress $\sigma_{\theta\theta}$ can be assumed to be independent of radius.

What will be the equivalent expression for the circumferential stress?

Solution

The relevant equation of equilibrium is

$$\frac{\partial \sigma_{rr}}{\partial r} + \frac{1}{r}\frac{\partial \sigma_{r\theta}}{\partial \theta} + \frac{\partial \sigma_{rz}}{\partial z} + \frac{(\sigma_{rr} - \sigma_{\theta\theta})}{r} + F_r = 0$$

Now for long cylinders plane strain conditions may be assumed,

i.e.

$$\frac{\partial \sigma_{rz}}{\partial z} = 0$$

By symmetry, the stress conditions are independent of θ,

$\therefore$

$$\frac{\partial \sigma_{r\theta}}{\partial \theta} = 0$$

and, in the absence of body forces,

$$F_r = 0$$

Thus the equilibrium equation reduces to

$$\frac{\partial \sigma_{rr}}{\partial r} + \frac{(\sigma_{rr} - \sigma_{\theta\theta})}{r} = 0$$

Since $\sigma_{\theta\theta}$ is independent of r this equation can be conveniently rearranged as follows:

$$\sigma_{rr} + r\frac{\partial \sigma_{rr}}{\partial r} = \sigma_{\theta\theta}$$

$$\frac{\partial}{\partial r}(r\sigma_{rr}) = \sigma_{\theta\theta}$$

Integrating,

$$r\sigma_{rr} = \sigma_{\theta\theta}r + C \qquad (1)$$

Now at $r = R$, $\sigma_{rr} = 0$

$\therefore$ substituting in (1), $\qquad 0 = R\sigma_{\theta\theta} + C$

$\therefore$ $\qquad C = -R\sigma_{\theta\theta} \qquad (2)$

Also at $r = R_0$, $\sigma_{rr} = -p$,

$$\therefore \qquad\qquad -R_0 p = R_0 \sigma_{\theta\theta} + C$$

$$= -(R - R_0)\sigma_{\theta\theta}$$

$$\therefore \qquad\qquad \sigma_{\theta\theta} = \frac{R_0 p}{(R - R_0)} \qquad\qquad (3)$$

Substituting in (1),

$$r\sigma_{rr} = \sigma_{\theta\theta} r - R\sigma_{\theta\theta} = -(R - r)\sigma_{\theta\theta}$$

$$\therefore \qquad\qquad \sigma_{rr} = -\frac{(R - r)}{r} \times \frac{R_0 p}{(R - R_0)}$$

$$\sigma_{rr} = -\frac{(T - t)}{r} p \frac{R_0}{T}$$

$$= -\frac{p R_0}{T} \frac{(T - t)}{(R_0 + t)}$$

and from (3)

$$\sigma_{\theta\theta} = \frac{R_0 p}{(R - R_0)} = \frac{R_0 p}{T}$$

Example 8.3

A three-dimensional complex stress system has principal stress values of 280 MN/m^2, 50 MN/m^2 and -120 MN/m^2. Determine (a) analytically and (b) graphically:

(i) the limiting value of the maximum shear stress;
(ii) the values of the octahedral normal and shear stresses.

Solution (a): Analytical

(i) The limiting value of the maximum shear stress is the greatest value obtained in any plane of the three-dimensional system. In terms of the principal stresses this is given by

$$\tau_{\max} = \tfrac{1}{2}(\sigma_1 - \sigma_3)$$

$$= \tfrac{1}{2}[280 - (-120) = \mathbf{200\ MN/m^2}$$

(ii) The octahedral normal stress is given by

$$\sigma_{\text{oct}} = \tfrac{1}{3}[\sigma_1 + \sigma_2 + \sigma_3]$$

$$= \tfrac{1}{3}[280 + 50 + (-120)] = \mathbf{70\ MN/m^2}$$

(iii) The octahedral shear stress is

$$\tau_{\text{oct}} = \tfrac{1}{3}\left[(\sigma_1 - \sigma_2)^2 + (\sigma_2 - \sigma_3)^2 + (\sigma_3 - \sigma_1)^2\right]^{1/2}$$

$$= \tfrac{1}{3}\left[(280 - 50)^2 + (50 + 120)^2 + (-120 - 280)^2\right]^{1/2}$$

$$= \tfrac{1}{3} [52900 + 28900 + 160000]^{1/2}$$

$$= \mathbf{163.9 \ MN/m^2}$$

Solution (b): Graphical

(i) The graphical solution is obtained by constructing the three-dimensional Mohr's representation of Fig. 8.43. The limiting value of the maximum shear stress is then equal to the radius of the principal circle.

i.e. $$\tau_{\max} = \mathbf{200 \ MN/m^2}$$

(ii) The direction cosines of the octahedral planes are

$$l = m = n = \frac{1}{\sqrt{3}} = 0.5774$$

i.e. $$\alpha = \beta = \gamma = \cos^{-1} 0.5774 = 54°52'$$

The values of the normal and shear stresses on these planes are then obtained using the procedures of §8.7.

By measurement, $$\sigma_{\text{oct}} = \mathbf{70 \ MN/m^2}$$

$$\tau_{\text{oct}} = \mathbf{164 \ MN/m^2}$$

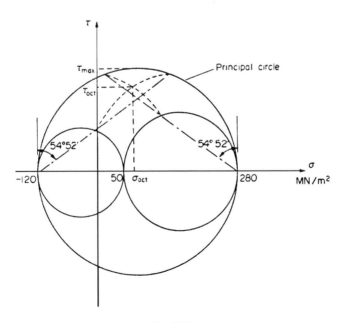

Fig. 8.43.

Example 8.4

A rectangular strain gauge rosette bonded at a point on the surface of an engineering component gave the following readings at peak load during test trials:

$$\varepsilon_0 = 1240 \times 10^{-6}, \quad \varepsilon_{45} = 400 \times 10^{-6}, \quad \varepsilon_{90} = 200 \times 10^{-6}$$

Determine the magnitude and direction of the principal stresses present at the point, and hence construct the full three-dimensional Mohr representations of the stress and strain systems present. $E = 210$ GN/m^2, $\nu = 0.3$.

Solution

The two-dimensional Mohr's strain circle representing strain conditions in the plane of the surface at the point in question is drawn using the procedure of §14.14[†] (Fig. 8.44).

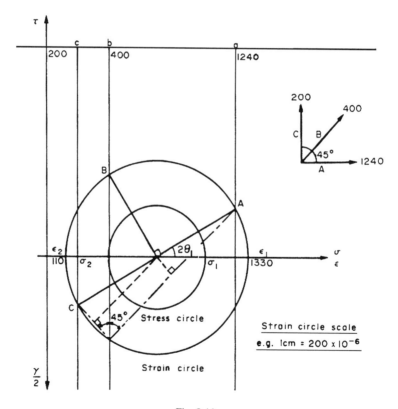

Fig. 8.44.

This establishes the values of the principal strains in the surface plane as 1330 $\mu\varepsilon$ and 110 $\mu\varepsilon$.

[†] E.J. Hearn, *Mechanics of Materials 1*, Butterworth-Heinemann, 1997.

The relevant two-dimensional *stress* circle can then be superimposed as described in §14.13 using the relationships:

$$\text{radius of stress circle} = \frac{(1-\nu)}{(1+\nu)} \times \text{radius of strain circle}$$

$$= \frac{0.7}{1.3} \times 3.05 = 1.64 \text{ cm}$$

$$\text{stress scale} = \frac{E}{(1-\nu)} \times \text{strain scale}$$

$$= \frac{210 \times 10^9}{0.7} \times 200 \times 10^{-6}$$

$$= 60 \text{ MN/m}^2$$

i.e. 1 cm on the stress diagram represents 60 MN/m^2.

The two principal stresses in the plane of the surface are then:

$$\sigma_1 (= 5.25 \text{ cm}) = \textbf{315 MN/m}^2$$

$$\sigma_2 (= 2.0 \ \text{ cm}) = \textbf{120 MN/m}^2$$

The third principal stress, normal to the free (unloaded) surface, is zero,

i.e. $$\sigma_3 = 0$$

The directions of the principal stresses are also obtained from the stress circle. With reference to the 0° gauge direction,

$$\sigma_1 \text{ lies at } \theta_1 = \textbf{15° clockwise}$$

$$\sigma_2 \text{ lies at } (15° + 90°) = \textbf{105° clockwise}$$

with σ_3 **normal to the surface** and hence to the plane of σ_1 and σ_2.

N.B. – These angles are the directions of the principal stresses (and strains) and they do not refer to the directions of the plane on which the stresses act, these being normal to the above directions.

It is now possible to determine the value of the third principal strain, i.e. that normal to the surface. This is given by eqn. (14.2) as

$$\varepsilon_3 = \frac{1}{E} [\sigma_3 - \nu\sigma_1 - \nu\sigma_2]$$

$$= \frac{1}{210 \times 10^9} [0 - 0.3(315 + 120)] \, 10^6$$

$$= -621 \times 10^{-6} = \textbf{--621 } \boldsymbol{\mu\varepsilon}$$

The complete Mohr's three-dimensional stress and strain representations can now be drawn as shown in Figs. 8.45 and 8.46.

[†] E.J. Hearn, *Mechanics of Materials 1*, Butterworth-Heinemann, 1997.

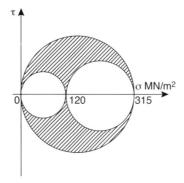

Fig. 8.45. Mohr stress circles.

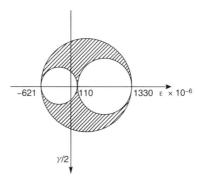

Fig. 8.46. Mohr strain circles.

Problems

8.1 (B). Given that the following strains exist at a point in a three-dimensional system determine the equivalent stresses which act at the point.

Take $E = 206$ GN/m^2 and $\nu = 0.3$.

$$\varepsilon_{xx} = 0.0010 \qquad \gamma_{xy} = 0.0002$$

$$\varepsilon_{yy} = 0.0005 \qquad \gamma_{zx} = 0.0008$$

$$\varepsilon_{zz} = 0.0007 \qquad \gamma_{yz} = 0.0010$$

[420, 340, 372, 15.8, 63.4, 79.2 MN/m^2.]

8.2 (B). The following cartesian stresses act at a point in a body subjected to a complex loading system. If $E = 206$ GN/m^2 and $\nu = 0.3$, determine the equivalent strains present.

$$\sigma_{xx} = 225 \text{ MN/m}^2 \qquad \sigma_{yy} = 75 \text{ MN/m}^2 \qquad \sigma_{zz} = 150 \text{ MN/m}^2$$

$$\tau_{xy} = 110 \text{ MN/m}^2 \qquad \tau_{yz} = 50 \text{ MN/m}^2 \qquad \tau_{zx} = 70 \text{ MN/m}^2$$

[764.6, 182, 291, 1388, 631, 883.5, all $\times 10^{-6}$.]

8.3 (B). Does a uniaxial stress field produce a uniaxial strain condition? Repeat Problem 8.2 for the following stress field:

$$\sigma_{xx} = 225 \text{ MN/m}^2$$

$$\sigma_{yy} = \sigma_{zz} = \tau_{xy} = \tau_{yz} = \tau_{zx} = 0$$

[No; 1092, −327.7, −327.7, 0, 0, 0, all $\times 10^{-6}$.]

8.4 (C). The state of stress at a point in a body is given by the following equations:

$$\sigma_{xx} = ax + by^2 + cz^3 \qquad \tau_{xy} = l + mz$$

$$\sigma_{yy} = dx + ey^2 + fz^3 \qquad \tau_{yz} = ny + pz$$

$$\sigma_{zz} = gx + hy^2 + kz^3 \qquad \tau_{zx} = qx^2 + sz^2$$

If equilibrium is to be achieved what equations must the body-force stresses X, Y and Z satisfy?

$$[-(a + 2sz); \ -(p + 2ey); \ -(n + 2qx + 3kz^2).]$$

8.5 (C). At a point the state of stress may be represented in standard form by the following:

$$
\begin{array}{ccc}
(3x^2 + 3y^2 - z) & (z - 6xy - \frac{3}{4}) & (x + y - \frac{3}{2}) \\[4pt]
(z - 6xy - \frac{3}{4}) & 3y^2 & 0 \\[4pt]
(x + y - \frac{3}{2}) & 0 & (3x + y - z + \frac{5}{4})
\end{array}
$$

Show that, if body forces are neglected, equilibrium exists.

8.6 (C). The plane stress distribution in a flat plate of unit thickness is given by:

$$\sigma_{xx} = yx^3 - 2axy + by$$

$$\sigma_{yy} = xy^3 - 2x^3y$$

$$\sigma_{xy} = -\frac{3}{2}x^2y^2 + ay^2 + \frac{x^4}{2} + c$$

Show that, in the absence of body forces, equilibrium exists. The load on the plate is specified by the following boundary conditions:

$$\text{At } x = \pm\frac{w}{2}, \quad \sigma_{xy} = 0$$

$$\text{At } x = -\frac{w}{2}, \quad \sigma_{xx} = 0$$

where w is the width of the plate.

If the length of the plate is L, determine the values of the constants a, b and c and determine the total load on the edge of the plate, $x = w/2$.

$$[B.P.] \ \left[\frac{3w^2}{8}, \ -\frac{w^3}{4}, \ -\frac{w^4}{32}, \ -\frac{w^3L^2}{4} \right]$$

8.7 (C). Derive the stress equations of equilibrium in cylindrical coordinates and show how these may be simplified for plane strain conditions.

A long, thin-walled cylinder of inside radius R and wall thickness T is subjected to an internal pressure p. Show that, if the hoop stresses are assumed independent of radius, the radial stress at any thickness t is given by

$$\sigma_{\mathrm{rr}} = \frac{pR}{(R + t)} \left[\frac{t}{T} - 1 \right]$$

8.8 (B). Prove that the following relationship exists between the direction cosines:

$$l^2 + m^2 + n^2 = 1$$

8.9 (C). The six cartesian stress components are given at a point P for three different loading cases as follows (all MN/m^2):

	Case 1	Case 2	Case 3
σ_{xx}	100	100	100
σ_{yy}	200	200	−200
σ_{zz}	300	100	100
τ_{xy}	0	300	200
τ_{yz}	0	100	300
τ_{zx}	0	200	300

Determine for each case the resultant stress at P on a plane through P whose normal is coincident with the X axis.

$$[100, 374, 374 \text{ MN/m}^2.]$$

8.10 (C). At a point in a material the stresses are:

$$\sigma_{xx} = 37.2 \text{ MN/m}^2 \qquad \sigma_{yy} = 78.4 \text{ MN/m}^2 \qquad \sigma_{zz} = 149 \text{ MN/m}^2$$

$$\sigma_{xy} = 68.0 \text{ MN/m}^2 \qquad \sigma_{yz} = -18.1 \text{ MN/m}^2 \qquad \sigma_{zx} = 32 \text{ MN/m}^2$$

Calculate the shear stress on a plane whose normal makes an angle of 48° with the X axis and 71° with the Y axis.

$$[41.3 \text{ MN/m}^2.]$$

8.11 (C). At a point in a stressed material the cartesian stress components are:

$$\sigma_{xx} = -40 \text{ MN/m}^2 \qquad \sigma_{yy} = 80 \text{ MN/m}^2 \qquad \sigma_{zz} = 120 \text{ MN/m}^2$$

$$\sigma_{xy} = 72 \text{ MN/m}^2 \qquad \sigma_{yz} = 46 \text{ MN/m}^2 \qquad \sigma_{zx} = 32 \text{ MN/m}^2$$

Calculate the normal, shear and resultant stresses on a plane whose normal makes an angle of 48° with the X axis and 61° with the Y axis.

$$[135, 86.6, 161 \text{ MN/m}^2.]$$

8.12 (C). Commencing from the equations defining the state of stress at a point, derive the general stress relationship for the normal stress on an inclined plane:

$$\sigma_n = \sigma_{xx}l^2 + \sigma_{zz}n^2 + \sigma_{yy}m^2 + 2\sigma_{xy}lm + 2\sigma_{yz}mn + 2\sigma_{zx}ln$$

Show that this relationship reduces for the plane stress system ($\sigma_{zz} = \sigma_{zx} = \sigma_{zy} = 0$) to the well-known equation

$$\sigma_n = \tfrac{1}{2}(\sigma_{xx} + \sigma_{yy}) + \tfrac{1}{2}(\sigma_{xx} - \sigma_{yy})\cos 2\theta + \sigma_{xy}\sin 2\theta$$

where $\cos\theta = l$.

8.13 (C). At a point in a material a resultant stress of value 14 MN/m^2 is acting in a direction making angles of 43°, 75° and 50°53′ with the coordinate axes X, Y and Z.

(a) Find the normal and shear stresses on an oblique plane whose normal makes angles of 67°13′, 30° and 71°34′, respectively, with the same coordinate axes.

(b) If $\sigma_{xy} = 1.5 \text{ MN/m}^2$, $\sigma_{yz} = -0.2 \text{ MN/m}^2$ and $\sigma_{xz} = 3.7 \text{ MN/m}^2$ determine σ_{xx}, σ_{yy} and σ_{zz}.

$$[10, 9.8, 19.9, 3.58, 23.5 \text{ MN/m}^2.]$$

8.14 (C). Three principal stresses of 250, 100 and -150 MN/m^2 act in a direction X, Y and Z respectively. Determine the normal, shear and resultant stresses which act on a plane whose normal is inclined at 30° to the Z axis, the projection of the normal on the XY plane being inclined at 55° to the XZ plane.

$$[-75.2, 134.5, 154.1 \text{ MN/m}^2.]$$

8.15 (C). The following cartesian stress components exist at a point in a body subjected to a three-dimensional complex stress system:

$$\sigma_{xx} = 97 \text{ MN/m}^2 \qquad \sigma_{yy} = 143 \text{ MN/m}^2 \qquad \sigma_{zz} = 173 \text{ MN/m}^2$$

$$\sigma_{xy} = 0 \qquad \sigma_{yz} = 0 \qquad \sigma_{zx} = 102 \text{ MN/m}^2$$

Determine the values of the principal stresses present at the point.

$$[233.8, 143.2, 35.8 \text{ MN/m}^2.]$$

8.16 (C). A certain stress system has principal stresses of 300 MN/m^2, 124 MN/m^2 and 56 MN/m^2.

(a) What will be the value of the maximum shear stress?

(b) Determine the values of the shear and normal stresses on the octahedral planes.

(c) If the yield stress of the material in simple tension is 240 MN/m^2, will the above stress system produce failure according to the distortion energy and maximum shear stress criteria?

$$[122 \text{ MN/m}^2; \; 104, 160 \text{ MN/m}^2; \text{ No, Yes.}]$$

8.17 (C). A pressure vessel is being tested at an internal pressure of 150 atmospheres (1 atmosphere = 1.013 bar). Strains are measured at a point on the inside surface adjacent to a branch connection by means of an equiangular strain rosette. The readings obtained are:

$$\varepsilon_0 = 0.23\% \qquad \varepsilon_{+120} = 0.145\% \qquad \varepsilon_{-120} = 0.103\%$$

Draw Mohr's circle to determine the magnitude and direction of the principal strains. $E = 208$ GN/m^2 and $\nu = 0.3$. Determine also the octahedral normal and shear strains at the point.

$$[0.235\%, 0.083\%, -0.142\%, 9°28'; \varepsilon_{\text{oct}} = 0.0589\%, \gamma_{\text{oct}} = 0.310\%.]$$

8.18 (C). At a point in a stressed body the principal stresses in the X, Y and Z directions are:

$$\sigma_1 = 49 \text{ MN/m}^2 \qquad \sigma_2 = 27.5 \text{ MN/m}^2 \qquad \sigma_3 = -6.3 \text{ MN/m}^2$$

Calculate the resultant stress on a plane whose normal has direction cosines $l = 0.73$, $m = 0.46$, $n = 0.506$. Draw Mohr's stress plane for the problem to check your answer. [38 MN/m^2.]

8.19 (C). For the data of Problem 8.18 determine graphically, and by calculation, the values of the normal and shear stresses on the given plane.
Determine also the values of the octahedral direct and shear stresses. [30.3, 23 MN/m^2; 23.4, 22.7 MN/m^2.]

8.20 (C). During tests on a welded pipe-tee, internal pressure and torque are applied and the resulting distortion at a point near the branch gives rise to shear components in the r, θ and z directions.
A rectangular strain gauge rosette mounted at the point in question yields the following strain values for an internal pressure of 16.7 MN/m^2:

$$\varepsilon_0 = 0.0013 \qquad \varepsilon_{45} = 0.00058 \qquad \varepsilon_{90} = 0.00187$$

Use the Mohr diagrams for stress and strain to determine the state of stress on the octahedral plane. $E = 208$ GN/m^2 and $\nu = 0.29$.
What is the direct stress component on planes normal to the direction of zero extension?

$$[\sigma_{\text{oct}} = 310 \text{ MN/m}^2; \tau_{\text{oct}} = 259 \text{ MN/m}^2; 530 \text{ MN/m}^2.]$$

8.21 (C). During service loading tests on a nuclear pressure vessel the distortions resulting near a stress concentration on the inside surface of the vessel give rise to shear components in the r, θ and z directions. A rectangular strain gauge rosette mounted at the point in question gives the following strain values for an internal pressure of 5 MN/m^2.

$$\varepsilon_0 = 150 \times 10^{-6}, \varepsilon_{45} = 220 \times 10^{-6} \text{ and } \varepsilon_{90} = 60 \times 10^{-6}$$

Use the Mohr diagrams for stress and strain to determine the principal stresses and the state of stress on the octahedral plane at the point. For the material of the pressure vessel $E = 210$ GN/m^2 and $\nu = 0.3$.
[B.P.] [52.5, 13.8, -5 MN/m^2; $\sigma_{\text{oct}} = 21$ MN/m^2, $\tau_{\text{oct}} = 24$ MN/m^2.]

8.22 (C). From the construction of the Mohr strain plane show that the ordinate $\frac{1}{2}\gamma$ for the case of $\alpha = \beta = \gamma$ (octahedral shear strain) is

$$\tfrac{1}{3}[(\varepsilon_1 - \varepsilon_2)^2 + (\varepsilon_2 - \varepsilon_3)^2 + (\varepsilon_3 - \varepsilon_1)^2]^{1/2}$$

8.23 (C). A stress system has three principal values:

$$\sigma_1 = 154 \text{ MN/m}^2 \qquad \sigma_2 = 113 \text{ MN/m}^2 \qquad \sigma_3 = 68 \text{ MN/m}^2$$

(a) Find the normal and shear stresses on a plane with direction cosines of $l = 0.732$, $m = 0.521$ with respect to the σ_1 and σ_2 directions.
(b) Determine the octahedral shear and normal stresses for this system. Check numerically.

$$[126, 33.4 \text{ MN/m}^2; 112, 35.1 \text{ MN/m}^2.]$$

8.24 (C). A plane has a normal stress of 63 MN/m^2 inclined at an angle of 38° to the greatest principal stress which is 126 MN/m^2. The shear stress on the plane is 92 MN/m^2 and a second principal stress is 53 MN/m^2. Find the value of the third principal stress and the angle of the normal of the plane to the direction of stress.
[-95 MN/m^2; 60°.]

8.25 (C). The normal stress σ_n on a plane has a direction cosine l and the shear stress on the plane is τ_n. If the two smaller principal stresses are equal show that

$$\sigma_1 = \sigma_n + \frac{\tau_n}{l}\sqrt{(1 - l^2)} \quad \text{and} \quad \sigma_2 = \sigma_3 = \sigma_n - \tau_n \frac{1}{\sqrt{(1 - l^2)}}$$

If $\tau_n = 75$ MN/m^2, $\sigma_n = 36$ MN/m^2 and $l = 0.75$, determine, graphically σ_1 and σ_2. [102, -48 MN/m^2.]

8.26 (C). If the strains at a point are $\varepsilon = 0.0063$ and $\gamma = 0.00481$, determine the value of the maximum principal strain ε_1 if it is known that the strain components make the following angles with the three principal strain directions:

$$\text{For } \varepsilon : \quad \alpha = 38.5° \qquad \beta = 56° \qquad \gamma = \text{positive}$$
$$\text{For } \gamma : \quad \alpha' = 128°32' \qquad \beta' = 45°10' \qquad \gamma' = \text{positive} \qquad\qquad [0.0075.]$$

8.27 (C). What is meant by the term deviatoric strain as related to a state of strain in three dimensions? Show that the sum of three deviatoric strains ε'_1, ε'_2 and ε'_3 is zero and also that they can be related to the principal strains $\varepsilon_1, \varepsilon_2$ and ε_3 as follows:

$$\varepsilon_1'^2 + \varepsilon_2'^2 + \varepsilon_3'^2 = \tfrac{1}{3}[(\varepsilon_1 - \varepsilon_2)^2 + (\varepsilon_2 - \varepsilon_1)^2 + (\varepsilon_3 - \varepsilon_1)^2] \qquad\qquad \text{[C.E.I.]}$$

8.28 (C). The readings from a rectangular strain gauge rosette bonded to the surface of a strained component are as follows:

$$\varepsilon_0 = 592 \times 10^{-6} \qquad \varepsilon_{45} = 308 \times 10^{-6} \qquad \varepsilon_{90} = -432 \times 10^{-6}$$

Draw the full three-dimensional Mohr's stress and strain circle representations and hence determine:
(a) the principal strains and their directions;
(b) the principal stresses;
(c) the maximum shear stress.

Take $E = 200 \text{ GN/m}^2$ and $\nu = 0.3$.

$$[640 \times 10^{-6}, -480 \times 10^{-6}; \text{ at } 12° \text{ and } 102° \text{ to } A, 109, -63.5, 86.25 \text{ MN/m}^2]$$

8.29 (C). For a rectangular beam, unit width and depth $2d$, simple beam theory gives the longitudinal stress $\sigma_{xx} = CMy/I$ where

$$y = \text{ordinate in depth direction (+ downwards)}$$

$$M = \text{BM in } yx \text{ plane (+ sagging)}$$

The shear force is Q and the shear stress τ_{xy} is to be taken as zero at top and bottom of the beam.
$\sigma_{yy} = 0$ at the bottom and $\sigma_{yy} = -w/$unit length, i.e. a distributed load, at the top.

$$\sigma_{zz} = \sigma_{zx} = \sigma_{zy} = 0$$

Using the equations of equilibrium in cartesian coordinates and without recourse to beam theory, find the distribution of σ_{yy} and σ_{xy}.

$$\text{[U.L.]} \quad \left[\sigma_{yy} = \frac{w}{2I}\left(d^2 y - \frac{y^3}{3} - \frac{2d^3}{3}\right), \quad \sigma_{xy} = -\frac{Q}{2I}(d^2 - y^2). \right]$$

8.30 (C). Determine whether the following strain fields are compatible:

$$\begin{array}{llll}
\text{(a)} & \varepsilon_{xx} = 2x^2 + 3y^2 + z + 1 & \text{(b)} & \varepsilon_{xx} = 3y^2 + xy \\
& \varepsilon_{yy} = 2y^2 + x^2 + 3z + 2 & & \varepsilon_{yy} = 2y + 4z + 3 \\
& \varepsilon_{zz} = 3x + 2y + z^2 + 1 & & \varepsilon_{zz} = 3zx + 2xy + 3yz + 2 \\
& \gamma_{xy} = 8xy & & \gamma_{xy} = 6xy \\
& \gamma_{yz} = 0 & & \gamma_{yz} = 2x \\
& \gamma_{zx} = 0 & & \gamma_{zx} = 2y \\
& \text{[Yes]} & & \text{[No]}
\end{array}$$

8.31 (C). The normal stress σ_n on a plane has a direction cosine l and the shear stress on the plane is τ. If the two smaller principal stresses are equal show that

$$\sigma_1 = \sigma_n + \frac{\tau}{l}\sqrt{(1 - l^2)} \quad \text{and} \quad \sigma_2 = \sigma_3 = \sigma_n - \frac{\tau l}{\sqrt{(1 - l^2)}}$$

8.32 (C). (i) A long thin-walled cylinder of internal radius R_0, external radius R and wall thickness T is subjected to an internal pressure p, the external pressure being zero. Show that if the circumferential stress ($\sigma_{\theta\theta}$) is independent of the radius r then the radial stress (σ_{rr}) at any thickness t is given by

$$\sigma_{rr} = -p\frac{R_0}{T}\frac{(T - t)}{(R_0 + t)}$$

The relevant equation of equilibrium which may be used is:

$$\frac{\partial \sigma_{rr}}{\partial r} + \frac{1}{r}\frac{\partial \sigma_{r\theta}}{\partial \theta} + \frac{\partial \sigma_{rz}}{\partial z} + \frac{(\sigma_{rr} - \sigma_{\theta\theta})}{r} + F_r = 0$$

(ii) Hence determine an expression for $\sigma_{\theta\theta}$ in terms of T.

(iii) What difference in approach would you adopt for a similar treatment in the case of a thick-walled cylinder?

[B.P.] [$R_0 p/T$.]

8.33 (C). Explain what is meant by the following terms and discuss their significance:

(a) Octahedral planes and stresses.

(b) Hydrostatic and deviatoric stresses.

(c) Plastic limit design.

(d) Compatibility.

(e) Principal and product second moments of area. [B.P.]

8.34 (C). At a point in a stressed material the cartesian stress components are:

$$\sigma_{xx} = -40 \text{ MN/m}^2 \qquad \sigma_{yy} = 80 \text{ MN/m}^2 \qquad \sigma_{zz} = 120 \text{ MN/m}^2$$

$$\sigma_{xy} = 72 \text{ MN/m}^2 \qquad \sigma_{xz} = 32 \text{ MN/m}^2 \qquad \sigma_{yz} = 46 \text{ MN/m}^2$$

Calculate the normal, shear and resultant stresses on a plane whose normal makes an angle of $48°$ with the X axis and $61°$ with the Y axis. [B.P.] [135.3, 86.6, 161 MN/m^2.]

8.35 (C). The Cartesian stress components at a point in a three-dimensional stress system are those given in problem 8.33 above.

(a) What will be the directions of the normal and shear stresses on the plane making angles of $48°$ and $61°$ with the X and Y axes respectively?

$$[l'm'n' = 0.1625, 0.7010, 0.6914; \; l_s m_s n_s = -0.7375, 0.5451, 0.4053]$$

(b) What will be the magnitude of the shear stress on the octahedral planes where $l = m = n = 1/\sqrt{3}$?

[10.7 MN/m^2]

8.36 (C). Given that the cartesian stress components at a point in a three-dimensional stress system are:

$$\sigma_{xx} = 20 \text{ MN/m}^2, \qquad \sigma_{yy} = 5 \text{ MN/m}^2, \qquad \sigma_{zz} = -50 \text{ MN/m}^2$$

$$\tau_{xy} = 0, \qquad \tau_{yz} = 20 \text{ MN/m}^2, \qquad \tau_{zx} = -40 \text{ MN/m}^2$$

(a) Determine the stresses on planes with direction cosines 0.8165, 0.4082 and 0.4082 relative to the X, Y and Z axes respectively. [$-14.2, 46.1, 43.8$ MN/m^2]

(b) Determine the shear stress on these planes in a direction with direction cosines of $0, -0.707, 0.707$.

[39 MN/m^2]

8.37 (C). In a finite element calculation of the stresses in a steel component, the stresses have been determined as follows, with respect to the reference directions X, Y and Z:

$$\sigma_{xx} = 10.9 \text{ MN/m}^2 \qquad \sigma_{yy} = 51.9 \text{ MN/m}^2 \qquad \sigma_{zz} = -27.8 \text{ MN/m}^2$$

$$\tau_{xy} = -41.3 \text{ MN/m}^2 \qquad \tau_{yz} = -8.9 \text{ MN/m}^2 \qquad \tau_{zx} = 38.5 \text{ MN/m}^2$$

It is proposed to change the material from steel to unidirectional glass-fibre reinforced polyester, and it is important that the direction of the fibres is the same as that of the maximum principal stress, so that the tensile stresses perpendicular to the fibres are kept to a minimum.

Determine the values of the three principal stresses, given that the value of the intermediate principal stress is 3.9 MN/m^2. [-53.8; 3.9; 84.9 MN/m^2]

Compare them with the safe design tensile stresses for the glass-reinforced polyester of: parallel to the fibres, 90 MN/m^2; perpendicular to the fibres, 10 MN/m^2.

Then take the direction cosines of the *major* principal stress as $l = 0.569, m = -0.781, n = 0.256$ and determine the maximum allowable misalignment of the fibres to avoid the risk of exceeding the safe design tensile stresses. (Hint: compression stresses can be ignored.) [15.9°]

8.38 (C). The stresses at a point in an isotropic material are:

$$\sigma_{xx} = 10 \text{ MN/m}^2 \qquad \sigma_{yy} = 25 \text{ MN/m}^2 \qquad \sigma_{zz} = 50 \text{ MN/m}^2$$

$$\tau_{xy} = 15 \text{ MN/m}^2 \qquad \tau_{yz} = 10 \text{ MN/m}^2 \qquad \tau_{zx} = 20 \text{ MN/m}^2$$

Determine the magnitudes of the maximum principal normal strain and the maximum principal shear strain at this point, if Young's modulus is 207 GN/m^2 and Poisson's ratio is 0.3. [280$\mu\varepsilon$; 419$\mu\varepsilon$]

8.39 (C). Determine the principal stresses in a three-dimensional stress system in which:

$$\sigma_{xx} = 40 \text{ MN/m}^2 \qquad \sigma_{yy} = 60 \text{ MN/m}^2 \qquad \sigma_{zz} = 50 \text{ MN/m}^2$$
$$\sigma_{xz} = 30 \text{ MN/m}^2 \qquad \sigma_{xy} = 20 \text{ MN/m}^2 \qquad \sigma_{yz} = 10 \text{ MN/m}^2$$

$$[90 \text{ MN/m}^2, 47.3 \text{ MN/m}^2, 12.7 \text{ MN/m}^2]$$

8.40 (C). If the stress tensor for a three-dimensional stress system is as given below and one of the principal stresses has a value of 40 MN/m² determine the values of the three eigen vectors.

$$\begin{bmatrix} 30 & 10 & 10 \\ 10 & 0 & 20 \\ 10 & 20 & 0 \end{bmatrix}$$

$$[0.816, 0.408, 0.408]$$

8.41 (C). Determine the values of the stress invariants and the principal stresses for the cartesian stress components given in Problem 8.2.   [450; 423.75; 556.25; 324.8; 109.5; 15.6 MN/m²]

8.42 (C). The stress tensor for a three-dimensional stress system is given below. Determine the magnitudes of the three principal stresses and determine the eigen vectors of the major principal stress.

$$\begin{bmatrix} 80 & 15 & 10 \\ 15 & 0 & 25 \\ 10 & 25 & 0 \end{bmatrix}$$

$$[85.3, 19.8, -25.1 \text{ MN/m}^2, 0.9592, 0.2206, 0.1771.]$$

8.43 (C). A hollow steel shaft is subjected to combined torque and internal pressure of unknown magnitudes. In order to assess the strength of the shaft under service conditions a rectangular strain gauge rosette is mounted on the outside surface of the shaft, the centre gauge being aligned with the shaft axis. The strain gauge readings recorded from this gauge are shown in Fig. 8.47.

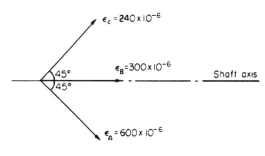

Fig. 8.47.

If E for the steel = 207 GN/m² and $\nu = 0.3$, determine:
(a) the principal strains and their directions;
(b) the principal stresses.

Draw complete Mohr's circle representations of the stress and strain systems present and hence determine the maximum shear stresses and maximum shear strain.
$$[636 \times 10^{-6} \text{ at } 16.8° \text{ to } A, -204 \times 10^{-6} \text{ at } 106.8° \text{ to } A, -360 \times 10^{-6} \text{ perp. to plane}; 159, -90, 0 \text{ MN/m}^2;$$
$$79.5 \text{ MN/m}^2, 996 \times 10^{-6}.]$$

8.44 (C). At a certain point in a material a resultant stress of 40 MN/m² acts in a direction making angles of 45°, 70° and 60° with the coordinate axes X, Y and Z. Determine the values of the normal and shear stresses on an oblique plane through the point given that the normal to the plane makes angles of 80°, 54° and 38° with the same coordinate axes.

If $\sigma_{xy} = 25 \text{ MN/m}^2$, $\sigma_{xz} = 18 \text{ MN/m}^2$ and $\sigma_{yz} = -10 \text{ MN/m}^2$, determine the values of σ_{xx}, σ_{yy} and σ_{zz} which act at the point.   [28.75, 27.7 MN/m²; $-3.5, 29.4, 28.9 \text{ MN/m}^2$.]

8.45 (C). The plane stress distribution in a flat plate of unit thickness is given by

$$\sigma_{xx} = x^3 y - 2y^3 x$$

$$\sigma_{yy} = y^3 x - 2pxy + qx$$

$$\sigma_{xy} = \frac{y^4}{2} - \frac{3}{2}x^2 y^2 + px^2 + s$$

If body forces are neglected, show that equilibrium exists.
The dimensions of the plate are given in Fig. 8.48 and the following boundary conditions apply:

$$\text{at } y = \pm\frac{b}{2} \qquad \sigma_{xy} = 0$$

and

$$\text{at } y = -\frac{b}{2} \qquad \sigma_{yy} = 0$$

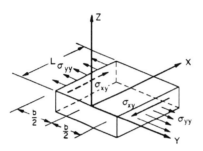

Fig. 8.48.

Determine:
(a) the values of the constants p, q and s;

(b) the total load on the edge $y = \pm b/2$. [B.P.] $\left[\dfrac{3b^2}{8}, \dfrac{-b^3}{4}, \dfrac{-b^4}{32}, \dfrac{b^3 L^2}{4} \right]$

8.46 (C). Derive the differential equation in cylindrical coordinates for radial equilibrium without body force of an element of a cylinder subjected to stresses σ_r, σ_θ.

A steel tube has an internal diameter of 25 mm and an external diameter of 50 mm. Another tube, of the same steel, is to be shrunk over the outside of the first so that the shrinkage stresses just produce a condition of yield at the inner surfaces of each tube. Determine the necessary difference in diameters of the mating surfaces before shrinking and the required external diameter of the outer tube. Assume that yielding occurs according to the maximum shear stress criterion and that no axial stresses are set up due to shrinking. The yield stress in simple tension or compression = 420 MN/m^2 and $E = 208$ GN/m^2. [C.E.I.] [0.126 mm, 100 mm.]

8.47 (C). For a particular plane strain problem the strain displacement equations in cylindrical coordinates are:

$$\varepsilon_r = \frac{\partial u}{\partial r}, \quad \varepsilon_\theta = \frac{u}{r}, \quad \varepsilon_z = \gamma_{r\theta} = \gamma_{\theta z} = \gamma_{zr} = 0$$

Show that the appropriate compatibility equation in terms of stresses is

$$vr\frac{\partial \sigma_r}{\partial r} - (1 - v)r\frac{\partial \sigma_\theta}{\partial r} + \sigma_r - \sigma_\theta = 0$$

where v is Poisson's ratio.
State the nature of a problem that the above equations can represent. [C.E.I.]

8.48 (C). A bar length L, depth d, thickness t is simply supported and loaded at each end by a couple C as shown in Fig. 8.49. Show that the stress function $\phi = Ay^3$ adequately represents this problem. Determine the value of the coefficient A in terms of the given symbols. [A = 2C/td^3]

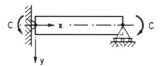

Fig. 8.49.

8.49 (C). A cantilever of unit width and depth $2d$ is loaded with forces at its free end as shown in Fig. 8.50. The stress function which satisfies the loading is found to be of the form:

$$\phi = ay^2 + by^3 + cxy^3 + exy$$

where the coordinates are as shown.

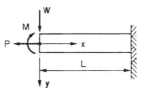

Fig. 8.50.

Determine the value of the constants a, b, c and e and hence show that the stresses are:

$$\sigma_{xx} = P/2d + 3My/2d^3 - 3Wxy/2d^3,$$

$$\sigma_{yy} = 0$$

$$\tau_{xy} = 3Wy^2/4d^3 - 3W/4d.$$

8.50 (C). A cantilever of unit width length L and depth $2a$ is loaded by a linearly distributed load as shown in Fig. 8.51, such that the load at distance x is qx per unit length. Proceeding from the sixth order polynomial derive the 25 constants using the boundary conditions, overall equilibrium and the biharmonic equation. Show that the stresses are:

$$\sigma_{xx} = \frac{qx^3y}{4a^3} + \frac{q}{4a^3}\left(-2xy^3 + \frac{6}{5}a^2xy\right)$$

$$\sigma_{yy} = -q\frac{x}{2} + qx\left(\frac{y^3}{4a^3} - \frac{3y}{4a}\right)$$

$$\tau_{xy} = \frac{3qx^2}{8a^3}(a^2 - y^2) - \frac{q}{8a^3}(a^4 - y^4) + \frac{3q}{20a}(a^2 - y^2)$$

Examine the state of stress at the free end ($x = 0$) and comment on the discrepancy of the shear stress. Compare the shear stress obtained from elementary theory, for $L/2a = 10$, with the more rigorous approach with the additional terms.

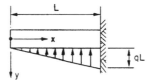

Fig. 8.51.

8.51 (C). Determine if the expression $\phi = (\cos^3 \theta)/r$ is a permissible Airy stress function, that is, make sure it satisfies the biharmonic equation. Determine the radial and shear stresses (σ_{rr} and $\tau_{r\theta}$) and sketch these on the periphery of a circle of radius a.

$$\left[\sigma_{rr} = \frac{2}{r^3} \cos \theta (3 - 5 \cos^2 \theta), \ \tau_{r\theta} = -\frac{6}{r^3} \cos^2 \theta \sin \theta. \right]$$

8.52 (C). The stress concentration factor due to a small circular hole in a flat plate subject to tension (or compression) in one direction is three. By superposition of the Kirsch solutions determine the stress concentration factors due to a hole in a flat plate subject to (a) pure shear, (b) two-dimensional hydrostatic tension. Show that the same result for case (b) can be obtained by considering the Lamé solution for a thick cylinder under external tension when the outside radius tends to infinity. [(a) 4; (b) 2.]

8.53 (C). Show that $\phi - Cr^2(\alpha - \theta + \sin \theta \cos \theta - \tan \alpha \cos^2 \theta)$ is a permissible Airy stress function and derive expressions for the corresponding stresses σ_{rr}, $\sigma_{\theta\theta}$ and $\tau_{r\theta}$.

These expressions may be used to solve the problem of a tapered cantilever beam of thickness carrying a uniformly distributed load q/unit length as shown in Fig. 8.52.

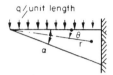

Fig. 8.52.

Show that the derived stresses satisfy every boundary condition along the edges $\theta = 0°$ and $\theta = \alpha$. Obtain a value for the constant C in terms of q and α and thus show that:

$$\sigma_{rr} = \frac{qr}{t(\tan \alpha - \alpha)} \quad \text{when } \theta = 0°$$

Compare this value with the longitudinal bending stress at $\theta = 0°$ obtained from the simple bending theory when $\alpha = 5°$ and $\alpha = 30°$. What is the percentage error when using simple bending?

$$\left[C = \frac{q}{2t(\tan \alpha - \alpha)}, \quad -0.2\% \text{ and } -7.6\% \text{ (simple bending is lower)} \right]$$

CHAPTER 9

INTRODUCTION TO THE FINITE ELEMENT METHOD

Introduction

So far in this text we have studied the means by which components can be analysed using so-called Mechanics of Materials approaches whereby, subject to making simplifying assumptions, solutions can be obtained by hand calculation. In the analysis of complex situations such an approach may not yield appropriate or adequate results and calls for other methods. In addition to experimental methods, numerical techniques using digital computers now provide a powerful alternative. Numerical techniques for structural analysis divides into three areas; the long established but limited capability *finite difference* method, the finite element method (developed from earlier structural matrix methods), which gained prominence from the 1950s with the advent of digital computers and, emerging over a decade later, the boundary element method. Attention in this chapter will be confined to the most popular finite element method and the coverage is intended to provide

- an insight into some of the basic concepts of the finite element method (fem.), and, hence, some basis of finite element (fe.), practice,
- the theoretical development associated with some relatively simple elements, enabling analysis of applications which can be solved with the aid of a simple calculator, and
- a range of worked examples to show typical applications and solutions.

It is recommended that the reader wishing for further coverage should consult the many excellent specialist texts on the subject.[1-10] This chapter does require some knowledge of matrix algebra, and again, students are directed to suitable texts on the subject.[11]

9.1. Basis of the finite element method

The fem. is a numerical technique in which the governing equations are represented in matrix form and as such are well suited to solution by digital computer. The solution region is represented, (discretised), as an assemblage (mesh), of small sub-regions called *finite elements*. These elements are connected at discrete points (at the extremities (corners), and in some cases also at intermediate points), known as nodes. Implicit with each element is its displacement function which, in terms of parameters to be determined, defines how the displacements of the nodes are interpolated over each element. This can be considered as an extension of the Rayleigh-Ritz process (used in Mechanics of Machines for analysing beam vibrations[6]). Instead of approximating the entire solution region by a single assumed displacement distribution, as with the Rayleigh-Ritz process, displacement distributions are assumed for each element of the assemblage. When applied to the analysis of a continuum (a solid or fluid through which the behavioural properties vary continuously), the discretisation becomes an assemblage of a number of elements each with a limited, i.e. *finite* number of degrees of freedom (dof). The *element* is the basic "building unit", with a predetermined number of dof., and can take various forms, e.g. one-dimensional rod or beam, two-dimensional membrane or plate, shell, and solid elements, see Fig. 9.1.

In stress applications, implicit with each element type is the nodal force/displacement relationship, namely the element stiffness property. With the most popular *displacement formulation* (discussed in §9.3), analysis requires the assembly and solution of a set of

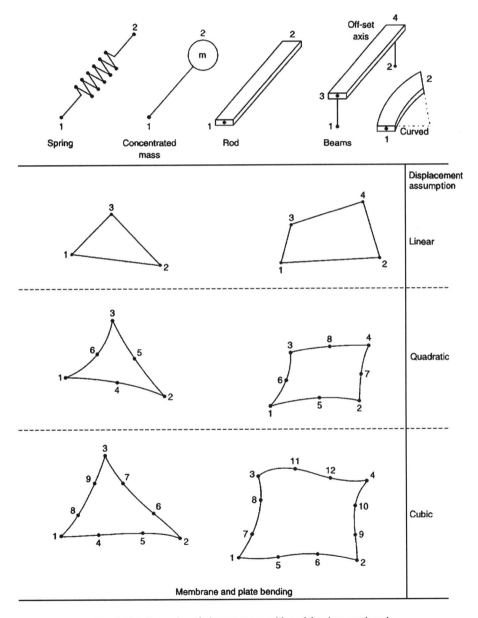

Fig. 9.1(a). Examples of element types with nodal points numbered.

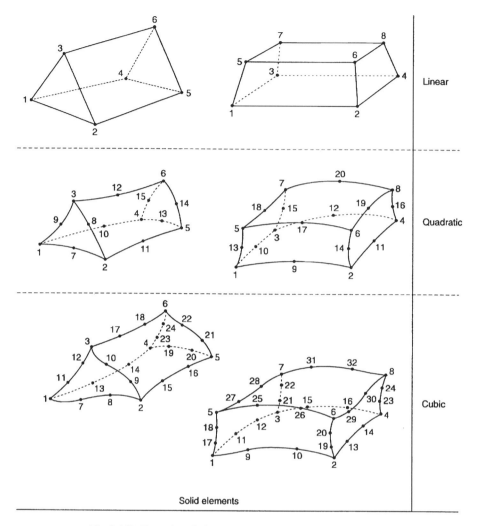

Fig. 9.1(b). Examples of element types with nodal points numbered.

simultaneous equations to provide the displacements for every node in the model. Once the displacement field is determined, the strains and hence the stresses can be derived, using strain-displacement and stress-strain relations, respectively.

9.2. Applicability of the finite element method

The fem. emerged essentially from the aerospace industry where the demand for extensive structural analyses was, arguably, the greatest. The general nature of the theory makes it applicable to a wide variety of boundary value problems (i.e. those in which a solution is required in a region of a body subject to satisfying prescribed boundary conditions, as encountered in equilibrium, eigenvalue and propagation or transient applications). Beyond the basic linear elastic/static stress analysis, finite element analysis (fea.), can provide solutions to non-linear material and/or geometry applications, creep, fracture mechanics, free and forced vibration. Furthermore, the method is not confined to solid mechanics, but is applied successfully to other disciplines such as heat conduction, fluid dynamics, seepage flow and electric and magnetic fields. However, attention in this text will be restricted to linearly elastic static stress applications, for which the assumption is made that the displacements are sufficiently small to allow calculations to be based on the undeformed condition.

9.3. Formulation of the finite element method

Even with restriction to solid mechanics applications, the fem. can be formulated in a variety of ways which broadly divides into 'differential equation', or 'variational' approaches. Of the differential equation approaches, the most important, most widely used and most extensively documented, is the *displacement, or stiffness, based fem.* Due to its simplicity, generality and good numerical properties, almost all major general purpose analysis programmes have been written using this formulation. Hence, only the displacement based fem. will be considered here, but it should be realised that many of the concepts are applicable to other formulations.

In §9.7, 9.8 and 9.9 the theory using the displacement method will be developed for a rod, simple beam and triangular membrane element, respectively. Before this, it is appropriate to consider here, a brief overview of the steps required in a fe. linearly elastic static stress analysis. Whilst it can be expected that there will be detail differences between various packages, the essential procedural steps will be common.

9.4. General procedure of the finite element method

The basic steps involved in a fea. are shown in the flow diagram of Fig. 9.2. Only a simple description of these steps is given below. The reader wishing for a more in-depth treatment is urged to consult some of numerous texts on the subject, referred to in the introduction.

9.4.1. Identification of the appropriateness of analysis by the finite element method

Engineering components, except in the simplest of cases, frequently have non-standard features such as those associated with the geometry, material behaviour, boundary conditions, or excitation (e.g. loading), for which classical solutions are seldom available. The analyst must therefore seek alternative approaches to a solution. One approach which can sometimes be very effective is to simplify the application grossly by making suitable approximations, leading to Mechanics of Materials solutions (the basis of the majority of this text). Allowance for the effects of local disturbances, e.g. rapid changes in geometry, can be achieved through the use of design charts, which provide a means of *local enhancement*. In current practice, many design engineers prefer to take advantage of high speed, large capacity, digital computers and use numerical techniques, in particular the fem. The range of application of the fem. has already been noted in §9.2. The versatility of the fem. combined with the avoidance, or reduction in the need for prototype manufacture and testing offer significant benefits. However, the purchase and maintenance of suitable fe. packages, provision of a computer platform with adequate performance and capacity, application of a suitably

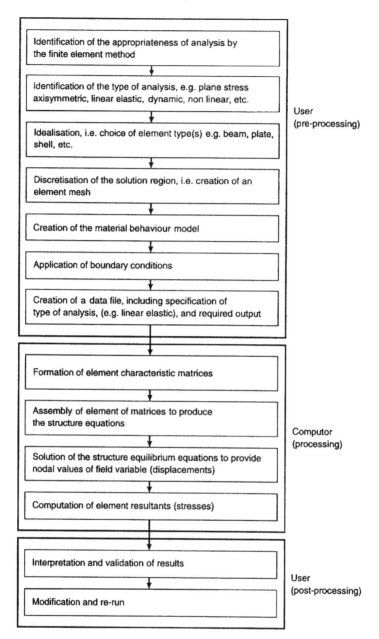

Fig. 9.2. Basic steps in the finite element method.

trained and experienced analyst and time for data preparation and processing should not be underestimated when selecting the most appropriate method. Experimental methods such as those described in Chapter 6 provide an effective alternative approach.

It is desirable that an analyst has access to all methods, i.e. analytical, numerical and experimental, and to not place reliance upon a single approach. This will allow essential validation of one technique by another and provide a degree of confidence in the results.

9.4.2. Identification of the type of analysis

The most appropriate type(s) of analysis to be employed needs to be identified in order that the component behaviour can best be represented. The assumption of either plane stress or plane strain is a common example. The high cost of a full three-dimensional analysis can be avoided if the assumption of both geometric and load symmetry can be made. If the application calls for elastic stress analysis, then the system equations will be linear and can be solved by a variety of methods, Gaussian elimination, Choleski factorisation or Gauss-Seidel procedure.[5]

For large displacement or post-yield material behaviour applications the system equations will be non-linear and iterative solution methods are required, such as that of Newton-Raphson.[5]

9.4.3. Idealisation

Commercially available finite element packages usually have a number of different elements available in the element library. For example, one such package, HKS ABAQUS[12] has nearly 400 different element variations. Examples of some of the commonly used elements have been given in Fig. 9.1.

Often the type of element to be employed will be evident from the physical application. For example, rod and beam elements can represent the behaviour of frames, whilst shell elements may be most appropriate for modelling a pressure vessel. Some regions which are actually three-dimensional can be described by only one or two independent coordinates, e.g. pistons, valves and nozzles, etc. Such regions can be idealised by using axisymmetric elements. Curved boundaries are best represented by elements having mid-side (or intermediate) nodes in addition to their corner nodes. Such elements are of higher *order* than linear elements (which can only represent straight boundaries) and include quadratic and cubic elements. The most popular elements belong to the so-called *isoparametric* family of elements, where the same parameters are used to define the geometry as define the displacement variation over the element. Therefore, those isoparametric elements of quadratic order, and above, are capable of representing curved sides and surfaces.

In situations where the type of elements to be used may not be apparent, the choice could be based on such considerations as

(a) number of dof.,
(b) accuracy required,
(c) computational effort,
(d) the degree to which the physical structure needs to be modelled.

Use of the elements with a quadratic displacement assumption are generally recommended as the best compromise between the relatively low cost but inferior performance of linear elements and the high cost but superior performance of cubic elements.

9.4.4. Discretisation of the solution region

This step is equivalent to replacing the actual structure or continuum having an infinite number of dof. by a system having a finite number of dof. This process, known as *discretisation*, calls for engineering judgement in order to model the region as closely as necessary. Having selected the element type, discretisation requires careful attention to *extent of the model* (i.e. location of model boundaries), *element size and grading, number of elements*, and factors influencing the *quality of the mesh*, to achieve adequately accurate results consistent with avoiding excessive computational effort and expense. These aspects are briefly considered below.

Extent of model

Reference has already been made above to applications which are axisymmetric, or those which can be idealised as such. Generally, advantage should be taken of geometric and loading symmetry wherever it exists, whether it be plane or axial. Appropriate boundary conditions need to be imposed to ensure the reduced portion is representative of the whole. For example, in the analysis of a semi-infinite tension plate with a central circular hole, shown in Fig. 9.3, only a quadrant need be modelled. However, in order that the quadrant is representative of the whole, respective v and u displacements must be prevented along the x and y direction symmetry axes, since there will be no such displacements in the full model/component.

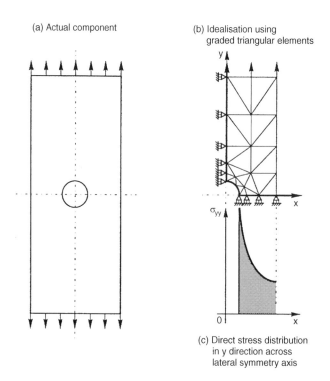

Fig. 9.3. Finite element analysis of a semi-infinite tension plate with a central circular hole, using triangular elements.

Further, it is known that disturbances to stress distributions due to rapid changes in geometry or load concentrations are only local in effect. Saint-Venant's principle states that the effect of stress concentrations essentially disappear within relatively small distances (approximately equal to the larger lateral dimension), from the position of the disturbance. Advantage can therefore be taken of this principle by reducing the necessary extent of a finite element model. A rule-of-thumb is that a model need only extend to one-and-a-half times the larger lateral dimension from a disturbance, see Fig. 9.4.

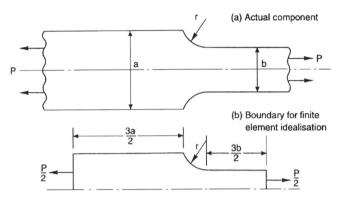

Fig. 9.4. Idealisation of a shouldered tension strip.

Element size and grading

The relative size of elements directly affects the quality of the solution. As the element size is reduced so the accuracy of solution can be expected to increase since there is better representation of the field variable, e.g. displacement, and/or better representation of the geometry. However, as the element size is reduced, so the number of elements increases with the accompanying penalty of increased computational effort. Needlessly small elements in regions with little variation in field variable or geometry will be wasteful. Equally, in regions where the stress variation is not of primary interest then a locally coarse mesh can be employed providing it is sufficiently far away from the region of interest and that it still provides an accurate stiffness representation. Therefore, element sizes should be graded in order to take account of anticipated stress/strain variations and geometry, and the results required. The example of stress analysis of a semi-infinite tension plate with a central circular hole, Fig. 9.3, serves to illustrate how the size of the elements can be graded from small-size elements surrounding the hole (where both the stress/strain and geometry are varying the most), to become coarser with increasing distance from the hole.

Number of elements

The number of elements is related to the previous matter of element size and, for a given element type, the number of elements will determine the total number of dof. of the model, and combined with the relative size determines the *mesh density*. An increase in the number of elements can result in an improvement in the accuracy of the solution, but a limit will be reached beyond which any further increase in the number of elements will not significantly improve the accuracy. This matter of *convergence* of solution is clearly important, and with experience a near optimal mesh may be achievable. As an alternative to increasing the number of elements, improvements in the model can be obtained by increasing the element order. This alternative form of *enrichment* can be performed manually (by substituting elements), or can be performed automatically, e.g. the commercial package RASNA has this capability. Clearly, any increase in the number of elements (or element order), and hence dof., will require greater computational effort, will put greater demands on available computer memory and increase cost.

Quality of the mesh

The quality of the fe. predictions (e.g. of displacements, temperatures, strains or stresses), will clearly be affected by the performance of the model and its constituent elements. The factors which determine quality[13] will now be explored briefly, namely

(a) coincident elements,
(b) free edges,
(c) poorly positioned "midside" nodes,
(d) interior angles which are too extreme,
(e) warping, and
(f) distortion.

(a) Coincident elements

Coincident elements refer to two or more elements which are overlaid and share some of the nodes, see Fig. 9.5. Such coincident elements should be deleted as part of cleaning-up of a mesh.

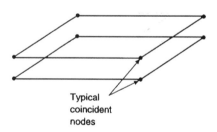

Fig. 9.5. Coincident elements.

(b) Free edges

A free edge should only exist as a model boundary. Neighbouring elements should share nodes along common inter-element boundaries. If they do not, then a free edge exists and will need correction, see Fig. 9.6.

(c) Poorly positioned "midside" nodes

Displacing an element's "midside" node from its mid-position will cause distortion in the mapping process associated with high order elements, and in extreme cases can significantly degrade an element's performance. There are two aspects to "midside" node displacement, namely, the relative position between the corner nodes, and the node's offset from a straight

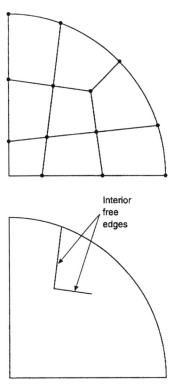

Fig. 9.6. Free edges.

line joining the corner nodes, see Fig. 9.7. The midside node's relative position should ideally be 50% of the side length for a parabolic element and 33.3% for a cubic element. An example of the effect of displacement of the "midside" node to the 25% position, is reported for a parabolic element[14] to result in a 15% error in the major stress prediction.

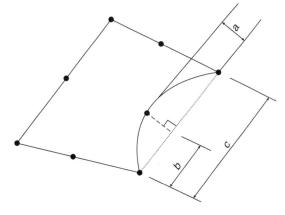

Percent displacement = 100 b/c
Offset = a/c

Fig. 9.7. "Midside" node displacement.

(d) Interior angles which are too extreme

Interior angles which are excessively small or large will, like displaced "midside" nodes, cause distortion in the mapping process. A re-entrant corner (i.e. an interior angle greater than 180°), see Fig. 9.8, will cause failure in the mapping as the *Jacobian matrix* (relating the derivatives with respect to curvilinear (r,s), coordinates, to those with respect to cartesian (x,y), coordinates), will not have an inverse (i.e. its determinate will be zero). For quadrilateral elements the ideal interior angle is 90°, and for triangular elements it is 60°.

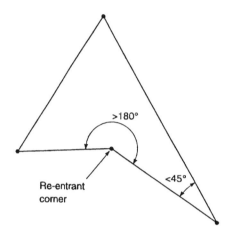

Fig. 9.8. Extreme interior angles.

(e) Warping

Warping refers to the deviation of the face of a "planar" element from being planar, see Fig. 9.9. The analogy of the three-legged milking stool (which is steady no-matter how uneven the surface is on which it is placed), to the triangular element serves to illustrate an advantage of this element over its quadrilateral counterpart.

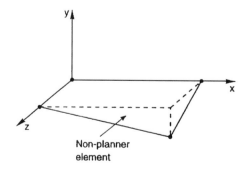

Fig. 9.9. Warping.

(f) Distortion

Distortion is the deviation of an element from its ideal shape, which corresponds to that in curvilinear coordinates. SDRC I-DEAS[13] gives two measures, namely

(1) the departure from the basic element shape which is known as *distortion*, see Fig. 9.10. Ideally, for a quadrilateral element, with regard to distortion, the shape should be a rectangle, and

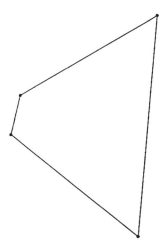

Fig. 9.10. Distortion.

(2) the amount of elongation suffered by an element which is known as *stretch*, or *aspect ratio distortion*, see Fig. 9.11. Ideally, for a quadrilateral element, with regard to stretch, the shape should be square.

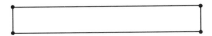

Fig. 9.11. Stretch.

Whilst small amounts of deviation of an element's shape from that of the parent element can, and must, be tolerated, unnecessary and excessive distortions and stretch, etc. must be avoided if degraded results are to be minimised. High order elements in gradually varying strain fields are most tolerant of shape deviation, whilst low order elements in severe strain fields are least tolerant.[5] There are automatic means by which element shape deviation can be measured, using information derived from the Jacobian matrix. Errors in a solution and the rate of convergence can be judged by computing so-called energy norms derived from successive solutions.[7] However, it is left to the judgement of the user to establish the degree of shape deviation which can be tolerated. Most packages offer quality checking facilities, which allows the user to interrogate the shape deviation of all, or a selection of, elements. I-DEAS provides a measure of element quality using a value with a range of −1 to +1, (where

+1 is the target value corresponding to zero distortion, and stretch, etc.). Negative values, which arise for example, from re-entrant corners, referred to above, will cause an attempted solution to fail, and hence need to be rectified. A distortion value above 0.7 can be considered acceptable, but errors will be incurred with any value below 1.0. However, circumstances may dictate acceptance of elements with a distortion value below 0.7. Similarly, as a rule-of-thumb, a stretch value above 0.5 can be considered acceptable, but again, errors will be incurred with any value below 1.0. Companies responsible for analyses may issue guidelines for quality, an example of which is shown in Table 9.1.

Table 9.1. Example of element quality guidelines.

Element	Interior angle°	Warpage	Distortion	Stretch
Triangle	30–90	N/A	0.35	0.3
Quadrilateral	45–135	0.2	0.60	0.3
Wedge	30–90	N/A	0.35	0.3
Tetrahedron	30–90	N/A	0.10	0.1
Hexahedron	45–135	0.2	0.5	0.3

9.4.5. Creation of the material model

The least material data required for a stress analysis is the empirical elastic modulus for the component under analysis describing the relevant stress/strain law. For a dynamic analysis, the material density must also be specified. Dependent upon the type of analysis, other properties may be required, including Poisson's ratio for two- and three-dimensional models and the coefficient of thermal expansion for thermal analyses. For analyses involving non-linear material behaviour then, as a minimum, the yield stress and yield criterion, e.g. von Mises, need to be defined. If the material within an element can be assumed to be isotropic and homogeneous, then there will be only one value of each material property. For non-isotropic material, i.e. orthotropic or anisotropic, then the material properties are direction and spacially dependent, respectively. In the extreme case of anisotropy, 21 independent values are required to define the material matrix.[5]

9.4.6. Node and element ordering

Before moving on to consider boundary conditions, it is appropriate to examine node and element ordering and its effect on efficiency of solution by briefly exploring the methods used. The formation of the element characteristic matrices (to be considered in §9.7, 9.8 and 9.9), and the subsequent solution are the two most computationally intensive steps in any fe. analysis. The computational effort and memory requirements of the solution are affected by the method employed, and are considered below.

It will be seen in Section 9.7, and subsequently, that the displacement based method involves the assembly of the *structural*, or *assembled, stiffness matrix* $[K]$, and the load and displacement column matrices, $\{P\}$ and $\{p\}$, respectively, to form the governing equation for stress analysis $\{P\} = [K]\{p\}$. With reference to §9.7, and subsequently, two features of the fem. will be seen to be that the assembled stiffness matrix $[K]$, is sparsely populated and is symmetric. Advantage can be taken of this in reducing the storage requirements of the computer. Two solution methods are used, namely, *banded* or *frontal*, the choice of which is dependent upon the number of dof. in the model.

Banded method of solution

The banded method is appropriate for small to medium size jobs (i.e. up to 10 000 dof.). By carefully ordering the dof. the assembled stiffness matrix $[K]$, can be banded with non-zero terms occurring only on the leading diagonal. Symmetry permits only half of the band to be stored, but storage requirements can still be high. It is advantageous therefore to minimise the *bandwidth*. A comparison of different node numbering schemes is provided by Figs. 9.12 and 9.13 in which a simple model comprising eight triangular linear elements is considered, and for further simplicity the nodal contributions are denoted as shaded squares, the empty squares denoting zeros.

The semi-bandwidth can be seen to depend on the node numbering scheme and the number of dof. per node and has a direct effect on the storage requirements and computational effort.

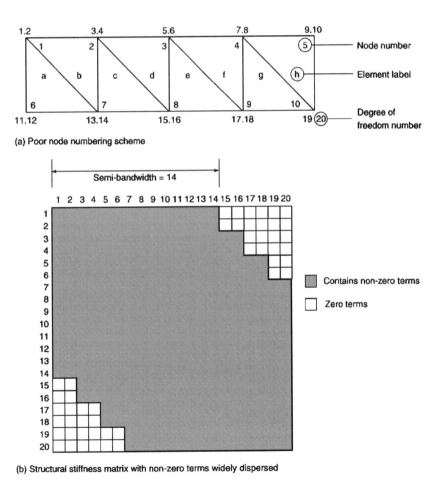

Fig. 9.12. Structural stiffness matrix corresponding to poor node ordering.

For a given number of dof. per node, which is generally fixed for each assemblage, the bandwidth can be minimised by using a proper node numbering scheme.

With reference to Figs. 9.12 and 9.13 there are a total of 20 dof. in the model (i.e. 10 nodes each with an assumed 2 dof.), and if the symmetry and bandedness is not taken advantage of, storage of the entire matrix would require $20^2 = 400$ locations. For the efficiently numbered model with a semi-bandwidth of 8, see Fig. 9.13, taking advantage of the symmetry and bandedness, the storage required for the upper, or lower, half-band is only $8 \times 20 = 160$ locations.

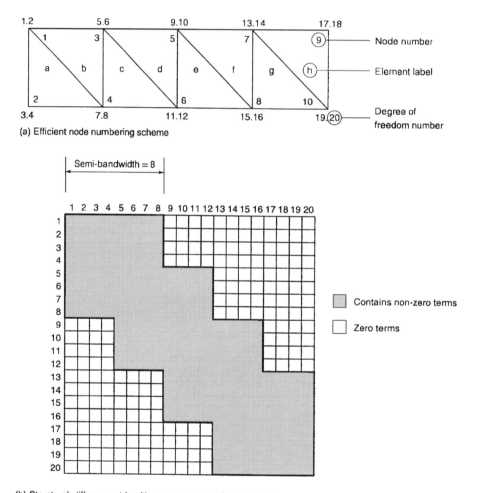

(a) Efficient node numbering scheme

(b) Structural stiffness matrix with non-zero terms closely banded

Fig. 9.13. Structural stiffness matrix corresponding to efficient node ordering.

From observation of Figs. 9.12 and 9.13 it can be deduced that the semi-bandwidth can be calculated from

$$\text{semi-bandwidth} = f(d+1)$$

where f is the number of dof. per node and d is the maximum largest difference in the node numbers for all elements in the assemblage. This expression is applicable to any type of finite element. It follows that to minimise the bandwidth, d must be minimised and this is achieved by simply numbering the nodes across the shortest dimension of the region.

For large jobs the capacity of computer memory can be exceeded using the above banded method, in which case a frontal solution is used.

Frontal method of solution

The frontal method is appropriate for medium to large size jobs (i.e. greater than 10 000 dof.). To illustrate the method, consider the simple two-dimensional mesh shown in Fig. 9.14. Nodal contributions are assembled in element order. With reference to Fig. 9.14, with the assembly of element number 1 terms (i.e. contributions from nodes 1, 2, 6 and 7), all information relating to node number 1 will be complete since this node is not common to any other element. Thus the dofs. for node 1 can be eliminated from the set of system equations. Element number 2 contributions are assembled next, and the system matrix will now contain contributions from nodes 2, 3, 6, 7 and 8. At this stage the dofs. for node number 2 can be eliminated. Further element contributions are merged and at each stage any nodes which do not appear in later elements are *reduced out*. The solution thus proceeds as a front through the system. As, for example, element number 14 is assembled, dofs. for the nodes indicated by line B are required, see Fig. 9.14. After eliminations which follow assembly of element number 14, dofs. associated with line C are needed. The solution front has thus moved from line A to C.

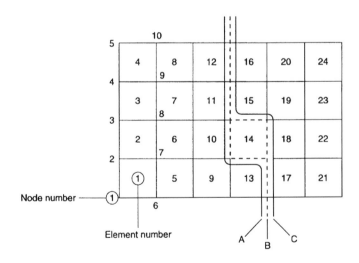

Fig. 9.14. Frontal method of solution.

To minimise memory requirements, which is especially important for jobs with large numbers of dof., the instantaneous width, i.e. *front size*, of the stiffness matrix during merging should be kept as small as possible. This is achieved by ensuring that elements are selected for merging in a specific order. Figures 9.15(a) and (b) serve to illustrate badly and well ordered elements, respectively, for a simple two-dimensional application. Front ordering facilities are available with some fe. packages which will automatically re-order the elements to minimise the front size.

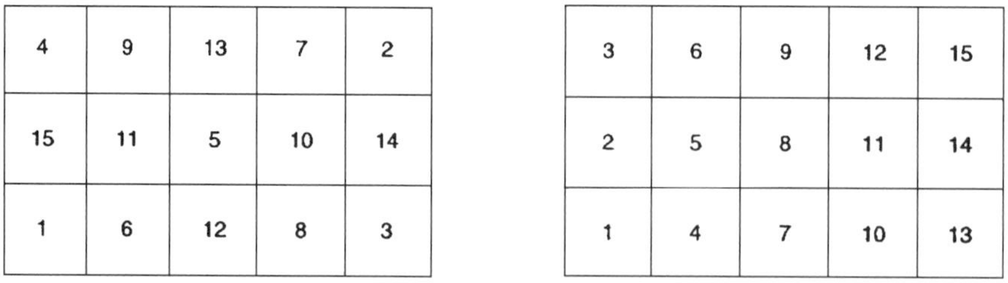

4	9	13	7	2
15	11	5	10	14
1	6	12	8	3

3	6	9	12	15
2	5	8	11	14
1	4	7	10	13

(a) Badly ordered elements   (b) Well ordered elements

Fig. 9.15. Examples of element ordering for frontal method.

9.4.7. Application of boundary conditions

Having created a mesh of finite elements and before the job is submitted for solution, it is necessary to enforce conditions on the boundaries of the model. Dependent upon the application, these can take the form of

- restraints,
- constraints,
- structural loads,
- heat transfer loads, or
- specification of active and inactive dof.

Attention will be restricted to a brief consideration of restraints and structural loads, which are sufficient conditions for a simple stress analysis. The reader wishing for further coverage is again urged to consult the many specialist texts.[1-10]

Restraints

Restraints, which can be applied to individual, or groups of nodes, involve defining the displacements to be applied to the possible six dof., or perhaps defining a temperature. As an example, reference to Fig. 9.3(b) shows the necessary restraints to impose symmetry conditions. It can be assumed that the elements chosen have only 2 dof. per node, namely u and v translations, in the x and y directions, respectively. The appropriate conditions are

along the x-axis, $v = 0$, and
along the y-axis, $u = 0$.

The usual symbol, representing a frictionless roller support, which is appropriate in this case, is shown in Fig. 9.16(a), and corresponds to zero normal displacement, i.e. $\delta_n = 0$, and zero tangential shear stress, i.e. $\tau_t = 0$, see Fig. 9.16(b).

In a static stress analysis, unless sufficient restraints are applied, the system equations (see §9.4.5), cannot be solved, since an inverse will not exist. The physical interpretation of this is that the loaded body is free to undergo unlimited *rigid body motion*. Restraints must be

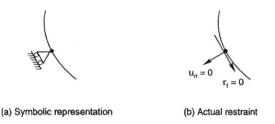

(a) Symbolic representation   (b) Actual restraint

Fig. 9.16. Boundary node with zero shear traction and zero normal displacement.

chosen to be sufficient, but not to create rigidity which does not exist in the actual component being modelled. This important matter of appropriate restraints can call for considerable engineering judgement, and the choice can significantly affect the behaviour of the model and hence the validity of the results.

Structural loads

Structural loads, which are applied to nodes can, through usual program facilities, be specified for application to groups of nodes, or to an entire model, and can take the form of loads, temperatures, pressures or accelerations. At the program level, only nodal loads are admissible, and hence when any form of distributed load needs to be applied, the nodal equivalent loads need to be computed, either manually or automatically. One approach is to simply define a set of statically equivalent loads, with the same resultant forces and moments as the actual loads. However, the most accurate method is to use kinematically equivalent loads[5] as simple statically equivalent loads do not give a satisfactory solution for other than the simplest element interpolation. Figure 9.17 illustrates the case of an element with a quadratic displacement interpolation. Here the distributed load of total value W, is replaced by three nodal loads which produce the same work done as that done by the actual distributed load.

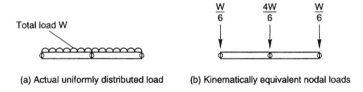

(a) Actual uniformly distributed load   (b) Kinematically equivalent nodal loads

Fig. 9.17. Structural load representation.

9.4.8. Creation of a data file

The data file, or *deck*, will need to be in precisely the format required by the particular program being used; although essentially all programs will require the same basic model data, i.e. nodal coordinates, element type(s) and connectivity, material properties and boundary conditions. The type of solution will need to be specified, e.g. linear elastic, normal modes, etc. The required output will also need to be specified, e.g. deformations, stresses, strains, strain energy, reactions, etc. Much of the tedium of producing a data file is removed if automatic data preparation is available. Such an aid is beneficial with regard to minimising the possibility of introducing data errors. The importance of avoiding errors cannot be over-emphasised, as the validity of the output is clearly dependent upon the correctness of the data. Any capability of a program to detect errors is to be welcomed. However, it should be realised that it is impossible for a program to detect all forms of error. e.g. incorrect but possible coordinates, incorrect physical or material properties, incompatible units, etc., can all go undetected. The user must, therefore, take every possible precaution to guard against errors. Displays of the mesh, including "shrunken" or "exploded" element views to reveal absent elements, restraints and loads should be scrutinised to ensure correctness before the solution stage is entered; the material and physical properties should also be examined.

9.4.9. Computer, processing, steps

The steps performed by the computer can best be followed by means of applications using particular elements, and this will be covered in §9.7 and subsequent sections.

9.4.10. Interpretation and validation of results

The numerical output following solution is often provided to a substantial number of decimal places which gives an aura of precision to the results. The user needs to be mindful that the fem. is numerical and hence is approximate. There are many potential sources of error, and a responsibility of the analyst is to ensure that errors are not significant. In addition to approximations in the model, significant errors can arise from round-off and truncation in the computation.

There are a number of checks that should be routine procedure following solution, and these are given below

- Ensure that any warning messages, given by the program, are pursued to ensure that the results are not affected. Error messages will usually accompany a failure in solution and clearly, will need corrective action.

- An obvious check is to examine the deformed geometry to ensure the model has behaved as expected, e.g. Poisson effect has occurred, slope continuity exists along axes of symmetry, etc.

- Ensure that equilibrium has been satisfied by checking that the applied loads and moments balance the reactions. Excessive out of balance indicates a poor mesh.

- Examine the smoothness of stress contours. Irregular boundaries indicate a poor mesh.

- Check inter-element stress discontinuities (stress jumps), as these give a measure of the quality of model. Large discontinuities indicate that the elements need to be enhanced.

- On traction-free boundaries the principal stress normal to the boundary should be zero. Any departure from this gives an indication of the quality to be expected in the other principal stress predictions for this point.

- Check that the directions of the principal stresses agree with those expected, e.g. normal and tangential to traction-free boundaries and axes of symmetry.

Results should always be assessed in the light of common-sense and engineering judgement. Manual calculations, using appropriate simplifications where necessary, should be carried out for comparison, as a matter of course.

9.4.11. Modification and re-run

Clearly, the need for design modification and subsequent fe. re-runs depends upon the particular circumstances. The computational burden may prohibit many re-runs. Indeed, for large jobs, (which may involve many thousands of dof. or many increments in the case of non-linear analyses), re-runs may not be feasible. The approach in such cases may be to run several exploratory crude models to gain some initial understanding how the component behaves, and hence aid final modelling.

9.5. Fundamental arguments

Regardless of the type of structure to be analysed, irrespective of whether the loading is static or dynamic, and whatever the material behaviour may be, there are only three types of argument to be invoked, namely, *equilibrium, compatibility and stress/strain law*. Whilst these arguments will be found throughout this text it is worthwhile giving them some explicit attention here as a sound understanding will help in following the theory of the fem. in the proceeding sections of this chapter.

9.5.1. Equilibrium

External nodal equilibrium

Static equilibrium requires that, with respect to some orthogonal coordinate system, the reactive forces and moments must balance the externally applied forces and moments. In fea. this argument extends to all nodes in the model. With reference to Fig. 9.18, some nodes may be subjected to applied forces and moments, (node number 4), and others may be support points (node numbers 1 and 6). There may be other nodes which appear to be neither of these (node numbers 2, 3, and 5), but are in fact nodes for which the applied force, or moment, is zero, whilst others provide support in one or two orthogonal directions and are loaded (or have zero load), in the remaining direction(s) (node number 6). Hence, for each node and with respect to appropriate orthogonal directions, satisfaction of external equilibrium requires

external loads or reactions = summation of internal, element, loads

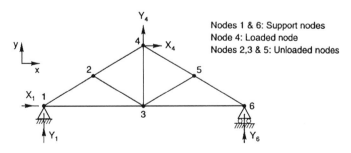

Fig. 9.18. Structural framework.

Then, for the jth node
$$\{P_j\} = \sum_{e=1}^{m} \{S^{(e)}\}$$

where the summation is of the internal loads at node j from all m elements joined at node j.

Use of this relation can be illustrated by considering the simple frame, idealised as planar with pin-joints and discretised as an assemblage of three elements, as shown in Fig. 9.19. The nodal force column matrix for the structure is

$$\{P\} = \{P_1 P_2 P_3\} = \{X_1 Y_1 X_2 Y_2 X_3 Y_3\}$$

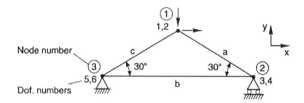

Fig. 9.19. Simple pin-jointed plane frame.

and the element force column matrix for the structure is

$$\{S\} = \{\{S^{(a)}\}\{S^{(b)}\}\{S^{(c)}\}\} = \{U_1 V_1 U_2 V_2, U_2 V_2 U_3 V_3, U_1 V_1 U_3 V_3\}$$

It follows from the above that external, nodal, equilibrium for the structure is satisfied by forming the relationship between the nodal and element forces as

$$
\begin{bmatrix} X_1 \\ Y_1 \\ X_2 \\ Y_2 \\ X_3 \\ Y_3 \end{bmatrix}
=
\begin{bmatrix}
1 & 0 & 0 & 0 & 0 & 0 & 0 & 0 & 1 & 0 & 0 & 0 \\
0 & 1 & 0 & 0 & 0 & 0 & 0 & 0 & 0 & 1 & 0 & 0 \\
0 & 0 & 1 & 0 & 1 & 0 & 0 & 0 & 0 & 0 & 0 & 0 \\
0 & 0 & 0 & 1 & 0 & 1 & 0 & 0 & 0 & 0 & 0 & 0 \\
0 & 0 & 0 & 0 & 0 & 0 & 1 & 0 & 0 & 0 & 1 & 0 \\
0 & 0 & 0 & 0 & 0 & 0 & 0 & 1 & 0 & 0 & 0 & 1
\end{bmatrix}
\begin{bmatrix} U_1 \\ V_1 \\ U_2 \\ V_2 \\ \overline{U_2} \\ V_2 \\ U_3 \\ V_3 \\ \overline{U_1} \\ V_1 \\ U_3 \\ V_3 \end{bmatrix}
$$

Or, more concisely, $$\{P\} = [a]^T \{S\}$$ (9.1)

which relates the nodal forces $\{P\}$ to the element forces $\{S\}$ for the whole structure.

Internal element equilibrium

Internal equilibrium can be explained most easily by considering an axial force element. For static equilibrium, the axial forces at each end will be equal in magnitude and opposite in direction. If the element is pin-ended and has a uniform cross-sectional area, A, then for equilibrium within the element

$$A\sigma_x = U, \tag{9.2}$$

in which the axial stress σ_x is taken to be constant over the cross-section.

9.5.2. Compatibility

External nodal compatibility

The physical interpretation of external compatibility is that any displacement pattern is not accompanied with voids or overlaps occurring between previously continuous members. In fea. this requirement is usually only satisfied at the nodes. Often it is only the displacement field which is continuous at the nodes, and not an element's first or higher order displacement derivatives. Figure 9.20 shows quadratically varying displacement fields for two adjoining quadrilateral elements and serves to illustrate these limitations.

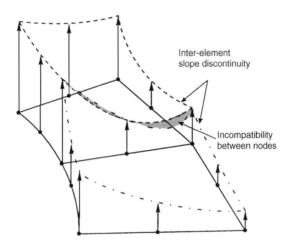

Fig. 9.20. Quadrilateral elements with quadratically varying displacement fields.

External, nodal, displacement compatibility will be shown to be automatically satisfied by a system of nodal displacements. For the simple frame shown in Fig. 9.19, the nodal displacement column matrix is

$$\{p\} = \{p_1 p_2 p_3\} = \{u_1 v_1, u_2 v_2, u_3 v_3\}$$

and the element displacement column matrix is

$$\{s\} = \{\{s^{(a)}\}\{s^{(b)}\}\{s^{(c)}\}\} = \{u_1 v_1 u_2 v_2, u_2 v_2 u_3 v_3, u_1 v_1 u_3 v_3\}$$

It follows from the above that external, nodal, compatibility is satisfied by forming the relationship between the element and nodal displacements as

$$
\begin{bmatrix} u_1 \\ v_1 \\ u_2 \\ v_2 \\ \overline{u_2} \\ v_2 \\ u_3 \\ v_3 \\ \overline{u_1} \\ v_1 \\ u_3 \\ v_3 \end{bmatrix}
=
\begin{bmatrix}
1 & 0 & 0 & 0 & 0 & 0 \\
0 & 1 & 0 & 0 & 0 & 0 \\
0 & 0 & 1 & 0 & 0 & 0 \\
0 & 0 & 0 & 1 & 0 & 0 \\
0 & 0 & 1 & 0 & 0 & 0 \\
0 & 0 & 0 & 1 & 0 & 0 \\
0 & 0 & 0 & 0 & 1 & 0 \\
0 & 0 & 0 & 0 & 0 & 1 \\
1 & 0 & 0 & 0 & 0 & 0 \\
0 & 1 & 0 & 0 & 0 & 0 \\
0 & 0 & 0 & 0 & 1 & 0 \\
0 & 0 & 0 & 0 & 0 & 1
\end{bmatrix}
\begin{bmatrix} u_1 \\ v_1 \\ u_2 \\ v_2 \\ u_3 \\ v_3 \end{bmatrix}
$$

Or, more concisely, $\{s\} = [a]\{p\}$ (9.3)

which relates all the element displacements $\{s\}$ to the nodal displacements $\{p\}$ for the whole structure.

Internal element compatibility

Again, for simplicity consider an axial force element. For the displacement within such an element not to introduce any voids or overlaps the displacement along the element, u, needs to be a continuous function of position, x. The compatibility condition is satisfied by

$$
\partial u / \partial x = \varepsilon_x \tag{9.4}
$$

9.5.3. Stress/strain law

Assuming for simplicity the material behaviour to be homogeneous, isotropic and linearly elastic, then Hooke's law applies giving, for a one-dimensional stress system in the absence of thermal strain,

$$
\varepsilon_x = \sigma_x / E \tag{9.5}
$$

in which E is the empirical modulus of elasticity.

9.5.4. Force/displacement relation

Combining eqns. (9.2), (9.4) and (9.5) and taking u to be a function of x only, gives

$$
U/A = \sigma_x = \sigma_x E = E \, du/dx
$$

Or, $U \, dx = AE \, du$

Integrating, and taking $u(0) = u_i$ and $u(L) = u_j$, corresponding to displacements at nodes i and j of an axial force element of length L, gives the force/displacement relationship

$$
U = AE(u_j - u_i)/L \tag{9.6}
$$

in which $(u_j - u_i)$ denotes the deformation of the element. Thus the force/displacement relationship for an axial force element has been derived from equilibrium, compatibility and stress/strain arguments.

9.6. The principle of virtual work

In the previous section the three basic arguments of equilibrium, compatibility and constitutive relations were invoked and, in the subsequent sections, it will be seen how these arguments can be used to derive rod and simple beam element equations. However, some situations, for example, may require elements of non-uniform cross-section or representation of complex geometry, and are not amenable to solution by this approach. In such situations, alternative approaches using energy principles are used, which allow the field variables to be represented by approximating functions whilst still satisfying the three fundamental arguments. Amongst the number of energy principles which can be used, the one known as the *principle of virtual work* will be considered here.

The equation of the principle of virtual work

Virtual work is produced by perturbing a system slightly from an equilibrium state. This can be achieved by allowing small, kinematically possible displacements, which are not necessarily real, and hence are *virtual displacements*. In the following brief treatment the corresponding equation of virtual work is derived by considering the linearly elastic, uniform cross-section, axial force element in Fig. 9.21. For a more rigorous treatment the reader is referred to Ref. 8 (p. 350). In Fig. 9.21 the nodes are shown detached to distinguish between the nodal and element quantities.

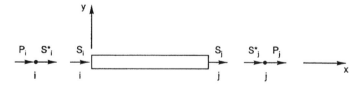

Fig. 9.21. Axial force element, shown with detached nodes.

Giving the nodal points virtual displacements $\bar{u}_i$ and $\bar{u}_j$, the virtual work for the two nodal points is

$$\overline{W} = (P_i + S_i^*)\bar{u}_i + (P_j + S_j^*)\bar{u}_j \tag{9.7}$$

This virtual work must be zero since the two nodal points are rigid bodies. It follows, since the virtual displacements are arbitrary and independent, that

$$P_i + S_i^* = 0 \quad \text{or} \quad P_i = -S_i^*$$

and

$$P_j + S_j^* = 0 \quad \text{or} \quad P_j = -S_j^*$$

which, for the single element, are the equations of external equilibrium.

Applying Newton's third law, the forces between the nodes and element are related as

$$S_i^* = -S_i \quad \text{and} \quad S_j^* = -S_j \tag{9.8}$$

Substituting eqns. (9.8) into eqn. (9.7) gives

$$\overline{W} = 0 = (P_i - S_i)\bar{u}_i + (P_j - S_j)\bar{u}_j$$
$$= (P_i\bar{u}_i + P_j\bar{u}_j) - (S_i\bar{u}_i + S_j\bar{u}_j) \tag{9.9}$$

The first quantity $(P_i\bar{u}_i + P_j\bar{u}_j)$, to first order approximation assuming linearly elastic behaviour represents the virtual work done by the applied external forces, denoted as $\overline{W}_e$. The second quantity, $(S_i\bar{u}_i + S_j\bar{u}_j)$, again to first order approximation represents the virtual work done by element internal forces, denoted as $\overline{W}_i$. Hence, eqn. (9.9) can be abbreviated to

$$0 = \overline{W}_e - \overline{W}_i \tag{9.10}$$

which is the equation of the principle of virtual work for a deformable body.

The external virtual work will be found from the product of external loads and corresponding virtual displacements, recognising that no work is done by reactions since they are associated with suppressed dof. The internal virtual work will be given by the strain energy, expressed using real stress and virtual strain (arising from virtual displacements), as

$$\overline{W}_i = \int_v \bar{\varepsilon}\sigma \, dv \tag{9.11}$$

which, for the case of a prismatic element with constant stress and strain over the volume, becomes

$$\overline{W}_i = \bar{\varepsilon}\sigma \, AL$$

9.7. A rod element

The formulation of a rod element will be considered using two approaches, namely the use of fundamental equations, based on equilibrium, compatibility and constitutive (i.e. stress/strain law), arguments and use of the principle of virtual work equation.

9.7.1. Formulation of a rod element using fundamental equations

Consider the structure shown in Fig. 9.18, for which the deformations (derived from the displacements), member forces, stresses and reactions are required. Idealising the structure such that all the members and loads are taken to be planar, and all the joints to act as frictionless hinges, i.e. pinned, and hence incapable of transmitting moments, the corresponding behaviour can be represented as an assemblage of rod finite elements. A typical element is shown in Fig. 9.22, for which the physical and material properties are taken to be constant throughout the element. Changes in properties, and load application are only admissible at nodal positions, which occur only at the extremities of the elements. Each node is considered to have two translatory freedoms, i.e. two dof., namely u and v displacements in the element x and y directions, respectively.

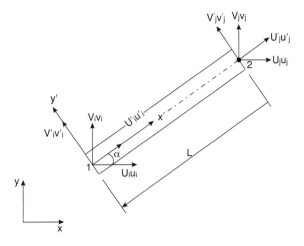

Fig. 9.22. An axial force rod element.

Element stiffness matrix in local coordinates

Quantities in the element coordinate directions are denoted with a prime ($'$), to distinguish from those with respect to the global coordinates. Displacements in the local element x' direction will cause an elongation of the element of $u'_j - u'_i$, with corresponding strain, $(u'_j - u'_i)/L$. Assuming Hookean behaviour, the element loads in the positive, local, x' direction will hence be given as

$$U'_i = AE(u'_i - u'_j)/L \quad \text{and} \quad U'_j = AE(u'_j - u'_i)/L$$

which are force/displacement relations similar to eqn. (9.6), and hence satisfy internal element equilibrium, compatibility and the appropriate stress/strain law.

In isolation, the element will not have any stiffness in the local y' direction. However, stiffness in this direction will arise from assembly with other elements with different inclinations. The element force/displacement relation can now be written in matrix form, as

$$
\begin{bmatrix} U'_i \\ V'_i \\ U'_j \\ V'_j \end{bmatrix} = \frac{AE}{L} \begin{bmatrix} 1 & 0 & -1 & 0 \\ 0 & 0 & 0 & 0 \\ -1 & 0 & 1 & 0 \\ 0 & 0 & 0 & 0 \end{bmatrix} \begin{bmatrix} u'_i \\ v'_i \\ u'_j \\ v'_j \end{bmatrix}
\tag{9.12}
$$

Or, more concisely, $\{S'^{(e)}\} = [k'^{(e)}]\{s'^{(e)}\}$ (9.13)

from which the *element stiffness* matrix with respect to local coordinates is:

$$
[k'^{(e)}] = \frac{AE}{L} \begin{bmatrix} 1 & 0 & -1 & 0 \\ 0 & 0 & 0 & 0 \\ -1 & 0 & 1 & 0 \\ 0 & 0 & 0 & 0 \end{bmatrix}
\tag{9.14}
$$

Element stress matrix in local coordinates

For a pin-jointed frame the only significant stress will be axial. Hence, with respect to local coordinates, the axial stress for a rod element will be given as:

$$\sigma^{(e)} = \frac{E}{L}[-1 \quad 0 \quad 1 \quad 0] \begin{bmatrix} u'_i \\ v'_i \\ u'_j \\ v'_j \end{bmatrix} \tag{9.15}$$

Or, more concisely, $\qquad \sigma^{(e)} = [H'^{(e)}]\{s'^{(e)}\}$ $\qquad\qquad\qquad$ (9.16)

from which the stress matrix with respect to local coordinates is:

$$[H'^{(e)}] = \frac{E}{L}[-1 \quad 0 \quad 1 \quad 0] \tag{9.17}$$

Transformation of displacements and forces

To enable assembly of contributions from each constituent element meeting at each joint, it is necessary to transform the force/displacement relationships to some global coordinate system, by means of a transformation matrix $[T^{(e)}]$. This matrix is derived by establishing the relationship between the displacements (or forces), in local coordinates x', y', and those in global coordinates x,y. Note that the element inclination α, is taken to be positive when acting clockwise viewed from the origin along the positive z-axis, and is measured from the positive global x-axis. With reference to Fig. 9.23, for node i,

$$u'_i = u_i \cos\alpha + v_i \sin\alpha, \quad \text{and} \quad v'_i = -u_i \sin\alpha + v_i \cos\alpha.$$

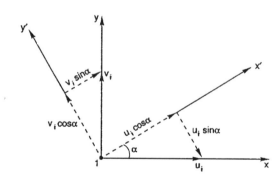

Fig. 9.23. Transformation of displacements.

Similarly, for node j,

$$u'_j = u_j \cos\alpha + v_j \sin\alpha, \quad \text{and} \quad v'_j = -u_j \sin\alpha + v_j \cos\alpha.$$

Writing in matrix form the above becomes:

$$\begin{bmatrix} u'_i \\ v'_i \\ u'_j \\ v'_j \end{bmatrix} = \begin{bmatrix} \cos\alpha & \sin\alpha & 0 & 0 \\ -\sin\alpha & \cos\alpha & 0 & 0 \\ 0 & 0 & \cos\alpha & \sin\alpha \\ 0 & 0 & -\sin\alpha & \cos\alpha \end{bmatrix} \begin{bmatrix} u_i \\ v_i \\ u_j \\ v_j \end{bmatrix}$$

Or, more concisely, $\qquad \{S'^{(e)}\} = [T^{(e)}]\{s^{(e)}\}$ $\qquad\qquad\qquad$ (9.18) in which the transformation matrix is:

$$[T^{(e)}] = \begin{bmatrix} \cos\alpha & \sin\alpha & 0 & 0 \\ -\sin\alpha & \cos\alpha & 0 & 0 \\ 0 & 0 & \cos\alpha & \sin\alpha \\ 0 & 0 & -\sin\alpha & \cos\alpha \end{bmatrix} \tag{9.19}$$

The same transformation enables the relationship between member loads in local and global coordinates to be written as:

$$\{S'^{(e)}\} = [T^{(e)}]\{S^{(e)}\} \tag{9.20}$$

Expressing eqn. (9.20) in terms of the member loads with respect to the required global coordinates, we obtain:

$$\{S^{(e)}\} = [T^{(e)}]^{-1}\{S'^{(e)}\}$$

Substituting from eqn. (9.13) gives:

$$\{S^{(e)}\} = [T^{(e)}]^{-1}[k'^{(e)}]\{s'^{(e)}\}$$

Further, substituting from eqn. (9.18) gives:

$$\{S^{(e)}\} = [T^{(e)}]^{-1}[k'^{(e)}][T^{(e)}]\{s^{(e)}\}$$

It can be shown, by equating work done in the local and global coordinates systems, that

$$[T^{(e)}]^{\mathrm{T}} = [T^{(e)}]^{-1}$$

(This property of the transformation matrix, $[T^{(e)}]$, whereby the inverse equals the transpose is known as orthogonality.) Hence, element loads are given by:

$$\{S^{(e)}\} = [T^{(e)}]^{\mathrm{T}}[k'^{(e)}][T^{(e)}]\{s^{(e)}\}$$

Or, more simply $$\{S^{(e)}\} = [k'^{(e)}]\{s^{(e)}\} \tag{9.21}$$

in which the element stiffness matrix in global coordinates is

$$[k^{(e)}] = [T^{(e)}]^{\mathrm{T}}[k'^{(e)}][T^{(e)}] \tag{9.22}$$

Element stiffness matrix in global coordinates

Substituting from eqns. (9.14) and (9.19) into eqn. (9.22), transposing the transformation matrix and performing the triple matrix product gives the element stiffness matrix in global coordinates as:

$$[k^{(e)}] = \frac{AE}{L} \begin{bmatrix} \cos^2\alpha & \cos\alpha\sin\alpha & -\cos^2\alpha & -\cos\alpha\sin\alpha \\ -\sin\alpha\cos\alpha & \sin^2\alpha & -\sin\alpha\cos\alpha & -\sin^2\alpha \\ -\cos^2\alpha & -\cos\alpha\sin\alpha & \cos^2\alpha & \cos\alpha\sin\alpha \\ -\sin\alpha\cos\alpha & -\sin^2\alpha & \sin\alpha\cos\alpha & \sin^2\alpha \end{bmatrix} \tag{9.23}$$

Element stress matrix in global coordinates

The element stress matrix found in local coordinates, eqn. (9.17), can be transformed to global coordinates by substituting from eqn. (9.18) into eqn. (9.16) to give

$$\sigma^{(e)} = [H'^{(e)}][T^{(e)}]\{s^{(e)}\} \tag{9.24}$$

in which the element stress matrix in global coordinates is

$$[H^{(e)}] = [H'^{(e)}][T^{(e)}] \tag{9.25}$$

Substituting from eqn. (9.17) and (9.19) into eqn. (9.25) gives the element stress matrix in global coordinates as

$$[H^{(e)}] = E^{(e)}[-\cos\alpha \quad -\sin\alpha \quad \cos\alpha \quad \sin\alpha]^{(e)}/L^{(e)} \tag{9.26}$$

Formation of structural governing equation and assembled stiffness matrix

With reference to § 9.5, external nodal equilibrium is satisfied by relating the nodal loads, $\{P\}$, to the element loads, $\{S\}$, via

$$\{P\} = [a]^{\mathrm{T}}\{S\} \tag{9.1}$$

Similarly, external, nodal, compatibility is satisfied by relating the element displacements, $\{s\}$, to the nodal displacements, $\{p\}$, via

$$\{s\} = [a]\{p\} \tag{9.3}$$

Substituting from eqn. (9.3) into eqn. (9.21) for all elements in the structure, gives:

$$\{S\} = [k][a]\{p\} \tag{9.27}$$

in which $[k]$ is the *unassembled stiffness* matrix. Premultiplying the above by $[a]^{\mathrm{T}}$ and substituting from eqn. (9.1) gives:

$$\{P\} = [a]^{\mathrm{T}}[k][a]\{p\}$$

Or, more simply $$\{P\} = [K]\{p\} \tag{9.28}$$

which is the *structural governing equation* for static stress analysis, relating the nodal forces $\{P\}$ to the nodal displacements $\{p\}$ for all the nodes in the structure, in which the *structural, or assembled stiffness* matrix

$$[K] = [a]^{\mathrm{T}}[k][a] \tag{9.29}$$

9.7.2. Formulation of a rod element using the principle of virtual work equation

Here, the principle of virtual work approach, described in § 9.6, will be used to formulate the equations for an axial force rod element. As described, the approach permits the displacement field to be represented by approximating functions, known as *interpolation* or *shape functions*, a brief description of which follows.

Shape functions

As the name suggests shape functions describe the way in which the displacements are interpolated throughout the element and often take the form of polynomials, which will be complete to some degree. The terms required to form complete linear, quadratic and cubic, etc., polynomials are given by Pascal's triangle and tetrahedron for two- and three dimensional elements, respectively. As well as completeness, there are other considerations to be made when choosing polynomial terms to ensure the element is well behaved, and the reader is urged to consult detailed texts.[6] One consideration, which will become apparent, is that the total number of terms in an interpolation polynomial should be equal to the number of dof. of the element.

Consider the axial force rod element shown in Fig. 9.24, for which the local and global axes have been taken to coincide. The purpose is to simplify the appearance of the equations by avoiding the need for the prime in denoting local coordinate dependent quantities. This element has only two nodes and each is taken to have only an axial dof. The total of only two dof. for this element limits the displacement interpolation function to a linear polynomial, namely

$$u(x) = \alpha_1 + \alpha_2 x$$

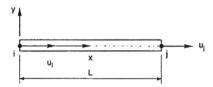

Fig. 9.24. Axial force rod element aligned with global x-axis.

where α_1 and α_2, to be determined, are known as *generalised coefficients*, and are dependent on the nodal displacements and coordinates.

Writing in matrix form

$$u(x) = \begin{bmatrix} 1 & x \end{bmatrix} \begin{bmatrix} \alpha_1 \\ \alpha_2 \end{bmatrix}$$

Or, more concisely,

$$u(x) = [x]\{\alpha\} \tag{9.30}$$

At the nodal points, $u(0) = u_i$ and $u(L) = u_j$.

Substituting into eqn. (9.30) gives

$$\begin{bmatrix} u_i \\ u_j \end{bmatrix} = \begin{bmatrix} 1 & 0 \\ 1 & L \end{bmatrix} \begin{bmatrix} \alpha_1 \\ \alpha_2 \end{bmatrix}$$

Or, more concisely

$$\{u\} = [A]\{\alpha\}$$

The column matrix of generalised coefficients, $\{\alpha\}$, is obtained by evaluating

$$\{\alpha\} = [A]^{-1}\{u\} \tag{9.31}$$

for which the required inverse of matrix $[A]$, i.e. $[A]^{-1}$ is obtained using standard matrix inversion whereby

$$[A]^{-1} = \text{adj } [A] / \det [A]$$

in which $\text{adj } [A] = [C]^{\mathrm{T}}$, where $[C]$ is the cofactor matrix of $[A]$

i.e. $\text{adj } [A] = \begin{bmatrix} L & -1 \\ 0 & 1 \end{bmatrix}^{\mathrm{T}} = \begin{bmatrix} L & 0 \\ -1 & 1 \end{bmatrix}$

and   det $[A] = 1 \times L - 0 \times 1 = L$

Hence,
$$[A]^{-1} = \frac{1}{L} \begin{bmatrix} L & 0 \\ -1 & 1 \end{bmatrix}$$

Substituting eqn. (9.31) into eqn. (9.30) and utilising the above result for $[A]^{-1}$, gives

$$u(x) = [1 \quad x]\frac{1}{L} \begin{bmatrix} L & 0 \\ -1 & 1 \end{bmatrix} \begin{bmatrix} u_i \\ u_j \end{bmatrix}$$

$$= \frac{1}{L}[L - x, x] \begin{bmatrix} u_i \\ u_j \end{bmatrix} = [1 - x/L, x/L] \begin{bmatrix} u_i \\ u_j \end{bmatrix}$$

$$= [N]\{u\} \tag{9.32}$$

in which $[N]$ is the matrix of shape functions. In this case, $N_1 = 1 - x/L$ and $N_2 = x/L$, and hence vary linearly over the element, as shown in Fig. 9.25. Note that the shape functions have the value unity at the node corresponding to the nodal displacement being interpolated and zero at all other nodes (in this case at the only other node), and is a property of all shape functions.

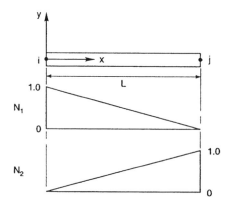

Fig. 9.25. Shape functions for the axial force rod element.

Element stiffness matrix in local coordinates

Consider the axial force element shown in Fig. 9.24. The only strain present will be a direct strain in the axial direction and is given by eqn. (9.4) as

$$\varepsilon_x = \varepsilon = \partial u / \partial x$$

Substituting from eqn. (9.32), gives

$$\varepsilon = \partial[N]\{u\}/\partial x = [B]\{u\} \tag{9.33}$$

and, taking the virtual strain to have a similar form to the real strain, gives

$$\bar{\varepsilon} = [B]\{\bar{u}\} \tag{9.34}$$

where

$$[B] = \partial[N]/\partial x$$

In the present case of the two-node linear rod element, eqn. (9.32) shows $[N] = \dfrac{1}{L}[L - x, x]$, and hence

$$[B] = \frac{1}{L}[-1 \ 1] \tag{9.35}$$

Note that in this case the derivative matrix $[B]$ contains only constants and does not involve functions of x and hence the strain, given by eqn. (9.33), will be constant along the length of the rod element.

Assuming Hookean behaviour and utilising eqn. (9.33)

$$\sigma = E[B]\{u\} \tag{9.36}$$

It follows for the linear rod element that the stress will be constant and is given as

$$\sigma = \frac{E}{L}[-1 \ 1] \begin{bmatrix} u_i \\ u_j \end{bmatrix} = \frac{E}{L}(u_j - u_i)$$

in which $(u_j - u_i)/L$ is the strain.

Substituting the expression for virtual strain from eqn. (9.34) and the real stress from eqn. (9.36) into eqn. (9.11) gives the internal virtual work as

$$\overline{W}_i = \int_v \overline{\varepsilon}\sigma \, dv = \int_v \{\overline{u}\}^{\mathrm{T}}[B]^{\mathrm{T}}E[B]\{u\} \, dv$$

Since the real and virtual displacements are constant, they can be taken outside the integral, to give

$$\overline{W}_i = \{\overline{u}\}^{\mathrm{T}} \int_v [B]^{\mathrm{T}}E[B] \, dv \, \{u\} \tag{9.37}$$

The external virtual work will be given by

$$\overline{W}_e = \{\overline{u}\}^{\mathrm{T}}\{P\} \tag{9.38}$$

Substituting from eqns. (9.37) and (9.38) into the equation of the principle of virtual work, eqn. (9.10) gives

$$0 = \{\overline{u}\}^{\mathrm{T}}\{P\} - \{\overline{u}\}^{\mathrm{T}} \int_v [B]^{\mathrm{T}}E[B] \, dv \, \{u\}$$

Or,

$$= \{\overline{u}\}^{\mathrm{T}}(\{P\} - \int_v [B]^{\mathrm{T}}E[B] \, dv \, \{u\})$$

The virtual displacements, $\{\overline{u}\}$, are arbitrary and nonzero, and hence the quantity in parentheses must be zero,

i.e.

$$\{P\} = \int_v [B]^{\mathrm{T}}E[B] \, dv \, \{u\} = [k'^{(e)}]\{u\} \tag{9.39}$$

where
$$[k'^{(e)}] = \int_v [B]^T E[B]\, dv \tag{9.40}$$

Evaluating the element stiffness matrix $[k'^{(e)}]$ for the linear rod element by substituting from eqns. (9.35) gives

$$[k'^{(e)}] = \int_v \frac{1}{L}\begin{bmatrix} -1 \\ 1 \end{bmatrix} E \frac{1}{L}[-1 \quad 1]\, dv$$

$$= \frac{E}{L^2}\begin{bmatrix} 1 & -1 \\ -1 & 1 \end{bmatrix}\int_v dv$$

For a prismatic element $\int_v dv = AL$, and the element stiffness matrix becomes

$$[k'^{(e)}] = \frac{AE}{L}\begin{bmatrix} 1 & -1 \\ -1 & 1 \end{bmatrix} \tag{9.41}$$

Expanding the force/displacement eqn. (9.39) to include terms associated with the y- direction, requires the insertion of zeros in the stiffness matrix of eqn. (9.41) and hence becomes identical to eqn. (9.14).

Element stress matrix in local coordinates

For the case of a linear rod element, substituting from eqn. (9.35) into eqn. (9.36) gives the element stress as

$$\sigma^{(e)} = \frac{E}{L}[-1 \ 1]\{u\} \tag{9.42}$$

Again, by inserting zeros in the matrix, to accommodate terms associated with the y- direction, eqn. (9.42) becomes identical to eqn. (9.15).

Transformation of element stiffness and stress matrices to global coordinates

The element stiffness and stress matrices obtained above can be transformed from local to global coordinates using the procedures of § 9.7.1 to give the results previously obtained, namely the stiffness matrix of eqn. (9.23) and stress matrix of eqn. (9.26).

Formation of structural governing equation and assembled stiffness matrix

Section 9.7.1 has covered the combination of individual element stiffness contributions, necessary to analyse an assemblage of elements representing a complete framework. Equilibrium and compatibility arguments were used to form the structural governing eqn. (9.28) and hence the assembled stiffness matrix, eqn. (9.29). Now, the alternative principle of virtual work will be used to derive the same equations.

Eqn. (9.37) gives the element internal virtual work in local coordinates as

$$W_i^{(e)} = \{u'^{(e)}\}^T \int_v [B^{(e)}]^T E^{(e)}[B^{(e)}]dv\{u'^{(e)}\}$$

Summing all such contributions for the entire structure of m elements, gives

$$\overline{W}_i = \sum_{e=1}^{m}(\{\overline{u}'^{(e)}\}^{\mathrm{T}}\int_{v}[B^{(e)}]^{\mathrm{T}}E^{(e)}[B^{(e)}]\,dv\,\{u'^{(e)}\}) \tag{9.43}$$

Summing the contributions over all n nodes, the external virtual work will be given by

$$\overline{W}_e = \sum_{i=1}^{n}\overline{p}_iP_i = \{\overline{p}\}^{\mathrm{T}}\{P\} \tag{9.44}$$

where $\{\overline{p}\}$ is the column matrix of all nodal virtual displacements for the structure and $\{P\}$ is the column matrix of all nodal forces. Substituting from eqns. (9.43) and (9.44) into the equation of the principle of virtual work, eqn. (9.10) gives

$$0 = \{\overline{p}\}^{\mathrm{T}}\{P\} - \sum_{e=1}^{m}(\{\overline{u}'^{(e)}\}^{\mathrm{T}}\int_{v}[B^{(e)}]^{\mathrm{T}}E^{(e)}[B^{(e)}]\,dv\,\{u'^{(e)}\}) \tag{9.45}$$

Relating the virtual displacements in local and global coordinates via the transformation matrix $[T^{(e)}]$, gives

$$\{\overline{u}^{(e)}\} = [T^{(e)}]\{\overline{p}^{(e)}\} \quad \text{and} \quad \{\overline{u}'^{(e)}\}^{\mathrm{T}} = \{\overline{p}^{(e)}\}^{\mathrm{T}}[T^{(e)}]^{\mathrm{T}}$$

Summing the contributions and recalling $\{\overline{p}\}$ denotes the nodal displacements for the entire structure, gives

$$\sum_{e=1}^{m}\{\overline{u}'^{(e)}\}^{\mathrm{T}} = \{\overline{p}\}^{\mathrm{T}}\sum_{e=1}^{m}[T^{(e)}]^{\mathrm{T}} \quad \text{and} \quad \sum_{e=1}^{m}\{\overline{u}'^{(e)}\} = \sum_{e=1}^{m}([T^{(e)}])\{p\}$$

Hence, eqn. (9.45) can be re-written as

$$\{\overline{p}\}^{\mathrm{T}}\{P\} = \{\overline{p}\}^{\mathrm{T}}(\sum_{e=1}^{m}[T^{(e)}]^{\mathrm{T}}\int_{v}[B^{(e)}]^{\mathrm{T}}E^{(e)}[B^{(e)}]\,dv\,[T^{(e)}])\{p\}$$

$$= \{\overline{p}\}^{\mathrm{T}}\sum_{e=1}^{m}[k^{(e)}]\{p\} = \{\overline{p}\}^{\mathrm{T}}[K]\{p\} \tag{9.46}$$

where

$$[k^{(e)}] = [T^{(e)}]^{\mathrm{T}}\int_{v}[B^{(e)}]^{\mathrm{T}}E^{(e)}[B^{(e)}]\,dv\,[T^{(e)}]$$

and the assembled stiffness matrix

$$[K] = \sum_{e=1}^{m}[k^{(e)}] \tag{9.47}$$

It follows from eqn. (9.46) since $\{\overline{p}\}$ is arbitrary and non-zero, that

$$\{P\} = [K]\{p\}$$

which is the structural governing equation and the same as eqn. (9.28), and implies nodal force equilibrium.

9.8. A simple beam element

As with the previous treatment of the rod element, the two approaches using fundamental equations and the principle of virtual work will be employed to formulate the necessary equations for a simple beam element.

9.8.1. Formulation of a simple beam element using fundamental equations

Consider the case, similar to §9.7.1, in which the deformations, member stresses and reactions are required for planar frames, excepting that the member joints are now taken to be rigid and hence capable of transmitting moments. The behaviour of such frames can be represented as an assemblage of beam finite elements. A typical simple beam element is shown in Fig. 9.26, the physical and material properties of which are taken to be constant throughout the element. As with the previous rod element, changes in properties and load application are only admissible at nodal positions. In addition to u and v translatory freedoms, each node has a rotational freedom, θ, about the z axis, giving three dof. per node. Hence, axial, shear and flexural deformations will be represented, whilst torsional deformations which are inappropriate for planar frames will be ignored.

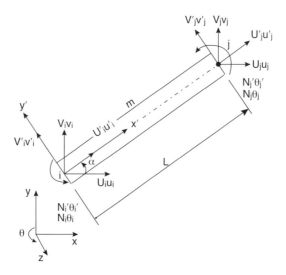

Fig. 9.26. A simple beam element.

Element stiffness matrix in local coordinates

The differential equation of flexure appropriate to a beam element can be written as

$$d^2v'/dx'^2 = N'/EI \qquad (9.48)$$

in which v' denotes the displacement in the local y' direction, N' is the element moment, E is the modulus of elasticity and I is the relevant second moment of area of the beam. The first derivative of the moment with respect to distance x' along a beam is known to give the shear force, V',

i.e.
$$dN'/dx' = V' \qquad (9.49)$$

Similarly, the first derivative of the shear force will give the loading intensity, ω',

i.e.
$$dV'/dx' = \omega' \qquad (9.50)$$

Differentiating eqn. (9.48) and utilising eqn. (9.49) gives

$$d^3v'/dx'^3 = V'/EI \qquad (9.51)$$

Differentiating again and utilising eqn. (9.50) gives

$$d^4v'/dx'^4 = \omega'/EI \qquad (9.52)$$

Integrating eqn. (9.52), recalling that loads can only be applied at the nodes, and hence $\omega' = 0$, gives

$$d^3v'/dx'^3 = C_1 = V'/EI, \quad \text{(from eqn. 9.51)} \qquad (9.53)$$

Further integration gives

$$d^2v'/dx'^2 = C_1x' + C_2 = N'/EI, \quad \text{(from eqn. 9.48)} \qquad (9.54)$$

$$dv'/dx' = C_1x'^2/2 + C_2x' + C_3 = \theta' \qquad (9.55)$$

$$\text{and} \quad v' = C_1x'^3/6 + C_2x'^2/2 + C_3x' + C_4 \qquad (9.56)$$

With reference to Fig. 9.26, it can be seen that

$$v'(0) = v'_i, \quad v'(L) = v'_j, \quad dv'/dx'(0) = \theta'_i, \quad dv'/dx'(L) = \theta'_j$$

It follows from eqn. (9.56) that
$$\qquad v'_i = C_4 \qquad (9.57)$$

from eqn. (9.55)
$$\qquad \theta'_i = C_3 \qquad (9.58)$$

from eqn. (9.56)

$$v'_j = C_1L^3/6 + C_2L^2/2 + C_3L + C_4 = C_1L^3/6 + C_2L^2/2 + \theta'_iL + v'_i \qquad (9.59)$$

and from eqn. (9.55)

$$\theta'_j = C_1L^2/2 + C_2L + C_3 \quad = C_1L^2/2 + C_2L + \theta'_i \qquad (9.60)$$

An expression for C_2 can now be obtained by multiplying eqn. (9.60) throughout by $L/3$ and subtracting the result from eqn. (9.59), (to eliminate C_1), to give

$$v'_j - \theta'_jL/3 = C_2(L^2/2 - L^2/3) + \theta'_i(L - L/3) + v'_i = C_2L^2/6 + \theta'_i2L/3 + v'_i$$

Rearranging, $C_2 \qquad = 6(v'_j - v'_i)/L^2 - 6(\theta'_iL/3 + \theta'_i2L/3)/L$

$$= 6(-v'_i + v'_j)/L^2 - 2(2\theta'_i + \theta'_j)/L \qquad (9.61)$$

Rearranging eqn. (9.60) and substituting from eqn. (9.61) gives

$$C_1 = (2/L^2)[(\theta'_j - \theta'_i) - 6(-v'_i + v'_j)/L + 2(2\theta'_i + \theta'_j)]$$

$$= 12(v'_i - v'_j)/L^3 + 6(\theta'_i + \theta'_j)/L^2 \qquad (9.62)$$

Substituting for constant C_1 from eqn. (9.62) into eqn. (9.53) gives shear force

$$V' = EIC_1 = 12EI(v'_i - v'_j)/L^3 + 6EI(\theta'_i + \theta'_j)/L^2 \qquad (9.63)$$

Substituting for constants C_1 and C_2 from eqns. (9.62) and (9.63) into eqn. (9.54) gives the moment

$$N' = EI(C_1x' + C_2)$$
$$= 6EI(2x' - L)(v'_i - v'_j)/L^3 + 6EIx'(\theta'_i + \theta'_j)/L^2 - 2EI(2\theta'_i + \theta'_j)/L \qquad (9.64)$$

Note that the shear force, eqn. (9.63) is independent of distance x' along the beam, i.e. constant, whilst the moment, eqn. (9.64) is linearly dependent on distance x', consistent with a beam subjected to concentrated forces.

It follows that the nodal shear force and moments are given as

$$V'(0) = V'(L) = 12EI(v'_i - v'_j)/L^3 + 6EI(\theta'_i + \theta'_j)/L^2 \qquad (9.65)$$

$$N'(0) = 6EI(-v'_i + v'_j)/L^2 - 2EI(2\theta'_i + \theta'_j)/L \qquad (9.66)$$

$$N'(L) = 6EI(v'_i - v'_j)/L^2 + 2EI(\theta'_i + 2\theta'_j)/L \qquad (9.67)$$

The shear force and moments given by eqns. (9.65)–(9.67) use a Mechanics of Materials sign convention, namely, a positive shear force produces a clockwise couple and a positive moment produces sagging. To conform with the sign convention shown in Fig. 9.26, the following changes are required:

$$V'_i = -V'_j = V'(0)$$
$$N'_i = -N'(0)$$
$$N'_j = N'(L)$$

Writing in matrix form, eqns. (9.65)–(9.67) become

$$\begin{bmatrix} V'_i \\ N'_i \\ V'_j \\ N'_j \end{bmatrix} = \frac{EI}{L} \begin{bmatrix} 12/L^2 & 6/L & -12/L^2 & 6/L \\ 6/L & 4 & -6/L & 2 \\ -12/L^2 & -6/L & 12/L^2 & -6/L \\ 6/L & 2 & -6/L & 4 \end{bmatrix} \begin{bmatrix} v'_i \\ \theta'_i \\ v'_j \\ \theta'_j \end{bmatrix} \qquad (9.68)$$

Combining eqn. (9.68) with eqn. (9.12) gives the matrix equation relating element axial and shear forces and moments to the element displacements as

$$\begin{bmatrix} U'_i \\ V'_i \\ N'_i \\ U'_j \\ V'_j \\ N'_j \end{bmatrix} = \begin{bmatrix} AE/L & 0 & 0 & -AE/L & 0 & 0 \\ 0 & 12EI/L^3 & 6EI/L^2 & 0 & -12EI/L^3 & 6EI/L^2 \\ 0 & 6EI/L^2 & 4EI/L & 0 & -6EI/L^2 & 2EI/L \\ -AE/L & 0 & 0 & AE/L & 0 & 0 \\ 0 & -12EI/L^3 & -6EI/L^2 & 0 & 12EI/L^3 & -6EI/L^2 \\ 0 & 6EI/L^2 & 2EI/L & 0 & -6EI/L^2 & 4EI/L \end{bmatrix} \begin{bmatrix} u'_1 \\ v'_i \\ \theta'_i \\ u'_j \\ v'_j \\ \theta'_j \end{bmatrix}$$

$$(9.69)$$

from which the element stiffness matrix with respect to local coordinates is:

$$[k^{(e)}] = \begin{bmatrix} AE/L & 0 & 0 & -AE/L & 0 & 0 \\ 0 & 12EI/L^3 & 6EI/L^2 & 0 & -12EI/L^3 & 6EI/L^2 \\ 0 & 6EI/L^2 & 4EI/L & 0 & -6EI/L^2 & 2EI/L \\ -AE/L & 0 & 0 & AE/L & 0 & 0 \\ 0 & -12EI/L^3 & -6EI/L^2 & 0 & 12EI/L^3 & -6EI/L^2 \\ 0 & 6EI/L^2 & 2EI/L & 0 & -6EI/L^2 & 4EI/L \end{bmatrix} \qquad (9.70)$$

Element stress matrix in local coordinates

Only bending and axial stresses will be considered, shear stresses being taken as insignificant. The points for stress calculation will be the extreme top and bottom fibres at each end of the element, which will always include the maximum stress point. With reference to Fig. 9.27, the beam element stress matrix will be

$$\{\sigma^{(e)}\} = \{\sigma_i^{\text{top}} \sigma_i^{\text{btm}} \sigma_j^{\text{top}} \sigma_j^{\text{btm}}\}$$

Fig. 9.27. Locations for beam element stress calculation.

Relating these stresses to the internal loads gives

$$\begin{bmatrix} \sigma_i^{\text{top}} \\ \sigma_i^{\text{btm}} \\ \sigma_j^{\text{top}} \\ \sigma_j^{\text{btm}} \end{bmatrix} = \begin{bmatrix} -1/A & 0 & t/I & 0 & 0 & 0 \\ -1/A & 0 & -b/I & 0 & 0 & 0 \\ 0 & 0 & 0 & 1/A & 0 & -t/I \\ 0 & 0 & 0 & 1/A & 0 & b/I \end{bmatrix} \begin{bmatrix} U_i' \\ V_i' \\ N_i' \\ U_j' \\ V_j' \\ N_j' \end{bmatrix} \qquad (9.71)$$

Or, more concisely, $$\{\sigma^{(e)}\} = [h^{(e)}]\{S'^{(e)}\} \qquad (9.72)$$

Substituting for the element loads column matrix using a relation of the form of eqn. (9.13) gives

$$\{\sigma^{(e)}\} = [h^{(e)}][k'^{(e)}]\{s'^{(e)}\} = [H'^{(e)}]\{s'^{(e)}\}$$

which is the same form as eqn. (9.16) and $[H'^{(e)}]$ is the stress matrix with respect to local coordinates. Evaluating $[h^{(e)}][k'^{(e)}]$ gives the stress matrix as

$$[H'^{(e)}] = \frac{E}{L} \begin{bmatrix} -1 & 6t/L & 4t & 1 & 6t/L & 2t \\ -1 & -6b/L & -4b & 1 & -6b/L & -2b \\ -1 & -6t/L & -2t & 1 & -6t/L & -4t \\ -1 & 6b/L & 2b & 1 & 6b/L & 4b \end{bmatrix} \qquad (9.73)$$

Transformation of displacements and loads

Relations of similar form to those of eqns. (9.18)–(9.23) but with additional rotational dof. terms, previously not included in the rod element transformation, will enable the above element stiffness and stress matrices to be transformed from local to global coordinates. The expanded form of eqn. (9.18) for the beam element will be given as

$$
\begin{bmatrix} u_i' \\ v_i' \\ \theta_i' \\ u_j' \\ v_j' \\ \theta_j' \end{bmatrix} =
\begin{bmatrix}
\cos\alpha & \sin\alpha & 0 & 0 & 0 & 0 \\
-\sin\alpha & \cos\alpha & 0 & 0 & 0 & 0 \\
0 & 0 & 1 & 0 & 0 & 0 \\
0 & 0 & 0 & \cos\alpha & \sin\alpha & 0 \\
0 & 0 & 0 & -\sin\alpha & \cos\alpha & 0 \\
0 & 0 & 0 & 0 & 0 & 1
\end{bmatrix}
\begin{bmatrix} u_i \\ v_i \\ \theta_i \\ u_j \\ v_j \\ \theta_j \end{bmatrix}
$$

in which the transformation matrix is:

$$
[T^{(e)}] =
\begin{bmatrix}
\cos\alpha & \sin\alpha & 0 & 0 & 0 & 0 \\
-\sin\alpha & \cos\alpha & 0 & 0 & 0 & 0 \\
0 & 0 & 1 & 0 & 0 & 0 \\
0 & 0 & 0 & \cos\alpha & \sin\alpha & 0 \\
0 & 0 & 0 & -\sin\alpha & \cos\alpha & 0 \\
0 & 0 & 0 & 0 & 0 & 1
\end{bmatrix}
\tag{9.74}
$$

Element stiffness matrix in global coordinates

$$
[k^{(e)}] = \frac{E}{L}
\begin{bmatrix}
A\cos^2\alpha + (12I\sin^2\alpha/L^2), & & \\
(A - 12I/L^2)\cos\alpha\sin\alpha, & A\sin^2\alpha + (12I\cos^2\alpha)/L^2, & \\
-(6I\sin\alpha)/L, & (6I\cos\alpha)/L, & 4I, \\
-A\cos^2\alpha - (12I\sin^2\alpha)/L^2, & (-A + 12I/L^2)\cos\alpha\sin\alpha, & (6I\sin\alpha)/L, \\
(-A + 12I/L^2)\cos\alpha\sin\alpha, & -A\sin^2\alpha - (12I\cos^2\alpha)/L^2, & -(6I\cos\alpha)/L, \\
-(6I\sin\alpha)/L, & (6I\cos\alpha)/L, & 2I,
\end{bmatrix}
$$

$$
\begin{array}{c}
symmetric \\
\begin{matrix}
A\cos^2\alpha + (12I\sin^2\alpha)/L^2, & & \\
(A - 12I/L^2)\cos\alpha\sin\alpha, & A\sin^2\alpha + (12I\cos^2\alpha)/L^2, & \\
(6I\sin\alpha)/L, & -(6I\cos\alpha)/L, & 4I
\end{matrix}
\end{array}
\tag{9.75}
$$

Element stress matrix in global coordinates

Substituting from eqns. (9.73) and (9.74) into eqn. (9.25) gives the element stress matrix in global coordinates as:

$$
[H^{(e)}] = \frac{E}{L}
\begin{bmatrix}
-\cos\alpha - 6t\sin(\alpha)/L & -\sin\alpha + 6t\cos(\alpha)/L & 4t \\
-\cos\alpha + 6b\sin(\alpha)/L & -\sin\alpha - 6b\cos(\alpha)/L & -4b \\
-\cos\alpha + 6t\sin(\alpha)/L & -\sin\alpha - 6t\cos(\alpha)/L & -2t \\
-\cos\alpha - 6b\sin(\alpha)/L & -\sin\alpha + 6b\cos(\alpha)/L & 2b
\end{bmatrix}
$$

$$
\begin{bmatrix}
\cos\alpha + 6t\sin(\alpha)/L & \sin\alpha - 6t\cos(\alpha)/L & 2t \\
\cos\alpha - 6b\sin(\alpha)/L & \sin\alpha + 6b\cos(\alpha)/L & -2b \\
\cos\alpha - 6t\sin(\alpha)/L & \sin\alpha + 6t\cos(\alpha)/L & -4t \\
\cos\alpha + 6b\sin(\alpha)/L & \sin\alpha - 6b\cos(\alpha)/L & 4b
\end{bmatrix}
\tag{9.76}
$$

Formation of structural governing equation and assembled stiffness matrix

Whilst the beam element matrices include rotational dof. terms, not present in the rod element matrices, the procedures of Section 9.7.1 still apply, and lead to the structural governing equation

$$\{P\} = [K]\{p\} \tag{9.28}$$

and the assembled stiffness matrix

$$[K] = [a]^{T}[k][a] \tag{9.29}$$

9.8.2. Formulation of a simple beam element using the principle of virtual work equation

As Section 9.7.2 the principle of virtual work equation will be invoked, this time to formulate the equations for a simple beam element.

Consider the simple beam element shown in Fig. 9.28, for which the local and global axes have again been taken to coincide to avoid need for the prime and hence to simplify the appearance of the equations. The two nodes are each taken to have only normal and rotational dof. The terms associated with the omitted axial dof. have already been derived for the linear rod element in §9.7 and will be incorporated once the other terms have been derived. The total of four dof. for this beam element permits the displacement to be interpolated by a fourth order polynomial, namely

$$v(x) = \alpha_1 + \alpha_2 x/L + \alpha_3 x^2/L^2 + \alpha_4 x^3/L^3$$

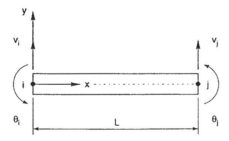

Fig. 9.28. Simple beam element, aligned with global *x*-axis.

where α_1 to α_4 are generalised coefficients to be determined. Utilisation of eqns. (9.48) and (9.49) shows this polynomial will provide for a linearly varying moment and constant shear force and hence will enable an exact solution for beams subjected to concentrated loads.

Writing in matrix form,

$$v(x) = [1, x/L, x^2/L^2, x^3/L^3] \begin{bmatrix} \alpha_1 \\ \alpha_2 \\ \alpha_3 \\ \alpha_4 \end{bmatrix}$$

or, more concisely

$$v(x) = [x]\{\alpha\} \tag{9.77}$$

At the nodal points, $v(0) = v_i$; $v(L) = v_j$;

and $dv/dx(0) = \theta_i$; $dv/dx(L) = \theta_j$

Substituting into eqn. (9.77) gives

$$
\begin{bmatrix} v_i \\ \theta_i \\ v_j \\ \theta_j \end{bmatrix} = \begin{bmatrix} 1 & 0 & 0 & 0 \\ 0 & 1/L & 0 & 0 \\ 1 & 1 & 1 & 1 \\ 0 & 1/L & 2/L & 3/L \end{bmatrix} \begin{bmatrix} \alpha_1 \\ \alpha_2 \\ \alpha_3 \\ \alpha_4 \end{bmatrix}
$$

Or, more concisely,

$$\{v\} = [A]\{\alpha\}$$

Expressing in terms of the column matrix of generalised coefficients, $\{\alpha\}$, gives

$$\{\alpha\} = [A]^{-1}\{v\} \tag{9.78}$$

Evaluation of eqn. (9.78) requires the inverse of matrix $[A]$ obtained from

$$
\text{adj } [A] = [C]^{\text{T}} = \begin{bmatrix} 1/L^2 & 0 & -3/L^2 & 2/L^2 \\ 0 & 1/L & -2/L & 1/L \\ 0 & 0 & 3/L^2 & -2/L^2 \\ 0 & 0 & -1/L & 1/L \end{bmatrix}^{\text{T}}
$$

$$
= \begin{bmatrix} 1/L^2 & 0 & 0 & 0 \\ 0 & 1/L & 0 & 0 \\ -3/L^2 & -2/L & 3/L^2 & -1/L \\ 2/L^2 & 1/L & -2/L^2 & 1/L \end{bmatrix}
$$

and $\det [A] = 1(1/L)(1 \times 3/L - 1 \times 2/L) \quad = 1/L^2$

Hence, $[A]^{-1} = L^2 \begin{bmatrix} 1/L^2 & 0 & 0 & 0 \\ 0 & 1/L & 0 & 0 \\ -3/L^2 & 2/L & 3/L^2 & -1/L \\ 2/L^2 & 1/L & -2/L^2 & 1/L \end{bmatrix}$

$$
= \begin{bmatrix} 1 & 0 & 0 & 0 \\ 3 & L & 0 & 0 \\ -3 & -2L & 3 & -L \\ 2 & L & -2 & L \end{bmatrix}
$$

Substituting eqn. (9.78) into eqn. (9.77) and utilising the above result for $[A]^{-1}$, gives

$$
v(x) = [1, x/L, x^2/L^2, x^3/L^3] \begin{bmatrix} 1 & 0 & 0 & 0 \\ 0 & L & 0 & 0 \\ -3 & -2L & 3 & -L \\ 2 & L & -2 & L \end{bmatrix} \begin{bmatrix} v_i \\ \theta_i \\ v_j \\ \theta_j \end{bmatrix}
$$

$$
= [1 - 3x^2/L^2 + 2x^3/L^3, \, x - 2x^2/L + x^3/L^2, \, 3x^2/L^2 - 2x^3/L^3, \, -x^2/L + x^3/L^2] \begin{bmatrix} v_i \\ \theta_i \\ v_j \\ \theta_j \end{bmatrix}
$$

$$\tag{9.79}$$

which has the same form as eqn. (9.32), in this case with shape functions

$$N_1 = 1 - 3x^2/L^2 + 2x^3/L^3$$

$$N_2 = x - 2x^2/L + x^3/L^2$$

$$N_3 = 3x^2/L^2 - 2x^3/L^3$$

$$N_4 = -x^2/L + x^3/L^2$$

the variation of which is shown in Fig. 9.29.

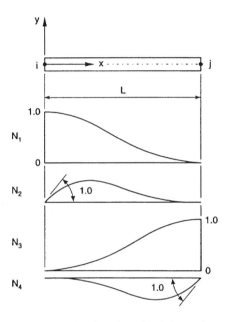

Fig. 9.29. Shape functions for a simple beam element.

Element stiffness matrix in local coordinates

Longitudinal bending stress is given by simple bending theory as

$$\sigma = M y/I$$

in which, from the differential equation of flexure, eqn. (9.48),

$$M = EI d^2v/dx^2$$

to give

$$\sigma = E y\, d^2v/dx^2 \tag{9.80}$$

Substituting eqn. (9.79) into eqn. (9.80) gives

$$\sigma = E y[(12x/L^3 - 6/L^2)(6x/L^2 - 4/L)(-12x/L^3 + 6/L^2)(6x/L^2 - 2/L)] \begin{bmatrix} v_i \\ \theta_i \\ v_j \\ \theta_j \end{bmatrix}$$

$$= E[B]\{u\} \tag{9.81}$$

which has the same form as eqn. (9.36), except in this case

$$[B] = yd^2[N]/dx^2 \tag{9.82}$$

Assuming Hookean behaviour, the longitudinal bending strain is obtained using eqn. (9.81), as

$$\varepsilon = \sigma/E = [B]\{u\}$$

which has the same form as eqn. (9.33).

Taking the virtual longitudinal strain to be given in a form similar to the real strain,

i.e. $$\varepsilon = [B]\{u\} \tag{9.83}$$

Substituting the real stress from eqn. (9.81) and the virtual strain from eqn. (9.83) into the equation of the principle of virtual work, (9.10), gives

$$0 = \{u\}^T\{P\} - \int_v \{u\}^T[B]^T E[B]\{u\}\, dv$$

The real and virtual displacements, being constant, can be taken outside the integral, to give

$$= \{u\}^T(\{P\} - \int_v [B]^T E[B]\, dv\, \{u\})$$

The virtual displacements, $\{u\}$, are arbitrary and nonzero, hence

$$\{P\} = \int_v [B]^T E[B]\, dv\, \{u\} = [k'^{(e)}]\{u\}$$

where $[k'^{(e)}] = \int_v [B]^T E[B]\, dv$, which is an identical result to eqn. (9.40). Substituting from eqn. (9.82) gives

$$[k'(e)] = \int_v y(d^2[N]/dx^2)^T E y(d^2[N]/dx^2)\, dv$$

$$= EI \int_0^L (d^2[N]/dx^2)^T (d^2[N]/dx^2)\, dx$$

$$= EI \int_0^L \begin{bmatrix} \dfrac{6}{L^2}\left(2\dfrac{x}{L}-1\right) \\[2mm] \dfrac{2}{L}\left(3\dfrac{x}{L}-2\right) \\[2mm] \dfrac{6}{L^2}\left(-2\dfrac{x}{L}+1\right) \\[2mm] \dfrac{2}{L}\left(3\dfrac{x}{L}-1\right) \end{bmatrix}$$

$$\times \left[\dfrac{6}{L^2}\left(2\dfrac{x}{L}-1\right) \quad \dfrac{2}{L}\left(3\dfrac{x}{L}-2\right) \quad \dfrac{6}{L^2}\left(-2\dfrac{x}{L}+1\right) \quad \dfrac{2}{L}\left(3\dfrac{x}{L}-1\right) \right] dx \tag{9.84}$$

The following gives examples of evaluating the integrals of eqn. (9.84) for two elements of the stiffness matrix, the rest are obtained by the same procedure.

$$k_{11} = EI \int_0^L \left[\dfrac{6}{L^2}\left(2\dfrac{x}{L}-1\right) \dfrac{6}{L^2}\left(2\dfrac{x}{L}-1\right) \right] dx = \dfrac{36EI}{L^4} \int_0^L \left(4\dfrac{x^2}{L^2} - 4\dfrac{x}{L} + 1\right) dx$$

$$= \dfrac{36EI}{L^4}\left[\dfrac{4x^3}{3L^2} - 2\dfrac{x^2}{L} + x\right] = \dfrac{36EI}{L^3}\left[\dfrac{4}{3} - 2 + 1\right] = \dfrac{12EI}{L^3}$$

and $k_{12} = k_{21} = EI \int_0^L \left[\frac{6}{L^2} \left(2\frac{x}{L} - 1 \right) \frac{2}{L} \left(3\frac{x}{L} - 2 \right) \right] dx = \frac{12EI}{L^3} \int_0^L \left(6\frac{x^2}{L^2} - 7\frac{x}{L} + 2 \right) dx$

$$= \frac{12EI}{L^3} \left[\frac{2x^3}{L^2} - \frac{7x^2}{2L} + 2x \right] = \frac{12EI}{L^2} \left[2 - \frac{7}{2} + 2 \right] = \frac{6EI}{L^2}$$

Evaluation of all the integrals of eqn. (9.84) leads to the beam element flexural stiffness matrix

$$[k'^{(e)}] = \frac{EI}{L} \begin{bmatrix} 12/L^2 & 6/L & -12/L^2 & 6/L \\ 6/L & 4 & -6/L & 2 \\ -12/L^2 & -6/L & 12/L^2 & -6/L \\ 6/L & 2 & -6/L & 4 \end{bmatrix}$$

which is identical to the stiffness matrix of eqn. (9.68) derived using fundamental equations. The same arguments made in Section 9.8.1 apply with regard to including axial terms to give the force/displacement relation, eqn. (9.69), and corresponding element stiffness matrix, eqn. (9.70).

Element stress matrix in local coordinates

Bending and axial stresses are obtained using the same relations as those in §9.8.1.

Transformation of element stiffness and stress matrices to global coordinates

The element stiffness and stress matrices are transformed from local to global coordinates using the procedures of §2.4.8.1 to give the stiffness matrix of eqn. (9.75) and stress matrix of eqn. (9.76).

Formation of structural governing equation and assembled stiffness matrix

The theorem of virtual work used in §9.7.2 to formulate rod element assemblages applies to the present beam elements. It follows, therefore, that the assembled stiffness matrix will be given by eqn. (9.47). The displacement column matrices will, for beams, include rotational dof., not present for rod elements. Further, at the nodes, moment equilibrium, as well as force equilibrium, is now implied by eqn. (9.28).

9.9. A simple triangular plane membrane element

The common occurrence of thin-walled structures merits devoting attention here to their analysis. Many applications are designed on the basis of in-plane loads only with resistance arising from membrane action rather than bending. Whilst thin plates can be curved to resist normal loads by membrane action, for simplicity only planar applications will be considered here. Membrane elements can have three or four edges, which can be straight or curvilinear, however, attention will be restricted here to the simplest, triangular, membrane element.

Unlike the previous rod and beam element formulations, with which displacement fields can be represented exactly and derived from fundamental arguments, the displacement fields represented by two-dimensional elements can only be approximate, and need to be derived using an energy principle. Here, the principle of virtual work will be invoked to derive the membrane element equations.

9.9.1. Formulation of a simple triangular plane membrane element using the principle of virtual work equation

With reference to Fig. 9.30, each node of the triangular membrane element has two dof., namely u and v displacements in the global x and y directions, respectively. The total of six dof. for the element limits the u and v displacement to linear interpolation. Hence

$$u(x, y) = \alpha_1 + \alpha_2 x + \alpha_3 y$$

and

$$v(x, y) = \alpha_4 + \alpha_5 x + \alpha_6 y$$

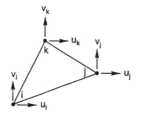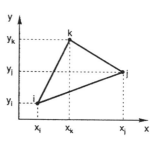

Fig. 9.30. Triangular plane membrane element.

or, in matrix form

$$
\begin{bmatrix} u(x, y) \\ v(x, y) \end{bmatrix} = \begin{bmatrix} 1 & x & y & 0 & 0 & 0 \\ 0 & 0 & 0 & 1 & x & y \end{bmatrix} \begin{bmatrix} \alpha_1 \\ \alpha_2 \\ \alpha_3 \\ \alpha_4 \\ \alpha_5 \\ \alpha_6 \end{bmatrix}
\tag{9.85}
$$

At the nodal points, $u(x_i, y_i) = u_i$, $v(x_i, y_i) = v_i$,

$u(x_j, y_j) = u_j$, $v(x_j, y_j) = v_j$,

and $u(x_k, y_k) = u_k$, $v(x_k, y_k) = v_k$

Substituting into eqn. (9.85) gives

$$
\begin{bmatrix} u_i \\ u_j \\ u_k \\ v_i \\ v_j \\ v_k \end{bmatrix} = \begin{bmatrix} 1 & x_i & y_i & 0 & 0 & 0 \\ 1 & x_j & y_j & 0 & 0 & 0 \\ 1 & x_k & y_k & 0 & 0 & 0 \\ 0 & 0 & 0 & 1 & x_i & y_i \\ 0 & 0 & 0 & 1 & x_j & y_j \\ 0 & 0 & 0 & 1 & x_k & y_k \end{bmatrix} \begin{bmatrix} \alpha_1 \\ \alpha_2 \\ \alpha_3 \\ \alpha_4 \\ \alpha_5 \\ \alpha_6 \end{bmatrix}
\tag{9.86}
$$

Or, more concisely, $\{p\} = [A]\{\alpha\}$

Similar to the previous sections, the column matrix of generalised coefficients, $\{\alpha\}$, is obtained by evaluating

$$\{\alpha\} = [A]^{-1}\{p\} \tag{9.87}$$

where, by arranging the dof. in the above sequence enables suitable partitioning of $[A]$ and minimises the effort required to obtain the inverse. Unlike the previous treatment of the rod and beam element, the evaluation of $[A]^{-1}$ is delayed until Example 9.5. The result, however, is given by eqn. (9.88). It is hoped this departure will enable the element formulation to be more easily assimilated.

$$[A]^{-1} = \frac{1}{2a} \begin{bmatrix} x_2y_3 - x_2y_2 & x_3y_1 - x_1y_3 & x_1y_2 - x_2y_1 & 0 & 0 & 0 \\ y_2 - y_3 & y_3 - y_1 & y_1 - y_2 & 0 & 0 & 0 \\ x_3 - x_2 & x_1 - x_3 & x_2 - x_1 & 0 & 0 & 0 \\ 0 & 0 & 0 & x_2y_3 - x_3y_2 & x_3y_1 - x_1y_3 & x_1y_2 - x_2y_1 \\ 0 & 0 & 0 & y_2 - y_3 & y_3 - y_1 & y_1 - y_2 \\ 0 & 0 & 0 & x_3 - x_2 & x_1 - x_3 & x_2 - x_1 \end{bmatrix} \tag{9.88}$$

where a is equal to the area of the element.

Substituting from eqn. (9.87) into eqn. (9.85), utilising the result for $[A]^{-1}$, i.e. eqn. (9.88), and writing concisely the result of the matrix multiplication, gives

$$\begin{bmatrix} u(x, y) \\ v(x, y) \end{bmatrix} = \begin{bmatrix} N_i & N_j & N_k & 0 & 0 & 0 \\ 0 & 0 & 0 & N_i & N_j & N_k \end{bmatrix} \begin{bmatrix} u_i \\ u_j \\ u_k \\ v_k \\ v_j \\ v_k \end{bmatrix}$$

where the shape functions are given as

$$N_i = \frac{1}{2a}[x_jy_k - x_ky_j + (y_j - y_k)x + (x_k - x_j)y]$$

$$N_j = \frac{1}{2a}[x_ky_i - x_iy_k + (y_k - y_i)x + (x_i - x_k)y] \tag{9.89}$$

$$N_k = \frac{1}{2a}[x_iy_j - x_jy_i + (y_i - y_j)x + (x_j - x_i)y]$$

Note that the shape functions of eqns. (9.89) are linear in x and y. Further, evaluation of eqns. (9.89) shows that shape function $N_i(x_i, y_i) = 1$ and $N_i(x, y) = 0$ at nodes j and k, and at all points on the line joining these nodes. Similarly, $N_j(x_j, y_j) = 1$ and $N_k(x_k, y_k) = 1$, and equal zero at, and on the line between, the other nodes.

Formulation of element stiffness matrix

For plane stress analysis, the strain/displacement relations are

$$\varepsilon_{xx} = \partial u/\partial x, \quad \varepsilon_{yy} = \partial v/\partial y, \quad \varepsilon_{xy} = \partial u/\partial y + \partial v/\partial x$$

where ε_{xx} and ε_{yy} are the direct strains parallel to the x and y axes, respectively, and ε_{xy} is the shear strain in the xy plane. Writing in matrix form gives

$$\{\varepsilon\} = \begin{bmatrix} \varepsilon_{xx} \\ \varepsilon_{yy} \\ \varepsilon_{xy} \end{bmatrix} = \begin{bmatrix} \partial/\partial x & 0 \\ 0 & \partial/\partial y \\ \partial/\partial y & \partial/\partial x \end{bmatrix} \begin{bmatrix} u \\ v \end{bmatrix}$$

Substituting from eqn. (9.85) and performing the partial differentiation, the above becomes

$$\{\varepsilon\} = \begin{bmatrix} 0 & 1 & 0 & 0 & 0 & 0 \\ 0 & 0 & 0 & 0 & 0 & 1 \\ 0 & 0 & 1 & 0 & 1 & 0 \end{bmatrix} \begin{bmatrix} \alpha_1 \\ \alpha_2 \\ \alpha_3 \\ \alpha_4 \\ \alpha_5 \\ \alpha_6 \end{bmatrix}$$

Substituting from eqn. (9.87) gives

$$\{\varepsilon\} = \begin{bmatrix} 0 & 1 & 0 & 0 & 0 & 0 \\ 0 & 0 & 0 & 0 & 0 & 1 \\ 0 & 0 & 1 & 0 & 1 & 0 \end{bmatrix} [A]^{-1} \begin{bmatrix} u_i \\ u_j \\ u_k \\ v_i \\ v_j \\ v_k \end{bmatrix}$$

Or, more concisely, $$\{\varepsilon\} = [B]\{u\} \tag{9.90}$$

where $$[B] = \begin{bmatrix} 0 & 1 & 0 & 0 & 0 & 0 \\ 0 & 0 & 0 & 0 & 0 & 1 \\ 0 & 0 & 1 & 0 & 1 & 0 \end{bmatrix} [A]^{-1} \tag{9.91}$$

Note that matrix $[B]$ is independent of position within the element with the consequence that the strain, and hence the stress, will be constant throughout the element.

For plane stress analysis ($\sigma_{zz} = \sigma_{xz} = \sigma_{yz} = 0$) with isotropic material behaviour, the stress/strain relations in matrix form are

$$\{\sigma\} = \begin{bmatrix} \sigma_{xx} \\ \sigma_{yy} \\ \sigma_{xy} \end{bmatrix} = \frac{E}{(1-v^2)} \begin{bmatrix} 1 & v & 0 \\ v & 1 & 0 \\ 0 & 0 & \dfrac{1-v}{2} \end{bmatrix} \{\varepsilon\}$$

Or, more concisely, $$\{\sigma\} = [D]\{\varepsilon\} \tag{9.92}$$

where σ_{xx} and σ_{yy} are the direct stresses parallel to x and y axes, respectively, σ_{xy} is the shear stress in the xy plane, and $[D]$ is known as the *elasticity matrix*.

Following the same arguments used in the rod and beam formulations, namely, taking the expression for virtual strain to have a similar form to the real strain, eqn. (9.90), and substituting this and the expression for real stress, eqn. (9.92), into the equation of the principle of virtual work, (9.10), gives the element stiffness matrix as

$$[k^{(e)}] = \int_v [B]^{\mathrm{T}}[D][B]\,dv \tag{9.93}$$

The only departure of eqn. (9.93) from the previous expressions is the replacement of the modulus of elasticity, E, by the elasticity matrix $[D]$, due to the change from a one- to a two-dimensional stress system.

Recalling, for the present case that the displacement fields are linearly varying, then matrix $[B]$ is independent of the x and y coordinates. The assumption of isotropic homogeneous material means that matrix $[D]$ is also independent of coordinates. It follows, assuming a constant thickness, t, throughout the element, of area, a, eqn. (9.93) can be integrated to give

$$[k^{(e)}] = at\,[B]^{\mathrm{T}}[D][B] \qquad (9.94)$$

Element stress matrix

The expression for the element direct and shear stresses is obtained by substituting from eqn. (9.90) into eqn. (9.92), to give

$$\{\sigma^{(e)}\} = [D][B]\{u\}$$

or, more fully,

$$\begin{bmatrix} \sigma_{xx} \\ \sigma_{yy} \\ \sigma_{xy} \end{bmatrix} = [D][B] \begin{bmatrix} u_i \\ u_j \\ u_k \\ v_i \\ v_j \\ v_k \end{bmatrix} \qquad (9.95)$$

These stresses are with respect to the global coordinate axes and are taken to act at the element centroid.

Formation of structural governing equation and assembled stiffness matrix

As Sections 9.7.2 and 9.8.2, the structural governing equation is given by eqn. (9.28) and the assembled stiffness matrix by eqn. (9.47).

9.10. Formation of assembled stiffness matrix by use of a dof. correspondence table

Element stiffness matrices given, for example, by eqn. (9.23), are formed for each element in the structure being analysed, and are combined to form the assembled stiffness matrix $[K]$. Where nodes are common to more than one element, the assembly process requires that appropriate stiffness contributions from all such elements are summed for each node. Execution of finite element programs will enable assembly of the element stiffness contributions by utilising, for example, eqn. (9.29) deriving matrix $[a]$, and hence $[a]^{\mathrm{T}}$, from the connectivity information provided by the element mesh. Alternatively, eqn. (9.47) can be used, the matrix summation requiring that all element stiffness matrices, $[k^{(e)}]$, are of the same order as the assembled stiffness matrix $[K]$. However, by efficient "housekeeping" only those rows and columns containing the non-zero terms need be stored.

For the purpose of performing hand calculations, the tedium of evaluating the triple matrix product of eqn. (9.29) can be avoided by summing the element stiffness contributions according to eqn. (9.47). The procedure to be adopted follows, and uses a so-called *dof. correspondence table*. Consider assembly of the element stiffness contributions for the simple pin-jointed plane frame idealised as three rod elements, shown in Fig. 9.19. The element stiffness matrices in global coordinates can be illustrated as:

$$[k^{(a)}] = \begin{bmatrix} a_{11} & a_{12} & a_{13} & a_{14} \\ a_{21} & a_{22} & a_{23} & a_{24} \\ a_{31} & a_{32} & a_{33} & a_{34} \\ a_{41} & a_{42} & a_{43} & a_{44} \end{bmatrix}$$

$$[k^{(b)}] = \begin{bmatrix} b_{11} & b_{12} & b_{13} & b_{14} \\ b_{21} & b_{22} & b_{23} & b_{24} \\ b_{31} & b_{32} & b_{33} & b_{34} \\ b_{41} & b_{42} & b_{43} & b_{44} \end{bmatrix}$$

$$[k^{(c)}] = \begin{bmatrix} c_{11} & c_{12} & c_{13} & c_{14} \\ c_{21} & c_{22} & c_{23} & c_{24} \\ c_{31} & c_{32} & c_{33} & c_{34} \\ c_{41} & c_{42} & c_{43} & c_{44} \end{bmatrix}$$

The procedure is as follows:

• Label a diagram of the frame with dof. numbers in node number sequence.

• Construct a dof. correspondence table, entering a set of dof. numbers for each node of every element. For the rod element there will be two dof. in each set, namely, u and v displacements, and two sets per element, one for each node. The sequence of the sets must correspond to progression along the local axis direction, i.e. along each positive x' direction. This is essential to maintain consistency with the element matrices, above, the terms of which have been shown in eqn. (9.23) to involve angle α, the value of which will correspond to the inclination of the element at the end chosen as the origin of its local axis. The sequence shown in Table 9.2 corresponds to $\alpha_a = 330°$, $\alpha_b = 180°$ and $\alpha_c = 210°$. The u and v dof. sequence within each set must be maintained.

• Choose an element for which the stiffness contributions are to be assembled.

• Assemble by either rows or columns according to the dof. correspondence table.

• Repeat for the remaining elements until all are assembled.

Table 9.2. Dof. correspondence table for assembly of structural stiffness matrix, $[K]$.

Row and/or column in element stiffness matrix, $[k^{(e)}]$	Row and/or column in assembled stiffness matrix, [K]		
	element a	element b	element c
1	1	3	1
2	2	4	2
3	3	5	5
4	4	6	6

For example, choosing to assemble element b contributions by rows, then the first and the "element b" columns of the dof. table, Table 9.2, are used. Start by inserting in row 3, columns 3, 4, 5, 6 of structural stiffness matrix $[K]$, the stiffness contributions respectively from row 1, columns 1, 2, 3, 4 of element stiffness matrix, $[k^{(b)}]$. Repeat for the remaining rows 4, 5, 6, inserting in columns 3, 4, 5, 6 of $[K]$, the respective contributions from rows

2, 3, 4, columns 1, 2, 3, 4 of $[k^{(b)}]$. Repeat for remaining elements a and c, to give finally:

$$[K] = \begin{bmatrix} a_{11} + c_{11} & a_{12} + c_{12} & a_{13} & a_{14} & c_{13} & c_{14} \\ a_{21} + c_{21} & a_{22} + c_{22} & a_{23} & a_{24} & c_{23} & c_{24} \\ a_{31} & a_{32} & a_{33} + b_{11} & a_{34} + b_{12} & b_{13} & b_{14} \\ a_{41} & a_{42} & a_{43} + b_{21} & a_{44} + b_{22} & b_{23} & b_{24} \\ c_{31} & c_{32} & b_{31} & b_{32} & b_{33} + c_{33} & b_{34} + c_{34} \\ c_{41} & c_{42} & b_{41} & b_{42} & b_{43} + c_{43} & b_{44} + c_{44} \end{bmatrix}$$

The above assembly procedure is generally applicable to any element, albeit with detail changes. In the case of the simple beam element, with its rotational, as well as translational dof., reference to § 9.8 shows that the element stiffness matrix is of order 6×6, and hence there will be two additional rows in the dof. correspondence table. A similar argument holds for the triangular membrane element, with its three nodes each having 2 dof. The Examples at the end of this chapter illustrate the assembly for rod, beam and membrane elements.

9.11. Application of boundary conditions and partitioning

With reference to §9.4.7, before the governing eqn. (9.28) can be solved to yield the unknown displacements, appropriate restraints need to be imposed. At some nodes the displacements will be prescribed, for example, at a fixed node the nodal displacements will be zero. Hence, some of the nodal displacements will be unknown, $\{p_\alpha\}$, and some will be prescribed, $\{p_\beta\}$. Following any necessary rearrangement to collect together equations relating to unknown, and those relating to prescribed, displacements, eqn. (9.28) can be partitioned into

$$\begin{bmatrix} \{P_\alpha\} \\ \overline{\{P_\beta\}} \end{bmatrix} = \begin{bmatrix} [K_{\alpha\alpha}] & [K_{\alpha\beta}] \\ \overline{[K_{\beta\alpha}]} & \overline{[K_{\beta\beta}]} \end{bmatrix} \begin{bmatrix} \{p_\alpha\} \\ \overline{\{p_\beta\}} \end{bmatrix} \qquad (9.96)$$

It will be found that where the loads are known, $\{P_\alpha\}$, [i.e. prescribed nodal forces (and moments, in beam applications)], the corresponding displacements will be unknown, $\{p_\alpha\}$, and where the displacements are known, $\{p_\beta\}$, (i.e. prescribed nodal displacements), the forces, $\{P_\beta\}$, (and moments, in beam applications), usually the reactions, will be unknown.

9.12. Solution for displacements and reactions

A solution for the unknown nodal displacements, $\{p_\alpha\}$, is obtained from the upper partition of eqn. (9.96)

$$\{P_\alpha\} = [K_{\alpha\alpha}]\{p_\alpha\} + [K_{\alpha\beta}]\{p_\beta\}$$

Rearranging

$$[K_{\alpha\alpha}]\{p_\alpha\} = \{P_\alpha\} - [K_{\alpha\beta}]\{p_\beta\}$$

To obtain a solution for the unknown nodal displacements, $\{p_\alpha\}$, it is only necessary to invert the submatrix $[K_{\alpha\alpha}]$. Pre-multiplying the above equation by $[K_{\alpha\alpha}]^{-1}$ (and using the matrix relation, $[K_{\alpha\alpha}]^{-1}[K_{\alpha\alpha}] = [I]$, the unit matrix), will yield the values of the unknown nodal displacements as

$$\{p_\alpha\} = [K_{\alpha\alpha}]^{-1}\{P_\alpha\} - [K_{\alpha\alpha}]^{-1}[K_{\alpha\beta}]\{p_\beta\} \qquad (9.97)$$

If all the prescribed displacements are zero, i.e. $\{p_\beta\} = \{0\}$, the above reduces to

$$\{p_\alpha\} = [K_{\alpha\alpha}]^{-1}\{P_\alpha\} \tag{9.98}$$

The unknown reactions, $\{P_\beta\}$, can be found from the lower partition of eqn. (9.96)

$$\{P_\beta\} = [K_{\beta\alpha}]\{p_\alpha\} + [K_{\beta\beta}]\{p_\beta\} \tag{9.99}$$

Again, if all the prescribed displacements are zero, the above reduces to

$$\{P_\beta\} = [K_{\beta\alpha}]\{p_\alpha\} \tag{9.100}$$

Bibliography

1. R.T. Fenner, *Finite Element Methods for Engineers*, Macmillan Press, London, 1996.
2. E. Hinton and D.R.J. Owen, *An Introduction to Finite Element Computations*, Pineridge Press, Swansea, 1979.
3. R.K. Liversley, *Finite Elements; An Introduction for Engineers*, Cambridge University Press, 1983.
4. NAFEMS, *Guidelines to Finite Element Practice*, DTI, NEL, Glasgow, 1992.
5. NAFEMS, *A Finite Element Primer*, DTI, NEL, Glasgow, 1992.
6. S.S. Rao, *The Finite Element Method in Engineering*, Pergamon Press, Oxford, 1989.
7. J.N. Reddy, *An Introduction to the Finite Element Method*, McGraw-Hill, 1993.
8. K.C. Rockey, *et al., The Finite Element Method; A Basic Introduction*, Crosby Lockwood Staples, London, 1975.
9. R.L. Sack, *Matrix Structural Analysis*, PWS-Kent, 1989.
10. O.C. Zienkiewicz, *The Finite Element Method*, McGraw-Hill, London, 1988.
11. C.A. Brebbia and A.J. Ferrante, *Computational Methods for the Solution of Engineering Problems*, Pentech Press, London, 1986.
12. HKS ABAQUS, User's Manual, 1995.
13. SDRC I-DEAS Finite Element Modelling, User's Guide, 1991.
14. K. Thomas, Effects of geometric distortion on the accuracy of plane quadratic isoparametric finite elements, Guidelines for finite element idealisation, Meeting Preprint 2504, ASCE, pp. 161–204, 1975.

Examples

Example 9.1

Figure 9.31 shows a planar steel support structure, all three members of which have the same axial stiffness, such that $AE/L = 20$ MN/m throughout. Using the displacement based finite element method and treating each member as a rod:

(a) assemble the necessary terms in the structural stiffness matrix;

(b) hence, determine, with respect to the global coordinates (i) the nodal displacements, and (ii) the reactions, showing the latter on a sketch of the structure and demonstrating that equilibrium is satisfied.

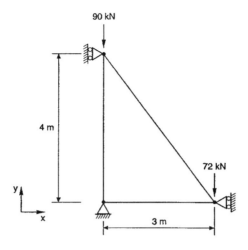

Fig. 9.31.

Solution

(a) Figure 9.32 shows suitable node, dof. and element labelling. Lack of symmetry prevents any advantage being taken to reduce the calculations. None of the members are redundant and hence the stiffness contributions of all three members need to be included.

All three elements will have the same stiffness matrix scalar,

i.e. $$(AE/L)^{(a)} = (AE/L)^{(b)} = (AE/L)^{(c)} = AE/L$$

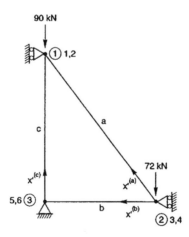

Fig. 9.32.

With reference to §9.7, the element stiffness matrix with respect to global coordinates is given by

$$[k^{(e)}] = \left(\frac{AE}{L}\right)^{(e)} \begin{bmatrix} \cos^2 \alpha \\ \sin \alpha \cos \alpha & \sin^2 \alpha & \text{symmetric} \\ -\cos^2 \alpha & -\sin \alpha \cos \alpha & \cos^2 \alpha \\ -\sin \alpha \cos \alpha & -\sin^2 \alpha & \sin \alpha \cos \alpha & \sin^2 \alpha \end{bmatrix}^{(e)}$$

Evaluating the stiffness matrix for each element:

Element a

$$\alpha^{(a)} = -\tan^{-1}(4/3), \quad \cos\alpha^{(a)} = -0.6, \quad \sin\alpha^{(a)} = 0.8$$

$$[k^{(a)}] = \frac{AE}{L}\begin{bmatrix} 0.36 & -0.48 & -0.36 & 0.48 \\ -0.48 & 0.64 & 0.48 & -0.64 \\ -0.36 & 0.48 & 0.36 & -0.48 \\ 0.48 & -0.64 & -0.48 & 0.64 \end{bmatrix}$$

Element b

$$\alpha^{(b)} = 180°, \quad \cos\alpha^{(b)} = -1, \quad \sin\alpha^{(b)} = 0$$

$$[k^{(b)}] = \frac{AE}{L}\begin{bmatrix} 1 & 0 & -1 & 0 \\ 0 & 0 & 0 & 0 \\ -1 & 0 & 1 & 0 \\ 0 & 0 & 0 & 0 \end{bmatrix}$$

Element c

$$\alpha^{(c)} = 90, \quad \cos\alpha^{(c)} = 0, \quad \sin\alpha^{(c)} = 1$$

$$[k^{(c)}] = \frac{AE}{L}\begin{bmatrix} 0 & 0 & 0 & 0 \\ 0 & 1 & 0 & -1 \\ 0 & 0 & 0 & 0 \\ 0 & -1 & 0 & 1 \end{bmatrix}$$

The structural stiffness matrix can now be assembled using a dof. correspondence table, (ref. §9.10). Observation of the highest dof. number, i.e. 6, gives the order (size), of the structural stiffness matrix, i.e. 6 × 6. The structural governing equations and hence the required structural stiffness matrix are therefore given as

Row/column in [k(e)]	Row/column in [K]		
	a	b	c
1	3	3	5
2	4	4	6
3	1	5	1
4	2	6	2

$$\begin{bmatrix} X_1 \\ Y_1 \\ X_2 \\ Y_2 \\ X_3 \\ Y_3 \end{bmatrix} = \frac{AE}{L}\begin{bmatrix} 0.36 & -0.48 & -0.36 & 0.48 & 0 & 0 \\ -0.48 & 0.64 & 0.48 & -0.64 & 0 & -1 \\ -0.36 & 0.48 & 0.36 & -0.48 & & \\ 0.48 & -0.64 & -0.48 & 0.64 & & \\ 0 & 0 & & & 1 & 0 \\ 0 & -1 & & & 0 & 1 \end{bmatrix}\begin{bmatrix} u_1 \\ v_1 \\ u_2 \\ v_2 \\ u_3 \\ v_3 \end{bmatrix}$$

(b) (i) Rearranging and partitioning, with $u_1 = u_2 = u_3 = v_3 = 0$, (i.e. $\{p_\beta\} = 0$)

$$\begin{bmatrix} Y_1 \\ Y_2 \end{bmatrix} = \begin{bmatrix} -90 \times 10^3 \\ -72 \times 10^3 \end{bmatrix} = \frac{AE}{L} \begin{bmatrix} 1.64 & -0.64 \\ -0.64 & 0.64 \end{bmatrix} \begin{bmatrix} v_1 \\ v_2 \end{bmatrix} \quad i.e.\{P_\alpha\} = [K_{\alpha\alpha}]\{p_\alpha\}$$

Inverting $[K_{\alpha\alpha}]$ to enable a solution for the displacements using $\{p_\alpha\} = [K_{\alpha\alpha}]^{-1}\{P_\alpha\}$

$$\text{adj } [K_{\alpha\alpha}] = \frac{AE}{L} \begin{bmatrix} 0.64 & 0.64 \\ 0.64 & 1.64 \end{bmatrix} \quad \text{and} \quad \det[K_{\alpha\alpha}] = 0.64(AE/L)^2$$

$$\text{Then } [K_{\alpha\alpha}]^{-1} = \frac{L}{AE} \begin{bmatrix} 1 & 1 \\ 1 & 2.5625 \end{bmatrix} \quad \text{Check:} \quad \frac{L}{AE} \begin{bmatrix} 1 & 1 \\ 1 & 2.5625 \end{bmatrix} \frac{AE}{L} \begin{bmatrix} 1.64 & -0.64 \\ -0.64 & 0.64 \end{bmatrix} = [I]$$

The required displacements are found from

$$\{p_\alpha\} = [K_{\alpha\alpha}]^{-1}\{P_\alpha\}$$

Substituting, $\begin{bmatrix} v_1 \\ v_2 \end{bmatrix} = \frac{L}{AE} \begin{bmatrix} 1 & 1 \\ 1 & 2.5625 \end{bmatrix} \begin{bmatrix} Y_1 \\ Y_2 \end{bmatrix} = -5 \times 10^{-8} \begin{bmatrix} 1 & 1 \\ 1 & 2.5625 \end{bmatrix} \begin{bmatrix} 90.10^3 \\ 72.10^3 \end{bmatrix}$

$$= \begin{bmatrix} -8.10 \\ -13.73 \end{bmatrix}_{mm}$$

The required nodal displacements are therefore $v_1 = -8.10$ mm and $v_2 = -13.73$ mm.
(ii) With reference to §9.12, nodal reactions are obtained from

$$\{P_\beta\} = [K_{\beta\alpha}]\{p_\alpha\}$$

Substituting gives

$$\begin{bmatrix} X_1 \\ X_2 \\ X_3 \\ Y_3 \end{bmatrix} = \frac{AE}{L} \begin{bmatrix} -0.48 & 0.48 \\ 0.48 & -0.48 \\ 0 & 0 \\ -1 & 0 \end{bmatrix} \begin{bmatrix} v_1 \\ v_2 \end{bmatrix} = 2 \times 10^7 \begin{bmatrix} -0.48 & 0.48 \\ 0.48 & -0.48 \\ 0 & 0 \\ -1 & 0 \end{bmatrix} \begin{bmatrix} -8.10 \times 10^{-3} \\ -13.725 \times 10^{-3} \end{bmatrix}$$

$$= \begin{bmatrix} -54 \\ 54 \\ 0 \\ 162 \end{bmatrix}_{kN}$$

The required nodal reactions are therefore $X_1 = -54$ kN, $X_2 = 54$ kN, $X_3 = 0$ and $Y_3 = 162$ kN.

Representing these reactions together with the applied forces on a sketch of the structure, Fig. 9.33, and considering force and moment equilibrium, gives

$$\sum F_x = (54 - 54) \text{ kN} \qquad = 0$$

$$\sum F_y = (162 - 90 - 72) \text{ kN} \qquad = 0$$

$$\sum M_3 = (54 \times 4 - 73 \times 3) \text{ kNm} = 0$$

Hence, equilibrium is satisfied by the system of forces.

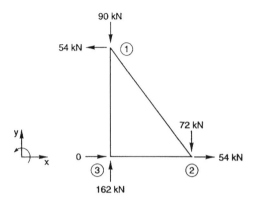

Fig. 9.33.

Example 9.2

Figure 9.34 shows the members and idealised support conditions for a roof truss. All three members of which are steel and have the same cross-sectional area such that $AE = 12$ MN throughout. Using the displacement based finite element method, treating the truss as a pin-jointed plane frame and each member as a rod:

(a) assemble the necessary terms in the structural stiffness matrix;

(b) hence, determine the nodal displacements with respect to the global coordinates, for the condition shown in Fig. 9.34.

(c) If, under load, the left support sinks by 5 mm, determine the resulting new nodal displacements, with respect to the global coordinates.

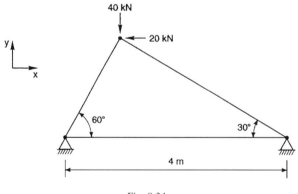

Fig. 9.34.

Solution

(a) Figure 9.35 shows suitable node, dof. and element labelling. Lack of symmetry prevents any advantage being taken to reduce the calculations. However, since both ends of the horizontal member are fixed it is redundant therefore and does not need to be considered further.

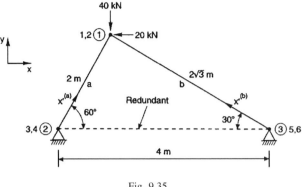

Fig. 9.35.

All three members have the same AE, hence

$$(AE)^{(a)} = (AE)^{(b)} = AE$$

With reference to §9.7, the element stiffness matrix with respect to global coordinates is given by

$$[k^{(e)}] = \left(\frac{AE}{L}\right)^{(e)} \begin{bmatrix} \cos^2 \alpha \\ \sin \alpha \cos \alpha & \sin^2 \alpha & \text{symmetric} \\ -\cos^2 \alpha & -\sin \alpha \cos \alpha & \cos^2 \alpha \\ -\sin \alpha \cos \alpha & -\sin^2 \alpha & \sin \alpha \cos \alpha & \sin^2 \alpha \end{bmatrix}^{(e)}$$

Evaluating the stiffness matrix for both elements:

Element a

$$L^{(a)} = 2\text{m}, \quad \alpha^{(a)} = 60°, \quad \cos \alpha^{(a)} = 1/2, \quad \sin \alpha^{(a)} = \sqrt{(3)}/2$$

$$[k^{(a)}] = \frac{AE}{8} \begin{bmatrix} 1 & \sqrt{3} & -1 & -\sqrt{3} \\ \sqrt{3} & 3 & -\sqrt{3} & -3 \\ -1 & -\sqrt{3} & 1 & \sqrt{3} \\ -\sqrt{3} & -3 & \sqrt{3} & 3 \end{bmatrix}$$

Element b

$$L^{(b)} = 2\sqrt{3}\text{m}, \quad \alpha^{(b)} = 150°, \quad \cos \alpha^{(b)} = -\sqrt{(3)}/2, \quad \sin \alpha^{(a)} = 1/2$$

$$[k^{(b)}] = \frac{AE}{8} \begin{bmatrix} \sqrt{3} & -1 & -\sqrt{3} & 1 \\ -1 & 1/\sqrt{3} & 1 & -1/\sqrt{3} \\ -\sqrt{3} & 1 & \sqrt{3} & -1 \\ 1 & -1/\sqrt{3} & -1 & 1/\sqrt{3} \end{bmatrix}$$

The structural stiffness matrix can now be multi-assembled using a dof. correspondence table, (ref. §9.10), and will be of order 6 × 6. Only the upper sub-matrices need to be completed, i.e. $[K_{\alpha\alpha}]$ and $[K_{\alpha\beta}]$, since the reactions are not required in this example. The necessary structural governing equations and hence the required structural stiffness matrix are therefore given as

Row/ column in $[k^{(e)}]$	Row/column in $[K]$	
	a	b
1	3	5
2	4	6
3	1	1
4	2	2

$$
\begin{bmatrix} X_1 \\ Y_1 \\ X_2 \\ Y_2 \\ X_3 \\ Y_3 \end{bmatrix}
= \frac{AE}{8}
\begin{bmatrix}
1 & \sqrt{3} & -1 & -\sqrt{3} & & u_1 \\
\sqrt{3} & & -1 & & -\sqrt{3} & 1 \\
\sqrt{3} & 3 & -\sqrt{3} & -3 & & \\
-1 & 1/\sqrt{3} & & & 1 & -1/\sqrt{3} \\
& & & & & \\
& & & & & \\
& & & & & \\
& & & & & \\
\end{bmatrix}
\begin{bmatrix} u_1 \\ v_1 \\ u_2 \\ v_2 \\ u_3 \\ v_3 \end{bmatrix}
$$

(b) Corresponding to $u_2 = v_2 = u_3 = v_3 = 0$, the partitioned equations reduce to:

$$
\begin{bmatrix} X_1 \\ Y_1 \end{bmatrix} = \begin{bmatrix} -20 \times 10^3 \\ -40 \times 10^3 \end{bmatrix} = \frac{AE}{8} \begin{bmatrix} 2.7321 & 0.7321 \\ 0.7321 & 3.5774 \end{bmatrix} \begin{bmatrix} u_1 \\ v_1 \end{bmatrix} \quad \text{i.e. } \{P_\alpha\} = [k_{\alpha\alpha}]\{p_\alpha\}
$$

Inverting $[K_{\alpha\alpha}]$ to enable a solution for the displacements from $\{p_\alpha\} = [K_{\alpha\alpha}]^{-1}\{P_\alpha\}$

$$
\text{adj } [K_{\alpha\alpha}] = \left(\frac{AE}{8}\right)^2 \begin{bmatrix} 3.5774 & -0.7321 \\ -0.7321 & 2.7321 \end{bmatrix} \text{ and det } [K_{\alpha\alpha}] = \left(\frac{AE}{8}\right)^2 9.2378
$$

Then $[K_{\alpha\alpha}]^{-1}$

$$
= \frac{8}{AE} \begin{bmatrix} 0.3873 & -0.07925 \\ -0.07925 & 0.2958 \end{bmatrix} \text{Check:} \frac{8}{AE} \begin{bmatrix} 0.3873 & -0.07925 \\ -0.07925 & 0.2958 \end{bmatrix} \frac{8}{AE} \begin{bmatrix} 2.7321 & 0.7321 \\ 0.7321 & 3.5774 \end{bmatrix} = [I]
$$

Hence, the required displacements are given by

$$
\{p_\alpha\} = [K_{\alpha\alpha}]^{-1}\{P_\alpha\}
$$

Substituting, $\begin{bmatrix} u_1 \\ v_1 \end{bmatrix} = \frac{8}{12 \times 10^6} \begin{bmatrix} 0.3873 & -0.07925 \\ -0.07925 & 0.2958 \end{bmatrix} \begin{bmatrix} -20.10^3 \\ -40.10^3 \end{bmatrix} = \begin{bmatrix} -3.05 \\ -6.83 \end{bmatrix}_{\text{mm}}$

The required nodal displacements are therefore $u_1 = -3.05$ mm and $v_1 = -6.83$ mm.

(c) With reference to §9.12, for non-zero prescribed displacements, i.e. $\{p_\beta\} \neq \{0\}$, the full partition of the governing equation is required, namely,

$$\{P_\alpha\} = [K_{\alpha\alpha}]\{p_\alpha\} + [K_{\alpha\beta}]\{p_\beta\}$$

Rearranging for the unknown displacements

$$\{p_\alpha\} = [K_{\alpha\alpha}]^{-1}\{P_\alpha\} - [K_{\alpha\alpha}]^{-1}[K_{\alpha\beta}]\{p_\beta\}$$

Evaluating

$$[K_{\alpha\alpha}]^{-1}[K_{\alpha\beta}] = \begin{bmatrix} 0.3873 & -0.07925 \\ -0.07925 & 0.2958 \end{bmatrix} \begin{bmatrix} -1 & -1.7321 & -1.7321 & 1 \\ -1.7321 & -3 & 1 & -0.5773 \end{bmatrix}$$

$$= \begin{bmatrix} -0.25 & -0.4331 & -0.75 & 0.4331 \\ -0.4331 & -0.75 & 0.4331 & -0.25 \end{bmatrix}$$

and $[K_{\alpha\alpha}]^{-1}[K_{\alpha\beta}]\{p_\beta\} = \begin{bmatrix} -0.25 & -0.4331 & -0.75 & 0.4331 \\ -0.4331 & -0.75 & 0.4331 & -0.25 \end{bmatrix} \begin{bmatrix} 0 \\ -5 \times 10^{-3} \\ 0 \\ 0 \end{bmatrix} = \begin{bmatrix} 2.1655 \\ 3.75 \end{bmatrix}_{mm}$

Recalling from part (b) that

$$[K_{\alpha\alpha}]^{-1}\{P_\alpha\} = \begin{bmatrix} -3.05 \\ -6.83 \end{bmatrix}_{mm}$$

and substituting into the above rearranged governing equation,

$$\text{i.e.} \quad \begin{bmatrix} u_1 \\ v_1 \end{bmatrix} = \begin{bmatrix} -3.05 \\ -6.83 \end{bmatrix} - \begin{bmatrix} 2.1655 \\ 3.75 \end{bmatrix} = \begin{bmatrix} -5.22 \\ -10.58 \end{bmatrix}_{mm}$$

yields the required new nodal displacements, namely, $u_1 = -5.22$ mm and $v_1 = -10.58$ mm.

Example 9.3

A steel beam is supported and loaded as shown in Fig. 9.36. The relevant second moments of area are such that $I^{(a)} = 2I^{(b)} = 2 \times 10^{-5}$ m⁴ and Young's modulus E for the beam material $= 200$ GN/m². Using the displacement based finite element method and representing each member by a simple beam element:

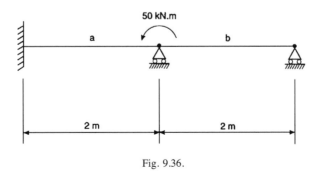

Fig. 9.36.

(a) determine the nodal displacements;
(b) hence, determine the nodal reactions, representing these on a sketch of the deformed geometry. Show that both force and moment equilibrium is satisfied.

Solution

(a) Figure 9.37 shows suitable node, dof. and element labelling. Lack of symmetry prevents any advantage being taken to reduce the calculations. There are no redundant members.

Employing two beam finite elements, (which is the least number in this case), both elements will have the same E/L, i.e.

$$(E/L)^{(a)} = (E/L)^{(b)} = E/L$$

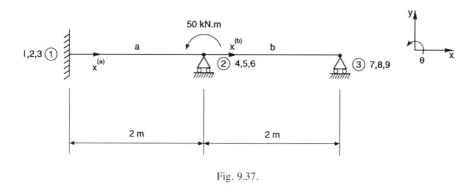

Fig. 9.37.

However, the second moments of area will be different, such that

$$I^{(a)} = 2I \text{ and } I^{(b)} = I$$

and will be the only difference between the two element stiffness matrices.

With reference to §9.8 and in the absence of axial forces, each element stiffness matrix with respect to local coordinates is given as

$$[k^{(e)}] = \left(\frac{EI}{L}\right)^{(e)} \begin{bmatrix} 12/L^2 & & & \\ 6/L & 4 & \text{symmetric} & \\ -12/L^2 & -6/L & 12/L^2 & \\ 6/L & 2 & -6/L & 4 \end{bmatrix}$$

The above local coordinate element stiffness matrix will, in this case, be identical to that with respect to global coordinates since the local and global axes coincide.

Substituting for both elements:

Element a

Recalling $I^{(a)} = 2I$

$$[k^{(a)}] = \left(\frac{EI}{L}\right) \begin{bmatrix} 24/L^2 & & & \\ 12/L & 8 & \text{symmetric} & \\ -24/L^2 & -12/L & 24/L^2 & \\ 12/L & 4 & -12/L & 8 \end{bmatrix}$$

Element b

Recalling $I^{(b)} = I$

$$[k^{(b)}] = \frac{EI}{L} \begin{bmatrix} 12/L^2 \\ 6/L & 4 & \text{symmetric} \\ -12/L^2 & -6/L & 12/L^2 \\ 6/L & 2 & -6/L & 4 \end{bmatrix}$$

The structural stiffness matrix can now be assembled. A dof. correspondence table can be used as an aid to assembly. However, observation of the relatively simple element connectivity, shows that the stiffness contributions for element a will occupy the upper left 4×4 locations, whilst those for element b will occupy the lower right 4×4 locations of the 6×6 structural stiffness matrix. The reduced structural stiffness matrix is due to the omission of axial terms, otherwise the matrix would have been of order 9×9. Hence, completing only those columns needed for the solution, gives

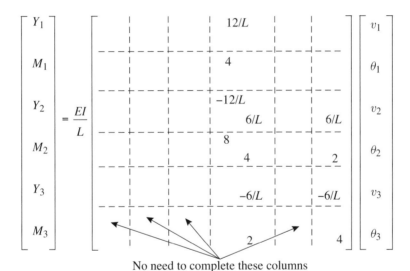

No need to complete these columns

Corresponding to $v_1 = \theta_1 = v_2 = v_3 = 0$ (by omitting axial terms it has already been taken that $u_1 = u_2 = u_3 = 0$), the partitioned equations reduce to

$$\begin{bmatrix} M_2 \\ M_3 \end{bmatrix} = \frac{EI}{L} \begin{bmatrix} 12 & 2 \\ 2 & 4 \end{bmatrix} \begin{bmatrix} \theta_2 \\ \theta_3 \end{bmatrix} \quad \text{i.e. } \{P_\alpha\} = [K_{\alpha\alpha}]\{p_\alpha\}$$

Inverting $[K_{\alpha\alpha}]$ to enable a solution for the displacements from $\{p_\alpha\} = [K_{\alpha\alpha}]^{-1}\{P_\alpha\}$

where adj $[K_{\alpha\alpha}] = \frac{EI}{L} \begin{bmatrix} 4 & -2 \\ -2 & 12 \end{bmatrix}$ and det $[K_{\alpha\alpha}] = 44(EI/L)^2$

Then, $[K_{\alpha\alpha}]^{-1} = \frac{L}{44EI} \begin{bmatrix} 4 & -2 \\ -2 & 12 \end{bmatrix}$ Check $\frac{L}{44EI} \begin{bmatrix} 4 & -2 \\ -2 & 12 \end{bmatrix} \frac{EI}{L} \begin{bmatrix} 12 & 2 \\ 2 & 4 \end{bmatrix} = [I]$

The required displacements are found from

$$\{p_\alpha\} = [K_{\alpha\alpha}]^{-1}\{P_\alpha\}$$

Substituting $\begin{bmatrix} \theta_2 \\ \theta_3 \end{bmatrix} = \dfrac{L}{44EI} \begin{bmatrix} 4 & -2 \\ -2 & 12 \end{bmatrix} \begin{bmatrix} M_2 \\ M_3 \end{bmatrix}$

$$= \frac{2}{44 \times 200 \times 10^9 \times 1 \times 10^{-5}} \begin{bmatrix} 4 & -2 \\ -2 & 12 \end{bmatrix} \begin{bmatrix} 5 \times 10^4 \\ 0 \end{bmatrix}$$

$$= 2.2727 \times 10^{-4} \begin{bmatrix} 20 \\ -10 \end{bmatrix} = \begin{bmatrix} 4.545.10^{-3} \\ -2.273.10^{-3} \end{bmatrix}_{\text{rad}} = \begin{bmatrix} 0.260 \\ -0.130 \end{bmatrix}_{\text{deg}}$$

The required nodal displacements are therefore $\theta_2 = 0.26°$ and $\theta_3 = -0.13°$.

(b) With reference to §9.12, nodal reactions are obtained from

$$\{P_\alpha\} = [K_{\alpha\beta}]\{p_\alpha\}$$

Substituting gives $\begin{bmatrix} Y_1 \\ M_1 \\ Y_2 \\ Y_3 \end{bmatrix} = \dfrac{EI}{L} \begin{bmatrix} 12/L & 0 \\ 4 & 0 \\ -6/L & 6/L \\ -6/L & -6/L \end{bmatrix} \begin{bmatrix} \theta_2 \\ \theta_3 \end{bmatrix}$

$$= \frac{200 \times 10^9 \times 1 \times 10^{-5}}{2} \begin{bmatrix} 6 & 0 \\ 4 & 0 \\ -3 & 3 \\ -3 & -3 \end{bmatrix} \begin{bmatrix} 4.545 \times 10^{-3} \\ -2.273 \times 10^{-3} \end{bmatrix}$$

$$= \begin{bmatrix} 27.27 \text{ kN} \\ 18.18 \text{ kNm} \\ -20.45 \text{ kN} \\ -6.82 \text{ kN} \end{bmatrix}$$

The required nodal reactions are therefore $Y_1 = 27.27$ kN, $M_1 = 18.18$ kNm, $Y_2 = -20.45$ kN and $Y_3 = -6.82$ kN.

Representing these reactions together with the applied moment on a sketch of the deformed beam, Fig. 9.38, and considering force and moment equilibrium, gives

$$\Sigma F_y = (27.27 - 20.45 - 6.82) \text{ kN} = 0$$

$$\Sigma M_1 = (18.18 + 50 - 20.45 \times 2 - 6.82 \times 4) \text{ kNm} = 0$$

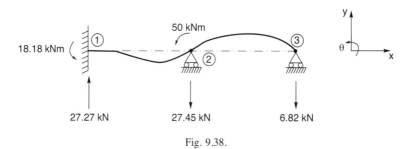

Fig. 9.38.

Example 9.4

The vehicle engine mounting bracket shown in Fig. 9.39 is made from uniform steel channel section for which Young's modulus, $E = 200$ GN/m^2. It can be assumed for both

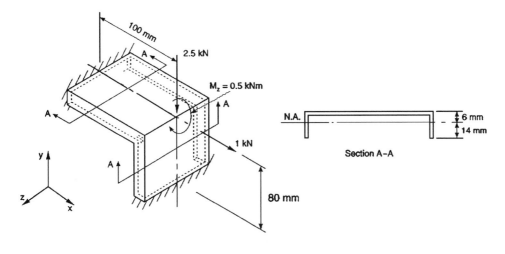

Fig. 9.39.

channels that the relevant second moment of area, $I = 2 \times 10^{-8}$ m^4 and cross-sectional area, $A = 4 \times 10^{-4}$ m^2. The bracket can be idealised as two beams, the common junction of which can be assumed to be infinitely stiff and the other ends to be fully restrained. Using the displacement based finite element method, and representing the constituent members as simple beam elements:

(a) assemble the necessary terms in the structural stiffness matrix;

(b) hence, determine for the condition shown in Fig. 9.39 (i) the nodal displacements with respect to the global coordinates, and (ii) the combined axial and bending extreme fibre stresses at the built-in ends and at the common junction.

Solution

(a) Figure 9.40 shows suitable node, dof. and element labelling. The structure does not have symmetry or redundant members. The least number of beam elements will be used to

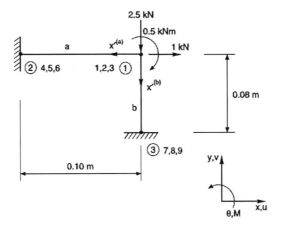

Fig. 9.40.

minimise the hand calculations which, in this example, is two.
Both elements will have the same A, E and I,

i.e. $$(A, E, I)^{(a)} = (A, E, I)^{(b)} = A, E, I,$$

but will have different lengths, i.e. $L^{(a)}$ and $L^{(b)}$.

With reference to §9.8, the element stiffness matrix inclusive of axial terms and in global coordinates is appropriate, namely:

$$[k^{(e)}] = \left(\frac{E}{L}\right)^{(e)} \begin{bmatrix} A\cos^2\alpha + (12I\sin^2\alpha)/L^2, \\ (A-12I/L^2)\cos\alpha\sin\alpha, & A\sin^2\alpha + (12I\cos^2\alpha)/L^2, \\ -(6I\sin\alpha)L, & (6I\cos\alpha)/L, & 4I, \\ -A\cos^2\alpha - (12I\sin^2\alpha)/L^2, & -(A-12I/L^2)\cos\alpha\sin\alpha, & (6I\sin\alpha)/L, \\ -(A-12I/L^2)\cos\alpha\sin\alpha, & -A\sin^2\alpha - (12I\cos^2\alpha)/L^2, & -(6I\cos\alpha)/L, \\ -(6I\sin\alpha)/L, & (6I\cos\alpha)/L, & 2I \end{bmatrix}$$

$$\begin{matrix} A\cos^2\alpha + (12I\sin^2\alpha)/L^2, & \text{symmetric} \\ (A-12I/L^2)\cos\alpha\sin\alpha, & A\sin^2\alpha + (12I\cos^2\alpha)/L^2, \\ (6I\sin\alpha)/L, & -(6I\cos\alpha)/L, & 4I \end{matrix}$$

Evaluating, for both elements, only those stiffness terms essential for the analysis:

Element a
$$L^{(a)} = 0.1m, \alpha^{(a)} = 180°, \cos\alpha^{(a)} = -1, \sin\alpha^{(a)} = 0$$

$$[k^{(a)}] = \frac{E}{L^{(a)}} \begin{bmatrix} A & 0 & 0 \\ 0 & 12I/L^2 & -6I/L \\ 0 & -6I/L & 4I \end{bmatrix}^{(a)} = E\times 10^{-4} \begin{bmatrix} 40 & 0 & 0 \\ 0 & 2.4 & -0.12 \\ 0 & -0.12 & 8\times10^{-3} \end{bmatrix}$$

No need to complete these rows and columns, for these examples

Element b
$$L^{(b)} = 0.08m, \quad \alpha^{(b)} = 270°, \quad \cos\alpha^{(b)} = 0, \quad \sin\alpha^{(b)} = -1$$

$$[k^{(b)}] = \frac{E}{L^{(b)}} \begin{bmatrix} 12I/L^2 & 0 & 6I/L \\ 0 & A & 0 \\ 6I/L & 0 & 4I \end{bmatrix}^{(b)} = E\times10^{-4} \begin{bmatrix} 4.6875 & 0 & 0.1875 \\ 0 & 50 & 0 \\ 0.1875 & 0 & 10.10^{-3} \end{bmatrix}$$

The structural stiffness matrix can now be assembled. Whilst the structure has a total of 9 dof., only 3 are active, the remaining 6 dof. are suppressed corresponding to the statement in the question regarding the ends being fully restrained. The node numbering adopted in Fig. 9.40 simplifies the stiffness assembly, whereby the first 3 × 3 submatrix terms for both elements are assembled in the first 3 × 3 locations of the structural stiffness matrix; these being the only terms associated with the active dofs. It follows that rearrangement is unnecessary, prior to partitioning. The necessary structural governing equations and hence the required structural stiffness matrix are therefore given as

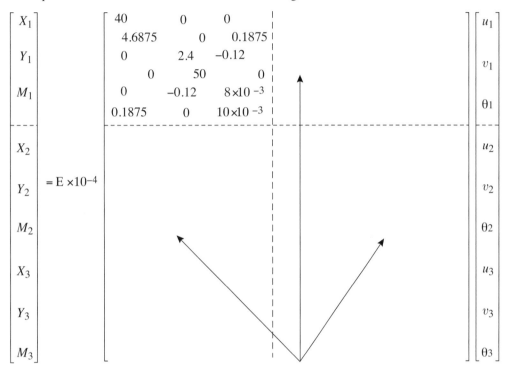

These submatrices are not required, for this example.

(b) (i) Corresponding to $u_2 = v_2 = \theta_2 = u_3 = v_3 = \theta_3 = 0$, the partitioned equations reduce to

$$\begin{bmatrix} X_1 \\ Y_1 \\ M_1 \end{bmatrix} = E \times 10^{-4} \begin{bmatrix} 44.6875 & 0 & 0.1875 \\ 0 & 52.4 & -0.12 \\ 0.1875 & -0.12 & 0.018 \end{bmatrix} \begin{bmatrix} u_1 \\ v_1 \\ \theta_1 \end{bmatrix}$$

$$= 10^7 \begin{bmatrix} 89.375 & 0 & 0.375 \\ 0 & 104.8 & -0.24 \\ 0.375 & -0.24 & 0.036 \end{bmatrix} \begin{bmatrix} u_1 \\ v_1 \\ \theta_1 \end{bmatrix}$$

i.e. $\{P_\alpha\} = [K_{\alpha\alpha}]\{p_\alpha\}$

Inverting $[K_{\alpha\alpha}]$ to enable a solution for the displacements from $\{p_\alpha\} = [K_{\alpha\alpha}]^{-1}\{P_\alpha\}$

where $\quad$ adj $[K_{\alpha\alpha}] = 10^{14} \begin{bmatrix} 3.7152 & -0.09 & -39.3 \\ -0.09 & 3.0769 & 21.45 \\ -39.3 & 21.45 & 9\,366.5 \end{bmatrix}$

and $\det [K_{\alpha\alpha}] = 10^{21}\{89.375[104.8 \times 0.036$

$$- (-0.24)(-0.24)] - 0 + 0.375(0 - 0.375 \times 104.8)\}$$

$$= 317.3085 \times 10^{21}$$

Then $[K_{\alpha\alpha}]^{-1} = 10^{-10} \begin{bmatrix} 11.7085 & -0.2836 & -123.8542 \\ -0.2836 & 9.6969 & 67.5998 \\ -123.8542 & 67.5998 & 29518.59 \end{bmatrix}$

The required displacements are found from

$$p_\alpha = [K_{\alpha\alpha}]^{-1}\{P_\alpha\}$$

Substituting $\begin{bmatrix} u_1 \\ v_1 \\ \theta_1 \end{bmatrix} = 10^{-10} \begin{bmatrix} 11.7085 & -0.2836 & -123.8542 \\ -0.2836 & 9.6969 & 67.5998 \\ -123.8542 & 67.5998 & 29\,518.59 \end{bmatrix} 10^3 \begin{bmatrix} 1 \\ -2.5 \\ -0.5 \end{bmatrix}$

$$= \begin{bmatrix} 7.434 \times 10^{-6} \text{ m} \\ -5.833 \times 10^{-6} \text{ m} \\ -1.505 \times 10^{-3} \text{ rad} \end{bmatrix}$$

The required nodal displacements are therefore $u_1 = 7.434 \times 10^{-6}$ m, $v_1 = -5.833 \times 10^{-6}$ m and $\theta_1 = -1.505 \times 10^{-3}$ rad.

(b) (ii) With reference to §9.8, the element stress matrix in global coordinates is given as

$$[H^{(e)}] = \frac{E}{L} \begin{bmatrix} -\cos\alpha - 6t\sin(\alpha)/L & -\sin\alpha + 6t\cos(\alpha)/L & 4t & \cos\alpha + 6t\sin(\alpha)/L & \sin\alpha - 6t\cos(\alpha)/L & 2t \\ -\cos\alpha + 6b\sin(\alpha)/L & -\sin\alpha - 6b\cos(\alpha)/L & -4b & \cos\alpha - 6b\sin(\alpha)/L & \sin\alpha + 6b\cos(\alpha)/L & -2b \\ -\cos\alpha + 6t\sin(\alpha)/L & -\sin\alpha - 6t\cos(\alpha)/L & -2t & \cos\alpha - 6t\sin(\alpha)/L & \sin\alpha + 6t\cos(\alpha)/L & -4t \\ -\cos\alpha - 6b\sin(\alpha)/L & -\sin\alpha + 6b\cos(\alpha)/L & 2b & \cos\alpha + 6b\sin(\alpha)/L & \sin\alpha - 6b\cos(\alpha)/L & 4b \end{bmatrix}$$

Evaluating, for both elements, only those terms essential for the analysis:

Element a

$t^{(a)} = 14 \times 10^{-3}$ m, $b^{(a)} = 6 \times 10^{-3}$ m, and recalling from part (a) $L^{(a)} = 0.1$ m, $\alpha^{(a)} = 180°$, $\cos\alpha^{(a)} = -1$, $\sin\alpha^{(a)} = 0$

$$[H^{(a)}] = \frac{200 \times 10^9}{0.1} \begin{bmatrix} 1 & -0.84 & 56 \times 10^{-3} & & & \\ 1 & 0.36 & -24 \times 10^{-3} & & & \\ 1 & 0.84 & -28 \times 10^{-3} & & & \\ 1 & -0.36 & 12 \times 10^{-3} & & & \end{bmatrix}$$

No need to complete these columns for this example
With reference to §9.7, the element stresses are obtained from

With reference to §9.7, the element stresses are obtained from

$$\{\sigma^{(e)}\} = [H^{(e)}]\{s^{(e)}\}$$

where, for element a, the displacement column matrix is

$$\{s^{(a)}\} = \{u_1 \ v_1 \ \theta_1 \ u_2 \ v_2 \ \theta_2\} \text{ in which } u_2 = v_2 = \theta_2 = 0 \text{ in this example.}$$

Substituting for element a and letting superscript i denote extreme inner fibres and superscript o denote extreme outer fibres, gives

$$\begin{bmatrix} \sigma_1^i \\ \sigma_1^o \\ \sigma_2^i \\ \sigma_2^o \end{bmatrix} = 2 \times 10^{12} \begin{bmatrix} 1 & -0.84 & 56 \times 10^{-3} & & & \\ 1 & 0.36 & -24 \times 10^{-3} & & & \\ 1 & 0.84 & -28 \times 10^{-3} & & & \\ 1 & -0.36 & 12 \times 10^{-3} & & & \end{bmatrix} \begin{bmatrix} 7.434 \times 10^{-6} \\ -5.833 \times 10^{-6} \\ 1.505 \times 10^{-3} \\ 0 \\ 0 \\ 0 \end{bmatrix} = \begin{bmatrix} -143.89 \times 10^6 \\ 82.91 \times 10^6 \\ 89.35 \times 10^6 \\ -17.05 \times 10^6 \end{bmatrix}$$

The required element stresses are therefore $\sigma_1^i = 143.89$ MN/m² (C), $\sigma_1^o = 82.91$ MN/m² (T), $\sigma_2^i = 89.35$ MN/m2(T) and $\sigma_2^o = 17.05$ MN/m² (C).

Element b

$t^{(b)} = 6 \times 10^{-3}$ m, $b^{(b)} = 14 \times 10^{-3}$ m, and recalling from part (a) $L^{(b)} = 0.08$ m, $\alpha^{(b)} = 270°$, $\cos \alpha^{(b)} = 0$, $\sin \alpha^{(b)} = -1$,

$$[H^{(b)}] = \frac{200 \times 10^9}{0.08} \begin{bmatrix} 0.45 & 1 & 24 \times 10^{-3} & & & \\ -1.05 & 1 & -56 \times 10^{-3} & & & \\ -0.45 & 1 & -12 \times 10^{-3} & & & \\ 1.05 & 1 & 28 \times 10^{-3} & & & \end{bmatrix}$$

Again, the element stresses are obtained from

$$\{\sigma^{(e)}\} = [H^{(e)}]\{s^{(e)}\}$$

where, for element b, the displacement column matrix is

$$\{s^{(b)}\} = \{u_1 \; v_1 \; \theta_1 \; u_3 \; v_3 \; \theta_3\} \text{ in which } u_3 = v_3 = \theta_3 = 0 \text{ in this example.}$$

Substituting for element b gives

$$\begin{bmatrix} \sigma_1^o \\ \sigma_1^i \\ \sigma_3^o \\ \sigma_3^i \end{bmatrix} = 2.5 \times 10^2 \begin{bmatrix} 0.45 & 1 & 24 \times 10^{-3} & & & \\ -1.05 & 1 & -56 \times 10^{-3} & & & \\ -0.45 & 1 & -12 \times 10^{-3} & & & \\ 1.05 & 1 & 28 \times 10^{-3} & & & \end{bmatrix} \begin{bmatrix} 7.434 \times 10^6 \\ -5.833 \times 10^6 \\ -1.505 \times 10^3 \\ 0 \\ 0 \\ 0 \end{bmatrix} = \begin{bmatrix} -96.52 \times 10^6 \\ 176.60 \times 10^6 \\ 22.20 \times 10^6 \\ -100.42 \times 10^6 \end{bmatrix}_{N/m^2}$$

The required element stresses are therefore $\sigma_1^i = 176.60$ MN/m² (T), $\sigma_1^o = 96.52$ MN/m² (C), $\sigma_3^i = 100.42$ MN/m²(C) and $\sigma_3^o = 22.20$ MN/m² (T).

Example 9.5

Derive the stiffness matrix in global coordinates for a three-node triangular membrane element for plane stress analysis. Assume that the elastic modulus, E, and thickness, t, are constant throughout, and that the displacement functions are

$$u(x, y) = \alpha_1 + \alpha_2 x + \alpha_3 y$$

$$v(x, y) = \alpha_4 + \alpha_5 x + \alpha_6 y$$

Solution

With reference to §9.9 and with respect to the node labelling shown in Fig. 9.41, matrix $[A]$ will be given as:

$$[A] = \begin{bmatrix} 1 & x_1 & y_1 & 0 & 0 & 0 \\ 1 & x_2 & y_2 & 0 & 0 & 0 \\ 1 & x_3 & y_3 & 0 & 0 & 0 \\ \hline 0 & 0 & 0 & 1 & x_1 & y_1 \\ 0 & 0 & 0 & 1 & x_2 & y_2 \\ 0 & 0 & 0 & 1 & x_3 & y_3 \end{bmatrix} = \begin{bmatrix} A_{\alpha\alpha} & A_{\alpha\beta} \\ \hline A_{\beta\alpha} & A_{\beta\alpha} \end{bmatrix}$$

Then $[A]^{-1} = \begin{bmatrix} [A_{\alpha\alpha}]^{-1} & 0 \\ 0 & [A_{\beta\beta}]^{-1} \end{bmatrix}$ where $[A_{\alpha\alpha}]^{-1} = [A_{\beta\beta}]^{-1}$, in this case

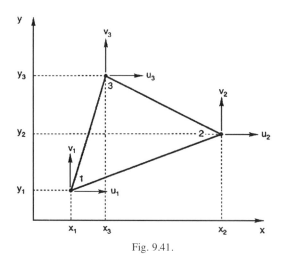

Fig. 9.41.

Obtaining the inverse of the partition

$$\text{adj } [A_{\alpha\alpha}] = [C_{\alpha\alpha}]^T = \begin{bmatrix} x_2 y_3 - x_3 y_2 & y_2 - y_3 & x_3 - x_2 \\ x_3 y_1 - x_1 y_3 & y_3 - y_1 & x_1 - x_3 \\ x_1 y_2 - x_2 y_1 & y_1 - y_2 & x_2 - x_1 \end{bmatrix}^T$$

$$= \begin{bmatrix} x_2 y_3 - x_3 y_2 & x_3 y_1 - x_1 y_3 & x_1 y_2 - x_2 y_1 \\ y_2 - y_3 & y_3 - y_1 & y_1 - y_2 \\ x_3 - x_2 & x_1 - x_3 & x_2 - x_1 \end{bmatrix}$$

and det $[A_{\alpha\alpha}] = (x_2 y_3 - x_3 y_2) - (x_1 y_3 - x_3 y_1) + (x_1 y_2 - x_2 y_1)$

$$= x_1(y_2 - y_3) + x_2(y_3 - y_1) + x_3(y_1 - y_2)$$

$$= 2 \times \text{area of element} = 2a, \text{ (see following derivation)}$$

Then $\quad [A_{\alpha\alpha}]^{-1} = \dfrac{\text{adj}[A_{\alpha\alpha}]}{\det[A_{\alpha\alpha}]} = \dfrac{1}{2a} \begin{bmatrix} x_2 y_3 - x_3 y_2 & x_3 y_1 - x_1 y_3 & x_1 y_2 - x_2 y_1 \\ y_2 - y_3 & y_3 - y_1 & y_1 - y_2 \\ x_3 - x_2 & x_1 - x_3 & x_2 - x_1 \end{bmatrix}$

Hence, $\quad [A]^{-1} = \dfrac{1}{2a} \begin{bmatrix} x_2 y_3 - x_3 y_2 & x_3 y_1 - x_1 y_3 & x_1 y_2 - x_2 y_1 & 0 & 0 & 0 \\ y_2 - y_3 & y_3 - y_1 & y_1 - y_2 & 0 & 0 & 0 \\ x_3 - x_2 & x_1 - x_3 & x_2 - x_1 & 0 & 0 & 0 \\ 0 & 0 & 0 & x_2 y_3 - x_3 y_2 & x_3 y_1 - x_1 y_3 & x_1 y_2 - x_2 y_1 \\ 0 & 0 & 0 & y_2 - y_3 & y_3 - y_1 & y_1 - y_2 \\ 0 & 0 & 0 & x_3 - x_2 & x_1 - x_3 & x_2 - x_1 \end{bmatrix}$

Area of element

With reference to Fig. 9.42, area of triangular element,

$a = $ area of enclosing rectangle $-$ (area of triangles b, c and d)

$$= (x_2 - x_1)(y_3 - y_1) - (1/2)(x_2 - x_1)(y_2 - y_1) - (1/2)(x_2 - x_3)(y_3 - y_2)$$

$$- (1/2)(x_3 - x_1)(y_3 - y_1)$$

$$= x_2 y_3 - x_2 y_1 - x_1 y_3 + x_1 y_1 - (1/2)[x_2 y_2 - x_2 y_1 - x_1 y_2 + x_1 y_1)$$

$$+ (x_2 y_3 - x_2 y_2 - x_3 y_3 + x_3 y_2) + (x_3 y_3 - x_3 y_1 - x_1 y_3 + x_1 y_1)]$$

$$= (1/2)(x_2 y_3 - x_2 y_1 - x_1 y_3 + x_1 y_2 - x_3 y_2 + x_3 y_1)$$

$$= (1/2)[x_1(y_2 - y_3) + x_2(y_3 - y_1) + x_3(y_1 - y_2)]$$

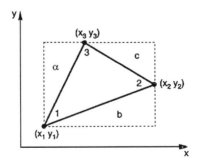

Fig. 9.42.

§9.9 gives matrix $[B]$ as

$$[B] = \begin{bmatrix} 0 & 1 & 0 & 0 & 0 & 0 \\ 0 & 0 & 0 & 0 & 0 & 1 \\ 0 & 0 & 1 & 0 & 1 & 0 \end{bmatrix} [A]^{-1}$$

Substituting for $[A]^{-1}$ from above and evaluating the product gives

$$[B] = \frac{1}{2a} \begin{bmatrix} y_2 - y_3 & y_3 - y_1 & y_1 - y_2 & 0 & 0 & 0 \\ 0 & 0 & 0 & x_3 - x_2 & x_1 - x_3 & x_2 - x_1 \\ x_3 - x_2 & x_1 - x_3 & x_2 - x_1 & y_2 - y_3 & y_3 - y_1 & y_1 - y_2 \end{bmatrix}$$

The required element stiffness matrix can now be found by substituting into the relation

$$[k] = at \, [B]^T [D][B]$$

$$= \frac{at}{2a} \begin{bmatrix} y_{23} & 0 & x_{32} \\ y_{31} & 0 & x_{13} \\ y_{12} & 0 & x_{21} \\ 0 & x_{32} & y_{23} \\ 0 & x_{13} & y_{31} \\ 0 & x_{21} & y_{12} \end{bmatrix} \frac{E}{1-v^2} \begin{bmatrix} 1 & v & 0 \\ v & 1 & 0 \\ 0 & 0 & (1-v)/2 \end{bmatrix} \frac{1}{2a} \begin{bmatrix} y_{23} & y_{31} & y_{12} & 0 & 0 & 0 \\ 0 & 0 & 0 & x_{32} & x_{13} & x_{21} \\ x_{32} & x_{13} & x_{21} & y_{23} & y_{31} & y_{12} \end{bmatrix}$$

where the abbreviation y_{23} denotes $y_2 - y_3$, etc.

Choosing to evaluate the product $[D][B]$ first, gives

$$[k] = \frac{Et}{4a(1-v^2)} \begin{bmatrix} y_{23} & 0 & x_{32} \\ y_{31} & 0 & x_{13} \\ y_{12} & 0 & x_{21} \\ 0 & x_{32} & y_{23} \\ 0 & x_{13} & y_{31} \\ 0 & x_{21} & y_{12} \end{bmatrix} \begin{bmatrix} y_{23} & y_{31} & y_{12} & vx_{32} & vx_{13} & vx_{21} \\ vy_{23} & vy_{31} & vy_{12} & x_{32} & x_{13} & x_{21} \\ \dfrac{(1-v)}{2}x_{32} & \dfrac{(1-v)}{2}x_{13} & \dfrac{(1-v)}{2}x_{21} & \dfrac{(1-v)}{2}y_{23} & \dfrac{(1-v)}{2}y_{31} & \dfrac{(1-v)}{2}y_{12} \end{bmatrix}$$

Completing the matrix multiplication, reversing the sequence of some of the coordinates so that all subscripts are in descending order, gives the required element stiffness matrix as

$$[k] = \frac{Et}{4a(1-v^2)} \begin{bmatrix} y_{32}^2 + x_{32}^2(1-v)/2, & & & \\ -y_{32}y_{31} - x_{31}x_{32}(1-v)/2, & y_{31}^2 + x_{31}^2(1-v)/2, & & \\ y_{21}y_{32} + x_{21}x_{32}(1-v)/2, & -y_{21}y_{31} - x_{21}x_{31}(1-v)/2, & y_{21}^2 + x_{21}^2(1-v)/2 & \\ -vx_{32}y_{32} - y_{32}x_{32}(1-v)/2, & vx_{32}y_{31} + y_{32}x_{31}(1-v)/2, & -vx_{32}y_{21} - y_{32}x_{21}(1-v)/2, \\ vx_{31}y_{32} + y_{31}x_{32}(1-v)/2, & -vx_{31}y_{31} - y_{31}x_{31}(1-v)/2, & vx_{31}y_{21} + y_{31}x_{21}(1-v)/2, \\ -vx_{21}y_{32} - y_{21}x_{32}(1-v)/2, & vx_{21}y_{31} + y_{21}x_{31}(1-v)/2, & -vx_{21}y_{21} - y_{21}x_{21}(1-v)/2, \end{bmatrix}$$

Symmetric

$$\begin{matrix} x_{32}^2 + y_{32}^2(1-v)/2, & & \\ -x_{31}x_{32} - y_{31}y_{32}(1-v)/2, & x_{31}^2 + y_{31}^2(1-v)/2, & \\ x_{21}x_{32} + y_{21}y_{32}(1-v)/2, & -x_{21}x_{31} - y_{21}y_{31}(1-v)/2, & x_{21}^2 + y_{21}^2(1-v)/2 \end{matrix}$$

Example 9.6

(a) Evaluate the element stiffness matrix, in global coordinates, for the three-node triangular membrane element, labelled a in Fig. 9.43. Assume plane stress conditions, Young's modulus, $E = 200$ GN/m^2, Poisson's ratio, $v = 0.3$, thickness, $t = 1$ mm, and the same displacement functions as Example 9.5.

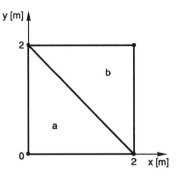

Fig. 9.43.

(b) Evaluate the element stiffness matrix for element b, assuming the same material properties and thickness as element a. Hence, evaluate the assembled stiffness matrix for the continuum.

Solution

(a) Figure 9.44 shows suitable node labelling for a single triangular membrane element. The resulting element stiffness matrix from the previous Example, 9.5, can be utilised. A specimen evaluation of an element stiffness term is given below for k_{11}. The rest are obtained by following the same procedure.

$$k_{11} = \frac{Et}{4a(1 - v^2)}[y_{32}^2 + x_{32}^2(1 - v)/2]$$

$$= \frac{Et}{4a(1 - v^2)}[(y_3 - y_2)^2 + (x_3 - x_2)^2(1 - v)/2]$$

Substituting $\quad = \dfrac{200 \times 10^9 \times 1 \times 10^{-3}}{4 \times 2(1 - 0.3^2)}[(2 - 0)^2 + (0 - 2)^2(1 - 0.3)/2]$

$$= 14.835 \times 10^7 \text{ N/m}$$

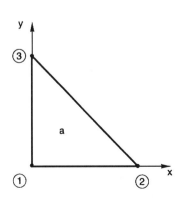

Fig. 9.44.

Evaluation of all the terms leads to the required triangular membrane element stiffness matrix for element a, namely

$$[k^{(a)}] = 10^7 [\text{N/m}] \begin{bmatrix} 14.835 & & & & & \\ -10.989 & 10.989 & & \text{symmetric} & & \\ -3.846 & 0 & 3.846 & & & \\ 7.143 & -3.297 & -3.846 & 14.835 & & \\ -3.846 & 0 & 3.846 & -3.846 & 3.846 & \\ -3.297 & 3.297 & 0 & -10.989 & 0 & 10.989 \end{bmatrix}$$

(b) Element b can temporarily also be labelled with node numbers 1, 2 and 3, as element a. To avoid confusion, this is best done with the elements shown "exploded", as in Fig. 9.45. The alternative is to re-number the subscripts in the element stiffness matrix result from Example 9.5.

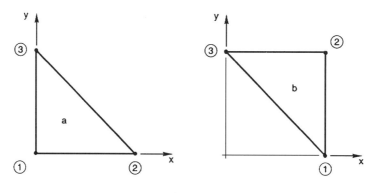

Fig. 9.45.

Performing the evaluations similar to part (a) leads to the required stiffness matrix for element b, namely

$$[k^{(b)}] = 10^7 [\text{N/m}] \begin{bmatrix} 3.846 & & & & & \\ -3.846 & 14.835 & & \text{symmetric} & & \\ 0 & -10.989 & 10.989 & & & \\ 0 & -3.297 & 3.297 & 10.989 & & \\ -3.846 & 7.143 & -3.297 & -10.989 & 14.835 & \\ 3.846 & -3.846 & 0 & 0 & -3.846 & 3.846 \end{bmatrix}$$

With reference to §9.10, the structural stiffness matrix can now be assembled using a dof. correspondence table. The order of the structural stiffness matrix will be 8×8, corresponding to four nodes, each having 2 dof. The dof. sequence, u_1, u_2, u_3, v_1, v_2, v_3, adopted for the convenience of inverting matrix $[A]$, covered in §9.9, can be converted to the more usual sequence, i.e. u_1, v_1, u_2, v_2, u_3, v_3, with the aid a dof. correspondence table. Whilst this re-sequencing is optional, the converted sequence is likely to result in less rearrangement of rows and columns, prior to partitioning the assembled stiffness matrix, than would otherwise be needed.

If row and column interchanges are to be avoided in solving the following Example, 9.7, and therefore save some effort, then the dof. labelling of Fig. 9.46 is recommended. This implies the final node numbering, also shown.

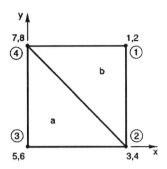

Fig. 9.46.

The dof. correspondence table will be as follows:

Row/column in $[k^{(e)}]$		1	2	3	4	5	6
Row/column in [K]	a	5	3	7	6	4	8
	b	3	1	7	4	2	8

Assembling the structural stiffness matrix, gives

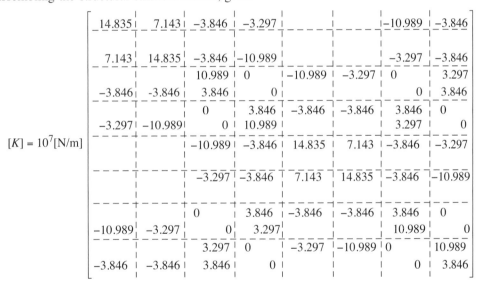

Summing the element stiffness contributions, and writing the structural governing equations, gives the result as

$$
\begin{bmatrix} X_1 \\ Y_1 \\ X_2 \\ Y_2 \\ X_3 \\ Y_3 \\ X_4 \\ Y_4 \end{bmatrix} = 10^7 \text{ [N/m]}
\begin{bmatrix}
14.835 & 7.143 & -3.846 & -3.297 & 0 & 0 & -10.989 & -3.846 \\
7.143 & 14.835 & -3.846 & -10.989 & 0 & 0 & -3.297 & -3.846 \\
-3.846 & -3.846 & 14.835 & 0 & -10.989 & -3.297 & 0 & 7.143 \\
-3.297 & -10.989 & 0 & 14.835 & -3.846 & -3.846 & 7.143 & 0 \\
0 & 0 & -10.989 & -3.846 & 14.835 & 7.143 & -3.846 & -3.297 \\
0 & 0 & -3.297 & -3.846 & 7.143 & 14.835 & -3.846 & -10.989 \\
-10.989 & -3.297 & 0 & 7.143 & -3.846 & -3.846 & 14.835 & 0 \\
-3.846 & -3.846 & 7.143 & 0 & -3.297 & -10.989 & 0 & 14.835
\end{bmatrix}
\begin{bmatrix} u_1 \\ v_1 \\ u_2 \\ v_2 \\ u_3 \\ v_3 \\ u_4 \\ v_4 \end{bmatrix}
$$

i.e. $\{P\} = [K]\{p\}$

where [K] is the required assembled stiffness matrix.

Example 9.7

Figure 9.47 shows a 1 mm thick sheet of steel, one edge of which is fully restrained whilst the opposite edge is subjected to a uniformly distributed tension of total value 40 kN. For the material Young's modulus, $E = 200$ GN/m^2 and Poisson's ratio, $v = 0.3$, and plane stress condition can be assumed.

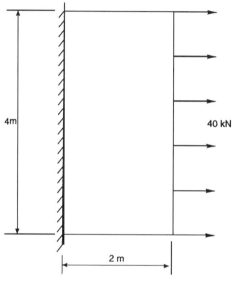

4m   40 kN

2 m

Fig. 9.47.

(a) Taking advantage of any symmetry, using two triangular membrane elements and hence the assembled stiffness matrix derived for the previous Example, 9.6, determine the nodal displacements in global coordinates.
(b) Determine the corresponding element principal stresses and their directions and illustrate these on a sketch of the continuum.

Solution

(a) Advantage can be taken of the single symmetry by modelling only half of the continuum. Figure 9.48 shows suitable node and dof. labelling, and division of the upper half of the continuum into two triangular membrane elements. Reference to the previous Example, 9.6, will reveal that the assembled stiffness matrix derived in answering this question can, conveniently, be utilised in solving the current example.

To simulate the clamped edge, dofs. 5 to 8 need to be suppressed, i.e. $u_3 = v_3 = u_4 = v_4 = 0$. Additionally, whilst node number 2 should be unrestrained in the x-direction, freedom in the y-direction needs to be suppressed to simulate the symmetry condition, i.e. $v_2 = 0$. Applying these boundary conditions and hence partitioning the structural stiffness matrix result from Example 9.6, gives the reduced equations as

$$\begin{bmatrix} X_1 \\ Y_1 \\ X_2 \end{bmatrix} = 10^7 [\text{N/m}] \begin{bmatrix} 14.835 & 7.143 & -3.846 \\ 7.143 & 14.835 & -3.846 \\ -3.846 & -3.846 & 14.835 \end{bmatrix} \begin{bmatrix} u_1 \\ v_1 \\ u_2 \end{bmatrix} \quad \text{i.e. } \{P_\alpha\} = [K_{\alpha\alpha}]\{p_\alpha\}$$

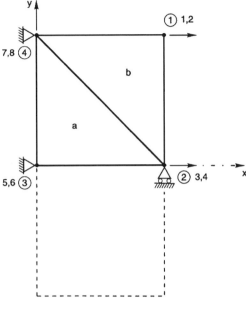

Fig. 9.48.

Inverting $[K_{\alpha\alpha}]$ to enable a solution for the displacements from $\{p_\alpha\} = [K_{\alpha\alpha}]^{-1}\{P_\alpha\}$

$$\text{where adj } [K_{\alpha\alpha}] = 10^{14} \begin{bmatrix} 205.286 & -91.175 & 29.583 \\ -91.175 & 205.286 & 29.583 \\ 29.583 & 29.583 & 169.055 \end{bmatrix}$$

and det $[K_{\alpha\alpha}] = 10^{21}[14.835(205.286) - 7.143(91.175) - 3.846(29.583)] = 2280.4 \times 10^{21}$

$$\text{Then } [K_{\alpha\alpha}]^{-1} = 10^{-10} \begin{bmatrix} 90.03 & -39.98 & 12.97 \\ -39.98 & 90.03 & 12.97 \\ 12.97 & 12.97 & 74.13 \end{bmatrix}$$

With reference to §9.4.7, the nodal load column matrix corresponding to a uniformly distributed load of 10 kN/m, will be given by

$$\{P_\alpha\} = \begin{bmatrix} X_1 \\ Y_1 \\ X_2 \end{bmatrix} = 10^3 \begin{bmatrix} 10 \\ 0 \\ 10 \end{bmatrix}_{[N]}$$

Hence, the nodal displacements are found from

$$\{p_\alpha\} = [K_{\alpha\alpha}]^{-1}\{P_\alpha\}$$

Substituting,

$$\begin{bmatrix} u_1 \\ v_1 \\ u_2 \end{bmatrix} = 10^{-10} \begin{bmatrix} 90.03 & -39.98 & 12.97 \\ -39.98 & 90.03 & 12.97 \\ 12.97 & 12.97 & 74.13 \end{bmatrix} 10^3 \begin{bmatrix} 10 \\ 0 \\ 10 \end{bmatrix} = 10^{-6} \begin{bmatrix} 103 \\ -27 \\ 87 \end{bmatrix}_{m} = \begin{bmatrix} 0.103 \\ -0.027 \\ 0.087 \end{bmatrix}_{mm}$$

The required nodal displacements are therefore $u_1 = 0.103$ mm, $v_1 = -0.027$ mm and $u_2 = 0.087$ mm.

(b) With reference to §9.9, element direct and shearing stresses are found from

$$
\begin{bmatrix} \sigma_{xx} \\ \sigma_{yy} \\ \sigma_{xy} \end{bmatrix} = [D][B] \begin{bmatrix} u_i \\ u_j \\ u_k \\ v_i \\ v_j \\ v_k \end{bmatrix}
$$

where, from Example 9.5,

$$
[D][B] = \frac{E}{2a(1-v^2)}
$$

$$
\begin{bmatrix} y_{23} & y_{31} & y_{12} & vx_{32} & vx_{13} & vx_{21} \\ vy_{23} & vy_{31} & vy_{12} & x_{32} & x_{13} & x_{21} \\ \dfrac{(l-v)}{2}x_{32} & \dfrac{(l-v)}{2}x_{13} & \dfrac{(l-v)}{2}x_{21} & \dfrac{(l-v)}{2}y_{23} & \dfrac{(l-v)}{2}y_{31} & \dfrac{(l-v)}{2}y_{12} \end{bmatrix}
$$

Evaluating the stresses for each element:

Element a

$$
\begin{bmatrix} \sigma_{xx} \\ \sigma_{yy} \\ \sigma_{xy} \end{bmatrix} = \frac{200 \times 10^9}{2 \times 2(1-0.3^2)} \begin{bmatrix} -2 & 2 & 0 & -0.6 & 0 & 0.6 \\ -0.6 & 0.6 & 0 & -2 & 0 & 2 \\ -0.7 & 0 & 0.7 & -0.7 & 0.7 & 0 \end{bmatrix} \begin{bmatrix} 0 \\ 87 \times 10^{-6} \\ 0 \\ 0 \\ 0 \\ 0 \end{bmatrix}
$$

$$
= \begin{bmatrix} 9.56 \times 10^6 \\ 2.87 \times 10^6 \\ 0 \end{bmatrix}_{N/m^2}
$$

The required principal stresses for element *a* are therefore $\sigma_1 = 9.56$ MN/m^2 (T) and $\sigma_2 = 2.87$ MN/m^2 (T), and are illustrated in Fig. 9.49.

Element b

$$
\begin{bmatrix} \sigma_{xx} \\ \sigma_{yy} \\ \sigma_{xy} \end{bmatrix} = \frac{200 \times 10^9}{2 \times 2(1-0.3^2)} \begin{bmatrix} 0 & 2 & -2 & -0.6 & 0.6 & 0 \\ 0 & 0.6 & -0.6 & -2 & 2 & 0 \\ -0.7 & 0.7 & 0 & 0 & 0.7 & -0.7 \end{bmatrix} \begin{bmatrix} 87 \times 10^{-6} \\ 103 \times 10^{-6} \\ 0 \\ 0 \\ -27 \times 10^{-6} \\ 0 \end{bmatrix}
$$

$$
= \begin{bmatrix} 10.43 \times 10^6 \\ 0.43 \times 10^6 \\ 2.92 \times 10^6 \end{bmatrix}_{N/m^2}
$$

The principal stresses are found from

$$
\sigma_1, \sigma_2 = \tfrac{1}{2}(\sigma_{xx} + \sigma_{yy}) \pm \tfrac{1}{2}\sqrt{[(\sigma_{xx} - \sigma_{yy})^2 + 4\sigma_{xy}^2]}
$$

Substituting gives

$$\sigma_1, \sigma_2 = \{\tfrac{1}{2}(10.43 + 0.43) \pm \tfrac{1}{2}\sqrt{[(10.43 - 0.43)^2 + 4 \times 2.92^2]}\}10^6 \ \text{N/m}^2$$

$$= (5.43 \pm 5.79)10^6 \ \text{N/m}^2$$

giving $\sigma_1 = 11.22 \ \text{MN/m}^2$ (T) and $\sigma_2 = 0.36 \ \text{MN/m}^2$ (C)
   The directions are found from

$$\theta = \tfrac{1}{2}\tan^{-1}[2\sigma_{xy}/(\sigma_{xx} - \sigma_{yy})]$$

substituting gives

$$\theta = \tfrac{1}{2}\tan^{-1}[2 \times 2.92/(10.43 - 0.43)] = 15.14°$$

The required principal stresses for element b are therefore $\sigma_1 = 11.22 \ \text{MN/m}^2$ (T) and $\sigma_2 = 0.36 \ \text{MN/m}^2$ (C) and are illustrated in Fig. 9.49.

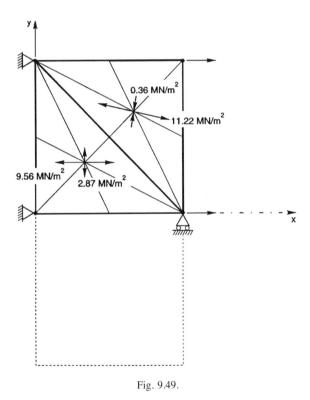

Fig. 9.49.

Problems

9.1 Figure 9.50 shows a support structure in the form of a pin-jointed plane frame, all three members of which are steel, of the same uniform cross-sectional area and length, such that $AE/L = 200 \ \text{kN/m}$, throughout.
(a) Using the displacement based finite element method and treating each member as a rod, determine the nodal displacements with respect to global coordinates for the frame shown in Fig. 9.50.

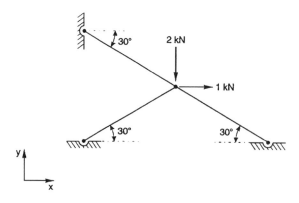

Fig. 9.50.

(b) Hence, determine the nodal reactions.

$$[-0.387, -13.557 \text{ mm}, -1.116, 0.644, 1.233, 0.712, -1.116, 0.644 \text{ kN}]$$

9.2 Figure 9.51 shows a roof truss, all members of which are made from steel, and have the same cross-sectional area, such that $AE = 10$ MN, throughout. For the purpose of analysis the truss can be treated as a pin-jointed plane frame. Using the displacement based finite element method, taking advantage of any symmetry and redundancies and treating each member as a rod element, determine the nodal displacements with respect to global coordinates.

$$[0.516, -2.280, -2.313 \text{ mm}]$$

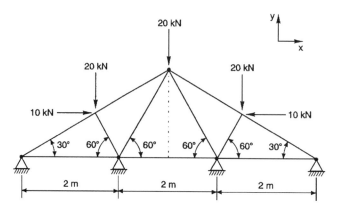

Fig. 9.51.

9.3 A recovery vehicle towing jib is shown in Fig. 9.52. It can be assumed that the jib can be idealised as a pin-jointed plane frame, with all three members made from steel, of the same uniform cross-sectional area, such that $AE = 40$ MN, throughout. Using the displacement based finite element method and treating each member as a rod, determine the maximum load P which can be exerted whilst limiting the resultant maximum deflection to 10 mm.

$$[36.5 \text{ kN}]$$

9.4 A hoist frame, arranged as shown in Fig. 9.53, comprises uniform steel members, each 1m long for which $AE = 200$ MN, throughout.

(a) Using the displacement based finite element method and assuming the frame members to be planar and pin-jointed, determine the nodal displacements with respect to global coordinates for the frame loaded as shown.

(b) Hence, determine the corresponding nodal reactions.

$$[0.5, 0.75, -0.722 \text{ mm}, 0, 86.6, -25.0, -43.3 \text{ kN}]$$

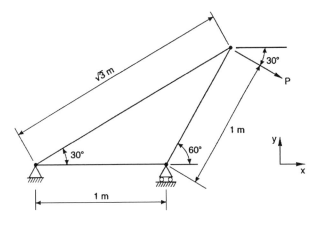

Fig. 9.52.

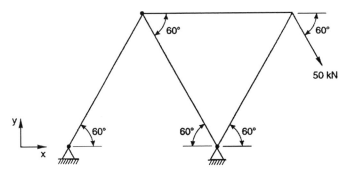

Fig. 9.53.

9.5 Figure 9.54 shows a towing "bracket" for a motor vehicle. Both members are made from uniform cylindrical section steel tubing, of outside diameter 40 mm, cross-sectional area 2.4×10^{-4} m^2, relevant second moment of area 4.3×10^{-8} m^4 and Young's modulus 200 GN/m^2. Using the displacement based finite element method, a simple beam element representation and assuming both members and load are coplanar:

(a) assemble the necessary terms in the structural stiffness matrix;
(b) hence, for the idealisation shown in Fig. 9.54, determine (i) the nodal displacements with respect to global coordinates, and (ii) the resultant maximum stresses at the built-in ends and at the common junction.

[0.421 mm, 0.103°, 131.34, 106.77, 127.20, 110.91 MN/m^2]

9.6 A stepped steel shaft supports a pulley, as shown in Fig. 9.55, is rigidly built-in at one end and is supported in a bearing at the position of the step. The bearing provides translational but not rotational restraint. Young's modulus for the material is 200 GN/m^2.

(a) Using the displacement based finite element method obtain expressions for the nodal displacements in global coordinates, using a two-beam model.
(b) Given that, because of a design requirement, the angular misalignment of the bearing cannot exceed 0.5°, determine the maximum load, P, that can be exerted on the pulley.
(c) Sketch the deformed geometry of the beam.

[$4.985 \times 10^{-8}P$, $-2.905 \times 10^{-8}P$, $8.475 \times 10^{-7}P$, 175 kN]

9.7 Figure 9.56 shows a chassis out-rigger which acts as a body support for an all-terrain vehicle. The out-rigger is constructed from steel channel section rigidly welded at the out-board edge and similarly welded to the vehicle chassis. For the channel material, Young's modulus, $E = 200$ GN/m^2, relevant second moment of area, $I = 2 \times 10^{-9}$ m^4 and cross-sectional area, $A = 4 \times 10^{-5}$ m^2. Using the displacement based finite element method, and representing the constituent members as simple beam elements

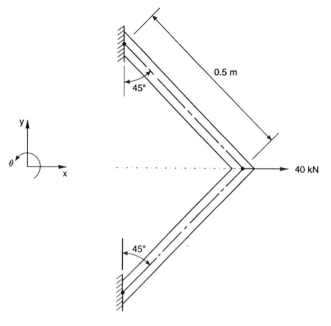

Fig. 9.54.

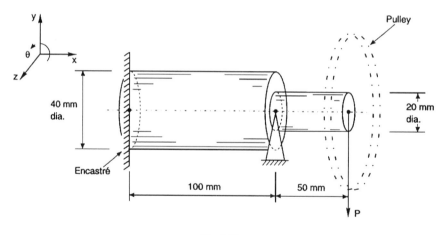

Fig. 9.55.

(a) determine the nodal displacements with respect to global coordinates.

(b) A modal analysis reveals that, to avoid resonance, the vertical stiffness of the out-rigger needs to be increased. Assuming only one of the members is to be stiffened, state which member and whether it should be the cross-sectional area or the second moment of area which should be increased, for most effect.

$$[0.1155, -0.4418 \text{ mm}, -0.322°, \text{ inclined member's csa.}]$$

9.8 The plane frame shown in Fig. 9.57 forms part of a steel support structure. The three members are rigidly connected at the common junction and are built-in at their opposite ends. All three members can be assumed to be axially rigid, and of constant cross-sectional area, A, relevant second moment of area, I, and Young's modulus, E. Using the displacement based finite element method and representing each member as a simple beam:

(a) show that the angular displacement of the common node, due to application of the moment, M, is given by $ML/8EI$;

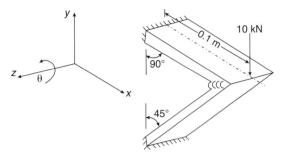

Fig. 9.56.

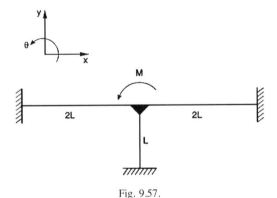

Fig. 9.57.

(b) determine all nodal reaction forces and moments due to this moment, and represent these reactions on a sketch of the frame. Show that both force and moment equilibrium is satisfied.

(c) If, due to a manufacturing defect, the joint at the lower end of the vertical member undergoes an angular displacement of $ML/4EI$, whilst all other properties remain unchanged, obtain a new expression for the angular displacement at the common junction.

$$[0, -3M/16L, M/8, -3M/4L, 0, 0, 3M/16L, M/8, 3M/4L, 0, M/4, ML/16EI]$$

9.9 Using the displacement based finite element method and a three-node triangular membrane element representation, determine the nodal displacements in global coordinates for the continuum shown in Fig. 9.58. Take advantage of any symmetry, assume plane stress conditions and use only two elements in the discretisation. For the material assume Young's modulus, $E = 200$ GN/m^2 and Poisson's ratio, $v = 0.3$.

$$[-3.00 \times 10^{-6}, 10.01 \times 10^{-6}, -3.00 \times 10^{-6}, 10.01 \times 10^{-6} \text{ m}]$$

9.10 A crude lifting device is fabricated from a triangular sheet of steel, 6 mm thick, as shown in Fig. 9.59. Assume for the material Young's modulus, $E = 200$ GN/m^2 and Poisson's ratio, $v = 0.3$, and that plane stress conditions are appropriate.

(a) Taking advantage of any symmetry, ignoring any instability and using only a single three-node triangular membrane element representation, use the displacement based finite element method to predict the nodal displacements in global coordinates.

(b) Determine the corresponding element principal stresses and their directions, and show these on a sketch of the element. $[-0.05, -0.17, -0.60$ mm, 134.85 MN/m^2 (T) at 31.7° from x-direction, 51.50 MN/m^2 (C)]

9.11 The web of a support structure, fabricated from steel sheet 1 mm thick, is shown in Fig. 9.60. Assume for the material Young's modulus, $E = 207$ GN/m^2 and Poisson's ratio, $v = 0.3$, and that plane stress conditions are appropriate.

(a) Neglecting any stiffening effects of adjoining members and any instability and using only a single three-node triangular membrane element representation, use the displacement based finite element method to predict the nodal displacements with respect to global coordinates.

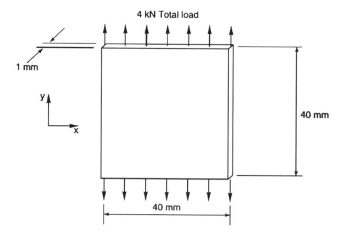

Fig. 9.58.

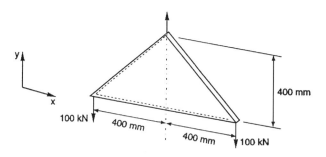

Fig. 9.59.

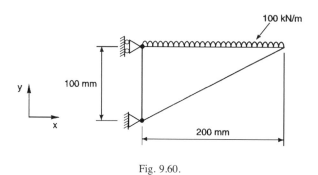

Fig. 9.60.

(b) Determine the corresponding element principal stresses and their directions, and show these on a sketch of the web.

$$[0.058, -0.60, -0.10 \text{ mm}, 123.6 \text{ MN/m}^2 \text{ (T) at } -31.7° \text{ from } x\text{-direction}, 323.6 \text{ MN/m}^2 \text{ (C)}]$$

9.12 Derive the stiffness matrix in global coordinates for a three-node triangular membrane element for plane *strain* conditions. Assume the displacement functions are the same as those of Example 9.5.

CHAPTER 10

CONTACT STRESS, RESIDUAL STRESS AND STRESS CONCENTRATIONS

Summary

The maximum pressure p_0 or compressive stress σ_c at the centre of contact between two curved surfaces is:

$$p_0 = -\sigma_c = \frac{3P}{2\pi ab}$$

where a and b are the major and minor axes of the Hertzian contact ellipse and P is the total load.

For *contacting parallel cylinders* of length L and radii R_1 and R_2,

maximum compressive stress, $\sigma_c = -0.591\sqrt{\dfrac{P\left(\dfrac{1}{R_1} + \dfrac{1}{R_2}\right)}{L\Delta}} = -p_0$

with $\Delta = \dfrac{1}{E_1}[1 - v_1^2] + \dfrac{1}{E_2}[1 - v_2^2]$

and the maximum shear stress, $\tau_{max} = 0.295\,p_0$ at a depth of $0.786b$ beneath the surface, with:

contact width, $\quad\quad b = 1.076\sqrt{\dfrac{P\Delta}{L\left(\dfrac{1}{R_1} + \dfrac{1}{R_2}\right)}}$

For *contacting spheres* of radii R_1 and R_2

maximum compressive stress, $\quad \sigma_c = -0.62\sqrt[3]{\dfrac{P}{\Delta^2}\left[\dfrac{1}{R_1} + \dfrac{1}{R_2}\right]^2} = -p_0$

maximum shear stress, $\quad \tau_{max} = 0.31\,p_0$ at a depth of $0.5b$ beneath the surface with:

contact width (circular) $\quad\quad b = 0.88\sqrt[3]{\dfrac{P\Delta}{\left(\dfrac{1}{R_1} + \dfrac{1}{R_2}\right)}}$

For a *sphere on a flat surface of the same material*

maximum compressive stress, $\quad\quad \sigma_c = -0.62\sqrt[3]{\dfrac{PE^2}{4R^2}}$

For a *sphere in a spherical seat* of the same material

$$\text{maximum compressive stress, } \sigma_c = -0.62\sqrt[3]{PE^2 \left[\frac{R_2 - R_1}{R_1 R_2}\right]^2}$$

For *spur gears*

$$\text{maximum contact stress, } \sigma_c = -0.475\sqrt{K}$$

with

$$K = \frac{W}{F_w d} \left[\frac{m+1}{m}\right]$$

with W = tangential driving load; F_w = face width; d = pinion pitch diameter; m = ratio of gear teeth to pinion teeth.

For *helical gears*

$$\text{maximum contact stress, } \sigma_c = -C\sqrt{\frac{K}{m_p}}$$

where m_p is the profile contact ratio and C a constant, both given in Table 10.2.

$$\text{Elastic stress concentration factor } K_t = \frac{\text{maximum stress, } \sigma_{\max}}{\text{nominal stress, } \sigma_{\text{nom}}}$$

$$\text{Fatigue stress concentration factor } K_f = \frac{S_n \text{ for the unnotched material}}{S_n \text{ for notched material}}$$

with S_n the endurance limit for n cycles of load.

$$\text{Notch sensitivity factor } q = \frac{K_f - 1}{K_t - 1}$$

or, in terms of a significant linear dimension (e.g. fillet radius) R and a material constant a

$$q = \frac{1}{(1 + a/R)}$$

$$\text{Strain concentration factor } K_\varepsilon = \frac{\text{max. strain at notch}}{\text{nominal strain at notch}}$$

Stress concentration factor K_p in presence of plastic flow is related to K_ε by Neuber's rule

$$K_p K_\varepsilon = K_t^2$$

10.1. Contact Stresses

Introduction

The design of components subjected to contact, i.e. local compressive stress, is extremely important in such engineering applications as bearings, gears, railway wheels and rails, cams, pin-jointed links, etc. Whilst in most other types of stress calculation it is usual to neglect local deflection at the loading point when deriving equations for stress distribution in general bodies, in contact situations, e.g. the case of a circular wheel on a flat rail, such an assumption would lead to infinite values of compressive stress (load ÷ "zero" area = infinity). This can only be avoided by local deflection, even yielding, of the material under the load to increase the bearing area and reduce the value of the compressive stress to some finite value.

Contact stresses between curved bodies in compression are often termed "Hertzian" contact stresses after the work on the subject by Hertz[1] in Germany in 1881. This work was concerned primarily with the evaluation of the maximum compressive stresses set up at the mating surfaces for various geometries of contacting body but it formed the basis for subsequent extension of consideration by other workers of stress conditions within the whole contact zone both at the surface and beneath it. It has now been shown that the strength and load-carrying capacity of engineering components subjected to contact conditions is not completely explained by the Hertz equations by themselves, but that further consideration of the following factors is an essential additional requirement:

(a) Local yielding and associated residual stresses

Yield has been shown to initiate sub-surface when the contact stress approaches $1.2 \, \sigma_y$ (σ_y being the yield stress of the contacting materials) with so-called "*uncontained plastic flow*" commencing when the stress reaches $2.8 \, \sigma_y$. Only at this point will material "escape" at the sides of the contact region. The ratio of loads to produce these two states is of the order of 350 although tangential (sliding) forces will reduce this figure significantly.

Unloading from any point between these two states produces a thin layer of residual tension at the surface and a sub-surface region of residual compression parallel to the surface. The residual stresses set up during an initial pass or passes of load can inhibit plastic flow in subsequent passes and a so-called "*shakedown*" situation is reached where additional plastic flow is totally prevented. Maximum contact pressure for shakedown is given by Johnson[14] as $1.6 \, \sigma_y$.

(b) Surface shear loading caused by mutual sliding of the mating surfaces

Pure rolling of parallel cylinders has been considered by Radzimovsky[5] whilst the effect of tangential shear loading has been studied by Deresiewicz[15], Johnson[16], Lubkin[17], Mindlin[18], Tomlinson[19] and Smith and Liu[20].

(c) Thermal stresses and associated material property changes resulting from the heat set up by sliding friction. (Local temperatures can rise to some 500°F above ambient).

A useful summary of the work carried out in this area is given by Lipson & Juvinal[21].

(d) The presence of lubrication – particularly hydrodynamic lubrication – which can greatly modify the loading and resulting stress distribution

The effects of hydrodynamic lubrication on the pressure distribution at contact (see Fig. 10.1) and resulting stresses have been considered by a number of investigators including Meldahl[22], M'Ewen[4], Dowson, Higginson and Whitaker[23], Crook[24], Dawson[25] and

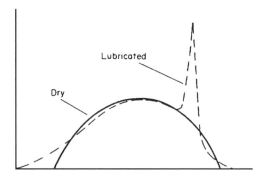

Fig. 10.1. Comparison of pressure distributions under dry and lubricated contact conditions.

Scott[26]. One important conclusion drawn by Dowson *et al.* is that at high load and not excessive speeds hydrodynamic pressure distribution can be taken to be basically Hertzian except for a high spike at the exit side.

(e) The presence of residual stresses at the surface of e.g. hardened components and their distribution with depth

In discussion of the effect of residually stressed layers on contact conditions, Sherratt[27] notes that whilst the magnitude of the residual stress is clearly important, the depth of the residually stressed layer is probably even more significant and the biaxiality of the residual stress pattern also has a pronounced effect. Considerable dispute exists even today about the origin of contact stress failures, particularly of surface hardened gearing, and the aspect is discussed further in §10.1.6 on gear contact stresses.

Muro[28], in X-ray studies of the residual stresses present in hardened steels due to rolling contact, identified a compressive residual stress peak at a depth corresponding to the depth of the maximum shear stress – a value related directly to the applied load. He therefore concluded that residual stress measurement could form a useful load-monitoring tool in the analysis of bearing failures.

Detailed consideration of these factors and even of the Hertzian stresses themselves is beyond the scope of this text. An attempt will therefore be made to summarise the essential formulae and behaviour mechanisms in order to provide an overall view of the problem without recourse to proof of the various equations which can be found in more advanced treatments such as those referred to below:-

The following special cases attracted special consideration:

  (i) *Contact of two parallel cylinders* – principally because of its application to roller bearings and similar components. Here the Hertzian contact area tends towards a long narrow rectangle and complete solutions of the stress distribution are available from Belajef[2], Foppl[3], M'Ewen[4] and Radzimovsky[5].

 (ii) *Spur and helical gears* – Buckingham[6] shows that the above case of contacting parallel cylinders can be used to fair accuracy for the contact of spur gears and whilst Walker[7] and Wellaver[8] show that helical gears are more accurately represented by contacting conical frustra, the parallel cylinder case is again fairly representative.

(iii) *Circular contact* – as arising in the case of contacting spheres or crossed cylinders. Full solutions are available by Foppl[3], Huber[9] Morton and Close[10] and Thomas and Hoersch[11].

(iv) *General elliptical contact*. Work on this more general case has been extensive and complete solutions exist for certain selected axes, e.g. the axes of the normal load. Authors include Belajef[2], Fessler and Ollerton[12], Thomas and Heorsch[11] and Ollerton[13].

Let us now consider the principal cases of contact loading:-

10.1.1. General case of contact between two curved surfaces

In his study of this general contact loading case, assuming elastic and isotropic material behaviour, Hertz showed that the intensity of pressure between the contacting surfaces could be represented by the elliptical (or, rather, semi-ellipsoid) construction shown in Fig. 10.2.

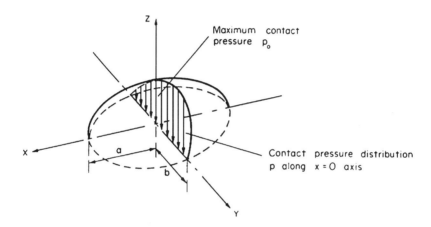

Fig. 10.2. Hertizian representation of pressure distribution between two curved bodies in contact.

If the maximum pressure at the centre of contact is denoted by p_0 then the **pressure at any other point within the contact region** was shown to be given by

$$p = p_0\sqrt{1 - \frac{x^2}{a^2} - \frac{x^2}{b^2}}$$ (10.1)

where a and b are the major and minor semi-axes, respectively.
The **total contact load** is then given by the volume of the semi-ellipsoid,

i.e. $$P = \frac{2}{3}\pi a b p_0$$ (10.2)

with the **maximum pressure** p_0 therefore given in terms of the applied load as

$$p_0 = \frac{3P}{2\pi ab} = \text{maximum compressive stress } \sigma_c$$ (10.3)

For any given contact load P it is necessary to determine the value of a and b before the maximum contact stress can be evaluated. These are found analytically from equations suggested by Timoshenko and Goodier[29] and adapted by Lipson and Juvinal[21].

i.e.
$$a = m \left[\frac{3P\Delta}{4A} \right]^{1/3} \quad \text{and} \quad b = n \left[\frac{3P\Delta}{4A} \right]^{1/3}$$

with
$$\Delta = \frac{1}{E_1}[1 - v_1^2] + \frac{1}{E_2}[1 - v_2^2]$$

a function of the elastic constants E and v of the contacting bodies and

$$A = \frac{1}{2} \left[\frac{1}{R_1} + \frac{1}{R_1'} + \frac{1}{R_2} + \frac{1}{R_2'} \right]$$

with R and R' the maximum and minimum radii of curvature of the unloaded contact surfaces in two perpendicular planes.

For flat-sided wheels R_1 will be the wheel radius and R_1' will be infinite. Similarly for railway lines with head radius R_2 the value of R_2' will be infinite to produce the flat length of rail.

$$B = \frac{1}{2} \left[\left(\frac{1}{R_1} - \frac{1}{R_1'} \right)^2 + \left(\frac{1}{R_2} - \frac{1}{R_2'} \right)^2 + 2 \left(\frac{1}{R_1} - \frac{1}{R_1'} \right) \left(\frac{1}{R_2} - \frac{1}{R_2'} \right) \cos 2\psi \right]^{1/2}$$

with ψ the angle between the planes containing curvatures $1/R_1$ and $1/R_2$.

Convex surfaces such as a sphere or roller are taken to be positive curvatures whilst internal surfaces of ball races are considered to be negative.

m and n are also functions of the geometry of the contact surfaces and their values are shown in Table 10.1 for various values of the term $\alpha = \cos^{-1}(B/A)$.

Table 10.1.

α degrees	20	30	35	40	45	50	55	60	65	70	75	80	85	90
m	3.778	2.731	2.397	2.136	1.926	1.754	1.611	1.486	1.378	1.284	1.202	1.128	1.061	1.000
n	0.408	0.493	0.530	0.567	0.604	0.641	0.678	0.717	0.759	0.802	0.846	0.893	0.944	1.000

10.1.2. Special case 1 – Contact of parallel cylinders

Consider the two parallel cylinders shown in Fig. 10.3(a) subjected to a contact load P producing a rectangular contact area of width $2b$ and length L. The contact stress distribution is indicated in Fig. 10.3(b).

The elliptical pressure distribution is given by the two-dimensional version of eqn (10.1)

i.e.
$$p = p_0 \sqrt{1 - \frac{y^2}{b^2}} \tag{10.5}$$

The **total load** P is then the volume of the prism

i.e.
$$P = \tfrac{1}{2}\pi b L p_0 \tag{10.6}$$

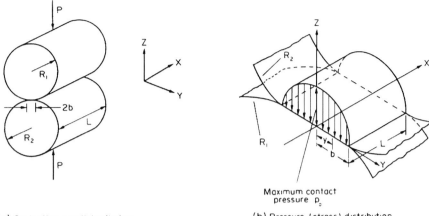

(a) Contacting parallel cylinders   (b) Pressure (stress) distribution

Fig. 10.3. (a) Contact of two parallel cylinders; (b) stress distribution for contacting parallel cylinders.

and the **maximum pressure or maximum compressive stress**

$$p_0 = \sigma_c = \frac{2p}{\pi bL} \tag{10.7}$$

The **contact width** can be related to the geometry of the contacting surfaces as follows:-

$$b = 1.076 \sqrt{\frac{P\Delta}{L\left(\dfrac{1}{R_1} + \dfrac{1}{R_2}\right)}} \tag{10.8}$$

giving the **maximum compressive stress** as:

$$\sigma_c = -p_0 = -0.591 \sqrt{\frac{P\left(\dfrac{1}{R_1} + \dfrac{1}{R_2}\right)}{L\Delta}} \tag{10.9}$$

(For a flat plate R_2 is infinite, for a cylinder in a cylindrical bearing R_2 is negative). **Stress conditions at the surface** on the load axis are then:

$$\sigma_z = \sigma_c = -p_0$$

$$\sigma_y = -p_0$$

$$\sigma_x = -2vp_0 \quad \text{(along cylinder length)}$$

The **maximum shear stress** is:

$$\tau_{\max} = 0.295p_0 \simeq 0.3p_0$$

occurring at a depth beneath the surface of $0.786\,b$ and on planes at $45°$ to the load axis.

In cases such as gears, bearings, cams, etc. which (as will be discussed later) can be likened to the contact of parallel cylinders, this shear stress will reduce gradually to zero as the rolling load passes the point in question and rise again to its maximum value as the next load contact is made. However, this will not be the greatest reversal of shear stress since there is another shear stress on planes parallel and perpendicular to the load axes known as the *"alternating"* or *"reversing"* shear stress, at a depth of 0.5 b and offset from the load axis by 0.85 b, which has a maximum value of 0.256 p_0 which changes from positive to negative as the load moves across contact.

The maximum shear stress on 45° planes thus varies between zero and 0.3 p_0 (approx) with an alternating component of 0.15 p_0 about a mean of 0.15 p_0. **The maximum alternating shear stress, however, has an alternating component of 0.256 p_0 about a mean of zero** – see Fig. 10.4. The latter is therefore considerably more significant from a fatigue viewpoint.

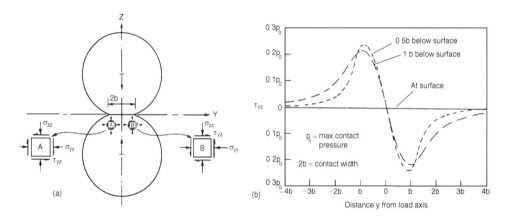

Fig. 10.4. Maximum alternating stress variation beneath contact surfaces.

N.B.: The above formulae assume the length of the cylinders to be very large in comparison with their radii. For short cylinders and/or cylinder/plate contacts with widths less than six times the contact area (or plate thickness less than six times the depth of the maximum shear stress) actual stresses can be significantly greater than those estimated by the given equations.

10.1.3. Combined normal and tangential loading

In normal contact conditions between contacting cylinders, gears, cams, etc. friction will be present reacting the sliding (or tendency to slide) of the mating surfaces. This will affect the stresses which are set up and it is usual in such cases to take the usual relationship between normal and tangential forces in the presence of friction

viz. $$F = \mu R \quad \text{or} \quad q = \mu p_0$$

where q is the tangential pressure distribution, assumed to be of the same form as that of the normal pressure. Smith and Liu[20] have shown that with such an assumption:

(a) A shear stress now exists on the surface at the contact point introducing principal stresses which are different from σ_x, σ_y and σ_z of the normal loading case.

(b) The maximum shear stress may exist either at the surface or beneath it depending on whether μ is greater than or less than 1/9 respectively.

(c) The stress range in the y direction is increased by almost 90% on the normal loading value and there is also a reversal of sign. A useful summary of stress distributions in graphical form is given by Lipson and Juvinal[21].

10.1.4. Special case 2 – Contacting spheres

For contacting spheres, eqns. (10.9) and (10.8) become
Maximum compressive stress (normal to surface)

$$\sigma_c = -p_0 = -0.62\sqrt[3]{\frac{P}{\Delta^2}\left[\frac{1}{R_1}+\frac{1}{R_2}\right]^2} \tag{10.10}$$

with a maximum value of

$$\sigma_c = -1.5P/\pi a^2 \tag{10.11}$$

Contact dimensions (circular)

$$a = b = 0.88\sqrt[3]{\frac{P\Delta}{\left[\dfrac{1}{R_1}+\dfrac{1}{R_2}\right]}} \tag{10.12}$$

As for the cylinder, if contact occurs between one sphere and a flat surface then R_2 is infinite, and if the sphere contacts inside a spherical seating then R_2 is negative.

The other two **principal stresses in the surface** plane are given by:

$$\sigma_x = \sigma_y = -\frac{(1+2\nu)}{2}p_0 \tag{10.13}$$

For steels with Poisson's ratio $\nu = 0.3$ the **maximum shear stress** is then:

$$\tau_{\text{max}} \mathrel{\hat=} 0.31p_0 \tag{10.14}$$

at a depth of half the radius of the contact surface.

The **maximum tensile stress** set up within the contact zone occurs at the edge of the contact zone in a radial direction with a value of:

$$\sigma_{t_{\text{max}}} = \frac{(1-2\nu)}{3}p_0 \tag{10.15}$$

The circumferential stress at the same point is equal in value, but compressive, whilst the stress normal to the surface is effectively zero since contact has ended. With equal and opposite principal stresses in the plane of the surface, therefore, the **material is effectively in a state of pure shear.**

The **maximum octahedral shear stress** which is also an important value in consideration of elastic failure, occurs at approximately the same depth below the surface as the maximum shear stress. Its value may be obtained from eqn (8.24) by substituting the appropriate values of σ_x, σ_y and σ_z found from Fig. 10.5 which shows their variation with depth beneath the surface.

The **relative displacement, e, of the centres of the two spheres** is given by:

$$e = 0.77\sqrt[3]{P^2\left(\frac{1}{E_1}+\frac{1}{E_2}\right)^2\left(\frac{1}{R_1}+\frac{1}{R_2}\right)} \tag{10.16}$$

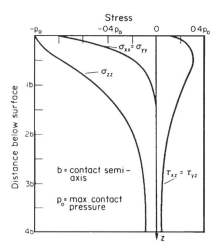

Fig. 10.5. Variation of stresses beneath the surface of contacting spheres.

For a *sphere contacting a flat surface* of the same material $R_2 = \infty$ and $E_1 = E_2 = E$. Substitution in eqns. (10.10) and (10.16) then yields

maximum compressive stress

$$\sigma_c = -0.62 \sqrt[3]{\frac{PE^2}{4R_1^2}} \qquad\qquad (10.17)$$

and relative displacement of centres

$$e = 1.54 \sqrt[3]{\frac{P^2}{2E^2 R_1}} \qquad\qquad (10.18)$$

For a *sphere on a spherical seat* of the same material

$$\sigma_c = -0.62 \sqrt[3]{PE^2 \left[\frac{R_2 - R_1}{R_1 R_2}\right]^2} \qquad\qquad (10.19)$$

with

$$e = 1.54 \sqrt[3]{\frac{P^2}{2E^2} \left[\frac{R_2 - R_1}{R_1 R_2}\right]} \qquad\qquad (10.20)$$

For *other, more general, loading cases* the reader is referred to a list of formulae presented by Roark and Young[33].

10.1.5. Design considerations

It should be evident from the preceding sections that the maximum Hertzian compressive stress is not, in itself, a valid criteria of failure for contacting members although it can be used as a valid design guide provided that more critical stress states which have a more direct influence on failure can be related directly to it. It has been shown, for example, that alternating shear stresses exist beneath the surface which are probably critical to fatigue life but these can be expressed as a simple proportion of the Hertzian pressure p_0 so that p_0 can be used as a simple index of contact load severity.

The contact situation is complicated under real service loading conditions by the presence of e.g. residual stresses in hardened surfaces, local yielding and associated additional residual stresses, friction forces and lubrication, thermal stresses and dynamic (including shock) load effects.

The failure of brittle materials under contact conditions correlates more closely with the maximum tensile stress at the surface rather than sub-surface shear stresses, whilst for static or very slow rolling operations failure normally arises as a result of excessive plastic flow producing indentation (*"brinelling"*) of the surface. In both cases, however, the Hertzian pressure remains a valuable design guide or reference.

By far the greatest number of failures of contacting components remains the surface or sub-surface fatigue initiated type variably known as *"pitting"*, *"spalling"*, *"onion-peel spalling"* or *"flaking"*. The principal service areas in which this type of failure occurs are gears and bearings.

10.1.6. Contact loading of gear teeth

Figure 10.6 shows the stress conditions which prevail in the region of a typical gear tooth contact. Immediately at the contact point, or centre of contact, there is the usual position of maximum compressive stress (p_0). Directly beneath this, and at a depth of approximately one-third of the contact width, is the maximum shear stress τ_{max} acting on planes at 45° to the load axis. Between these two positions lies the maximum alternating or reversed shear stress τ_{alt} acting on planes perpendicular and parallel to the surface. Whilst τ_{alt} is numerically smaller than τ_{max} it alternates between positive and negative values as the tooth proceeds

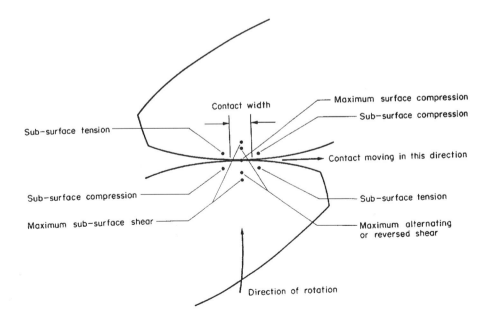

Fig. 10.6. Stress conditions in the region of gear tooth contact.

through mesh giving a stress range greater than that of τ_{max} which ranges between a single value and zero. It is argued by many that, for this reason, τ_{alt} is probably more significant to fatigue life than τ_{max} – particularly if its depth relates closely to that of peak residual stresses or case–core junctions of hardened gears.

As the gears rotate there is a combination of rolling and sliding motions, the latter causing additional surface stresses not shown in Fig. 8.6. Ahead of the contact area there is a narrow band of compression and behind the contact area a narrow band of tension. A single point on the surface of a gear tooth therefore passes through a complex variety of stress conditions as it goes through its meshing cycle. Both the surface and alternating stress change sign and other sub-surface stresses change from zero to their maximum value. Add to these fatigue situations the effects of residual stress, lubrication, thermal stresses and dynamic loading and it is not surprising that gears may fail in one of a number of ways either at the surface or sub-surface.

The majority of gear tooth failures are surface failures due to "pitting", "spalling", "flaking", "wear", etc. the three former modes referring to the fracture and shedding of pieces of various size from the surface. Considerable speculation and diverse views exist even among leading workers as to the true point of origin of some of these failures and considerable evidence has been produced of, apparently, both surface and sub-surface crack initiation. The logical conclusion would therefore seem to be that both types of initiation are possible depending on precisely the type of loading and contact conditions.

A strong body of opinion supports the suggestion of Johnson[14,16] and Almen[30] who attribute contact stress failures to local plastic flow at inclusions or flaws in the material, particularly in situations where a known overload has occurred at some time prior to failure. The overload is sufficient to produce the initial plastic flow and successive cycles then extend the region of plasticity and crack propagation commences. Dawson[25] and Akaoka[31] found evidence of sub-surface cracks running parallel to the surface, some breaking through to the surface, others completely unconnected with it. These were attributed to the fatigue action of the maximum alternating (reversed) shear stress. Undoubtedly, from the evidence presented by other authors, cracks can also initiate at the surface probably producing a "pitting" type of failure, i.e. smaller depth of damage. These cracks are suggested to initiate at positions of maximum tensile stress in the contact surface and subsequent propagation is then influenced by the presence (or otherwise) of lubricant.

In the case of helical gears, three-dimensional photoelastic tests undertaken by the author[32] indicate that maximum sub-surface stresses are considerably greater than those predicted by standard design procedures based on Hertzian contact and uniform loading along the contact line. Considerable non-uniformity of load was demonstrated which, together with dynamic effects, can cause maximum loads and stresses many times above the predicted nominal values. The tests showed the considerable benefit to be gained on the load distribution and resulting maximum stress values by the use of tip and end relief of the helical gear tooth profile.

10.1.7. Contact stresses in spur and helical gearing

Whilst the radius of an involute gear tooth will change slightly across the width of contact with a mating tooth it is normal to ignore this and take the contact of spur gear teeth as equivalent to the contact of parallel cylinders with the same radius of curvature at the point of contact. The Hertzian eqns. (10.8) and (10.9) can thus be applied to **spur gears** and, for typical steel elastic constant values of $\nu = 0.3$ and $E = 206.8$ GN/m^2, the **maximum contact** stress becomes

$$\sigma_c = -p_o = -0.475\sqrt{K} \text{ MN/m}^2 \qquad (10.21)$$

where

$$K = \frac{W}{F_w d} \left[\frac{m+1}{m} \right]$$

with W = tangential driving load = pinion torque ÷ pinion pitch radius
$\quad F_w$ = face width
$\quad d$ = pinion pitch diameter
$\quad m$ = ratio of gear teeth to pinion teeth; the pinion taken to be the
$\qquad$ smaller of the two mating teeth.

For **helical gears, the maximum contact stress** is given by

$$\sigma_c = -p_o = -C\sqrt{\frac{K}{m_p}} \qquad (10.22)$$

where K is the same factor as for spur gears
$\quad m_p$ is the profile contact ratio
$\quad C$ is a constant

the values of m_p and C being found in Table 10.2, for various helix angles and pressure angles.

Table 10.2. Typical values of C and m_p for helical gears.

Pressure angle	Spur		15° Helix		30° Helix		45° Helix	
	C	m_p	C	m_p	C	m_p	C	m_p
$14\frac{1}{2}°$	0.546	2.10	0.528	2.01	0.473	1.71	0.386	1.26
$17\frac{1}{2}°$	0.502	1.88	0.485	1.79	0.435	1.53	0.355	1.13
$20°$	0.474	1.73	0.458	1.65	0.410	1.41	0.335	1.05
$25°$	0.434	1.52	0.420	1.45	0.376	1.25	0.307	0.949

10.1.8. Bearing failures

Considerable care is necessary in the design of bearings when selecting appropriate ball and bearing race radii. If the radii are too similar the area of contact is large and excessive wear and thermal stress (from frictional heating) results. If the radii are too dissimilar then the contact area is very small, local compressive stresses become very high and the load capacity of the bearing is reduced. As a compromise between these extremes the radius of the race is normally taken to be between 1.03 and 1.08 times the ball radius.

Fatigue life tests and service history then indicate that the life of ball bearings varies approximately as the cube of the applied load whereas, for roller bearings, a 10/3 power relationship is more appropriate. These relationships can only be used as a rough "rule of thumb", however, since commercially produced bearings, even under nominally similar and controlled production conditions, are notorious for the wide scatter of fatigue life results.

As noted previously, the majority of bearing failures are by spalling of the surface and most of the comments given in §10.1.6 relating to gear failures are equally relevant to bearing failures.

10.2. Residual Stresses

Introduction

It is probably true to say that all engineering components contain stresses (of variable magnitude and sign) before being subjected to service loading conditions owing to the history of the material prior to such service. These stresses, produced as a result of mechanical working of the material, heat treatment, chemical treatment, joining procedure, etc., are termed *residual stresses* and they can have a very significant effect on the fatigue life of components. These residual stresses are "locked into" the component in the absence of external loading and represent a datum stress over which the service load stresses are subsequently superimposed. If, by fortune or design, the residual stresses are of opposite sign to the service stresses then part of the service load goes to reduce the residual stress to zero before the combined stress can again rise towards any likely failure value; such residual stresses are thus extremely beneficial to the strength of the component and significantly higher fatigue strengths can result. If, however, the residual stresses are of the same sign as the applied stress, e.g. both tensile, then a smaller service load is required to produce failure than would have been the case for a component with a zero stress level initially; the strength and fatigue life in this case is thus reduced. Thus, both the magnitude and sign of residual stresses are important to fatigue life considerations, and methods for determining these quantities are introduced below.

It should be noted that whilst preceding chapters have been concerned with situations where it has been assumed that stresses are zero at zero load this is not often the case in practice, and great care must be exercised to either fully evaluate the levels of residual stress present and establish their effect on the strength of the design, or steps must be taken to reduce them to a minimum.

Bearing in mind that most loading applications in engineering practice involve fatigue to a greater or less degree it is relevant to note that surface residual stresses are the most critical as far as fatigue life is concerned since, almost invariably, fatigue cracks form at the surface. The work of §11.1.3 indicates that whilst tensile mean stresses promote fatigue crack initiation and propagation, compressive mean stresses are beneficial in that they impede fatigue failure. Compressive residual stresses are thus generally to be preferred (and there is not always a choice of course) if fatigue lives of components are to be enhanced. Indeed, compressive stresses are often deliberately introduced into the surface of components, e.g. by chemical methods which will be introduced below, in order to increase fatigue lives. There are situations, however, where compressive residual stress can be most undesirable; these include potential buckling situations where compressive surface stresses could lead to premature buckling failure, and operating conditions where the service loading stresses are also compressive. In the latter case the combined service and residual stresses may reach a sufficiently high value to exceed yield in compression and produce local plasticity on the first cycle of loading. On unloading, tensile residual stress "pockets" will be formed and these can act as local stress concentrations and potential fatigue crack initiation positions. Such a situation arises in high-temperature applications such as steam turbines and nuclear plant, and in contact load applications.

Whilst it has been indicated above that tensile residual stresses are generally deleterious to fatigue life there are again exceptions to this "rule", and very significant ones at that! It is now quite common to deliberately overload structures and components during proof testing to produce plastic flow at discontinuities and other stress concentrations to reduce their stress concentration effect on subsequent loading cycles. Other important techniques which involve the deliberate overloading of components in order to produce residual stress distribution favourable to subsequent loading cycles include "autofrettage" of thick cylinders (see §3.20(a)), "overspeeding" of rotating discs (see §3.20(b)) and pre-stressing of springs (see §3.8).

Whilst engineers have been aware of residual stresses for many years it is only recently that substantial efforts have been made to investigate their magnitudes and distributions with depth in components and hence their influence on performance and service life. This is probably due to the conservatism of old design procedures which generally incorporated sufficiently large safety factors to mask the effects of residual stresses on component integrity. However, with current drives for economy of manufacture coupled with enhanced product safety and reliability, design procedures have become far more stringent and residual stress effects can no longer be ignored. Principally, the designer needs to consider the effect of residual stress on structural or component failure but there is also need for detailed consideration of distortion and stability factors which are also closely related to residual stress levels.

10.2.1. Reasons for residual stresses

Residual stresses generally arise when conditions in the outer layer of a material differ from those internally. This can arise by one of three principal mechanisms: (a) mechanical processes, (b) chemical treatment, (c) heat treatment, although other mechanism are also discussed in the subsequent text.

(a) Mechanical processes

The most significant mechanical processes which induce surface residual stresses are those which involve plastic yielding and hence "cold-working" of the material such as rolling, shot-peening and forging. Practically all other standard machining procedures such as grinding, turning, polishing, etc., also involve local yielding (to a lesser extent perhaps) and also induce residual stresses. Reference should also be made to §3.9 and §3.10 which indicate how residual stresses can be introduced due to bending or torsion beyond the elastic limit.

Cold working

Shot peening is a very popular method for the introduction of favourable compressive residual stresses in the surface of components in order to increase their fatigue life. It is a process whereby small balls of iron or steel shot are bombarded at the component surface at high velocity from a rotating nozzle or wheel. It is applicable virtually to all metals and all component geometries and so is probably the most versatile of all the mechanical working processes. The bombardment tends to compress the surface layer and thus laterally try to expand it. This lateral expansion at the surface is resisted by the core material and residual compression results, its magnitude depending on the size of shot used and the peening velocity. Typically, residual stresses of the order of half the yield strength of the material are readily obtained, with peak values slightly sub-surface. However, special procedures such as "strain peening" which bombard the surface whilst applying external tensile loads can produce residuals approaching the full yield strength.

The major benefit of shot peening arises in areas of small fillet radii, notches or other high stress gradient situations and on poor surface finishes such as those obtained after rough machining or decarburisation. It is widely used in machine parts produced from high-strength steels and on gears, springs, structural components, engine con-rods and other motor vehicle components when fatigue lives have been shown to have been increased by factors in excess of 100%.

A number of different peening procedures exist in addition to standard shot peening with spherical shot, e.g. needle peening (bombardment by long needles with rounded ends), hammer peening (surface indented with radiused tool), roller-burnishing (rolling of under-sized hole to required diameter), roto peening (impact of shot-coated flexible flaps). Figure 10.7 shows a typical residual stress distribution produced by shot peening, the maximum residual stress attainable being given by the following "rule of thumb" estimate

$$\sigma_m \simeq 500 + (0.2 \times \text{tensile strength})$$

for steels with a tensile strength between 650 MN/m^2 and 2 GN/m^2.

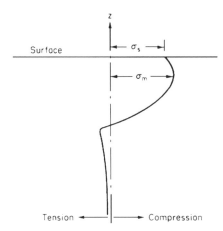

Fig. 10.7. Typical residual stress distribution with depth for the shot-peening process.

For lower-strength steels and alloys σ_m can initially reach the yield stress or 0.1% proof stress but this will fade under cyclic loading.

Cold rolling of threads, crankpins and axles relies on similar principles to those outlined above with, in this case, continuous pressure of the rollers producing controlled amounts of cold working. Further examples of cold working are the bending of pipes and conduits, cold shaping of brackets and clips and cold drawing of bars and tubes – sometimes of complex cross-section.

In some of the above applications the stress gradient into the material can be quite severe and a measurement technique which can produce results over reasonable depth is essential if residual stress–fatigue life relationships and mechanisms are to be fully understood.

Machining

It has been mentioned above that plastic deformation is almost invariably present in any machining process and the extent of the plastically deformed layer, and hence of the residually stressed region, will depend on the depth of cut, sharpness of tool, rates of speed and feed and the machineability of the material. With sharp tools, the heat generated at the tip of the tool will not have great influence and the residually stressed layer is likely to be compressive and relatively highly localised near the surface. With blunt tools or multi-tipped tools, particularly grinding, much more heat will be generated and if cooling is not sufficient this will produce thermally induced compressive stresses which can easily exceed the tensile stresses applied by the mechanical action of the tool. If they are large enough to exceed yield then tensile residual stresses may arise on cooling and care may need to be exercised in the type and level of service stress to which the component is then subjected. The depth of the residually stressed layer will depend upon the maximum temperatures reached during the machining operation and upon the thermal expansion coefficient of the material but it is likely that it will exceed that due to machining plastic deformation alone.

Residual stresses in manufactured components can often be very high; in grinding, for example, it is quite possible for the tensile residual stresses to produce cracking, particularly sub-surface, and etching techniques are sometimes employed after the grinding of e.g. bearings to remove a small layer on the surface in order to check for grinding damage. Distortion is another product of high residual stresses, produced particularly in welding and other heat treatment processes.

(b) Chemical treatment

The principal chemical treatments which are used to provide components with surface residual stress layers favourable to subsequent service fatigue loading conditions are nitriding, tufftriding and carburising.

Nitriding

Nitriding is a process whereby certain alloy steels are heated to about 550°C in an ammonia atmosphere for periods between 10 and 100 hours. Nitrides form in the surface of the steel with an associated volume increase. The core material resists this expansion and, as a result, residual compressive stresses are set up which can be very high (see Fig. 10.8). The surface layer, which typically is of the order of 0.5 mm thick, is extremely hard and the combination of this with the high surface residual compressive stresses make nitrided components exceptionally resistant to stress concentrations such as surface notches; fatigue lives of nitrided components are thus considerably enhanced over those of the parent material.

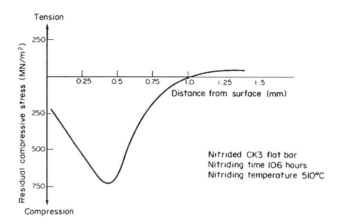

Fig. 10.8. Typical residual stress results for nitrided steel bar using the X-ray technique.

Minimal distortion or warping is produced by the nitriding process and no quenching is required.

Tufftriding

A special version of nitriding known as "tough nitriding" or, simply, "Tufftriding" consists of the heating of steel in a molten cyanic salt bath for approximately 90 minutes to allow nitrogen to diffuse into the steel surface and combine with the iron carbide formed in the outer skin when carbon is also released from the cyanic bath. The product of this combination is carbon-bearing epsilon-iron-nitride which forms a very tough but thin, wear-resistant layer, typically 0.1 mm thick. The process is found to be particularly appropriate for plain medium-carbon steels with little advantage over normal nitriding for the higher-strength alloy steels.

Carburising

Introduction of carbon into surface layers to produce so-called carburising may be carried out by solid, liquid or gaseous media. In each case the parent material contained in the selected medium such as charcoal, liquid sodium cyanide plus soda ash, or neutral gas enriched with propane, is heated to produce diffusion of the carbon into the surface. The depth of hardened case resulting varies from, typically, 0.25 mm on small articles to 0.37 mm on bearings (see Fig. 10.9).

(c) Heat treatment

Unlike chemical treatments, heat treatment procedures do not alter the chemical composition at the surface but simply modify the metallurgical structure of the parent material. Principal heat treatment procedures which induce favourable residual stress layers are induction hardening and flame hardening, although many other processes can also be considered within this category such as flame cutting, welding, quenching and even hot rolling or

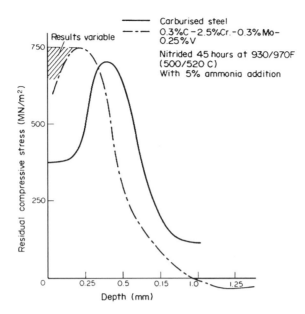

Fig. 10.9. Comparison of residual stress pattern present in nitrided chromium−molybdenum−vanadium steel and in a carburised steel − results obtained using the X-ray technique.

forging. In the two latter cases, however, chemical composition effects are included since carbon is removed from the surface by oxidation. This "decarburising" process produces surface layers with physical properties generally lower than those of the core and it is thus considered as a weakening process.

Returning to the more conventional heat treatment processes of flame and induction hardening, these again have a major effect at the surface where temperature gradients are the most severe. They produce both surface hardening and high compressive residual stresses with associated fatigue life improvements of up to 100%. There is some evidence of weakening at the case to core transition region but the process remains valuable for components with sharp stress gradients around their profile or in the presence of surface notches.

In both cases the surface is heated above some critical temperature and rapidly cooled and it is essential that the parent material has sufficient carbon or alloys to produce the required hardening by quenching. Heating either takes place under a gas flame or by electric induction heating caused by eddy currents generated in the surface layers. Typically, flame hardening is used for such components as gears and cams whilst induction hardening is applied to crankshaft journals and universal joints.

Many other components ranging from small shafts and bearings up to large forgings, fabricated structures and castings are also subjected to some form of heat treatment. Occasionally this may take the form of simple stress-relief operations aiming to reduce the level of residual stresses produced by prior manufacturing processes. Often, however, the treatment may be applied in order to effect some metallurgical improvement such as the normalising of large castings and forgings to improve their high-temperature creep characteristics or the surface hardening of gears, shafts and bearings. The required phase change of such processes usually entails the rapid cooling of components from some elevated temperature and it is this cooling which induces thermal gradients and, if these are sufficiently large (i.e. above yield), residual stresses. The component surfaces tend to cool more rapidly, introducing tensile stresses in the outer layers which are resisted by the greater bulk of the core material and result in residual compressive stress. As stated earlier, the stress gradient with depth into the material will depend upon the temperatures involved, the coefficient of thermal expansion of the material and the method of cooling. Particularly severe stress gradients can be produced by rapid quenching in water or oil.

Differential thermal expansion is another area in which residual stresses–or stress systems which can be regarded as residual stresses–arise. In cases where components constructed from materials with different coefficients of linear expansion are subjected to uniform temperature rise, or in situations such as heat exchangers or turbine casings where one material is subjected to different temperatures in different areas, free expansions do not take place. One part of the component attempts to expand at a faster rate and is constrained from doing so by an adjacent part which is either cooler or has a lower coefficient of expansion. Residual stresses will most definitely occur on cooling if the differential expansion stresses at elevated temperatures exceed yield.

When dealing with the quenching of heated parts, as mentioned above, a simple rule is useful to remember: *"What cools last is in tension"*. Thus the surface which generally cools first ends up in residual biaxial compression whilst the inner core is left in a state of triaxial tension. An exception to this is the quenching of normal through-hardened components when residual tensile stresses are produced at the surface unless a special process introduced by the General Motors Corporation of the U.S.A. termed "Marstressing" is used. This probably explains why surface-hardened parts generally have a much greater fatigue life than corresponding through-hardened items.

Should residual tensile stresses be achieved in a surface and be considered inappropriate then they can be relieved by tempering, although care must be taken to achieve the correct balance of ductility and strength after completion of the tempering process.

(d) Welds

One of the most common locations of fatigue failures resulting from residual stresses is at welded joints. Any weld junction can be considered to have three different regions; (a) the parent metal, (b) the weld metal, and (c) the heat-affected-zone (H.A.Z.), each with their own different physical properties including expansion coefficients. Residual stresses are then produced by the restraint of the parent metal on the shrinkage of the hot weld metal when it cools, and by differences in phase transformation behaviour of the three regions.

The magnitude and distribution of the residual stresses will depend upon the degree of preheat of the surfaces prior to welding, the heat input during welding, the number of weld passes, the match of the parent and weld metal and the skill of the operator. Even though the residuals can often be reduced by subsequent heat treatment this is not always effective owing to the different thermal expansions of the three zones. Differences between other physical properties in the three regions can also mean that failures need not always be associated with the region or part which is most highly stressed. Generally it is the heat-affected zones which contain sharp peaks of residual stress.

In welded structures, longitudinal shrinkages causes a weld and some parent material on either side to be in a state of residual tension often as high as the yield stress. This is balanced in the remainder of the cross-section by a residual compression which, typically, varies between 20 and 100 MN/m^2. When service load compressive stresses are applied to the members, premature yielding occurs in the regions of residual compressive stress, the stiffness of the member is reduced and there is an increased tendency for the component to buckle. In addition to this longitudinal "tendon force" effect there are also transverse effects in welds known as "pull-in" and "wrap-up" effects (see Fig. 10.10) again dictated by the level of residual stress set up.

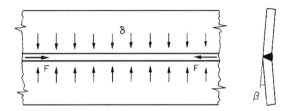

Fig. 10.10. The three basic parameters used to describe global weld shrinkage: F = tendon force; δ = pull-in; β-wrap-up.

The control of distortion is a major problem associated with large-scale welding. This can sometimes be minimised by clamping parts during welding to some pre-form curvatures or templates so that on release, after welding, they spring back to the required shape. Alternatively, components can be stretched or subjected to heat in order to redistribute the residual stress pattern and remove the distortion. In both cases, care needs to be exercised that unfavourable compressive stresses are not set up in regions which are critical to buckling failure.

(e) Castings

Another common problem area involving thermal effects and associated residual stresses is that of large castings. Whilst the full explanation for the source of residuals remains unclear (even after 70 years of research) it is clear that at least two mechanisms exist. Firstly, there are the physical restrictions placed on contraction of the casting, as it cools, by the mould itself and the differential thermal effects produced by different rates of cooling in different sections of e.g. different thickness. Secondly, there are metallurgical effects which arise largely as a result of differential cooling rates. Metal phase transformations and associated volume changes therefore occur at different positions at different times and rates. It is also suspected that different rates of cooling through the transformation range may create different material structures with different thermal coefficients. It is likely of course that the residual stress distribution produced will be as a result of a combination of these, and perhaps other, effects. It is certainly true, however, that whatever the cause there is frequently a need to subject large castings to some form of stress-relieving operation and any additional process such as this implies additional cost. It is therefore to be hoped that recent advances in measurement techniques, notably in the hole drilling method, will lead to a substantially enhanced understanding of residual stress mechanisms and to the development, for example, of suitable casting procedures which may avoid the need for additional stress-relieving operations.

10.2.2. The influence of residual stress on failure

It has been shown that residual stresses can be accommodated within the elastic failure theories quite simply by combining the residual and service load stresses (taking due account of sign) and inserting the combined stress value into the appropriate yield criterion. This is particularly true for ductile materials when both the Von Mises distortion energy theory and the Tresca maximum shear stress theory produce good correlation with experimental results. Should yielding in fact occur, there will normally be a change in the residual stress magnitude (usually a reduction) and distribution. The reduction of residual stress in this way is known as *"fading"*.

It is appropriate to mention here another type of failure phenomenon which is related directly to residual stress termed *"stress corrosion cracking"*. This occurs in metals which are subjected to corrosive environments whilst stressed, the cracks appearing in the surface layers.

Another source of potential failure is that of residual stress systems induced by the assembly of components with an initial lack of fit. This includes situations where the lack of fit is by deliberate design, e.g. shrinking or force-fit of compound cylinders or hubs on shafts and those where insufficient clearance or tolerances have been specified on mating components which, therefore, have to be forced together on assembly. This is, of course, a totally different situation to most of the cases listed above where residual stresses arise within a single member; it can nevertheless represent a potentially severe situation.

In contact loading situations such as in gearing or bearings, consideration should be given to the relationship between the distribution of residual stress with depth in the, typically, hardened surface and the depth at which the peak alternating shear stress occurs under the contact load. It is possible that the coincidence of the peak alternating value with the peak residual stress could explain the sub-surface initiation of cracks in spalling failures of such components. The hardness distribution with depth should also be considered in a similar way to monitor the strength/stress ratio, the lowest value of which can also initiate failure.

10.2.3. Measurement of residual stresses

The following methods have been used for residual stress investigations:

(1) Progressive turning or boring – Sach's method[34]
(2) Sectioning
(3) Layer removal – Rosenthal and Newton[35]
                        – Waisman and Phillips[36]
(4) Hole-drilling   – Mather[37]
                        – Bathgate[38]
                        – Procter and Beaney[39, 40, 41]
(5) Trepanning or ring method – Milbradt[42]
(6) Chemical etch – Waisman and Phillips[36]
(7) Stresscoat brittle lacquer drilling – Durelli and Tsao[43]
(8) X-ray – French and Macdonald[44]
              – Kirk[45, 46]
              – Andrews et al[47]
(9) Magnetic method – Abuki and Cullety[48]

(10) Hardness studies – Sines and Carlson[49]
(11) Ultrasonics – Noranha and West[50]
                       – Kino[51]
(12) Modified layer removal – Hearn and Golsby[52]
                                            – Spark machining–Denton[53]
(13) Photoelasticity   – Hearn and Golsby[52]

Of these techniques, the most frequently applied are the layer removal (either mechanically or chemically), the hole-drilling and the X-ray measurement procedures. Occasionally the larger scale sectioning of a component after, e.g. initially coating the surface with a photoelastic coating, a brittle lacquer or marking a grid, is useful for the semi-quantitative assessment of the type and level of residual stresses present. In each case the relaxed stresses are transferred to the coating or grid and are capable of interpretation. In the case of the brittle lacquer method the surface is coated with a layer of a brittle lacquer such as "Tenslac" or "Stress coat" and, after drying, is then drilled with a small hole at the point of interest. The relieved residual stresses, if of sufficient magnitude, will then produce a crack pattern in the lacquer which can be readily evaluated in terms of the stress magnitude and type.

The layer removal, progressive turning or boring, trepanning, chemical etch, modified layer removal and hole-drilling methods all rely on basically the same principle. The component is either machined, etched or drilled in stages so that the residual stresses are released producing relaxation deformations or strains which can be measured by mechanical methods or electrical resistance strain gauges and, after certain corrections, related to the initial residual stresses. Apart from the hole-drilling technique which is discussed in detail below, the other techniques of metal removal type are classed as destructive since the component cannot generally be used after the measurement procedure has been completed.

Most layer removal techniques rely on procedures for metal removal which themselves introduce or affect the residual stress distribution and associated measurement by the generation of heat or as a result of mechanical working of the surface – or both. Conventional machining procedures including grinding, milling and polishing all produce significant effects. Of the 'mechanical' processes, spark erosion has been shown to be the least damaging process and the only one to have an acceptably low effect on the measured stresses. Regrettably, however, it is not always available and it may prove impractical in certain situations, e.g. site measurement. In such cases, either chemical etching procedures are used or, if these too are impractical, then standard machining techniques have to be employed with suitable corrections applied to the results.

X-ray techniques are well established and will also be covered in detail below; they are, however, generally limited to the measurement of strains at, or very near to, the surface and require very sophisticated equipment if reasonably accurate results are to be achieved.

Ultrasonic and magneto-elastic methods until recently have not received much attention despite the promise which they show. Grain orientation and other metallurgical inhomogeneities affect the velocity and attenuation of ultrasonic waves and further development of the technique is required in order to effectively separate these effects from the changes due to residual stress. A sample of stress-free material is also required for calibration of the method for quantitative results. Considerable further development is also required in the case of the magneto-elastic procedure which relies on the changes which occur in magnetic flux densities in ferromagnetic materials with changing stress.

The attempts to relate residual stress levels to the hardness of surfaces again appear to indicate considerable promise since they would give an alternative non-destructive technique which is simple to apply and relatively inexpensive. Unfortunately, however, the proposals do not seem to have achieved acceptance to date and do not therefore represent any significant challenge to the three "popular" methods.

Let us now consider in greater detail the two most popular procedures, namely hole-drilling and X-ray methods.

The hole-drilling technique

The hole-drilling method of measurement of residual stresses was initially proposed by Mathar[37] in 1933 and involves the drilling of a small hole (i.e. small diameter and depth) normal to the surface at the point of interest and measurement of the resulting local surface deformations or strains. The radial stress at the edge of the hole must be zero from simple equilibrium conditions so that local redistribution of stress or "relaxation" must occur. At the time the technique was first proposed, the method of measurement of the relaxations was by mechanical extensometers and the accuracy of the technique was limited. Subsequent workers, and particularly those in recent years,[38−41] have used electric resistance strain gauges and much more refined procedures of hole drilling metal removal as described below.

The particular advantages of the hole drilling technique are that it is accurate, can be made portable and is the least "destructive" of the metal removal techniques, the small holes involved generally not preventing further use of the component under test−although care should be exercised in any such decision and may depend upon the level of stress present. Stress values are obtained at a point and their variation with depth can also be established. This is important with surface-hardening chemical treatments such as nitriding or carburising where substantial stress variation and stress reversals can take place beneath the surface – see Fig. 10.9.

Whichever method of hole drilling is proposed, the procedure now normally adopted is the bonding of a three-element strain gauge rosette at the point under investigation and the drilling of a hole at the gauge centre in order to release the residual stresses and allowing the recording of the three strains ε_1, ε_2 and ε_3 in the three gauge element directions. Beaney[39] then quotes the formula which may be used for evaluation of the principal residual stresses σ_1 and σ_2 in the following form:

$$\left.\begin{array}{c}\sigma_1\\\sigma_2\end{array}\right\} = -\frac{1}{K_1}\cdot\frac{E}{2}\left\{\frac{(\varepsilon_1+\varepsilon_3)}{1-v(K_2/K_1)}\pm\frac{1}{1+v(K_2/K_1)}\sqrt{(\varepsilon_3-\varepsilon_1)^2+[(\varepsilon_1+\varepsilon_3)^2-2\varepsilon_2]^2}\right\}$$

Values of K_1 and K_2 are found by calibration, the value of K_1 and hence the "sensitivity" depending on the geometry of the hole and the position of the gauges relative to the hole-close control of these parameters are therefore important. For hole depths of approx. one hole diameter little error is introduced for steels by assuming the "modified" Poissons ratio term $v(K_2/K_1)$ to be constant at 0.3.

It should be noted that the drilled hole will act as a stress raiser with a stress concentration factor of at least 2. Thus, if residual stress levels are over half the yield stress of the material in question then some local plasticity will arise at the edge of the hole and the above formula will over-estimate the level of stress. However, the over-estimation is predictable and can be calibrated and in any case is negligible for residuals up to 70% of the yield stress.

Scaramangas *et al.*[55] show how simple correction factors can be applied to allow for variations of stress with depth, for the effects of surface preparation when mounting the gauges and for plastic yielding at the hole edge.

Methods of hole-drilling: (a) high-speed drill or router

Until recently, the 'standard' method of hole-drilling has been the utilisation of a small diameter, high-speed, tipped drill (similar to that used by dentists) fitted into a centring device which can be accurately located over the gauge centre using a removable eyepiece and fixed rigidly to the surface (see Fig. 10.11). Having located the fixture accurately over the required drilling position using cross-hairs the eye-piece is then removed and replaced by the drilling head. Since flat-bottomed holes were assumed in the derivation of the theoretical expressions it is common to use end-milling cutters of between $\frac{1}{8}$ in and $\frac{1}{4}$ in (3 mm to 6 mm) diameter. Unfortunately, the drilling operation itself introduces machining stresses into the component, of variable magnitude depending on the speed and condition of the tool,

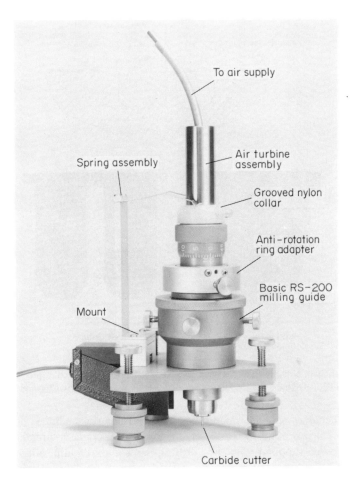

Fig. 10.11. Equipment used for the hole-drilling technique of residual stress measurement.

and these cannot readily be separated. Unless drilling is very carefully controlled, therefore, errors can arise in the measured strain values and the alternative "stress-free" machining technique outlined below is recommended.

(b) Air-abrasive machining

In this process the conventional drilling head is replaced by a device which directs a stream of air containing fine abrasive particles onto the surface causing controlled erosion of the material – see Fig. 10.12(a). The type of hole produced – see Fig. 10.12(b) – does not have a rectangular axial section but can be trepanned to produce axisymmetric, parallel-sided, holes

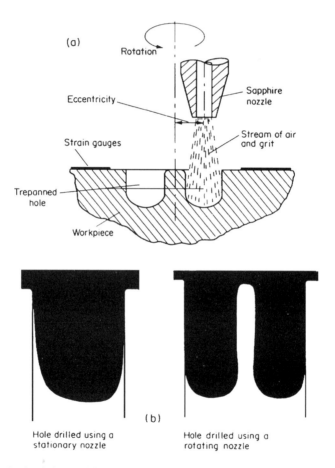

Fig. 10.12. (a) Air abrasive hole machining using a rotating nozzle: (b) types of holes produced: (i) hole drilled using a stationary nozzle, (ii) hole drilled using a rotating nozzle (trepanning)

with repeatable accuracy. Square-sided holes often aid this repeatability, but rotation of the nozzle on an eccentric around the gauge centre axis will produce a circular trepanned hole.

The optical device supplied with the "drilling" unit allows initial alignment of the unit with respect to the gauge centre and also measurement of the hole depth to enable stress distributions with depth to be plotted. It is also important to ascertain that the drilled depth is at least 1.65 mm since this has been shown to be the minimum depth necessary to produce full stress relaxation in most applications.

Whilst the strain gauge rosettes originally designed for this work were for hole sizes of 1.59 mm, larger holes have now been found to give greater sensitivity and accuracy. Optimum hole size is now stated to be between 2 and 2.2 mm.

X-ray diffraction

The X-ray technique is probably the most highly developed non-destructive measurement technique available today. Unfortunately, however, although semi-portable units do exist, it is still essentially a laboratory tool and the high precision equipment involved is rather expensive. Because the technique is essentially concerned with the measurement of stresses in the surface it is important that the very thin layer which is examined is totally representative of the conditions required.

Two principal X-ray procedures are in general use: the *diffractometer approach* and the *film technique*. In the first of these a diffractometer is used to measure the relative shift of X-ray diffraction lines produced on the irradiated surface. The individual crystals within any polycrystalline material are made up of families of identical planes of atoms, with a fairly uniform interplanar spacing d. The so-called lattice strain normal to the crystal planes is then $\Delta d/d$ and at certain angles of incidence (known as Bragg angles) X-ray beams will be diffracted from a given family of planes as if they were being reflected. The diffraction is governed by the Bragg equation:

$$n\lambda = 2d \sin \theta \qquad (10.23)$$

where n is an integer corresponding to the order of diffraction
$\qquad \lambda$ is the wavelength of the X-ray radiation
$\qquad \theta$ is the angle of incidence of the crystal plane.

Any change in applied or residual stress caused, e.g. by removal of a layer of material from the rear face of the specimen, will produce a change in the angle of reflection obtained by differentiating the above equation

i.e. $$\Delta\theta = -\frac{\Delta d}{d} \cdot \tan \theta \qquad (10.24)$$

This equation relates the change in angle of incidence to the lattice strain at the surface. Typically, $\Delta\theta$ ranges between 0.3° and 0.02° depending on the initial value of θ used.

Two experimental procedures can be used to evaluate the stresses in the surface: (a) the $\sin^2 \psi$ technique and (b) the two-exposure technique, and full details of these procedures are given by Kirk[45]. Certain problems exist in the interpretation of the results, such as the elastic anisotropy which is exhibited by metals with respect to their crystallographic directions. The appropriate values of E and ν have thus to be established by separate X-ray experiments on tensile test bars or four-point bending beams.

In applications where successive layers of the metal are removed in order to determine the values of sub-surface stresses (to a very limited depth), the material removal produces a redistribution of the stresses to be measured and corrections have to be applied. Standard formulae[54] exist for this process.

An alternative method for determination of the Bragg angle θ is to use the film technique with a so-called *"back reflection"* procedure. Here the surface is coated with a thin layer of powder from a standard substance such as silver which gives a diffraction ring near to that from the material under investigation. Measurements of the ring diameter and film–specimen distance are then used in either a single-exposure or double-exposure procedure to establish the required stress values.[47]

The X-ray technique is valid only for measurement in materials which are elastic, homogeneous and isotropic. Fortunately most polycrystaline metals satisfy this requirement to a fair degree of accuracy but, nevertheless, it does represent a constraint on wider application of the technique.

10.2.4. Summary of the principal effects of residual stress

(1) Considerable improvement in the fatigue life of components can be obtained by processes which introduce residual stress of appropriate sign in the surface layer. Compressive residual stresses are particularly beneficial in areas of potential fatigue (tensile stress) failure.

(2) Pre-loading of components beyond yield in the same direction as future service loading will produce residual stress systems which strengthen the components.

(3) Surface-hardened components have a greater fatigue resistance than through-hardened parts.

(4) Residual stresses have their greatest influence on parts which are expected to undergo high numbers of loading cycles (i.e. low strain–high cycle fatigue); they are not so effective under high strain–low cycle fatigue conditions.

(5) Considerable benefit can be obtained by local strengthening procedures at, e.g. stress concentrations, using shot peening or other localised procedures.

(6) Failures always occur at positions where the ratio of strength to stress is least favourable. This is particularly important in welding applications.

(7) Consideration of the influence of residual stresses must be part of the design process for all structures and components.

10.3. Stress concentrations

Introduction

In practically all the other chapters of this text loading conditions and components have been analysed in which stresses have been assumed, or shown, to be uniform or smoothly varying. In practice, however, this rarely happens owing to the presence of grooves, fillets, threads, holes, keyways, points of concentrated loading, material flaws, etc. In each of these cases, and many others too numerous to mention, the stress at the "discontinuity" is likely to be significantly greater than the assumed or nominally calculated figure and such discontinuities are therefore termed *stress raisers or stress concentrations.*

Most failures of structural members or engineering components occur at stress concentrations so that it is important that designers understand their significance and the magnitude of their effect since it is practically impossible to design any component without some form of stress raiser. In fatigue loading conditions, for example, virtually all failures occur at stress concentrations and it is therefore necessary to be able to develop a procedure which will take them into account during design strength calculations.

Geometric discontinuities such as holes, sharp fillet radii, keyways, etc. are probably the most prevalent causes of failure and typical examples of failure are shown in Figs. 10.13 and 10.14.

Fig. 10.13. Combined bending and tension fatigue load failure at a sharp fillet radius stress concentration position on a large retaining bolt of a heavy-duty extrusion press.

Fig. 10.14. Another failure emanating from a sharp fillet stress concentration position.

In order to be able to understand the stress concentration mechanism consider the simple example of the tensile bar shown in Figs. 10.15(a) and 10.15(b). In Fig. 10.15(a) the bar is solid and the tensile stress is nominally uniform at $\sigma = P/A = (P/bt)$ across the section.

In Fig. 10.15(b), however, the bar is drilled with a transverse hole of diameter d. Away from the hole the stress remains uniform across the section at $\sigma = P/bt$ and, using a similar calculation, the stress at the section through the centre of the hole should be $\sigma_{nom} = P/(b-d)t$ and uniform.

The sketch shows, however, that the stress at the edge of the hole is, in fact, much greater than this, indeed it is nearly 3 times as great (depending on the diameter d).

The ratio of the actual maximum stress σ_{max} and the nominal value σ_{nom} is then termed the *stress concentration factor* for the hole.

$$\textbf{Stress concentration factor } \mathbf{K}_t = \frac{\textbf{maximum stress}}{\textbf{nominal stress}} = \frac{\sigma_{max}}{\sigma_{nom}} \qquad (10.25)$$

For a small hole $K_t \simeq 3$.

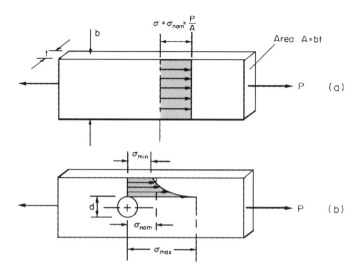

Fig. 10.15. Stress concentration effect of a hole in a tension bar.

It should also be observed that whilst the stress local to the hole is greater than the nominal stress, at distances greater than about one hole diameter away from the edge of the hole the stresses are less than the nominal value. This must be true from simple equilibrium condition since the sum of (stress × area) across the section must balance the applied force; if the stress is greater than the nominal or average stress at one point it must therefore be less in another.

It should be evident that even had a safety factor of, say, 2.8 been used in the stress calculations for the tensile bar in question the bar would have failed since the stress concentration factor exceeds this and it is important not to rely on safety factors to cover stress concentration effects which can generally be estimated quite well, as will be discussed later. Safety factors should be reserved for allowing for uncertainties in service load conditions which cannot be estimated or anticipated with any confidence.

The cause of the stress concentration phenomenon is perhaps best understood by the use of a few analogies; firstly, that of the flow of liquid through a channel. It can be shown that the distribution of stress through a material is analogous to that of fluid flow through a channel, the cross-section of which varies in the same way as that of the material cross-section. Thus Fig. 10.16 shows the experimentally obtained flow lines for a fluid flowing round a pin of diameter d in a channel of width b, i.e. the same geometry as that of the tensile bar. It will be observed that the flow lines crowd together as the fluid passes the pin and the velocity of flow increases significantly in order that the same quantity of fluid can pass per second through the reduced gap. This is directly analogous to the increased stress at the hole in the tensile bar. Any other geometrical discontinuity will have a similar effect see Fig. 10.17.

An alternative analogy is to consider the bar without the hole as a series of stretched rubber bands parallel to each other as are the flow lines of Fig. 10.16(a). Again inserting a pin to represent the hole in the bar produces a distortion of the bands and pressure on the pin at its top and bottom diameter extremities – again directly analogous to the increased "pressure" or stress felt by the bar at the edge of the hole.

It is appropriate to mention here that the stress concentration factor calculation of eqn. (10.25) only applies while stresses remain in the elastic range. If stresses are increased

Fig. 10.16. Flow lines (a) without and (b) with discontinuity.

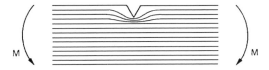

Fig. 10.17. Flow lines around a notch in a beam subjected to bending.

beyond the elastic region then local yielding takes place at the stress concentration and stresses will be redistributed as a result. In most cases this can reduce the level of the maximum stress which would be estimated by the stress concentration factor calculation. In the case of a notch or sharp-tipped crack, for example, the local plastic region forms to blunt the crack tip and reduce the stress-concentration effect for subsequent load increases. This local yielding represents a limiting factor on the maximum realistic value of stress concentration factor which can be obtained for most structural engineering materials. For very brittle materials such as glass, however, the high stress concentrations associated with very sharp notches or scratches can readily produce fracture in the absence of any significant plasticity. This, after all, is the principle of glass cutting!

The ductile flow or local yielding at stress concentrations is termed a *notch-strengthening effect* and stress concentration factors, although defined in the same way, become *plastic stress concentration factors* K_p. For most ductile materials, as the maximum stress in the component is increased up to the maximum tensile strength of the material, the value of K_p tends towards unity thus indicating that the *static* strength of the component has not been reduced significantly by the presence of the stress concentration. This is not the case for impact, fatigue or brittle fracture conditions where stress concentrations play a very significant part.

In complete contrast, stress concentrations of the types mentioned above are relatively inconsequential to the strength of heterogeneous brittle materials such as cast iron because of the high incidence of "natural" internal stress raisers within even the un-notched material, e.g. internal material flaws or impurities.

It has been shown above that the magnitude of the local increase in stress in the tensile bar caused by the stress concentration, i.e. the value of the stress concentration factor, is related to the geometry of both the bar and the hole since both b and d appear in the calculation of eqn (10.25).

Figure 10.18 shows the way in which the stress concentration value changes with different hole/bar geometries. It will be noted that the most severe effect (when related to the nominal area left after drilling the hole) is obtained when the hole diameter is smallest, producing a stress concentration factor (s.c.f.) of approximately 3. Whilst this is the largest s.c.f. value it does not mean, of course, that the bar is weaker the smaller the hole. Clearly a very large hole leaves very little material to carry the tensile load and the nominal stress will increase to produce failure. It is the combination of the nominal stress and the stress concentration factor which gives the value of the maximum stress that eventually produces

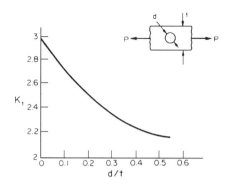

Fig. 10.18. Variation of elastic stress concentration factor K_t for a hole in a tensile bar with varying d/t ratios.

failure – both must therefore be considered.

i.e. Maximum stress = nominal stress × stress concentration factor.

If load on the bar is increased sufficiently then failure will occur, the crack emanating from the peak stress position at the edge of the hole across the section to the outside (see Fig. 10.19).

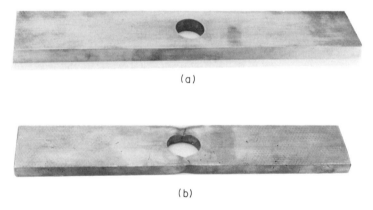

Fig. 10.19. Tensile bar loaded to destruction – crack initiates at peak stress concentration position at the hole edge.

Other geometric factors will affect the stress-concentration effect of discontinuities such as the hole, e.g. its shape. Figure 10.20 shows the effect of various hole shapes on the s.c.f. achieved in the tensile plate for which it can be shown that, approximately, $K_t = 1 + 2(A/B)$ where A and B are the major and minor axis dimensions of the elliptical holes perpendicular and parallel to the axis of the applied stress respectively. When $A = B$, the ellipse becomes the circular hole considered previously and $K_t \simeq 3$.

For large values of B, i.e. long elliptical slots parallel to the applied stress axis, stress concentration effects are reduced below 3 but for large A values, i.e. long elliptical slots perpendicular to the stress axis, s.c.f.'s rise dramatically and the potentially severe effect of slender slots or cracks such as this can readily be seen.

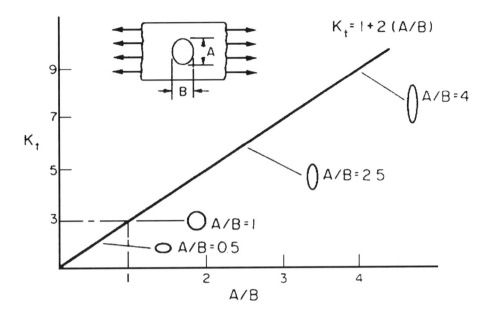

Fig. 10.20. Effect of shape of hole on the stress concentration factor for a bar with a transverse hole.

This is, of course, the theory of the perforated toilet paper roll which should tear at the perforation every time–which only goes to prove that theory very rarely applies perfectly in every situation!! (Closer consideration of the mode of loading and material used in this case helps to defend the theory, however.)

10.3.1. Evaluation of stress concentration factors

As stated earlier, the majority of the work in this text is devoted to consideration of stress situations where stress concentration effects are not present, i.e. to the calculation of nominal stresses. Before resulting stress levels can be applied to design situations, therefore, it is necessary for the designer to be able to estimate or predict the stress concentration factors associated with his particular design geometry and nominal stresses. In some cases these have been obtained analytically but in most cases graphs have been produced for standard geometric discontinuity configurations using experimental test procedures such as photoelasticity, or more recently, using finite element computer analysis.

Figures 10.21 to 10.30 give stress concentration factors for fillets, grooves and holes under various types of loading based upon a highly recommended reference volume[57]. Many other geometrical forms and loading conditions are considered in this and other reference texts[60] but for non-standard cases the application of the photoelastic technique is also highly recommended (see §6.12).

The reference texts give stress concentration factors not only for two-dimensional plane stress situations such as the tensile plate but also for triaxial stress systems such as the common case of a shaft with a transverse hole or circumferential groove subjected to tension, bending or torsion.

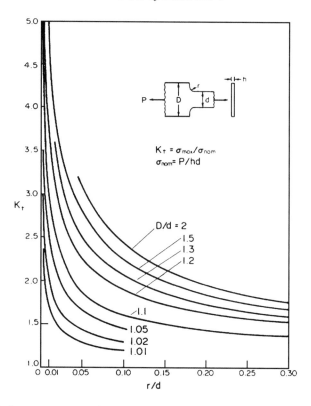

Fig. 10.21. Stress concentration factor K_t for a stepped flat tension bar with shoulder fillets.

Figures 10.31, 10.32 and 10.34 indicate the ease with which stress concentration positions can be identified within photoelastic models as the points at which the fringes are greatest in number and closest together. It should be noted that:

(1) Stress concentration factors are different for a single geometry subjected to different types of loading. Appropriate K_t values must therefore be obtained for each type of loading. Figure 10.33 shows the way in which the stress concentration factors associated with a groove in a circular bar change with the type of applied load.

(2) Care must be taken that stress concentration factors are applied to nominal stresses calculated on the same basis as that of the s.c.f. calculation itself, i.e. the same cross-sectional area must be used—usually the net section left after the concentration has been removed. In the case of the tensile bar of Fig. 10.15 for example, σ_{nom} has been taken as $P/(b-d)t$. An alternative system would have been to base the nominal stress σ_{nom} upon the full 'un-notched' cross-sectional area i.e. $\sigma_{nom} = P/t$. Clearly, the stress concentration factors resulting from this approach would be very different, particularly as the size of the hole increases.

(3) In the case of combined loading, the stress calculated under each type of load must be multiplied by its own stress concentration factor. In combined bending and axial load, for example, the bending stress ($\sigma_b = My/I$) should be multiplied by the bending s.c.f. and the axial stress ($\sigma_d = P/A$) multiplied by the s.c.f. in tension.

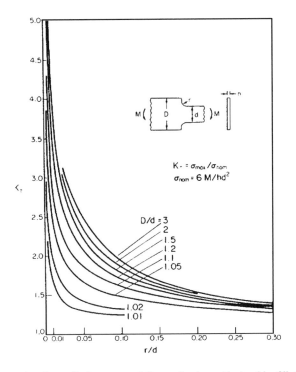

Fig. 10.22. Stress concentration factor K_t for a stepped flat tension bar with shoulder fillets subjected to bending.

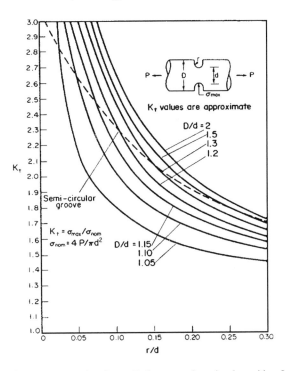

Fig. 10.23. Stress concentration factor K_t for a round tension bar with a U groove.

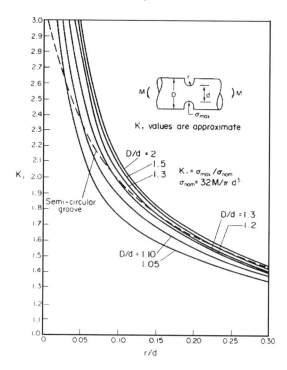

Fig. 10.24. Stress concentration factor K_t for a round bar with a U groove subjected to bending.

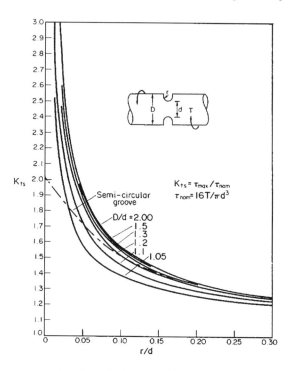

Fig. 10.25. Stress concentration factor K_t for a round bar with a U groove subjected to torsion.

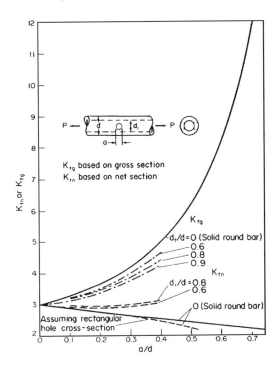

Fig. 10.26. Stress concentration factor K_t for a round bar or tube with a transverse hole subjected to tension.

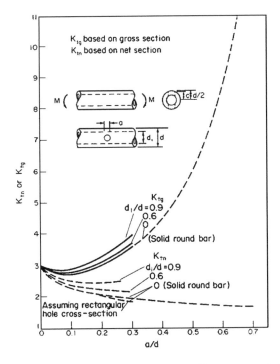

Fig. 10.27. Stress concentration factor K_t for a round bar or tube with a transverse hole subjected to bending.

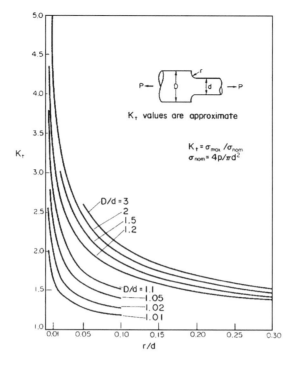

Fig. 10.28. Stress concentration factor K_t for a round bar with shoulder fillet subjected to tension.

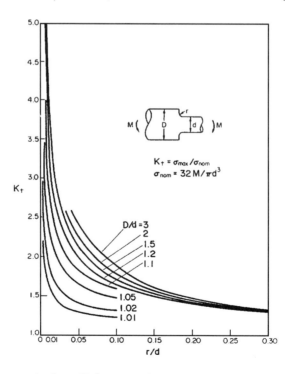

Fig. 10.29. Stress concentration factor K_t for a stepped round bar with shoulder fillet subjected to bending.

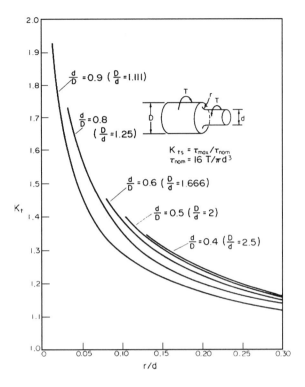

Fig. 10.30. Stress concentration factor K_t for a stepped round bar with shoulder fillet subjected to torsion.

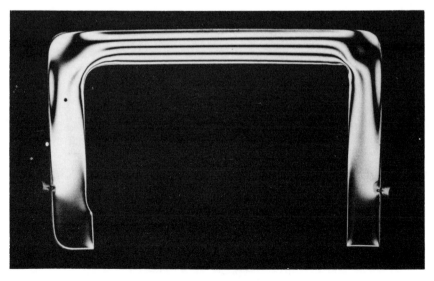

Fig. 10.31. Photoelastic fringe pattern of a portal frame showing stress concentration at the corner blend radii (different blend radii produce different stress concentration factors)

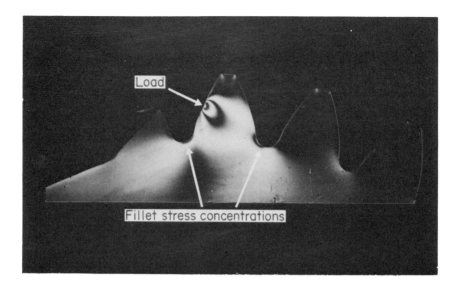

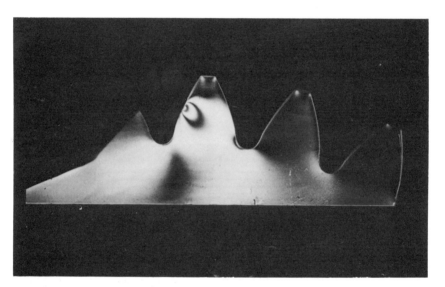

Fig. 10.32. Photoelastic fringe pattern of stress distribution in a gear tooth showing stress concentration at the loading point on the tooth flank and at the root fillet radii (higher concentration on the compressive fillet). Refer also to Fig. 10.45.

10.3.2. Saint-Venant's principle

The general problem of stress concentration was studied analytically by Saint-Venant who produced the following statement of principle: "If the forces acting on a small area of a body are replaced by a statically equivalent system of forces acting on the same area, there will be considerable changes in the local stress distribution but the effect at distances large compared with the area on which the forces act will be negligible". The effect of this principle is best demonstrated with reference to the photoelastic fringe pattern obtained in a model of a beam subjected to four-point bending, i.e. bending into a circular arc between the central

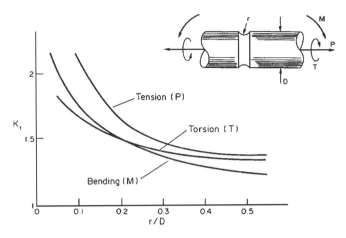

Fig. 10.33. Variation of stress concentration factors for a grooved shaft depending on the type of loading.

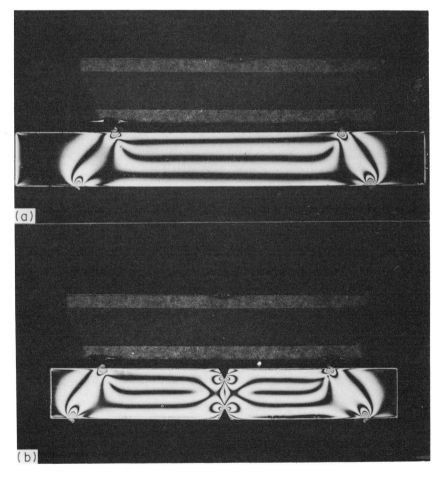

Fig. 10.34. (a) Photoelastic fringe pattern in a model of a beam subjected to four-point bending (i.e. circular arc bending between central supports): (b) as above but with a central notch.

supports – see Fig. 10.34(a). If the moment could have been applied by some other means so as to avoid the contact at the loading points then the fringe pattern would have been a series of parallel fringes, the centre one being the neutral axis. The stress concentrations due to the loading points are clearly visible as is the effect of these on the distribution of the fringes and hence stress. In particular, note the curvature of the neutral axis towards the inner loading points and the absence of the expected parallel fringe distribution both near to and outside the loading points. However, for points at least one depth of beam away from the stress concentrations (St. Venant) the fringe pattern is unaffected, the parallel fringes remain undisturbed and simple bending theory applies. If either the beam length is reduced or further stress concentrations (such as the notch of Fig. 10.34(b)) are introduced so that every part of the beam is within "one depth" of a stress concentration then at no point will simple theory apply and analysis of the fringe pattern is required for stress evaluation–there is no simple analytical procedure.

Similarly, in a round tension bar the stresses at the ends will be dependent upon the method of gripping or load application but within the main part of the bar, at least one diameter away from the loading point, stresses can again be obtained from simple theory. To the other extreme comes the case of a screw thread. The maximum s.c.f. arises at the first contacting thread at the plane of the bearing face of the head or nut and up to 70% of the load is carried by the first two or three threads. In such a case, simple theory cannot be applied anywhere within the component and the reader is referred to the appropriate B.S. Code of Practice and/or the work of Brown and Hickson[59].

10.3.3. Theoretical considerations of stress concentrations due to concentrated loads

A full treatment of the local stress distribution at points of application of concentrated load is beyond the scope of this text. Two particular cases will be introduced briefly, however, in order that the relevant useful equations can be presented.

(a) Concentrated load on the edge of an infinite plate

Work by St. Venant, Boussinesq and Flamant (see §8.7.9) has led to the development of a theory based upon the replacement of the concentrated load by a radial distribution of loads around a semi-circular groove (which replaces the local area of yielding beneath the concentrated load) (see Fig. 10.35). Elements in the material are then, according to Flamant, subjected to a radial compression of

$$\sigma_r = \frac{2P\cos\theta}{\pi b r} \quad \text{with } b = \text{ width of plate} \tag{10.26}$$

This produces element cartesian stresses of:

$$\sigma_{xx} = \sigma_r \sin^2\theta = -\frac{2P\cos\theta\sin^2\theta}{\pi b r} = -\frac{2Px^2 y}{\pi b(x^2 + y^2)^2} \tag{10.27}$$

$$\sigma_{yy} = \sigma_r \cos^2\theta = \frac{-2P\cos^3\theta}{\pi b r} = \frac{-2Py^3}{\pi b(x^2 + y^2)^2} \tag{10.28}$$

$$\tau_{xy} = \sigma_r \sin\theta\cos^2\theta = -\frac{2P\sin\theta\cos^2\theta}{\pi b r} = -\frac{2Pxy^2}{b(x^2 + y^2)^2} \tag{10.29}$$

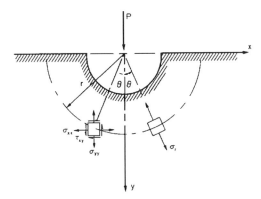

Fig. 10.35. Elemental stresses due to concentrated load P on the edge of an infinite plate.

(b) Concentrated load on the edge of a beam in bending

In this case a similar procedure is applied but, with a finite beam, consideration must be given to the horizontal forces set up within the groove which result in longitudinal stresses additional to the bending effects.

The total stress across the vertical section through the loading point (or groove) is then given by the so-called "*Wilson–Stokes equation*".

$$\sigma_{xx} = \frac{P}{\pi bd} \pm \left[\frac{L}{4} - \frac{d}{2\pi}\right] \frac{2Py}{bd^3} \tag{10.30}$$

where d is the depth of the beam, b the breadth and L the span.

This form of expression can be shown to indicate that the maximum longitudinal stresses set up are, in fact, less than those obtained from the simple bending theory alone (in the absence of the stress concentration).

10.3.4. Fatigue stress concentration factor

As noted above, the plastic flow which develops at positions of high stress concentration in ductile materials has a stress-relieving effect which significantly nullifies the effect of the stress raiser under static load conditions. Even under cyclic or fatigue loading there is a marked reduction in stress concentration effect and this is recognised by the use of a fatigue *stress concentration factor* K_f.

In the absence of any stress concentration (i.e. for $K_t = 1$) materials exhibit an "*endurance limit*" or "*fatigue limit*" – a defined stress amplitude below which the material can withstand an indefinitely large (sometimes infinite) number of repeated load cycles. This is often referred to as the un-notched fatigue limit – see Fig. 10.36.

For a totally brittle material in which the elastic stress concentration factor K_t might be assumed to have its full effect, e.g. $K_t = 2$, the fatigue life or notched endurance limit would be reduced accordingly. For materials with varying plastic flow capabilities, the effect of stress-raisers produces notched endurance limits somewhere between the un-notched value and that of the 'theoretical' value given by the full K_t – see Fig. 10.36, i.e. the fatigue stress concentration factor lies somewhere between the full K_t value and unity.

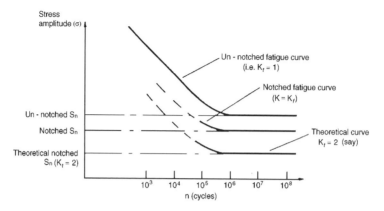

Fig. 10.36. Notched and un-notched fatigue curves.

If the endurance limit for a given number of cycles, n, is denoted by S_n then the fatigue stress concentration factor is defined as:

$$K_f = \frac{S_n \text{ for unnotched material}}{S_n \text{ for notched material}} \tag{10.31}$$

K_f is sometimes referred to by the alternative titles of "*fatigue strength reduction factor*" or, simply, the "*fatigue notch factor*".

The value of K_f is normally obtained from fatigue tests on identical specimens both with and without the notch or stress-raiser for which the stress concentration effect is required.

It is well known (and discussed in detail in Chapter 11) that the fatigue life of components is affected by a great number of variables such as mean stress, stress range, environment, size effect, surface condition, etc., and many different approaches have been proposed to allow realistic estimations of life under real working conditions as opposed to the controlled laboratory conditions under which most fatigue tests are carried out. One approach which is relevant to the present discussion is that proposed by Lipson & Juvinal[60] which utilises fatigue stress concentration factors, K_f, suitably modified by various coefficients to take account of the above-mentioned variables.

10.3.5. Notch sensitivity

A useful relationship between the elastic stress concentration factor K_t and the fatigue notch factor K_f introduces a *notch sensitivity* q defined as follows:

$$q = \frac{K_f - 1}{K_t - 1} \quad \text{or, in shear,} \quad q = \frac{K_{fs} - 1}{K_{ts} - 1}$$

which may be re-written in terms of the fatigue notch factor as:

$$K_f = 1 + q(K_t - 1) \quad \text{with } 0 \leqslant q \leqslant 1 \tag{10.32}$$

It will be seen that, at the extreme values of q, eqn. (10.32) is valid since when $q = 1$ the full effect of the elastic stress concentration factor K_t applies and $K_f = K_t$; similarly when $q = 0$ and full ductility applies there is, in effect, no stress concentration and $K_f = 1$ with the material behaving in an unnotched fashion.

The value of the notch sensitivity for stress raisers with a significant linear dimension (e.g. fillet radius) R and a material constant "a" is given by:

$$q = \frac{1}{\left(1 + \dfrac{a}{R}\right)} \tag{10.33}$$

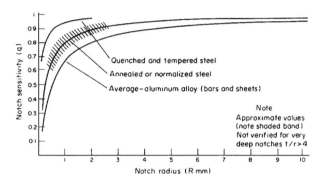

Fig. 10.37. Average fatigue notch sensitivity q for various notch radii and materials.

Typically, $a = 0.01$ for annealed or normalised steel, 0.0025 for quenched and tempered steel and 0.02 for aluminium alloy. However, values of "a" are not readily available for a wide range of materials and reference should be made to graphs of q versus R given by both Peterson[57] and Lipson and Juvinal[60].

The stress and strain distribution in a tensile bar containing a "through-hole" concentration are shown in Fig. 10.38 where the elastic stress concentration factor predictions are compared with those taking into account local yielding and associated stress redistribution.

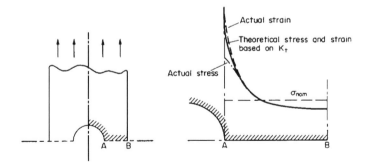

Fig. 10.38. Effect of a local yielding and associated stress re-distribution on the stress and strain concentration at the edge of a hole in a tensile bar.

10.3.6. Strain concentration – Neuber's rule

Within the elastic range, the concentration factor expressed in terms of strain rather than stress is equal to the stress concentration factor K_t. In the presence of plastic flow, however, the elastic stress concentration factor is reduced to the plastic factor K_p but local strains clearly exceed those predicted by elastic considerations – see Fig. 10.39.

A strain concentration factor can thus be defined as:

$$\mathbf{K}_\varepsilon = \frac{\textbf{maximum strain at the notch}}{\textbf{nominal strain at the notch}}$$

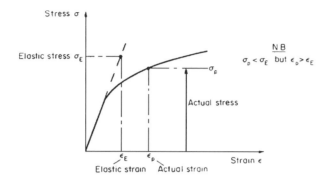

Fig. 10.39. Comparison of elastic and plastic stresses and strains.

the value of K_ε increasing as the value of K_p decreases. One attempt to relate the two factors is known as "*Neuber's Rule*", viz.

$$K_p K_\varepsilon = K_t^2 \qquad\qquad (10.34)$$

It is appropriate here to observe that recent research in the fatigue behaviour of materials indicates that the strain range of fatigue loading may be more readily related to fatigue life than the stress range which formed the basis of much early fatigue study. This is said to be particularly true of low-cycle fatigue where, in particular, the plastic strain range is shown to be critical.

10.3.7. Designing to reduce stress concentrations

From the foregoing discussion it should now be evident that stress concentrations are critical to the life of engineering components and that fatigue failures, for example, almost invariably originate at such positions. It is essential, therefore, for any design to be successful that detailed consideration is given to the reduction of stress concentration effects to an absolute minimum.

One important rule in this respect is concerned with the initial placement of the stress concentration. Assuming that some freedom exists as to the position of e.g. oil-holes, keyways, grooves, etc., then it is essential that these be located at positions where the nominal stress is as low as possible. The resultant magnitude of stress concentration factor × nominal stress is then also a minimum for a particular geometry of stress raiser.

In situations where no flexibility exists as to the position of the stress raiser then one of the procedures outlined below should be considered. In many cases a qualitative assessment of the benefits, or otherwise, of design changes is readily obtained by sketching the lines of stress flow through the component as in Fig. 10.17. Sharp changes in flow direction indicate high stress concentration factors, smooth changes in flow direction are the optimum solution.

The following standard stress concentration situations are common in engineering applications and procedures for reduction of the associated stress concentration factors are introduced for each case. The procedures, either individually or in combination, can then often be applied to produce beneficial stress reduction in other non-standard design situations.

(a) Fillet radius

Probably the most common form of stress concentration is that arising at the junction of two parts of a component of different shape, diameter, or other dimension. In almost every shaft, spindle, or axle design, for example, the component consists of a number of different diameter sections connected by shoulders and associated fillets.

If Fig. 10.40(a) is taken to be either the longitudinal section of a shaft or simply a flat plate, then the transition from one dimension to another via the right-angle junction is exceptionally bad design since the stress concentration associated with the sharp corner is exceedingly high. In practice, however, either naturally due to the fact that the machining tool has a finite radius, or by design, the junction is formed via a fillet radius and the wise designer employs the highest possible radius of fillet consistent with the function of the component in order to keep the s.c.f. as low as possible. Whilst, historically, circular arcs have generally been used for fillets, other types of blend geometry have been shown to produce even further reduction of s.c.f. notably elliptical and streamline fillets[61], the latter following similar contours to those of a fluid when it flows out of a hole in the bottom of a tank. Fig. 10.41 shows the effect of elliptical fillets on the s.c.f. values.

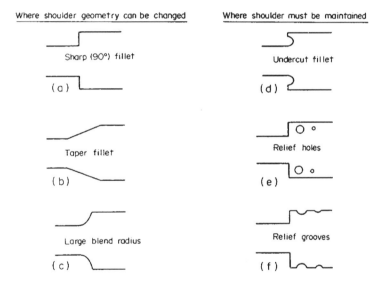

Fig. 10.40. Various methods for reduction of stress concentration factor at the junction of two parts of a component of different depth/diameter.

There are occasions, however, where the perpendicular faces at the junction need to be maintained and only a relatively small fillet radius can be allowed e.g. for retention of bearings or wheel hubs. A number of alternative solutions for reduction of the s.c.f's are shown in Fig. 10.40(d) to (f) and Fig. 10.42.

(b) Keyways or splines

It is common to use keyways or splines in shaft applications to provide transfer of torque between components. Gears or pulleys are commonly keyed to shafts, for example, by square

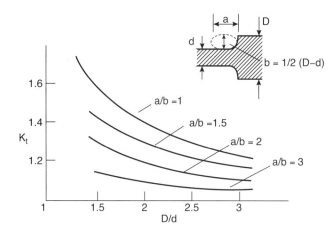

Fig. 10.41. Variation of elliptical fillet stress concentration factor with ellipse geometry.

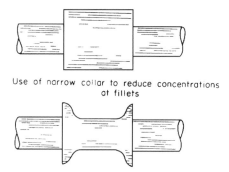

Use of narrow collar to reduce concentrations
at fillets

Fig. 10.42. Use of narrow collar to reduce stress concentration at fillet radii in shafts.

keys with side dimensions approximately equal to one-quarter of the shaft diameter with the depth of the keyway, therefore, one-eighth of the shaft diameter.

Analytical solutions for such a case have been carried out by both Leven[63] and Neuber[65] each considering the keyway without a key present. Neuber gives the following formula for stress concentration factor (based on shear stresses):

$$K_{t_s} = 1 + \sqrt{\frac{h}{r}} \qquad (10.35)$$

where h = keyway depth and r = radius at the base of the groove or keyway (see Fig. 10.43). For a semi-circular groove $K_{t_s} = 2$.

Leven, considering the square keyway specifically, observes that the s.c.f. is a function of the keyway corner radius and the shaft diameter. For a practical corner radius of about one-tenth the keyway depth $K_{t_s} \simeq 3$.

If fillet radii cannot be reduced then s.c.f.'s can be reduced by drilling holes adjacent to the keyway as shown in Fig. 10.43(b).

The presence of a key and its associated fit (or lack of) has a significant effect on the stress distribution and no general solution exists. Each situation strictly requires its own solution via practical testing such as photoelasticity.

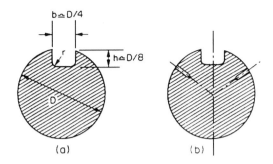

Fig. 10.43. Key-way dimensions and stress reduction procedure.

(c) Grooves and notches

Circumferential grooves or notches (particularly U-shaped notches) occur frequently in engineering design in such applications as C-ring retainer grooves, oil grooves, shoulder or grinding relief grooves, seal retainers, etc; even threads may be considered as multi-groove applications.

Most of the available s.c.f. data available for grooves or notches refers to U-shaped grooves and circular fillet radii and covers both plane stress and three-dimensional situations such as shafts with circumferential grooves. In general, the higher the blend radius, the lower the s.c.f; the optimum value being $K_t = 2$ for a semi-circular groove as calculated by Neuber's equation (10.35) above.

Some data exists for other forms of groove such as V notches and hyperbolic fillets but, particularly in bending and tension, the latter have little advantage over circular arcs and V notches only show significant advantage for included angles greater than 120°. In cases where s.c.f. data for a particular geometry of notch are not readily available recourse can be made to standard factor data for plates with a central hole.

Stress concentrations at notches and grooves can be reduced by the "metal removal – stiffness reduction" technique utilising any procedure which improves the stress flow, e.g. multiple notches of *U* grooves or selected hole drilling as shown in Fig. 10.44. Reductions of the order of 30% can be obtained.

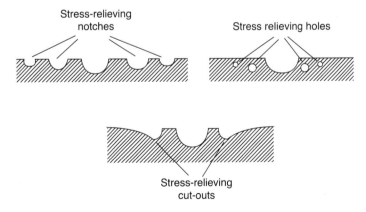

Fig. 10.44. Various procedures for the reduction of stress concentrations at notches or grooves.

This procedure of introducing secondary stress concentrations deliberately to reduce the local stiffness of the material adjacent to a stress concentration is a very powerful stress reduction technique. In effect, it causes more of the stiffer central region of the component to carry the load and persuades the stress lines to follow a path removed from the effect of the single, sharp concentration. Figures 10.40(d) to (f), 10.42 and 10.43 are all examples of the application of this technique, sometimes referred to as an "interference effect" the individual concentrations interfering with each other to mutual advantage.

(d) Gear teeth

The full analysis of the stress distribution in gear teeth is a highly complex problem. The reader is only referred in this section to the stress concentrations associated with the fillet radii at the base of the teeth – see Fig. 10.45.

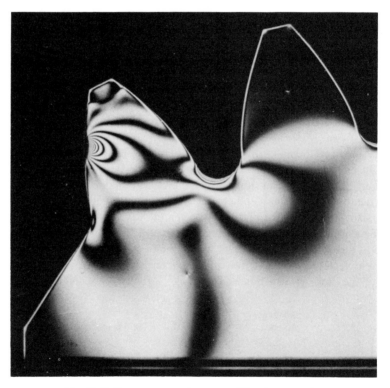

Fig. 10.45. Stress concentration at root fillet of gear tooth.

The loading on the tooth produces both direct stress and bending components on the root section and Dolan and Broghammer[68] in early studies of the problem gave the following formula for the combined stress concentration effect (for $20°$ pressure angle gears)

$$K_t = 0.18 + \frac{1}{\left(\dfrac{r}{t}\right)^{0.15} \left(\dfrac{h}{t}\right)^{0.45}}$$

Later work by Jacobson[69], again for 20° pressure angle gears, produced a series of charts of strength factors and more recently Hearn[66,67] has carried out photoelastic studies of both two-dimensional involute tooth forms and three-dimensional helical gears which introduce new considerations of stress concentration factors, notably their variation in both magnitude and position as the load moves up and down the tooth flank.

(e) Holes

From much of the previous discussion it should now be evident that holes represent very significant stress raisers, be they in two-dimensional plates or three-dimensional bars. Fortunately, a correspondingly high amount of information and data is available, e.g. Peterson[57], covering almost every foreseeable geometry and loading situation. This includes not only individual holes but rows and groups of holes, pin-joints, internally pressurised holes and intersecting holes.

(f) Oil holes

The use of transverse and longitudinal holes as passages for lubricating oil is common in shafting, gearing, gear couplings and other dynamic mechanisms. Occasionally similar holes are also used for the passage of cooling fluids.

In the case of circular shafts, no problem arises when longitudinal holes are bored through the centre of the shaft since the nominal torsional stress at this location is very small and the effect on the overall strength of the shaft is minimal. A transverse hole, however, is a significant source of stress concentration in any mode of loading, i.e. bending, torsion or axial load, and the relevant s.c.f. values must be evaluated from standard reference texts[57, 60]. Whatever the type of loading, the value of K_t increases as the size of the hole increases for a given shaft diameter, with minimum values for very small holes of 2 for torsion and 3 for bending and tension.

In cases of combined loading, a conservative estimate[58] of the stress concentration may be obtained from values of K_t given by either Peterson[57] or Lipson[60] for an infinite plate containing a transverse hole and subjected to an equivalent biaxial stress condition.

One procedure for the reduction of the stress concentration at the point where transverse holes cut the surface of shafts is shown in Fig. 10.46.

(g) Screw threads

Again the stress distribution in screw threads is extremely complex, values of the stress concentration factors associated with each thread being dependent upon the tooth form, the fit between the nut and the bolt, the nut geometry, the presence or not of a bolt shank and the load system applied. Pre-tensioning also has a considerable effect. However, from numerous photoelastic studies carried out by the author and others[59, 61, 62] it is clear that the greatest stress most often occurs at the first mating thread, generally at the mating face of the head of the nut with the bearing surface, with practically all the load shared between the first few threads. (One estimate of the source of bolt failures shows 65% in the thread at the nut face compared with 20% at the end of the thread and 15% directly under the bolt

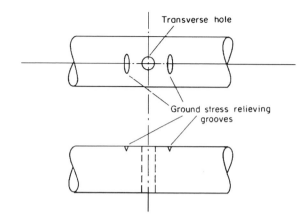

Fig. 10.46. Procedure for reduction of stress concentration at exit points of transverse holes in shafts.

head). Alternative designs of nut geometry can be introduced to spread the load distribution a little more evenly as shown in Figs. 10.47 and tapering of the thread is a very effective load-distribution mechanism.

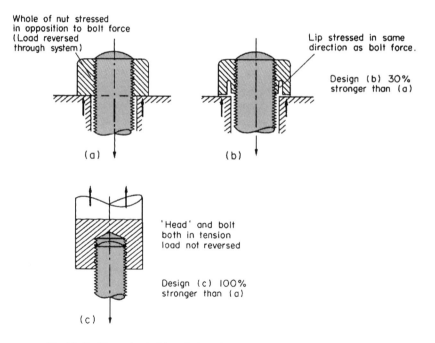

Fig. 10.47. Alternative bolt/nut designs for reduction of stress concentrations.

Reduction in diameter of the bolt shank and a correspondingly larger fillet radius under the bolt head also produces a substantial improvement as does the use of a material with a lower modulus of elasticity for the nut compared with the bolt; fatigue tests have shown strength improvements of between 35 and 60% for this technique.

Stress concentration data for various nut and bolt configurations are given by Hetenyi[62], again based on photoelastic studies. As an example of the severity of loading at the first thread, stress concentration factors of the order of 13 are readily obtained in conventional nut designs and even using the modified designs noted above s.c.f.'s of up to 9 are quite common. It is not perhaps surprising, therefore, that one of the most common causes of machinery or plant failure is that of stud or bolt fracture.

(h) Press or shrink fit members

There are some applications where discontinuity of component profile caused by two contacting members represents a substantial stress raiser effectively as great as a right-angle fillet. These include shrink or press-fit applications such as collars, gears, wheels, pulleys, etc., mounted on their drive shafts and even simple compressive loading of rectangular faces on wider support plates – see Fig. 10.48(a).

(a)

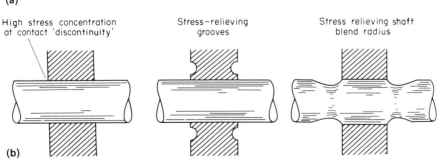

(b)

Fig. 10.48. (a) Photoelastic fringe pattern showing stress concentrations produced at contact discontinuities such as the loading of rectangular plates on a flat surface (equivalent to cross-section of cylindrical roller bearing on its support surface); (b) the reduction of stress concentration at press and shrink fits.

Significant stress concentration reductions can be obtained by introducing stress-relieving grooves or a blending fillet (or taper) in the press-fit member or the shaft – see Fig. 10.48.

10.3.8. Use of stress concentration factors with yield criteria

Whilst stress concentration factors are defined in terms of the maximum individual stress at the stress raiser it could be argued that, since stress conditions there are normally biaxial, it would be more appropriate to express them in terms of some "equivalent stress" employing one of the yield criteria introduced in Chapter 15.[†]

Since the maximum shear strain energy (distortion energy) theory of Von Mises is usually considered to be the most applicable to both static and dynamic conditions in ductile materials then, for a biaxial state the Von Mises equivalent stress can be defined as:

$$\sigma_e = \sqrt{\sigma_1^2 - \sigma_1\sigma_2 + \sigma_2^2} \tag{10.36}$$

and, since there is always a direct relationship between σ_1 and σ_2 within the elastic range for biaxial states (i.e. $\sigma_1 = k\sigma_2$) then

$$\sigma_e = \left[\sigma_1^2 - \frac{\sigma_1^2}{k} - \frac{\sigma_1^2}{k^2} \right]^{1/2}$$

$$= \sigma_1 \left[1 - \left(\frac{1}{k} \right) - \left(\frac{1}{k^2} \right) \right]^{1/2}$$

Then the stress concentration factor expressed in terms of this equivalent stress will be

$$K_e = \frac{\sigma_e}{\sigma_{\text{nom}}} = \frac{\sigma_1}{\sigma_{\text{nom}}} \left[1 - \left(\frac{1}{k} \right) - \left(\frac{1}{k^2} \right) \right]^{1/2} \tag{10.37}$$

Except for the special case of equal bi-axial stress conditions when $\sigma_1 = \sigma_2$ and $K = 1$ the value of K_e is always less than K_t.

A full treatment of the design procedures to be adopted for both ductile and brittle materials incorporating both yield criteria (Von Mises and Mohr) and stress concentration factors is carried out by Peterson[57] with consideration of static, alternating and combined static and alternating stress conditions.

10.3.9. Design procedure

The following procedure should be adopted for the design of components in order that the effect of stress concentration is minimised and for the component to operate safely and reliably throughout its intended service life.

(1) Prepare a draft design incorporating the principal features and requirements of the component. The dimensions at this stage will be obtained with reference to the nominal stresses calculated on the basis of known or estimated service loads.

(2) Identify the potential stress concentration locations.

[†] E.J. Hearn, *Mechanics of Materials 1*, Butterworth-Heinemann, 1997.

(3) Undertake the procedures outlined in §10.3.7 to reduce the stress concentration factors at these locations by:

    (a) streamlining the design where possible to avoid sharp changes in geometry and producing gradual fillet transitions between adjacent parts of different shape and size.

    (b) If fillet changes cannot be effected owing to design constraints, of e.g. bearing surfaces, undertake other modifications to the design to produce smoother "flow" of the stresses through the component.

    (c) Where appropriate, reduce the stiffness of the material adjacent to the stress concentration positions to allow greater flexibility and a reduction in the associated stress concentration factor. This is probably best achieved by removal of material as discussed earlier.

(4) Evaluate the stress concentration factors for the modified design using standard tables[57, 60] or experimental test procedures such as photoelasticity. Depending on the material and the loading conditions either K_t or K_f may be appropriate.

(5) Ensure that the maximum stress in the component taking into account both the stress concentration factors and an additional safety factor to account for service uncertainties, does not exceed the safe working stress for the material concerned.

References

1. Hertz, H. "On the contact of elastic solids", *J. reine angew Math.* (1881), **92**, 56.
2. Belajef, N. M. "On the problem of contact stresses", *Bull. Eng. Ways of Communication"*, St. Petersburg, 1917.
3. Foppl, L. "Zertschrift fur Angew", *Math und Mech.* (1936), **16**, 165.
4. M'Ewen, E. "The load-carrying capacity of the oil film between gear teeth". *The Engineer*, (1948), **186**, 234–235.
5. Radzimovsky, E. I. "Stress distribution of two rolling cylinders pressed together," *Bulletin* 408, Eng. Exp. Sta. Univ. Ill., 1953.
6. Buckingham, E. "Dynamic loads on gear teeth", ASME Special Committee on Strength of Gear Teeth. New York, 1931.
7. Walker, H. "A laboratory testing machine for helical gear tooth action", *The Engineer*, (6 June 1947), **183**, 486–488.
8. Wellauer, E. J. "Surface durability of helical and herringbone gears", *Machine Design*, May 1964.
9. Huber, M. T. *Ann. Phys. Paris* (1904), **14**, 153.
10. Morton W. B. and Close, L. J. "Notes on Hertz's theory of the contact of elastic bodies" *Phil. Mag.* (1922), 6th series, **43**, 320.
11. Thomas H. R. and Hoersch, V. A. "Stresses due to the pressure of one elastic solid on another" *Univ. Illinois Eng. Exp. Sta. Bull.* (1930), **212**.
12. Fessler, H. and Ollerton, E. "Contact stresses in toroids under radial loads", *Brit. J. Appl. Phys.* (1957), **8**, 387.
13. Ollerton, C. "Stresses in the contact zone". Convention on Adhesion. I. Mech. E., London, 1964.
14. Johnson, K. L. "A review of the theory of rolling contact stresses", O.E.C.D. Sub-group on Rolling Wear. Delft, April, 1965. *Wear* (1966), **9**, 4–19.
15. Deresiewicz, H. "Contact of elastic spheres under an oscillating torsional couple". *Ibid.* (1954), **76**, 52.
16. Johnson, K. L. "Plastic contact stresses" BSSM Conference. Sub-surface stresses." Nov. 1970, unpublished.
17. Lubkin, J. L. "The torsion of elastic spheres in contact" *J. Appl. Mech. Trans ASME* (1951), **73**, 183.
18. Mindlin, R. D. and Deresiewicz, H. "Elastic spheres in contact under varying oblique forces", *J. Appl. Mech. Trans ASME* (1953), **75**, 327.
19. Tomlinson, G. A. and Gough, H. J. "An investigation of the fretting corrosion of closely fitting surfaces". *Proc. I Mech. E.* (1939), **141**, 223.
20. Smith, J. O. and Liu, C. K. "Stresses due to tangential and normal loads on an elastic solid with application to some contact stress problems", *J. App. Mech.*, June 1953.
21. Lipson, C. and Juvinal, R. C. *Handbook of Stress and Strength.* Macmillan, New York, 1963.
22. Meldahl, A. "Contribution to the Theory of Lubrication of Gears and of the Stressing of the Lubricated Flanks of Gear Teeth" Brown Boveri Review. Vol. 28, No. 11, Nov. 1941.

23. Dowson, D., Higginson, G. R. and Whitaker, A. V. "Stress distribution in lubricated rolling contacts", *I. Mech. E. Fatigue in Rolling Contact* (1964).

24. Crook, A. W. "Lubrication of Rollers". Parts I–IV. *Phil. Trans. Roy. Soc.*-Ser. A250 (1958); *Ibid.*, 254 (1962) and 255 (1963).

25. Dawson, P. H. "Rolling contact fatigue crack initiation in a 0.3% carbon steel", *Proc. I. Mech. E.* (1968/69), **183**, Part 1.

26. Scott, D. "The effect of materials properties, lubricant and environment in rolling contact fatigue", *I. Mech. E. Fatigue in Rolling Contact* (1964).

27. Sherratt, F. "The influence of shot-peening and similar surface treatments on the fatigue properties of metals", B.S.S.M. Sub-Surface Stresses, 1970 (unpublished).

28. Muro, H. "Changes of residual stress due to rolling contacts", *J. Soc. Mat. Sci. Japan* (July 1969), **18**, 615–619.

29. Timoshenko, S. P. and Goodier, J. N. *Theory of Elasticity*, 2nd edn. McGraw-Hill, 1951.

30. Almen, J. O. "Surface deterioration of gear teeth", *Trans. A.S.M.* (Jan. 1950).

31. Akaoka, J. "Some considerations relating to plastic deformation under rolling contact, *Rolling Contact Phenomena*. Elsevier.

32. Hearn E. J. "A three-dimensional photoelastic analysis of the stress distribution in double helical epicyclic gears", *Strain* (July 1979).

33. Roark, R. J. and Young, W. C. *"Formulas for Stress and Strain"*, 5th edn., McGraw-Hill, 19

34. Heindlhofer, K. *Evaluation of Residual Stresses*. McGraw-Hill, 1948.

35. Rosenthal, and Norton, "A method of measuring tri-axial residual stresses in plates", *Journal Welding Society*, (Welding supplement, 1945), **24**, 295–307.

36. Waisman, and Phillips, "Simplified measurement of residual stresses", *S.E.S.A.*, **II** (2), 29–44.

37. Mathar, J. "Determination of initial stresses by measuring the deformations around drilled holes", *Trans A.S.M.E.*, **56** (4), (1934), 249–254.

38. Bathgate, R. G. "Measurement of non-uniform bi-axial residual stresses by the hole drilling method", *Strain.* **4** (2) (1968), 20–27.

39. Beaney, E. M. "The Air Abrasive Centre-Hole Technique for the Measurement of Residual Stresses". B.S.S.M. Int'l Conference: "Product Liability and Reliability", September 1980 (Birmingham).

40. Procter, E. "An Introduction to Residual Stresses, their Measurement and Reliability Aspects". B.S.S.M. Int'l Conference: "Product Liability and Reliability", September 1980 (Birmingham).

41. Beaney, E. M. and Procter, E. "A critical evaluation of the centre-hole technique for the measurement of residual stresses". C.E.G.B. report (UK), RD/B/N2492, Nov. 1972.

42. Milbradt, K. P. "Ring method determination of residual stresses" *S.E.S.A.*, **9** (1), 63–74.

43. Durelli, A. J. and Tsao, C. H. "Quantitative evaluation of residual stresses by the stresscoat drilling technique" *S.E.S.A.*, **9** (1), 195–207.

44. French, D. N. and Macdonald, B. A. "Experimental methods of X-ray analysis", *S.E.S.A.*, **26** (2), 456–462.

45. Kirk, D. "Theoretical aspects of residual stress measurement by X-ray diffractometry", *Strain*, **6** (2) (1970).

46. Kirk, D. "Experimental features of residual stress measurement by X-ray diffractometry", *Strain*, **7** (1) (1971).

47. Andrews, K. W. *et al.* "Stress measurement by X-ray diffraction using film techniques", *Strain*, **10** (3) (July 1974), 111–116.

48. Abuki, S. and Cullity, B. D. "A magnetic method for the determination of residual stresses", *S.E.S.A.*, **28** (1), 217–223.

49. Sines, G. and Carlson, R. "Hardness measurements for the determination of residual stresses", *A.S.T.M. Bull.*, no. 180 (February 1952), pp. 35–37.

50. Noranha, P. J. and Wert, J. J. "An ultrasonic technique for the measurement of residual stress", *Journal Testing and Evaluation*, **3** (2) (1975), 147–172.

51. Kino, G. S. *et al.* "Acoustic techniques for measuring stress regions in materials", Electric Power Research Institute NP-1043 Project 609-1. Interim Report Apr. 1979.

52. Hearn, E. J. and Golsby, R. J. "Residual stress investigation of a noryl telecommunication co-ordinate selector switch", *S.E.S.A.*, 3rd Int. Congress Experimental Mechanics, Los Angeles, 1973.

53. Denton, A. A. "The use of spark machining in the determination of local residual stress" *Strain*, **3** (3) (July 1967), 19–24.

54. Moore, M. G. and Evans, W. D. "Mathematical correction for stress in removed layers in X-ray diffraction residual stress analysis", *S.A.E. Trans.*, **66** (1958), 340–345.

55. Scaramangas, A. A. *et al.* "On the correction of residual stress measurements obtained using the centre-hole method", *Strain*, **18** (3) (August 1982), 88–96.

56. Huang, T. C. *Bibliography on Residual Stresses*. Soc. Auto. Eng., New York, 1954.

57. Peterson, R. E. *Stress Concentration Design Factors*, John Wiley & Sons, New York, 1953.

58. Perry, C. C. "Stress and strain concentration", Vishay Lecture and Series, Vishay Research and Education, Michigan, U.S.A.

59. Brown, A. F. C. and Hickson, V. M. "A photoelastic study of stresses in screw threads", *Proc. I. Mech. E.*, **1B**, 1952–3.

60. Lipson, C. and Juvinal, R. C. *Handbook of Stress and Strength*. Macmillan, New York, 1963.

61. Heywood, R. B. *Photoelasticity for Designers.* Pergamon, Oxford, 1969.
62. Hetenyi, M. "A photoelastic study of bolt and nut fastenings", *Trans. A.S.M.E. J. Appl. Mechs*, **65** (1943).
63. Leven, M. M. "Stresses in keyways by photoelastic methods and comparison with numerical solution", *Proc. S.E.S.A.*, **7** (2) (1949).
64. Roark, R. J. *Formulas for Stress and Strain*, 5th edition. McGraw-Hill, New York, 1975.
65. Neuber, N. *Theory of Notch Stresses.* Edwards, Michigan, 1946.
66. Hearn, E. J. "A new look at the bending strength of gear teeth", *Experimental Mechanics S.E.S.A.*, October 1980.
67. Hearn, E. J. "A new design procedure for helical gears", *Engineering*, October 1978.
68. Dolan and Broghamer
69. Jacobson, M. A. "Bending stresses in spur gear teeth: proposed new design factors. ...", *Proc. I. Mech. E.*, 1955.

Examples

Example 10.1

(a) Two parallel steel cylinders of radii 50mm and 100mm are brought into contact under a load of 2 kN. If the cylinders have a common length of 150 mm and elastic constants of $E = 208$ GN/m^3 and $v = 0.3$ determine the value of the maximum contact pressure. What will then be the magnitude and position of the maximum shear stress?

(b) How would the values change if the larger cylinder were replaced by a flat surface?

Solution (a)

For contacting parallel cylinders eqn. (10.9) gives the value of the maximum contact pressure (or compressive stress) as

$$\sigma_c = -p_0 = -0.591\sqrt{\frac{P}{L\Delta}\left(\frac{1}{R_1} + \frac{1}{R_2}\right)}$$

where

$$\Delta = \frac{1}{E_1}[1 - v_1^2] + \frac{1}{E_2}[1 - v_2^2]$$

$$= \frac{2}{E}[1 - v^2] \text{ for similar materials}$$

$$= \frac{2 \times 0.91}{208 \times 10^9}$$

∴ Max. contact pressure

$$p_0 = 0.591\sqrt{\frac{2 \times 10^3 \times 208 \times 10^9}{150 \times 10^{-3} \times 2 \times 0.91}\left(\frac{1}{50} + \frac{1}{100}\right)10^3}$$

$$= 0.591 \times 21.38 \times 10^7$$

$$= \mathbf{126.4 \ MN/m^2}$$

Maximum shear stress $= 0.295\,p_0 = \mathbf{37.3 \ MN/m^2}$

occurring at a depth $\mathbf{d = 0.786b}$

with (from eqn. (10.8)) $b = 1.076 \sqrt{\dfrac{P\Delta}{L\left(\dfrac{1}{R_1} + \dfrac{1}{R_2}\right)}}$

$$= 1.076 \sqrt{\frac{2 \times 10^3 \times 2 \times 0.91}{150 \times 10^{-3} \times 208 \times 10^9 \times 30}}$$

$$= 1.076 \times 0.624 \times 10^{-4}$$

$$= 0.067 \text{ mm}$$

∴   Depth of max shear stress $= 0.786 \times 0.067 = \mathbf{0.053\ mm}$

(*b*) Replacing the 100 mm cylinder by a flat surface makes $\dfrac{1}{R_2} = 0$ and

$$\text{contact pressure } p_0 = 0.591 \sqrt{\frac{2 \times 10^3 \times 208 \times 10^9}{150 \times 10^{-3} \times 2 \times 0.91}\left(\frac{1}{50}\right)} \, 10^3$$

$$= 0.591 \times 17.48 \times 10^7$$

$$= \mathbf{103.2\ MN/m^2}$$

with max shear stress $= 0.295 \times 103.2 = \mathbf{30.4\ MN/n^2}$

and $b = 0.082$ mm

∴   Depth of max shear stress $= 0.786 \times 0.082 = \mathbf{0.064\ mm}$.

Example 10.2

(a) What will be the maximum compressive stress set up when two spur gears transmit a torque of 250 N m? One gear has 150 teeth on a pitch circle diameter of 200 mm whilst the second gear has 200 teeth. Both gears have a common face-width of 200 mm. Assume $E = 208$ GN/m^2 and $\nu = 0.3$ for both gears.

(b) How will this value change if the spur gears are replaced by helical gears of $17\frac{1}{2}^\circ$ pressure angle and 30° helix?

Solution (a)

(*a*) From eqn. (10.21) the maximum compressive stress at contact is

$$\sigma_c = -p_0 = -0.475\sqrt{K}$$

with $K = \dfrac{W}{F_w d}\left[\dfrac{m+1}{m}\right]$

$$= \frac{250}{100 \times 10^{-3}} \times \frac{1}{200 \times 10^{-3} \times 200 \times 10^{-3}} \left[\frac{\dfrac{200}{150} + 1}{\dfrac{200}{150}}\right]$$

$$= 109375$$

$$\therefore \qquad \sigma_c = -0.475\sqrt{109375} = -\textbf{157.1 MN/m}^2$$

the negative sign indicating a compressive stress value.

Solution (b)

For the helical gears, eqn. (10.22) gives

$$\sigma_c = -p_0 = C\sqrt{\frac{K}{m_p}}$$

and for the given pressure angle and helix values Table 10.2 gives values of

$$C = 0.435 \text{ and } m_p = 1.53$$

$$\therefore \sigma_c = -0.435\sqrt{\frac{109375}{1.52}} = -\textbf{116.3 MN/m}^2$$

Example 10.3

A rectangular bar with shoulder fillet is subjected to a uniform bending moment of 100 Nm. Its dimensions are as follows (see Fig. 10.22) $D = 50$ mm; $d = 25$ mm; $r = 2.5$ mm; $h = 10$ mm.

Determine the maximum stress present in the bar for static load conditions. How would the value change if (a) the moment were replaced by a tensile load of 20 kN, (b) the moment and the tensile load are applied together.

Solution

For applied moment

From simple bending theory, nominal stress (related to smaller part of the bar) is:

$$\sigma_{\text{nom}} = \frac{M y}{I} = M \times \frac{d}{2} \times \frac{12}{hd^3} = \frac{6M}{hd^2}$$

$$= \frac{6 \times 100}{10 \times 10^{-3} \times (25 \times 10^{-3})^2} = 96 \text{ MN/m}^2$$

Now from Fig. 10.22 the elastic stress concentration factor for $D/d = 2$ and $r/d = 0.1$ is:

$$K_t = 1.85$$

$$\therefore \qquad \text{Maximum stress} = 1.85 \times 96 = \textbf{177.6 MN/m}^2.$$

(a) For tensile load

Again for smallest part of the bar

$$\sigma_{\text{nom}} = \frac{P}{hd} = \frac{20 \times 10^3}{10 \times 10^{-3} \times 25 \times 10^{-3}} = 80 \text{ MN/m}^2$$

and from Fig. 10.2, $\qquad K_t = 2.44$

$$\therefore \qquad \text{Maximum stress} = 2.44 \times 80 = \textbf{195.2 MN/m}^2.$$

(b) For combined bending and tensile load

Since the maximum stresses arising from both the above conditions will be direct stresses in the fillet radius then the effects may be added directly, i.e. the most adverse stress condition will arise in the bending tensile fillet when the maximum stress due to combined tension and bending will be:

$$\sigma_{max} = K_t \sigma_{b_{nom}} + K'_t \sigma_{d_{nom}}$$

$$= 177.6 + 195.2 = \textbf{372.8 MN/m}^2$$

Example 10.4

A semi-circular groove of radius 3 mm is machined in a 50 mm diameter shaft which is then subjected to the following combined loading system:

(a) a direct tensile load of 50 kN,
(b) a bending moment of 150 Nm,
(c) a torque of 320 Nm.

Determine the maximum value of the stress produced by each loading separately and hence estimate the likely maximum stress value under the combined loading.

Solution

For the shaft dimensions given, $D/d = 50/(50 - 6) = 1.14$ and $r/d = 3/44 = 0.068$.

(a) For tensile load

$$\text{Nominal stress } \sigma_{nom} = \frac{P}{A} = \frac{50 \times 10^3}{\pi \times (22 \times 10^{-3})^2} = 32.9 \text{ MN/m}^2.$$

From Fig. 10.23 $K_t = 2.51$

∴   Maximum stress $= 2.51 \times 32.9 = \textbf{82.6 MN/m}^2$.

(b) For bending

$$\text{Nominal stress } \sigma_{nom} = \frac{32M}{\pi d^3} = \frac{32 \times 150}{\pi \times (44 \times 10^{-3})^3} = \textbf{18 MN/m}^2$$

and from Fig. 10.24, $K_t = 2.24$

∴   Maximum stress $= 2.24 \times 18 = \textbf{40.3 MN/m}^2$.

(c) For torsion

$$\text{Nominal stress } \tau_{nom} = \frac{16T}{\pi d^3} = \frac{16 \times 320}{\pi \times (44 \times 10^{-3})^3} = 19.1 \text{ MN/m}^2$$

and from Fig. 10.25, $Kt_s = 1.65$

∴   Maximum stress $= 1.65 \times 19.1 = \textbf{31.5 MN/m}^2$

(d) For the combined loading the direct stresses due to bending and tension add to give a total maximum direct stress of $82.6 + 40.3 = 122.9$ MN/m² which will then act in conjunction with the shear stress of 31.5 MN/m² as shown on the element of Fig. 10.49.

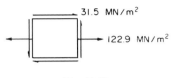

Fig. 10.49.

Then either by Mohrs circle or the use of eqn. (13.11)[†] the maximum principal stress will be

$$\sigma_1 = 130.5 \text{ MN/m}^2.$$

With a maximum shear stress of $\tau_{\text{max}} = 69 \text{ MN/m}^2$.

Example 10.5

Estimate the bending strength of the shaft shown in Fig. 10.50 for two materials

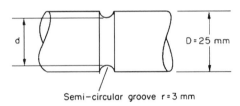

Semi-circular groove r = 3 mm

Fig. 10.50.

(a) Normalised 0.4% C steel with an unnotched endurance limit of 206 MN/m²
(b) Heat-treated $3\frac{1}{2}$% Nickel steel with an unnotched endurance limit of 480 MN/m².

Solution
From the dimension of the figure

$$\frac{D}{d} = \frac{25}{19} = 1.316 \quad \text{and} \quad \frac{r}{d} = \frac{3}{19} = 0.158$$

$$\therefore \text{ From Fig. 10.24} \qquad K_t = 1.75$$

From Fig. 10.37 for notch radius of 3 mm

$$q = 0.93 \text{ for normalised steel}$$

$$q = 0.97 \text{ for nickel steel (heat-treated)}$$

$\therefore$ From eqn. (10.32) for the normalised steel

$$K_f = 1 + q(K_t - 1)$$

$$= 1 + 0.93(1.75 - 1) = 1.698$$

[†] E.J. Hearn, *Mechanics of Materials 1*, Butterworth-Heinemann, 1997.

and the fatigue strength

$$\sigma_f = \frac{206}{1.698} = \textbf{12.3 MN/m}^2$$

and for the nickel steel

$$K_f = 1 + 0.97(1.75 - 1) = 1.728$$

and the fatigue strength

$$\sigma_f = \frac{480}{1.728} = \textbf{277.8 MN/m}^2$$

N.B. Safety factors should then be applied to these figures to allow for service loading conditions, etc.

Problems

10.1 (B). Two parallel steel cylinders of radii 100 mm and 150 mm are required to operate under service conditions which produce a maximum load capacity of 3000 N. If the cylinders have a common length of 200 mm and, for steel, $E = 208$ GN/m^2 and $v = 0.3$ determine:

(a) the maximum contact stress under peak load;
(b) the maximum shear stress and its location also under peak load.

[99.9 MN/m^2; 29.5 MN/m^2; 0.075 mm]

10.2 (B). How would the answers for problem 10.1 change if the 150 mm radius cylinder were replaced by a flat steel surface?   [77.4 MN/m^2; 22.8 MN/m^2; 0.097 mm]

10.3 (B). The 150 mm cylinder of problem 10.1 is now replaced by an aluminium cylinder of the same size. What percentage change of results is obtained?
For aluminium $E = 70$ GN/m^2 and $v = 0.27$. [−29.5%; −29.5%; +41.9%]

10.4 (B). A railway wheel of 400 mm radius exerts a force of 4500 N on a horizontal rail with a head radius of 300 mm. If $E = 208$ GN/m^2 and $v = 0.3$ for both the wheel and rail determine the maximum contact pressure and the area of contact.

[456 MN/m^2; 14.8 mm^2]

10.5 (B). What will be the contact area and maximum compressive stress when two steel spheres of radius 200 mm and 150 mm are brought into contact under a force of 1 kN? Take $E = 208$ GN/m^2 and $v = 0.3$.

[751 MN/m^2; 2.01 mm^2]

10.6 (B). Determine the maximum compressive stress set up in two spur gears transmitting a pinion torque of 160 Nm. The pinion has 100 teeth on a pitch circle diameter of 130 mm; the gear has 200 teeth and there is a common face-width of 130 mm. Take $E = 208$ GN/m^2 and $v = 0.3$.   [222 MN/m^2]

10.7 (B). Assuming the data of problem 10.6 now relate to a pair of helical gears of 30 helix and 20° pressure angle what will now be the maximum compressive stress?

[161.4 MN/m^2]

CHAPTER 11

FATIGUE, CREEP AND FRACTURE

Summary

Fatigue loading is generally defined by the following parameters

$$\text{stress range, } \sigma_r = 2\sigma_a$$

$$\text{mean stress, } \sigma_m = \tfrac{1}{2}(\sigma_{\max} + \sigma_{\min})$$

$$\text{alternating stress amplitude, } \sigma_a = \tfrac{1}{2}(\sigma_{\max} - \sigma_{\min})$$

When the mean stress is not zero

$$\text{stress ratio, } R_s = \frac{\sigma_{\min}}{\sigma_{\max}}$$

The *fatigue strength* σ_N for N cycles under zero mean stress is related to that σ_a under a condition of mean stress σ_m by the following alternative formulae:

$$\sigma_a = \sigma_N[1 - (\sigma_m/\sigma_{TS})] \quad \text{(Goodman)}$$

$$\sigma_a = \sigma_N[1 - (\sigma_m/\sigma_{TS})^2] \quad \text{(Geber)}$$

$$\sigma_a = \sigma_N[1 - (\sigma_m/\sigma_y)] \quad \text{(Soderberg)}$$

where σ_{TS} = tensile strength and σ_y = yield strength of the material concerned. Applying a factor of safety F to the Soderberg relationship gives

$$\sigma_a = \frac{\sigma_N}{F}\left[1 - \left(\frac{(\sigma_m \cdot F)}{\sigma_y}\right)\right]$$

Theoretical elastic stress concentration factor for elliptical crack of major and minor axes A and B is

$$K_t = 1 + 2A/B$$

The relationship between any given number of cycles n at one particular stress level to that required to break the component at the same stress level N is termed the "*stress ratio*" (n/N). *Miner's law* then states that for cumulative damage actions at various stress levels:

$$\frac{n_1}{N_1} + \frac{n_2}{N_2} + \frac{n_3}{N_3} + \cdots + \text{etc.} = 1$$

The *Coffin–Manson law* relating the plastic strain range $\Delta\varepsilon_p$ to the number of cycles to failure N_f is:

$$\Delta\varepsilon_p = K(N_f)^{-b}$$

or

$$\Delta\varepsilon_p = \left(\frac{N_f}{D}\right)^{-b}$$

where D is the ductility, defined in terms of the reduction in area r during a tensile test as

$$D = l_n \left(\frac{1}{1-r} \right)$$

The total strain range = elastic + plastic strain ranges

i.e. $\Delta\varepsilon_t = \Delta\varepsilon_e + \Delta\varepsilon_p$

the elastic range being given by *Basquin's law*

$$\Delta\varepsilon_e = \frac{3.5\sigma_{TS}}{E} \cdot N_f^{-0.12}$$

Under creep conditions the *secondary creep rate* ε_s^0 is given by the *Arrhenius equation*

$$\varepsilon_s^0 = Ae^{\left(-\frac{H}{RT}\right)}$$

where H is the activation energy, R the universal gas constant, T the absolute temperature and A a constant.

Under increasing stress the power law equation gives the secondary creep rate as

$$\varepsilon_s^0 = \beta\sigma^n$$

with β and n both being constants.

The latter two equations can then be combined to give

$$\varepsilon_s^0 = K\sigma^n e^{\left(-\frac{H}{RT}\right)}$$

The *Larson–Miller parameter* for life prediction under creep conditions is

$$P_1 = T(\log_{10} t_r + C)$$

The *Sherby–Dorn parameter* is

$$P_2 = \log_{10} t_r - \frac{\alpha}{T}$$

and the *Manson–Haferd parameter*

$$P_3 = \frac{T - T_a}{\log_{10} t_r - \log_{10} t_a}$$

where t_r = time to rupture and T_a and $\log_{10} t_a$ are the coordinates of the point at which graphs of T against $\log_{10} t_r$ converge. C and α are constants.

For *stress relaxation under constant strain*

$$\frac{1}{\sigma^{n-1}} = \frac{1}{\sigma_i^{n-1}} + \beta E(n-1)t$$

where σ is the instantaneous stress, σ_i the initial stress, β and n the constants of the power law equation, E is Young's modulus and t the time interval.

Griffith predicts that fracture will occur at a fracture stress σ_f given by

$$\sigma_f^2 = \frac{2bE\gamma}{\pi a(1-v^2)} \quad \text{for plane strain}$$

or
$$\sigma_f^2 = \frac{2bE\gamma}{\pi a} \quad \text{for plane stress}$$

where $2a$ = initial crack length (in an infinite sheet)

b = sheet thickness

γ = surface energy of crack faces.

Irwin's expressions for the cartesian components of stress at a crack tip are, in terms of polar coordinates;

$$\sigma_{yy} = \frac{K}{\sqrt{2\pi r}} \cos\frac{\theta}{2} \left[1 + \sin\frac{\theta}{2}\sin\frac{3\theta}{2} \right]$$

$$\sigma_{xx} = \frac{K}{\sqrt{2\pi r}} \cos\frac{\theta}{2} \left[1 - \sin\frac{\theta}{2}\sin\frac{3\theta}{2} \right]$$

$$\sigma_{xy} = \frac{K}{\sqrt{2\pi r}} \cos\frac{\theta}{2} \sin\frac{\theta}{2}\cos\frac{3\theta}{2}$$

where K is the *stress intensity factor* $= \sigma\sqrt{\pi a}$

or, for an edge-crack in a semi-infinite sheet

$$K = 1.12\sigma\sqrt{\pi a}$$

For *finite size components* with cracks generally growing from a free surface the *stress intensity factor* is modified to

$$K = \sigma Y \sqrt{a}$$

where Y is a *compliance function* of the form

$$Y = A\left(\frac{a}{W}\right)^{1/2} - B\left(\frac{a}{W}\right)^{3/2} + C\left(\frac{a}{W}\right)^{5/2} - D\left(\frac{a}{W}\right)^{7/2} + E\left(\frac{a}{W}\right)^{9/2}$$

In terms of load P, thickness b and width W

$$K = \frac{P}{bW^{1/2}} \cdot Y$$

For *elastic-plastic conditions* the plastic zone size is given by

$$r_p = \frac{K^2}{\pi\sigma_y^2} \quad \text{for plane stress}$$

and
$$r_p = \frac{K^2}{3\pi\sigma_y^2} \quad \text{for plane strain}$$

r_p being the extent of the plastic zone along the crack axis measured from the crack tip.

Mode II crack growth is described by the *Paris–Erdogan Law*

$$\frac{da}{dN} = C(\Delta K)^m$$

where C and m are material coefficients.

11.1. Fatigue

Introduction

Fracture of components due to fatigue is the most common cause of service failure, particularly in shafts, axles, aircraft wings, etc., where cyclic stressing is taking place. With static loading of a ductile material, plastic flow precedes final fracture, the specimen necks and the fractured surface reveals a fibrous structure, but with fatigue, the crack is initiated from points of high stress concentration on the surface of the component such as sharp changes in cross-section, slag inclusions, tool marks, etc., and then spreads or propagates under the influence of the load cycles until it reaches a critical size when fast fracture of the remaining cross-section takes place. The surface of a typical fatigue-failed component shows three areas, the small point of initiation and then, spreading out from this point, a smaller glass-like area containing shell-like markings called "*arrest lines*" or "*conchoidal markings*" and, finally, the crystalline area of rupture.

Fatigue failures can and often do occur under loading conditions where the fluctuating stress is below the tensile strength and, in some materials, even below the elastic limit. Because of its importance, the subject has been extensively researched over the last one hundred years but even today one still occasionally hears of a disaster in which fatigue is a prime contributing factor.

11.1.1. The S/N curve

Fatigue tests are usually carried out under conditions of rotating – bending and with a zero mean stress as obtained by means of a Wohler machine.

From Fig. 11.1, it can be seen that the top surface of the specimen, held "cantilever fashion" in the machine, is in tension, whilst the bottom surface is in compression. As the specimen rotates, the top surface moves to the bottom and hence each segment of the surface moves continuously from tension to compression producing a stress-cycle curve as shown in Fig. 11.2.

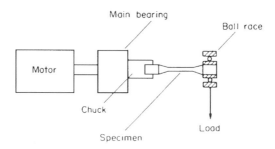

Fig. 11.1. Single point load arrangement in a Wohler machine for zero mean stress fatigue testing.

In order to understand certain terms in common usage, let us consider a stress-cycle curve where there is a positive tensile mean stress as may be obtained using other types of fatigue machines such as a Haigh "push-pull" machine.

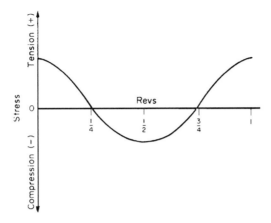

Fig. 11.2. Simple sinusoidal (zero mean) stress fatigue curve, "reversed-symmetrical".

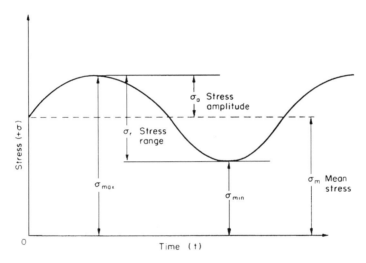

Fig. 11.3. Fluctuating tension stress cycle producing positive mean stress.

The stress-cycle curve is shown in Fig. 11.3, and from this diagram it can be seen that:

$$\textbf{Stress range, } \sigma_r = 2\sigma_a. \tag{11.1}$$

$$\textbf{Mean stress, } \sigma_m = \frac{\sigma_{max} + \sigma_{min}}{2} \tag{11.2}$$

$$\textbf{Alternating stress amplitude, } \sigma_a = \frac{\sigma_{max} - \sigma_{min}}{2} \tag{11.3}$$

If the mean stress is not zero, we sometimes make use of the "stress ratio" R_s where

$$R_s = \frac{\sigma_{min}}{\sigma_{max}} \tag{11.4}$$

The most general method of presenting the results of a fatigue test is to plot a graph of the stress amplitude as ordinate against the corresponding number of cycles to failure as abscissa, the amplitude being varied for each new specimen until sufficient data have been obtained. This results in the production of the well-known *S/N curve* – Fig. 11.4.

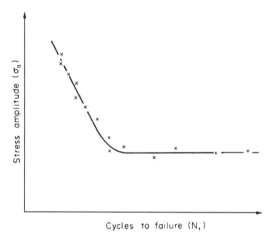

Fig. 11.4. Typical S/N curve fatigue life curve.

In using the S/N curve for design purposes it may be advantageous to express the relationship between σ_a and N_f, the number of cycles to failure. Various empirical relationships have been proposed but, provided the stress applied does not produce plastic deformation, the following relationship is most often used:

$$\sigma_r^a N_f = K \qquad (11.5)$$

Where a is a constant which varies from 8 to 15 and K is a second constant depending on the material – see Example 11.1.

From the S/N curve the "fatigue limit" or "endurance limit" may be ascertained. The *"fatigue limit"* is the stress condition below which a material may endure an infinite number of cycles prior to failure. Ferrous metal specimens often produce S/N curves which exhibit fatigue limits as indicated in Fig. 11.5(a). The *"fatigue strength"* or *"endurance limit"*, is the stress condition under which a specimen would have a fatigue life of N cycles as shown in

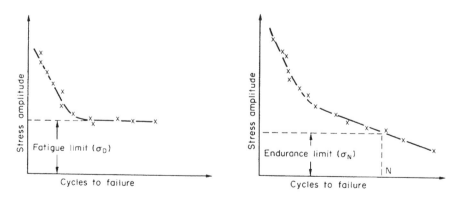

Fig. 11.5. S/N curve showing (a) fatigue limit, (b) endurance limit.

Fig. 10.5(b). Non-ferrous metal specimens show this type of curve and hence components made from aluminium, copper and nickel, etc., must always be designed for a finite life.

Another important fact to note is that the results of laboratory experiments utilising plain, polished, test pieces cannot be applied directly to structures and components without modification of the intrinsic values obtained. Allowance will have to be made for many differences between the component in its working environment and in the laboratory test such as the surface finish, size, type of loading and effect of stress concentrations. These factors will reduce the intrinsic (i.e. plain specimen) fatigue strength value thus,

$$\sigma'_N = \frac{\sigma_N}{K_f}[C_a \cdot C_b \cdot C_c] \tag{11.6}$$

where σ'_N is the "modified fatigue strength" or "modified fatigue limit", σ_N is the intrinsic value, K_f is the fatigue strength reduction factor (see § 11.1.4) and C_a, C_b and C_c are factors allowing for size, surface finish, type of loading, etc.

The types of fatigue loading in common usage include direct stress, where the material is repeatedly loaded in its axial direction; plane bending, where the material is bent about its neutral plane; rotating bending, where the specimen is being rotated and at the same time subjected to a bending moment; torsion, where the specimen is subjected to conditions which produce reversed or fluctuating torsional stresses and, finally, combined stress conditions, where two or more of the previous types of loading are operating simultaneously. It is therefore important that the method of stressing and type of machine used to carry out the fatigue test should always be quoted.

Within a fairly wide range of approximately 100 cycles/min to 6000 cycles/min, the effect of speed of testing (i.e. frequency of load cycling) on the fatigue strength of metals is small but, nevertheless, frequency may be important, particularly in polymers and other materials which show a large hysteresis loss. Test details should, therefore, always include the frequency of the stress cycle, this being chosen so as not to affect the result obtained (depending upon the material under test) the form of test piece and the type of machine used. Further details regarding fatigue testing procedure are given in BS3518: Parts 1 to 5.

Most fatigue tests are carried out at room temperature but often tests are also carried out at elevated or sub-zero temperatures depending upon the expected environmental operating conditions. At low temperatures the fatigue strength of metals show no deterioration and may even show a slight improvement, however, with increase in temperature, the fatigue strength decreases as creep effects are added to those of fatigue and this is revealed by a more pronounced effect of frequency of cycling and of mean stress since creep is both stress- and time-dependent.

When carrying out elevated temperature tests in air, oxidation of the sample may take place producing a condition similar to corrosion fatigue. Under the action of the cyclic stress, protective oxide films are cracked allowing further and more severe attack by the corrosive media. Thus fatigue and corrosion together ensure continuous propagation of cracks, and materials which show a definite fatigue limit at room temperature will not do so at elevated temperatures or at ambient temperatures under corrosive conditions – see Fig. 11.6.

11.1.2. P/S/N curves

The fatigue life of a component as determined at a particular stress level is a very variable quantity so that seemingly identical specimens may give widely differing results. This scatter

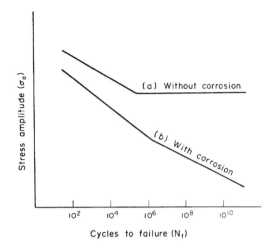

Fig. 11.6. The effect of corrosion on fatigue life. S/N Curve for (a) material showing fatigue limit; (b) same material under corrosion conditions.

arises from many sources including variations in material composition and heterogeneity, variations in surface finish, variations in axiality of loading, etc.

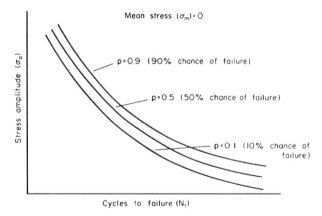

Fig. 11.7. P/S/N curves indicating percentage chance of failure for given stress level after known number of cycles (zero mean stress)

To overcome this problem, a number of test pieces should be tested at several different stresses and then an estimate of the life at a particular stress level for a given probability can be made. If the probability of 50% chance of failure is required then a P/S/N curve can be drawn through the median value of the fatigue life at the stress levels used in the test. It should be noted that this 50% ($p = 0.5$) probability curve is the curve often displayed in textbooks as *the* S/N curve for a particular material and if less probability of failure is required then the fatigue limit value will need to be reduced.

11.1.3. Effect of mean stress

If the fatigue test is carried out under conditions such that the mean stress is tensile (Fig. 11.3), then, in order that the specimen will fail in the same number of cycles as a similar specimen tested under zero mean stress conditions, the stress amplitude in the former case will have to be reduced. The fact that an increasing tensile mean stress lowers the fatigue or endurance limit is important, and all S/N curves should contain information regarding the test conditions (Fig. 11.8).

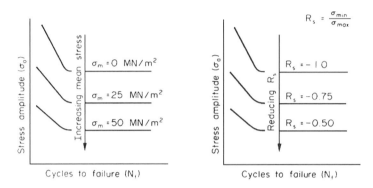

Fig. 11.8. Effect of mean stress on the S/N curve expressed in alternative ways.

A number of investigations have been made of the quantitative effect of *tensile mean stress* resulting in the following equations:

$$\text{Goodman}^{(1)} \quad \sigma_a = \sigma_N \left[1 - \left(\frac{\sigma_m}{\sigma_{TS}} \right) \right] \tag{11.7}$$

$$\text{Geber}^{(2)} \quad \sigma_a = \sigma_N \left[1 - \left(\frac{\sigma_m}{\sigma_{TS}} \right)^2 \right] \tag{11.8}$$

$$\text{Soderberg}^{(3)} \quad \sigma_a = \sigma_N \left[1 - \left(\frac{\sigma_m}{\sigma_y} \right) \right] \tag{11.9}$$

where σ_N = the fatigue strength for N cycles under zero mean stress conditions.
  σ_a = the fatigue strength for N cycles under condition of mean stress σ_m.
  σ_{TS} = tensile strength of the material.
  σ_y = yield strength of the material.

The above equations may be shown in graphical form (Fig. 11.9) and in actual practice it has been found that most test results fall within the envelope formed by the parabolic curve of Geber and the straight line of Goodman. However, because the use of Soderberg gives an additional margin of safety, this is the equation often preferred – see Example 11.2.
Even when using the Soderberg equation it is usual to apply a factor of safety F to both the alternating and the steady component of stress, in which case eqn. (11.9) becomes:

$$\sigma_a = \frac{\sigma_N}{F} \left(1 - \frac{\sigma_m \times F}{\sigma_y} \right) \tag{11.10}$$

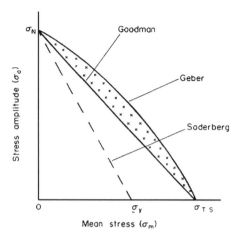

Fig. 11.9. Amplitude/mean stress relationships as per Goodman. Geber and Soderberg.

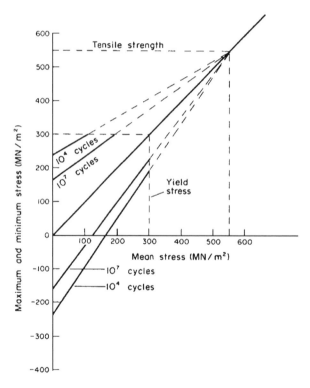

Fig. 11.10. Smith diagram.

The interrelationship of mean stress and alternating stress amplitude is often shown in diagrammatic form frequently collectively called Goodman diagrams. One example is shown in Fig. 11.10, and includes the experimentally derived curves for endurance limits of a specific steel. This is called a Smith diagram. Many alternative forms of presentation of data are possible including the Haigh diagram shown in Fig. 11.11, and when understood

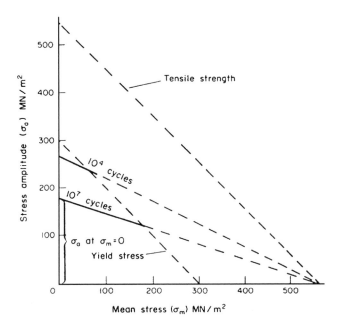

Fig. 11.11. Haigh diagram.

by the engineer these diagrams can be used to predict the fatigue life of a component under a particular stress regime. If the reader wishes to gain further information about the use of these diagrams it is recommended that other texts be consulted.

The effect of a *compressive mean stress* upon the life of a component is not so well documented or understood as that of a tensile mean stress but in general most materials do not become any worse and may even show an improved performance under a compressive mean stress. In calculations it is usual therefore to take the mean stress as zero under these conditions.

11.1.4. Effect of stress concentration

The influence of stress concentration (see §10.3) can be illustrated by consideration of an elliptical crack in a plate subjected to a tensile stress. Provided that the plate is very large, the "theoretical stress concentration" factor K_t is given by:

$$K_t = 1 + \frac{2A}{B} \qquad (11.11)$$

where "A" and "B" are the crack dimensions as shown in Fig. 11.12.

If the crack is perpendicular to the direction of stress, then A is large compared with B and hence K_t will be large. If the crack is parallel to the direction of stress, then A is very small compared with B and hence $K_t = 1$. If the dimensions of A and B are equal such that the crack becomes a round hole, then $K_t = 3$ and a maximum stress of $3\sigma_{nom}$ acts at the sides of the hole.

The effect of sudden changes of section, notches or defects upon the fatigue performance of a component may be indicated by the "*fatigue notch*" or "*fatigue strength reduction*" factor

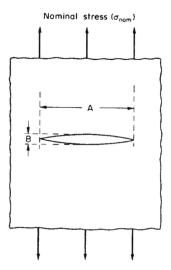

Fig. 11.12. Elliptical crack in semi-infinite plate.

K_f, which is the *ratio of the stress amplitude at the fatigue limit of an un-notched specimen, to that of a notched specimen under the same loading conditions.*

K_f is always less than the static theoretical stress concentration factor referred to above because under the compressive part of a tensile–compressive fatigue cycle, a fatigue crack is unlikely to grow. Also the ratio of K_f/K_t decreases as K_t increases, sharp notches having less effect upon fatigue life than would be expected. The extent to which the stress concentration effect under fatigue conditions approaches that for static conditions is given by the "*notch sensitivity factor*" q, and the relationship between them may be simply expressed by:

$$q = \frac{K_f - 1}{K_t - 1} \tag{11.12}$$

thus **q** is always less than 1. See also §10.3.5.

Notch sensitivity is a very complex factor depending not only upon the material but also upon the grain size, a finer grain size resulting in a higher value of q than a coarse grain size. It also increases with section size and tensile strength (thus under some circumstances it is possible to decrease the fatigue life by increasing tensile strength!) and, as has already been mentioned, it depends upon the severity of notch and type of loading.

In dealing with a ductile material it is usual to apply the factor K_f only to the fluctuating or alternating component of the applied stress. Equation (11.10) then becomes:

$$\sigma_a = \frac{\sigma_N}{F \cdot K_f} \left[1 - \left(\frac{\sigma_m \cdot F}{\sigma_y} \right) \right] \tag{11.13}$$

A typical application of this formula is given in Example 11.3.

11.1.5. Cumulative damage

In everyday, true-life situations, for example a car travelling over varying types of roads or an aeroplane passing through various weather conditions on its flight, stresses will not generally be constant but will vary according to prevailing conditions.

Several attempts have been made to predict the fatigue strength for such variable stresses using S/N curves for constant mean stress conditions. Some of the predictive methods available are very complex but the simplest and most well known is "*Miner's Law.*"

Miner[7] postulated that whilst a component was being fatigued, internal damage was taking place. The nature of the damage is difficult to specify but it may help to regard damage as the slow internal spreading of a crack, although this should not be taken too literally. He also stated that the extent of the damage was directly proportional to the number of cycles for a particular stress level, and quantified this by adding, "*The fraction of the total damage occurring under one series of cycles at a particular stress level, is given by the ratio of the number of cycles actually endured n to the number of cycles N required to break the component at the same stress level*". The ratio n/N is called the "*cycle ratio*" and Miner proposed that failure takes place when the sum of the cycle ratios equals unity.

i.e. when $\Sigma n/N = 1$

or $$\frac{n_1}{N_1} + \frac{n_2}{N_2} + \frac{n_3}{N_3} + \cdots + \textbf{etc} = 1 \qquad (11.14)$$

If equation (11.14) is merely treated as an algebraic expression then it should be unimportant whether we put n_3/N_3 before n_1/N_1 etc., but experience has shown that the order of application of the stress is a matter of considerable importance and that the application of a higher stress amplitude first has a more damaging effect on fatigue performance than the application of an initial low stress amplitude. Thus the cycle ratios rarely add up to 1, the sum varying between 0.5 and 2.5, but it does approach unity if the number of cycles applied at any given period of time for a particular stress amplitude is kept relatively small and frequent changes of stress amplitude are carried out, i.e. one approaches random loading conditions. A simple application of Miner's rule is given in Example 11.4.

11.1.6. Cyclic stress–strain

Whilst many components such as axle shafts, etc., have to withstand an almost infinite number of stress reversals in their lifetime, the stress amplitudes are relatively small and usually do not exceed the elastic limit. On the other hand, there are a growing number of structures such as aeroplane cabins and pressure vessels where the interval between stress cycles is large and where the stresses applied are very high such that plastic deformation may occur. Under these latter conditions, although the period in time may be long, the number of cycles to failure will be small and in recent years interest has been growing in this "*low cycle fatigue*".

If, during fatigue testing under these high stress cycle conditions, stress and strain are continually monitored, a hysteresis loop develops characteristic of each cycle.

Figure 11.13 shows typical loops under constant stress amplitude conditions, each loop being displaced to the right for the sake of clarity. It will be observed that with each cycle, because of work hardening, the width of the loop is reduced, eventually the loop narrowing to a straight line under conditions of total elastic deformation.

The relationship between the loop width W and the number of cycles N is given by:

$$W = AN^{-h} \qquad (11.15)$$

where A is a constant and h the measure of the rate of work hardening.

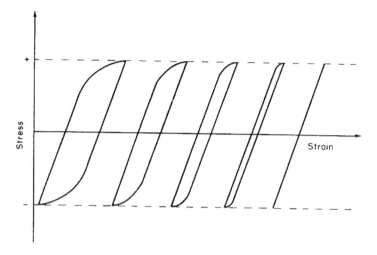

Fig. 11.13. Cyclic stress–strain under constant stress conditions – successive loading loops displaced to right for clarity. Hysteresis effects achieved under low cycle, high strain (constant stress amplitude) fatigue.

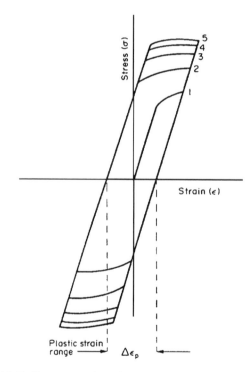

Fig. 11.14. Cyclic stress–strain under constant strain amplitude conditions.

If instead of using constant *stress* amplitude conditions, one uses constant *strain* amplitude conditions then the form of loop is indicated in Fig. 11.14. Under these conditions the stress range increases with the number of cycles but the extent of the increase reduces with each cycle such that after about 20% of the life of the component the loop becomes constant.

If now a graph is drawn (using logarithmic scales) of the plastic strain range against the number of cycles to failure a straight line results (Fig. 11.15). From this graph we obtain the following equation for the plastic strain range $\Delta\varepsilon_p$ which is known as the Coffin–Manson Law.[8]

$$\Delta\varepsilon_p = K(N_f)^{-b} \qquad (11.16)$$

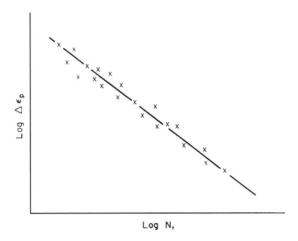

Fig. 11.15. Relationship between plastic strain and cycles to failure in low cycle fatigue.

The value of b varies between 0.5 and 0.6 for most metals, whilst the constant K can be related to the ductility of the metal. Equation (11.16) can also be expressed as:

$$\Delta\varepsilon_p = \left(\frac{N_f}{D}\right)^{-b} \qquad (11.17)$$

where D is the ductility as determined by the reduction in area r in a tensile test.

i.e.

$$D = l_n\left(\frac{1}{1-r}\right)$$

In many applications, the total strain range may be known but it may be difficult to separate it into plastic and elastic components; thus a combined equation may be more useful.

$$\Delta\varepsilon_1 = \Delta\varepsilon_e + \Delta\varepsilon_p$$

Where $\Delta\varepsilon_t$, $\Delta\varepsilon_e$ and $\Delta\varepsilon_p$ stand for total, elastic and plastic strain ranges respectively. Relationships between $\Delta\varepsilon_p$ and N_f are given above but $\Delta\varepsilon_e$ may be related to N_f by the following modified form of *Basquin's Law*.[9]

$$\Delta\varepsilon_e = 3.5 \times \frac{\sigma_{TS}}{E} \times N_f^{-0.12} \qquad (11.18)$$

If a graph is plotted (Fig. 11.16) of strain range against number of cycles to failure, it can be seen that the beginning part of the curve closely fits the slope of Coffin's equation while the latter part fits the modified Basquin's equation, the cross-over point being at about 10^5

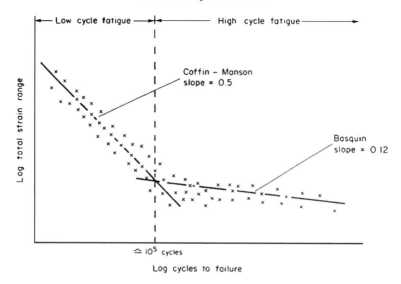

Fig. 11.16. Relationship between total strain and cycles to failure in low and high cycle fatigue.

cycles. Therefore, it can be said that up to this figure fatigue performance is a function of the material's ductility, whilst for cycles in excess of this, life is a function of the strength of the material.

11.1.7. Combating fatigue

When selecting a material for use under fatigue conditions it may be better to select one which shows a fatigue limit, e.g. steel, rather than one which exhibits an endurance limit, e.g. aluminium. This has the advantage of enabling the designer to design for an infinite life provided that the working stresses are kept to a suitably low level, whereas if the latter material is selected then design must be based upon a finite life.

In general, for most steels, the fatigue limit is about 0.5 of the tensile strength, there-fore, by selecting a high-strength material the allowable working stresses may be increased. Figure 11.17.

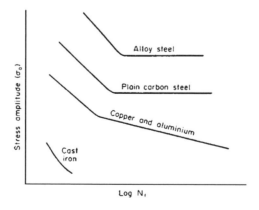

Fig. 11.17. Relative performance of various materials under fatigue conditions.

Following on the above, any process that increases tensile strength should raise the fatigue limit and one possible method of accomplishing this with steels is to carry out heat treatment. The general effect of heat treatment on a particular steel is shown in Fig. 11.18.

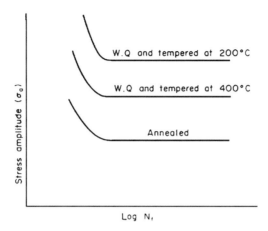

Fig. 11.18. Effect of heat treatment upon the fatigue limit of steel.

Sharp changes in cross-section will severely reduce the fatigue limit (see §10.3.4), and therefore generous radii can be used to advantage in design. Likewise, surface finish will also have a marked effect and it must be borne in mind that fatigue data obtained in laboratory tests are often based upon highly polished, notch-free, samples whilst in practice the component is likely to have a machined surface and many section changes. The sensitivity of a material to notches tends to increase with increase in tensile strength and decrease with increase in plasticity, thus, in design situations, a compromise between these opposing factors must be reached.

Figure 11.19 shows the fatigue limits of typical steels in service expressed as a percentage of the fatigue limits obtained for the same steels in the laboratory and it will be noticed that

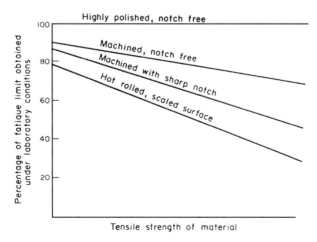

Fig. 11.19. Effect of surface conditions on the fatigue strength of materials.

the fatigue limit of a low-strength steel is not affected to the same extent as the high-strength steel. i.e. the former is less notch-sensitive (another factor to be taken account of when looking at the relative cost of the basic material). However, it must be pointed out that it may be poor economy to overspecify surface finish, particularly where stress levels are relatively low.

Because fatigue cracks generally initiate at the surface of a component under tensile stress conditions, certain processes, both chemical and mechanical, which introduce residual surface compressive stresses may be utilised to improve fatigue properties (see §10.2). However, the extent of the improvement is difficult to assess quantitatively at this juncture of time. Among the chemical treatments, the two most commonly employed are *carburising* and *nitriding* which bring about an expansion of the lattice at the metal surface by the introduction of carbon and nitrogen atoms respectively. Figure 11.20 shows the effect upon fatigue limit.

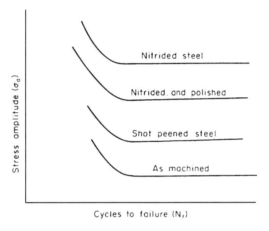

Fig. 11.20. Effect of processes which introduce surface residual stresses upon the fatigue strength of a steel.

The most popular mechanical method of improving fatigue limits is *shot peening*, the surface of the material being subjected to bombardment by small pellets or shot of suitable material. In this manner, compressive residual stresses are induced but only to a limited depth, roughly 0.25 mm. Other mechanical methods involve improving fatigue properties around holes by pushing through balls which are slightly over-sized – a process called "*ballising*," and the use of balls or a roller to cold work shoulders on fillets – a process called "*rolling*".

11.1.8. Slip bands and fatigue

The onset of fatigue is usually characterised by the appearance on the surface of the specimen of slip bands which, after about 5% of the fatigue life, become permanent and cannot be removed by electropolishing. With increase in the number of load cycles these bands deepen until eventually a crack is formed.

Using electron microscopical techniques Forsyth[10] observed *extrusions* and *intrusions* from well-defined slip bands and Cottrell[11] proposed a theory of cross-slip or slip on alternate slip planes whereby, during the tensile half of the stress cycle, slip occurs on each plane in turn to produce two surface steps which on the compressive half of the cycle are converted into an intrusion and an extrusion (see Fig. 11.21). Although an intrusion is only very small, being approximately 1 μm deep, it nevertheless can act as a stress raiser and initiate the formation of a true fatigue crack.

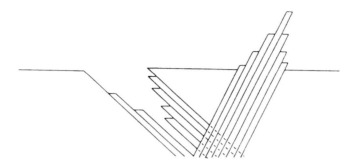

Fig. 11.21. Diagrammatic representation of the formation of intrusions and extrusions.

Fatigue endurance is commonly divided into two periods: (i) the *"crack initiation"* period; (ii) the *"crack growth"* or *"propagation"* period. It is now accepted that the fatigue crack is initiated by the deepening of the slip band grooves by dislocation movement into crevices and finally cracks, but this makes it very difficult to distinguish between crack initiation and crack propagation and therefore a division of the fatigue based upon mode of crack growth is often more convenient.

Initially the cracks will form in the surface grains and develop along the active slip plane as mentioned briefly above. These cracks are likely to be aligned with the direction of maximum shear within the component, i.e. at 45° to the maximum tensile stress. This is often referred to as *Stage I growth* and is favoured by zero mean stress and low cyclic stress conditions.

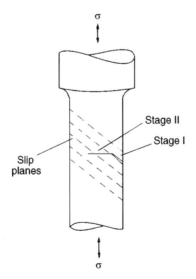

Fig. 11.22. Stage I and II fatigue crack propagation.

At some point, usually when the crack encounters a grain boundary. Stage I is replaced by *Stage II growth* in which the crack is normal to the maximum principal tensile stress. This stage is favoured by a tensile mean stress and high cyclic stress conditions. Close examination of the fractured surface shows that over that part associated with Stage II, there are a large number of fine lines called "*striations*", each line being produced by one fatigue cycle and by measuring the distance between a certain number of striations the fatigue crack growth rate can be calculated.

Once the fatigue crack has reached some critical length such that the energy for further growth can be obtained from the elastic energy of the surrounding metal, catastrophic failure takes place. This final fracture area is rougher than the fatigue growth area and in mild steel is frequently crystalline in appearance. Sometimes it may show evidence of plastic deformation before final separation occurred. Further discussion of fatigue crack growth is introduced in §11.3.7.

11.2. Creep

Introduction

Creep is the time-dependent deformation which accompanies the application of stress to a material. At room temperatures, apart from the low-melting-point metals such as lead, most metallic materials show only very small creep rates which can be ignored. With increase in temperature, however, the creep rate also increases and above approximately $0.4 \, T_m$, where T_m is the melting point on the Kelvin scale, creep becomes very significant. In high-temperature engineering situations related to gas turbine engines, furnaces and steam turbines, etc., deformation caused by creep can be very significant and must be taken into account.

11.2.1. The creep test

The creep test is usually carried out at a constant temperature and under constant load conditions rather than at constant stress conditions. This is acceptable because it is more representative of service conditions. A typical creep testing machine is shown in Fig. 11.23. Each end of the specimen is screwed into the specimen holder which is made of a creep-resisting alloy and thermocouples and accurate extensometers are fixed to the specimen in order to measure temperature and strain. The electric furnace is then lowered into place and when all is ready and the specimen is at the desired temperature, the load is applied by adding weights to the lower arm and readings are taken at periodic intervals of extension against time. It is important that accurate control of temperature is possible and to facilitate this the equipment is often housed in a temperature-controlled room.

The results from the creep test are plotted in graphical form to produce a typical curve as shown in Fig. 11.24. After the initial extension *OA* which is produced as soon as the test load is applied, and which is not part of the creep process proper (but which nevertheless should not be ignored), the curve can be divided into three stages. In the first or *primary* stage *AB*, the movement of dislocations is very rapid, any barriers to movement caused by work-hardening being overcome by the recovery processes, albeit at a decreasing rate. Thus the initial *creep strain rate* is high but it rapidly decreases to a constant value. In the

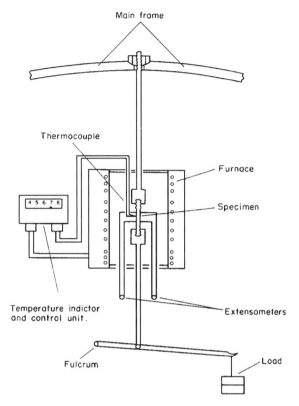

Fig. 11.23. Schematic diagram of a typical creep testing machine.

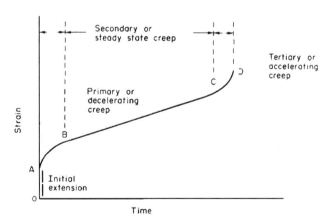

Fig. 11.24. Typical creep curve.

secondary stage BC, the work-hardening process of "dislocation pile-up" and "entanglement" are balanced by the recovery processes of "dislocation climb" and "cross-slip", to give a straight-line relationship and the slope of the graph in this steady-state portion of the curve is equal to the secondary creep rate. Since, generally, the primary and tertiary stages occur quickly, it is the secondary creep rate which is of prime importance to the design engineer.

The third or *tertiary* stage *CD* coincides with the formation of internal voids within the specimen and this leads to "necking", causing the stress to increase and rapid failure to result.

The shape of the creep curve for any material will depend upon the temperature of the test and the stress at any time since these are the main factors controlling the work-hardening and recovery processes. With increase in temperature, the creep rate increases because the softening processes such as "dislocation climb" can take place more easily, being diffusion-controlled and hence a thermally activated process.

It is expected, therefore, that the creep rate is closely related to the *Arrhenius equation*, viz.:

$$\varepsilon_s^0 = A e^{-H/RT} \tag{11.19}$$

where ε_s^0 is the *secondary creep rate*, H is the *activation energy* for creep for the material under test, R is the universal gas constant, T is the absolute temperature and A is a constant. It should be noted that both A and H are not true constants, their values depending upon stress, temperature range and metallurgical variables.

The secondary creep rate also increases with increasing stress, the relationship being most commonly expressed by the *power law equation*:

$$\varepsilon_s^0 = \beta \sigma^n \tag{11.20}$$

where β and n are constants, the value of n usually varying between 3 and 8.

Equations (11.19) and (11.20) may be combined to give:

$$\varepsilon_s^0 = K \sigma^n e^{-H/RT} \tag{11.21}$$

Figure 11.25 illustrates the effect of increasing stress or temperature upon the creep curve and it can be seen that increasing either of these two variables results in a similar change of creep behaviour, that is, an increase in the secondary or minimum creep rate, a shortening of the secondary creep stage, and the earlier onset of tertiary creep and fracture.

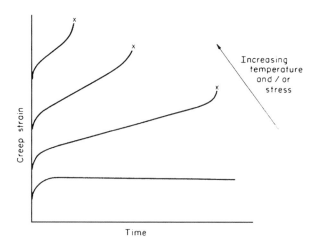

Fig. 11.25. Creep curves showing effect of increasing temperature or stress.

11.2.2. Presentation of creep data

When dealing with problems in which creep is important, the design engineer may wish to know whether the creep strain over the period of expected life of the component is tolerable, or he may wish to know the value of the maximum operating stress if the creep strain is not to exceed a specified figure over a given period of time.

In order to assist in the answering of these questions, creep data are often published in other forms than the standard strain–time curve. Figure 11.26 shows a number of fixed strain curves presented in the form of an *isometric stress–time* diagram which relates strain, stress and time for a fixed, specified, temperature and material, while Fig. 11.27 is an *isometric strain–time diagram*.

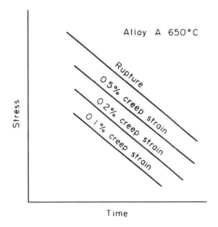

Fig. 11.26. Isometric stress–time diagrams.

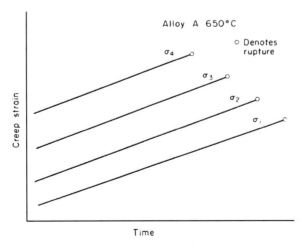

Fig. 11.27. Isometric strain–time diagram.

Sometimes, instead of presenting data relating to a fixed temperature, the strain may be constant and curves of equal time called *isochronous* stress–temperature curves. Fig. 11.28 may be given. Such curves can be used for comparing the properties of various alloys and Fig. 11.29 shows relations for a creep strain of 0.2% in 3000 hours. Such information might be applicable to an industrial gas turbine used intermittently.

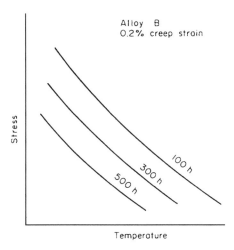

Fig. 11.28. Isochronous stress–temperature curves.

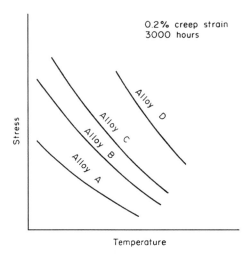

Fig. 11.29. 3000 hours, 0.2% creep–strain curves for various alloys.

11.2.3. The stress–rupture test

Where creep strain is the important design factor and fracture may be expected to take a very long time, the test is often terminated during the steady state of creep when sufficient information has been obtained to produce a sufficiently accurate value of the secondary creep rate. Where life is the important design parameter, then the test is carried out to destruction and this is known as a *stress–rupture test*.

Because the total strain in a rupture test is much higher than in a creep test, the equipment can be less sophisticated. The loads used are generally higher, and thus the time of test shorter, than for creep. The percentage elongation and percentage reduction in area at fracture may be determined but the principal information obtained is the time to failure at a fixed temperature under nominal stress conditions.

A graph (Fig. 11.30), is plotted of time to rupture against stress on a log–log basis, and often a straight line results for each test temperature. Any change in slope of this stress–rupture line may be due to change in the mechanism of creep rupture within the material.

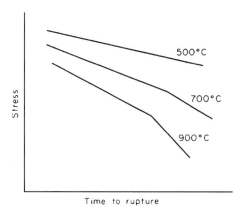

Fig. 11.30. Stress–rupture time curves at various temperatures.

11.2.4. Parameter methods

Very often, engineers have to confirm to customers that a particular component will withstand usage at elevated temperatures for a particular life-time which, in the case of furnace equipment or steam applications, may be a considerable number of years. It is impracticable to test such a component, for example, for twenty years before supplying the customer and therefore some method of extrapolation is required. The simplest method is to test at the temperature of proposed usage, calculate the minimum creep rate and assume that this will continue for the desired life-time and then ascertain whether the creep strain is acceptable. The obvious disadvantage of this method is that it does not allow for tertiary creep and sudden failure (which the creep curve shows will take place at some time in the future but at a point which cannot be determined because of time limitations).

In order to overcome this difficulty a number of workers have proposed methods involving accelerated creep tests, whereby the test is carried out at a higher temperature than that used in practice and the results used to predict creep-life or creep-strain over a longer period of time at a lower temperature.

The most well-known method is that of *Larson and Miller* and is based upon the Arrhenius equation (eqn. 11.19) which can be rewritten, in terms of $\log_{10}$ as in eqn. (11.22) or in terms of l_n without the constant 0.4343.

$$\log_{10} t_r = \log_{10} G + 0.4343 \cdot \frac{H}{R} \cdot \frac{1}{T} \tag{11.22}$$

where t_r is the time to rupture, G is a constant, T is the **absolute temperature**, R is the universal gas constant, and H is the activation energy for creep and is assumed to be stress-dependent.

$$\therefore \qquad\qquad \log_{10} t_r + C = m \cdot \frac{1}{T}$$

where m is a function of stress.

$$\therefore \qquad\qquad T(\log_{10} t_r + C) = m$$

this can be re-written as:

$$P_1 = f(\sigma)$$

where the *Larson–Miller parameter*

$$\boldsymbol{P_1 = T(\log_{10} t_r + C)} \tag{11.23}$$

the value of the constant C can be obtained from the intercept when $\log_{10} t_r$ is plotted against $1/T$. For ferrous metals it usually lies between 15 and 30. If a test is carried out under a certain value of stress and temperature, the value of t_r can be determined and, if repeated for other stress and temperature values, the results can be plotted on a *master curve* (Fig. 11.31).

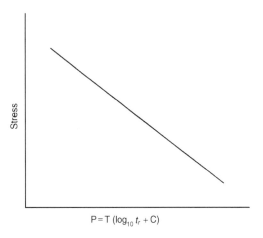

$$P = T\,(\log_{10} t_r + C)$$

Fig. 11.31. Larson–Miller master curve.

The value of the parameter P is the same for a wide variety of combinations of t_r and temperature, ranging from short times and high temperatures representing test conditions, to long times at lower temperatures representing service conditions.

Results obtained by other workers, notably Sherby and Dorn[14], suggest that G in the above equation (eqn. 11.22) is not a true constant but varies with stress whilst E is essentially constant. If 0.4343 E/R in eqn. (11.22) is replaced by α and $\log_{10} G$ by ϕ then eqn. (11.22) can be written as:

$$\log_{10} t_r - \frac{\alpha}{T} = \phi$$

or

$$P_2 = f(\sigma)$$

where the *Sherby–Dorn parameter*

$$P_2 = \log_{10} t_r - \frac{\alpha}{T} \tag{11.24}$$

the constant α being determined from the common slope of a plot of $\log_{10} t_r$ versus $1/T$. After a series of creep tests, a master curve can then be plotted and used in the same manner as for the Larson–Miller parameter.

Another parameter was suggested by Manson and Haferd[13] who found that, for a given material under different stress and temperature conditions, a family of lines was obtained which intercepted at a point when $\log t_r$ was plotted against T. The family of lines of this kind could be represented by the equation:

$$T - T_a = m(\log_{10} t_r - \log_{10} t_a) \tag{11.25}$$

where the slope m is a function of stress and T_a and $\log_{10} t_a$ are the coordinates of the converging point (Fig. 11.32).

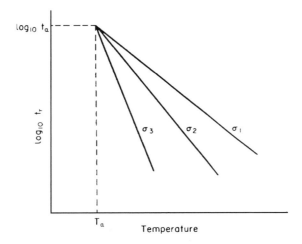

Fig. 11.32. Manson and Haferd curves.

The *Manson–Haferd parameter* can then be stated as:

$$P_3 = \frac{T - T_a}{\log_{10} t_r - \log_{10} t_a} \tag{11.26}$$

where $P_3 = m = f(\sigma)$.

A master curve can then be plotted in a similar manner to the other methods.

When using any of the above methods, certain facts should be borne in mind with regard to their limitations. Firstly, since the different methods give slightly differing results, this casts doubts on the validity of all three methods and, in general, although the Manson–Haferd parameter has been found to produce more accurate predictions, it is difficult to determine the exact location of the point of convergence of the lines. Secondly, if one is using higher test temperatures than operating temperatures then the mechanisms of creep may be different and unrelated one to the other. Thirdly, the mechanism of creep failure may change with temperature; at lower temperatures, failure is usually transcrystalline whilst at higher temperatures it is intercrystalline, the change-over point being the *equi-cohesive temperature*. Results above this temperature are difficult to correlate with those obtained below this temperature.

11.2.5. Stress relaxation

So far we have been concerned with the study of material behaviour under constant loading or *constant stress* conditions where increase in strain is taking place and may eventually lead to failure. However, there are important engineering situations involving cylinder-head bolts, rivets in pressure vessels operating at elevated temperatures, etc., where we consider the *strain* to be *constant* and then we need to evaluate the decrease in stress which may take place. This time-dependent decrease in stress under constant strain conditions is called "stress relaxation".

Consider two plates held together by a bolt deformed by a stress σ_i producing an initial strain ε_i which is all elastic.

Then
$$\varepsilon_i = \varepsilon_e = \sigma_i/E \tag{1}$$

At elevated temperatures and under conditions of steady-state creep, this bolt will tend to elongate at a rate ε^0 dictated by the power law:

$$\varepsilon^0 = \frac{d\varepsilon_c}{dt} = \beta\sigma^n \tag{2}$$

and, assuming the thickness of the plates remain constant, the strain caused by creep ε_c simply reduces the elastic part ε_e of the initial strain,

i.e.
$$\varepsilon_e = \varepsilon_i - \varepsilon_c \tag{3}$$

But, since the creep strain decreases the elastic component of the initial strain, a corresponding decrease in stress must also result from eqn. (1).

Since ε_i is constant, if we differentiate eqn. (3) with respect to time we obtain:

$$\frac{d\varepsilon_e}{dt} = -\frac{d\varepsilon_c}{dt} \tag{4}$$

but $\varepsilon_e = \sigma E$ where σ is the instantaneous stress, therefore the LHS of eqn. (4) can be replaced by $(1/E) \cdot (d\sigma/dt)$ whilst, from eqn. (2), the RHS of eqn. (4) can be replaced by $\beta\sigma^n$.

Therefore, eqn. (4) can be rewritten:

$$\frac{1}{E} \cdot \frac{d\sigma}{dt} = -\beta\sigma^n \tag{5}$$

$$\therefore \qquad \int \frac{d\sigma}{\sigma^n} = -E\beta \int dt$$

$$\therefore \qquad -\frac{1}{(n-1)\sigma^{n-1}} = -E\beta t + C \qquad (6)$$

To find C, consider the time $t = 0$, when the stress would be the initial stress σ_i

Then
$$C = -\frac{1}{(n-1)\sigma_i^{n-1}} \qquad (7)$$

substituting for C in eqn. (6), multiplying through by $(n-1)$ and re-arranging, gives:

$$\frac{1}{\sigma^{n-1}} = \frac{1}{\sigma_i^{n-1}} + \beta E(n-1)t \qquad (11.27)$$

11.2.6. Creep-resistant alloys

The time-dependent deformation called "creep", as with all deformation processes, is largely dependent upon dislocation movement and, therefore, the development of alloys with a high resistance to creep involves producing a material in which movement of dislocations only takes place with difficulty.

Since creep only becomes an engineering problem above about $0.4 \times$ melting point temperature on the Kelvin scale T_m, the higher the melting point of the major alloy constituent the better. However, there are practical limitations; for instance, some high-melting-point metals e.g. tungsten (M.Pt 3377° C) are difficult to machine, some Molybdenum (M.Pt 2607° C) form volatile oxides and some others, e.g. Osmium (M.Pt 3027 C) are very expensive and therefore Nickel (M.Pt 1453) and Cobalt (M.Pt 1492° C) are used extensively at the moment.

The movement of dislocations will be hindered to a greater extent in an alloy rather than in a pure metal and alloying elements such as chromium and cobalt are added therefore to produce a solid-solution causing "*solid-solution-hardening*". Best results are obtained by rising an alloying element whose atomic size and valency are largely different from those of the parent metal, but this limits the amount that may be added. Also, the greater the amount of alloying element, the lower is likely to be the melting range of the alloy. Thus, the benefits of solution-hardening which hinders the dislocation movement may be outweighed at higher temperatures by a close approach to the solidus temperature.

Apart from solution-hardening, most creep-resisting alloys are further strengthened by *precipitation hardening* which uses carbides, oxides, nickel-titanium-aluminium, and nitride particles to block dislocation movement. Further deformation can then only take place by the dislocation rising above or "climbing" over the precipitate in its path and this is a diffusion-controlled process. Thus, metals with a low rate of self-diffusion e.g. face-centered-metals such as nickel are preferred to body-centred-metals.

Finally, *cold-working* is another method of increasing the high-temperature strength of an alloy and hindering dislocation movement but, since cold work lowers the re-crystallization temperature, for best results it is limited to about 15–20%. The use of alloying elements which raise the re-crystallization temperature in these circumstances will be beneficial.

All the methods above have their limitations. In solid-solution-hardening, a temperature increase will produce a corresponding increase in the mobility of the solute atoms which tend to lock the dislocations, thus making dislocation movement easier. With precipitation hardening, the increase in temperature may produce "*over-ageing*", resulting in a coarsening of the precipitate or even a complete solution of the precipitate, both effects resulting in a softening and decrease in creep resistance. It may be possible, however, to arrange for a second precipitate to form which may strengthen the alloy. The effects of cold work are completely nullified when the temperature rises above the re-crystallisation temperature, hence the application of this technique is very limited.

At room temperature, grain boundaries are normally stronger than the grain material but, with increase in temperature, the strength of the boundary decreases at a faster rate than does the strength of the grain interior such that above the "*equi-cohesive temperature*" (Fig. 11.33), a coarse-grain material will have higher strength than a fine-grain material since the latter is associated with an increase in the amount of grain boundary region. It should be noted that T_e is not fixed, but dependent upon stress, being higher at high stresses than at low stresses.

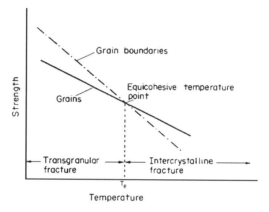

Fig. 11.33. Diagram showing concept of equi-cohesive temperature (T_e)

It can be shown also that creep rate is inversely proportional to the square of the grain size and dictated by the following formula:

$$\textbf{Creep rate} = \frac{K\sigma D}{d^2 T} \tag{11.28}$$

where K is a constant, σ is the applied stress, D is the coefficient of self diffusion, d is the grain size and T is the temperature. Thus, *grain boundary strengthening*, by the introduction of grain boundary carbide precipitates which help to prevent grain boundary sliding, and the control of grain size are important. Better still, the component may be produced from a single crystal such as the RB211 intermediate pressure turbine blade.

Apart from high creep resistance, alloys for use at high temperatures generally require other properties such as high oxidation resistance, toughness, high thermal fatigue resistance and low density, the importance of these factors depending upon the application of the material, and it is doubtful if any single test would provide a simple or accurate index of the qualities most desired.

11.3. Fracture mechanics

Introduction

The use of stress analysis in modern design procedures ensures that in normal service very few engineering components fail because they are overloaded. However, weakening of the component by such mechanisms as corrosion or fatigue-cracking may produce a catastrophic fracture and in some instances, such as in the design of motorcycle crash helmets, the fracture properties of the component are the most important consideration. The study of how materials fracture is known as *fracture mechanics* and the resistance of a material to fracture is colloquially known as its *"toughness"*.

No structure is entirely free of defects and even on a microscopic scale these defects act as stress-raisers which initiate the growth of cracks. The theory of fracture mechanics therefore assumes the pre-existence of cracks and develops criteria for the catastrophic growth of these cracks. The designer must then ensure that no such criteria can be met in the structure.

In a stressed body, a crack can propagate in a combination of the three opening modes shown in Fig. 11.34. *Mode I* represents opening in a purely tensile field while *modes II* and *III* are in-plane and anti-plane shear modes respectively. The most commonly found failures are due to cracks propagating predominantly in mode I, and for this reason materials are generally characterised by their resistance to fracture in that mode. The theories examined in the following sections will therefore consider mode I only but many of the conclusions will also apply to modes II and III.

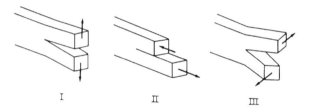

Fig. 11.34. The three opening modes, associated with crack growth: mode I–tensile; mode II–in-plane shear; mode III–anti-plane shear.

11.3.1. Energy variation in cracked bodies

A basic premise in thermodynamic theory is that a system will move to a state where the free energy of the system is lower. From this premise a simple criterion for crack growth can be formulated. It is assumed that a crack will only grow if there is a decrease in the free energy of the system which comprises the cracked body and the loading mechanism. The first usable criterion for fracture was developed from this assumption by Griffith[15], whose theory is described in detail in §11.3.2.

For a clearer understanding of Griffith's theory it is necessary to examine the changes in stored elastic energy as a crack grows. Consider, therefore, the simple case of a strip containing an edge crack of length a under uniaxial tension as shown in Fig. 11.35. If load W is applied gradually, the load points will move a distance x and the strain energy, U, stored in the body will be given by

$$U = \tfrac{1}{2}Wx$$

for purely elastic deformation.

The load and displacement are related by the *"compliance"* C,

i.e. $x = CW$ (11.29)

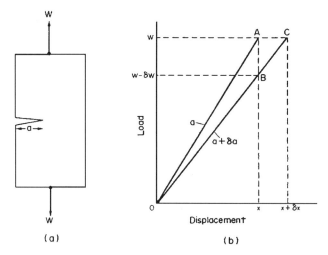

Fig. 11.35. (a) Cracked body under tensile load W; (b) force-displacement curves for a body with crack lengths a and $a + da$.

The compliance is itself a function of the crack length but the exact relationship varies with the geometry of the cracked body. However, if the crack length increases, the body will become less stiff and the compliance will increase.

There are two limiting conditions to be considered depending on whether the cracked body is maintained at (a) constant displacement or (b) constant loading. Generally a crack will grow with both changing loads and displacement but these two conditions represent the extreme constraints.

(a) Constant displacement

Consider the case shown in Fig. 11.35(b). If the body is taken to be perfectly elastic then the load–displacement relationship will be linear. With an initial crack length a loading will take place along the line OA. If the crack extends a small distance δa while the points of application of the load remain fixed, there will be a small increase in the compliance resulting in a decrease in the load of δW. The load and displacement are then given by the point B. The change in stored energy will then be given by

$$\delta U_x = \tfrac{1}{2}(W - \delta W)x - \tfrac{1}{2}Wx$$

$$\delta U_x = -\tfrac{1}{2}\delta Wx \tag{11.30}$$

(b) Constant loading

In this case, if the crack again extends a small distance δa the loading points must move through an additional displacement δx in order to keep the load constant. The load and displacement are then represented by the point C.

There would appear to be an increase in stored energy given by

$$\delta U = \tfrac{1}{2}W(x + \delta x) - \tfrac{1}{2}Wx$$

$$= \tfrac{1}{2}W\delta x$$

However, the load has supplied an amount of energy

$$= W\,\delta x$$

This has to be obtained from external sources so that there is a total reduction in the potential energy of the system of

$$\delta U_W = \tfrac{1}{2} W\,\delta x - W\,\delta x$$

$$\boldsymbol{\delta U_W = -\tfrac{1}{2} W\,\delta x} \tag{11.31}$$

For infinitesimally small increases in crack length the compliance C remains essentially constant so that

$$\delta x = C\,\delta W$$

Substituting in eqn. (11.31)

$$\delta U_W = -\tfrac{1}{2} W C\,\delta W = -\tfrac{1}{2} x\,\delta W$$

Comparison with eqn. (11.30) shows that, for small increases in crack length,

$$\delta U_w = \delta U_x$$

It is therefore evident that for small increases in crack length there is a similar decrease in potential energy no matter what the loading conditions. For large changes in crack length there is no equality but, generally, we are interested in the onset of crack growth since for monotonic (continuously increasing) loading catastrophic failure commonly follows crack initiation.

   If there is a decrease in potential energy when a crack grows then there must be an energy requirement for the production of a crack – otherwise all cracked bodies would fracture instantaneously. The following section examines the most commonly used fracture criterion based on a net decrease in energy.

11.3.2. Linear elastic fracture mechanics (L.E.F.M.)

(a) Griffith's criterion for fracture

Griffith's thermodynamics approach was the first to produce a usable theory of fracture mechanics.[12] His theoretical model shown in Fig. 11.36 was of an infinite sheet under a remotely applied uniaxial stress σ and containing a central crack of length $2a$. The preceding section has shown that when a crack grows there is a decrease in potential energy. Griffith, by a more mathematically rigorous treatment, was able to show that if that decrease in energy is greater than the energy required to produce new crack faces then there will be a net decrease in energy and the crack will propagate. For an increase in crack length of δa.

$$\delta U = 2\gamma b\,\delta a$$

γ is the surface energy of the crack faces;
b is the thickness of the sheet.
   At the onset of crack growth, δa is small and we have

$$\frac{dU}{da} = 2b\gamma$$

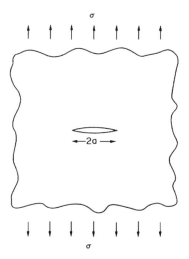

Fig. 11.36. Mathematical model for Griffith's analysis.

The expression on the left-hand side of the above equation is termed the *"critical strain energy release"* (with respect to crack length) and is usually denoted as G_c, i.e., at the onset of fracture,

$$G_c = \frac{\partial U}{\partial a} = 2b\gamma \tag{11.32}$$

This is the *Griffith criterion for fracture*.

   Griffith's analysis gives G_c in terms of the fracture stress σ_f

$$G_c = \frac{\sigma_f^2 \pi a}{E} \quad \text{in plane stress} \tag{11.33}$$

$$G_c = \frac{\sigma_f^2 \pi a}{E}(1 - v^2) \quad \text{in plane strain.} \tag{11.34}$$

For finite bodies and those with edge cracks, correction factors must be introduced. Usually this involves replacing the factor π by some dimensionless function of the cracked body's geometry.

   From eqns. (11.32) and (11.34) we can predict that, *for plane strain*, the fracture stress should be given by

$$\sigma_f^2 = \frac{2bE\gamma}{\pi a(1 - v^2)} \tag{11.35}$$

or, *for plane stress*:

$$\sigma_f^2 = \frac{2bE\gamma}{\pi a}$$

Griffith tested his theory on inorganic glasses and found a reasonable correlation between predicted and observed values of fracture stress. However, inorganic glasses are extremely brittle and when more ductile materials are examined it is found that the predicted values are far less than those observed. It is now known that even in apparently brittle fractures a ductile material will produce a localised plastic zone at the crack tip which effectively blunts the crack. This has not prevented some workers measuring G_c experimentally and using it as a means of comparing materials but it is then understood that the energy required to propagate the crack includes the energy to produce the plastic zone.

(b) Stress intensity factor

Griffith's criterion is an energy-based theory which ignores the actual stress distribution near the crack tip. In this respect the theory is somewhat inflexible. An alternative treatment of the elastic crack was developed by Irwin[16], who used a similar mathematical model to that employed by Griffith except in this case the remotely applied stress is biaxial – (see Fig. 11.37). Irwin's theory obtained expressions for the stress components near the crack tip. The most elegant expression of the stress field is obtained by relating the cartesian components of stress to polar coordinates based at the crack tip as shown in Fig. 11.38.

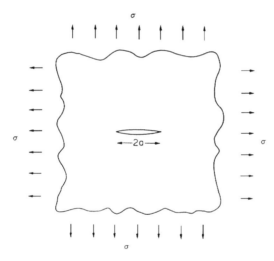

Fig. 11.37. Mathematical model for Irwin's analysis.

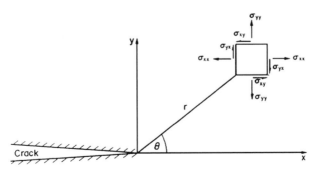

Fig. 11.38. Coordinate system for stress components in Irwin's analysis.

Then we have:

$$
\left.
\begin{aligned}
\sigma_{yy} &= \frac{K}{\sqrt{2\pi r}} \cos \frac{\theta}{2} \left[1 + \sin \frac{\theta}{2} \sin \frac{3\theta}{2} \right] \\[1em]
\sigma_{xx} &= \frac{K}{\sqrt{2\pi r}} \cos \frac{\theta}{2} \left[1 - \sin \frac{\theta}{2} \sin \frac{3\theta}{2} \right] \\[1em]
\sigma_{xy} &= \frac{K}{\sqrt{2\pi r}} \cos \frac{\theta}{2} \sin \frac{\theta}{2} \cos \frac{3\theta}{2}
\end{aligned}
\right\}
\tag{11.36}
$$

With, for plane stress,

$$
\sigma_{zz} = 0
$$

or, for plane strain,

$$
\sigma_{zz} = \nu(\sigma_{xx} + \sigma_{yy})
$$

with

$$
\sigma_{zx} = \sigma_{zy} = 0 \quad \text{for both cases.}
$$

The expressions on the right-hand side of the above equations are the first terms in series expansions but for regions near the crack tip where $r/a \gg 1$ the other terms can be neglected.

It is evident that each stress component is a function of the parameter K and the polar coordinates determining the point of measurement. The parameter K, which is termed the "*stress intensity factor*", therefore uniquely determines the stress field near the crack tip. If we base our criterion for fracture on the stresses near the crack tip then we are implying that the value of K determines whether the crack will propagate or not. The stress intensity factor K is simply a function of the remotely applied stress and crack length.

If more than one crack opening mode is to be considered then K sometimes carries the suffix I, II or III corresponding to the three modes shown in Fig. 11.34. However since this text is restricted to consideration of mode I crack propagation only, the formulae have been simplified by adopting the symbol K without its suffix. Other texts may use the full symbol K_I in development of similar formulae.

For Irwin's model, K is given by

$$
K = \sigma \sqrt{\pi a}
\tag{11.37a}
$$

For an edge crack in a semi-infinite sheet

$$
K = 1.12 \sigma \sqrt{\pi a}
\tag{11.37b}
$$

To accommodate different crack geometries a flaw shape parameter Q is sometimes introduced thus

$$
K = \sigma \sqrt{\frac{\pi a}{Q}}
\tag{11.37c}
$$

or, for an edge crack

$$
K = 1.12 \sigma \sqrt{\frac{\pi a}{Q}}
\tag{11.37d}
$$

Values of Q for various aspect (depth to width) ratios of crack can be obtained from standard texts*, but, typically, they range from 1.0 for an aspect ratio of zero to 2.0 for an aspect ratio of 0.4.

* Knott and Elliot, *Worked Examples in Fracture Mechanics*, Inst. met.

Normal manufactured components are of finite size and generally cracks grow from a free surface. However, for regions near the crack tip it is found that eqn (11.36) is a good approximation to the stress field if the stress intensity factor is modified to

$$K = \sigma Y \sqrt{a} \qquad (11.38)$$

where a is the length of the edge crack and Y is a dimensionless correction factor often termed a "*compliance function*". Y is a polynomial of the ratio a/W where W is the uncracked width in the crack plane (see Table 11.1).

Table 11.1. Table of compliance functions (Y).

Compliance function $Y = A\left(\dfrac{a}{W}\right)^{1/2} - B\left(\dfrac{a}{W}\right)^{3/2} + C\left(\dfrac{a}{W}\right)^{5/2} - D\left(\dfrac{a}{W}\right)^{7/2} + E\left(\dfrac{a}{W}\right)^{9/2}$

with W = uncracked specimen width; a = length of edge crack; b = specimen thickness; P = total load; L = distance between loading points

Specimen geometry	Specimen nomenclature	Equation for K	Compliance function constants				
			A	B	C	D	E
	Single edge notched (S.E.N.)	$K = \dfrac{P}{bW^{1/2}} \cdot Y$	1.99	0.41	18.70	38.48	53.85
	Three-point bend ($L = 4W$)	$K = \dfrac{3PL}{bW^{3/2}} \cdot Y$	1.93	3.07	14.53	25.11	25.80
	Four-point bend	$K = \dfrac{3PL}{bW^{3/2}} \cdot Y$	1.99	2.47	12.97	23.17	24.80
	Compact tension (C.T.S.)	$K = \dfrac{P}{bW^{1/2}} \cdot Y$	29.60	185.50	655.70	1017.0	638.90

It is common practice to express K in the directly measurable quantities of load P, thickness b, and width W. The effect of crack length is then totally incorporated into Y.

i.e.
$$K = \frac{P}{bW^{1/2}} \cdot Y \qquad (11.39)$$

In practice, values of K can be determined for any geometry and for different types of loading. Table 11.1 gives the expressions derived for common laboratory specimen geometries.

Photoelastic determination of stress intensity factors

The stress intensity factor K defined in §11.3.2 is an important and useful parameter because it uniquely describes the stress field around a crack tip under tension. In a two-dimensional system the stress field around the crack tip has been defined, in polar co-ordinates, by eqns. (11.36) as

$$\left.\begin{array}{l} \sigma_{xx} = \dfrac{K}{\sqrt{2\pi r}} \cos \dfrac{\theta}{2} \left[1 - \sin \dfrac{\theta}{2} \sin \dfrac{3\theta}{2} \right] \\[12pt] \sigma_{yy} = \dfrac{K}{\sqrt{2\pi r}} \cos \dfrac{\theta}{2} \left[1 + \sin \dfrac{\theta}{2} \sin \dfrac{3\theta}{2} \right] \\[12pt] \tau_{xy} = \dfrac{K}{\sqrt{2\pi r}} \cos \dfrac{\theta}{2} \sin \dfrac{\theta}{2} \cos \dfrac{3\theta}{2} \end{array}\right\} \qquad (11.36)\text{bis}$$

with the crack plane running along the line $\theta = \pi$, see Fig. 11.38.

From §6.15 photoelastic fringes or "isochromatics" are contours of equal maximum shear stress τ_{max} and from Mohr circle proportions, or from eqn. (13.13),[†]

$$\tau_{max} = \tfrac{1}{2}\sqrt{(\sigma_{xx} - \sigma_{yy})^2 + 4\tau_{xy}^2}$$

Substituting eqns. (11.36) gives

$$\tau_{max} = \frac{K \sin \theta}{2\sqrt{2\pi r}} \qquad (11.40)$$

If, therefore, a photoelastic model is constructed of the particular crack geometry under consideration (see Fig. 11.39) a plot of τ_{max} against $1/\sqrt{r}$ from the crack tip will produce a straight line graph from the slope of which K can be evaluated.

It is usual to record fringe order (and hence τ_{max} values) along the line $\theta = 90°$ when r will be at a maximum r_{max}. Then:

$$\tau_{max} = \frac{K}{2\sqrt{2\pi}} \cdot \frac{1}{\sqrt{r_{max}}}$$

i.e. $$\text{slope of graph} = \frac{K}{2\sqrt{2\pi}}$$

Correction factors

Equations (11.36), above, were derived for mathematical convenience for an infinite sheet containing a central crack of length $2a$ loaded under biaxial tension. For these conditions it is found that

$$K = \sigma\sqrt{\pi a}$$

In order to allow for the fact that the loading in the photoelastic test is uniaxial and that the model is of finite, limited size, a correction factor needs to be applied to the above K value in order that the "theoretical" value can be obtained and compared with the result obtained from the photoelastic test. This is normally written in terms of the function (a/w) where $a = $ edge crack length and $w = $ plate width. Then:

$$K = \sigma\sqrt{a}\left[1.99 - 0.41 \left(\frac{a}{w}\right) + 18.7 \left(\frac{a}{w}\right)^2 - 38.48 \left(\frac{a}{w}\right)^3 + 53.85 \left(\frac{a}{w}\right)^4 \right]$$

[†] E.J. Hearn, *Mechanics of Materials 1*, Butterworth-Heinemann, 1997.

Fig. 11.39. Photoelastic fringe pattern associated with a particular crack geometry (relevant stress component nomenclature given in Fig. 11.38)

11.3.3. Elastic–plastic fracture mechanics (E.P.F.M.)

Irwin's description of the stress components near an elastic crack can be summarised as

$$\text{Stress} \; \propto \; \frac{K}{r^{1/2}} \times \text{a function of } \theta - \text{see eqn. 11.36}$$

which implies that each stress component rises to infinity as the crack tip is approached and as r nears zero.

In particular, the vertical stress in the crack plane where $\theta = 0$ is given by

$$\sigma_{yy} = \frac{K}{(2\pi x)^{1/2}} \tag{11.41}$$

which is represented by the dotted line shown in Fig. 11.40(a).

In a ductile material then, at some point the stress will exceed the yield stress and the material will yield. By following Knott's analysis[17] we can estimate the extent of the plastic deformation.

If we consider *plane stress conditions* then σ_{yy} is the maximum and σ_{zz} the minimum ($= 0$) principal stress. Then, by the Tresca criterion, the material will shear in the yz plane

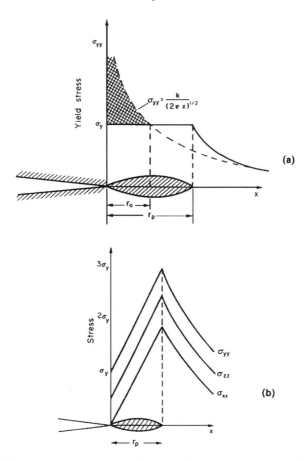

Fig. 11.40. Schematic representation of crack tip plasticity in; (a) plane stress; (b) plane strain.

and at 45 to the y and z axes when

$$\sigma_{yy} - \sigma_{zz} = \sigma_y$$

or $$\sigma_{yy} = \sigma_y$$

where σ_y is the yield stress in uniaxial tension.

At some distance r_0 from the crack tip $\sigma_{yy} = \sigma_y$ as shown in Fig. 11.40(a). By simple integration, the area under the curve between the crack tip and r_0 is equal to $2\sigma_y r_0$. The shaded area in the figure must therefore have an area $\sigma_y r_0$. It is conventional to assume that the higher stress levels associated with the shaded area are redistributed so that the static zone extends a distance r_p where

$$r_p = 2r_0 = \frac{K^2}{\pi\sigma_y^2} \tag{11.42}$$

Figure 11.40(a) shows the behaviour of $\sigma_{yy}(x, 0)$, i.e. the variation of σ_{yy} with x at $y = 0$, for the case of plane stress. This is only a first approximation, but the estimate of the plastic zone size differs only by a small numerical factor from refined treatments. *In the case of plane strain* σ_{zz} is non-zero and σ_{xx} is the smallest principal stress in the vicinity of the crack tip. Again, by using the Tresca criterion, we have

$$\sigma_{yy} - \sigma_{xx} = \sigma_y$$

or

$$\sigma_{yy} = \sigma_y + \sigma_{xx}$$

In the plastic zone the difference between σ_{yy} and σ_{xx} must be maintained at σ_y. As x increases from the crack tip σ_{xx} rises from zero and so σ_{yy} must rise above σ_y. The normal stress σ_{zz} must also increase and a schematic representation is shown in Fig. 11.40(b). The stress configuration is one of triaxial tension and the constraints on the material produce stresses higher than the uniaxial yield stress. The maximum stress under these conditions is often conveniently taken to be $3\sigma_y$. The plane strain plastic zone size is therefore taken to be one-third of the plane stress plastic zone

i.e.
$$r_p = \frac{K^2}{3\pi\sigma_y^2} \qquad (11.43)$$

In both states of stress it is seen that the square of the stress intensity factor determines the size of the plastic zone. This would seem paradoxical as K is derived from a perfectly elastic model. However, if the plastic zone is small, the elastic stress field in the region around the plastic zone will still be described by eqn. (11.43). The plasticity is then termed "*well-contained*" or "*K-controlled*" and we have an *elastic–plastic stress distribution*. A typical criterion for well-contained plasticity is that the plastic zone size should be less than one-fiftieth of the uncracked specimen ligament.

11.3.4. Fracture toughness

A fracture criterion for brittle and elastic–plastic cracks can be based on functions of the elastic stress components near the crack tip. No matter what function is assumed it is implied that K reaches some critical value since each stress component is uniquely determined by K. In other words the crack will become unstable when K reaches a value K_{IC}, the *Critical stress intensity factor* in mode I. K_{IC} is now almost universally denoted as the "*fracture toughness*", and is used extensively to classify and compare materials which fracture under plane strain conditions.

The fracture toughness is measured by increasing the load on a pre-cracked laboratory specimen which usually has one of the geometries shown in Table 11.1. When the onset of crack growth is detected then the load at that point is used to calculate K_{IC}.

In brittle materials, the onset of crack growth is generally followed by a catastrophic failure whereas ductile materials may withstand a period of stable crack growth before the final fracture. The start of the stable growth is usually detected by changes in the compliance of the specimen and a clip-gauge mounted across the mouth of the crack produces a sensitive method of detecting changes in compliance. It is important that the crack is sharp and that its length is known. In soft materials a razor edge may suffice, but in metals the crack is generally grown by fatigue from a machined notch. The crack length can be found after the final fracture by examining the fracture surfaces when the boundary between the two types of growth is usually visible. Typical values of the fracture toughness of some common materials are given in Table 11.2.

Table 11.2. Typical K_{IC} values.

Material	K_{IC} (MN/m$^{3/2}$)
Concrete (dependent on mix and void content)	0.1–0.15
Epoxy resin	0.5–2.0
Polymethylmethacrylate	2–3
Aluminium	20–30
Low alloy steel	40–60

11.3.5. Plane strain and plane stress fracture modes

Generally, in plane stress conditions, the plastic zone crack tip is produced by shear deformation through the thickness of the specimen. Such deformation is enhanced if the thickness of the specimen is reduced. If, however, the specimen thickness is increased then the additional constraint on through-thickness yielding produces a triaxial stress distribution so that approximate plane strain deformation occurs with shear in the xy plane. There is usually a transition from plane stress to plane strain conditions as the thickness is increased. As K_{IC} values are generally quoted for plane strain, it is important that this condition prevails during fracture toughness testing.

A well-established criterion for plane strain conditions is that the thickness B should obey the following:

$$B \geqslant 2.5 \frac{(K_{IC})^2}{\sigma_y^2} \qquad (11.44)$$

It should be noted that, even on the thickest specimens, a region of plane stress yielding is always present on the side surfaces because no triaxial stress can exist there. The greater plasticity associated with the plane stress deformation produces the characteristic "*shear lips*" often seen on the edges of fracture surfaces. In some instances the plane stress regions on the surfaces may be comparable in size with nominally plane strain regions and a mixed-mode failure is observed. However, many materials show a definite transition from plane stress to plane strain.

11.3.6. General yielding fracture mechanics

When the extent of plasticity which accompanies the growth of a crack becomes compa-rable with the crack length and the specimen dimensions we cannot apply linear elastic fracture mechanics (LEFM) and other theories have to be sought. It is beyond the scope of this book to review all the possible attempts to provide a unified theory. We will, however, examine the J integral developed by Rice[18] because this has found the greatest favour in recent years amongst researchers in this field. In its simplest form the J integral can be defined as

$$J = \frac{\partial U^*}{\partial a} \qquad (11.45)$$

where the asterisk denotes that this energy release rate includes both linear elastic and non-linear elastic strain energies. For linear elasticity J is equivalent to G.

The theory of the J integral was developed for non-linear *elastic* behaviour but, in the absence of any rival theory, the J integral is also used when the extent of *plasticity* produces a non-linear force-displacement curve.

As the crack propagates and the crack tip passes an element of the material, the element will partially unload. In cases of general yielding the elements adjacent to the crack tip will have been plastically deformed and will not unload reversibly, and the strain energy released will not be as great as for reversible non-linear elastic behaviour. At the initiation of growth no elements will have unloaded so that if we are looking for a criterion for crack growth then the difference between plastic and nonlinear elastic deformation may not be significant. By analogy with Griffith's definition of G_c in eqn. (11.32) we can define

$$J_c = \left(\frac{\partial U^*}{\partial a} \right)_c \tag{11.46}$$

the critical strain energy release rate for crack growth. Here the energy required to extend the crack is dominated by the requirement to extend the plastic zone as the crack grows. The surface energy of the new crack faces is negligible in comparison. Experiments on mild steel[16] show that J_c is reasonably constant for the initiation of crack growth in different specimen geometries.

Plastic deformation in many materials is a time-dependent process so that, at normal rates of loading, the growth of cracks through structures with gross yielding can be stable and may be arrested by removing the load.

Calculation of J

Several methods of calculating J exist, but the simplest method using normal laboratory equipment is that developed by Begley and Landes.[19] Several similar specimens of any suitable geometry are notched or pre-cracked to various lengths. The specimens are then extended while the force–displacement curves are recorded. Two typical traces where the crack length a_2 is greater than a_1 are shown in Fig. 11.41. At any one displacement x, the area under the $W-x$ curve gives U^*. For any given displacement, a graph can be plotted of

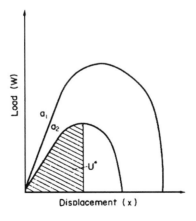

Fig. 11.41. Force–displacement curves for cracked bodies exhibiting general yielding (crack length $a_1 < a_2$)

U^* against crack length (Fig. 11.42). The slopes of these curves give J for any given combination of crack length and displacement, and can be plotted as a function of displacement (Fig. 11.43). By noting the displacement at the onset of crack growth, J_c can be assessed.

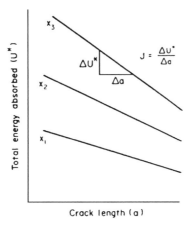

Fig. 11.42. Total energy absorbed as a function of crack length and at constant displacement ($x_3 > x_2 > x_1$)

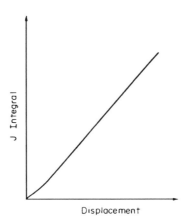

Fig. 11.43. The J integral as a function of displacement.

11.3.7. Fatigue crack growth

The failure of engineering components most commonly occurs at stress levels far below the maximum design stress. Also, components become apparently more likely to fail as their service life increases. This phenomenon, commonly termed fatigue, see §11.1, involves the growth of small defects into macroscopic cracks which grow until K_{IC} is exceeded and catastrophic failure occurs. One of the earliest observations of fatigue failure was that the amplitude of fluctuations in the applied stress had a greater influence on the fatigue life of a component than the mean stress level. In fact if there is no fluctuation in loading then fatigue failure cannot occur, whatever magnitude of static stress is applied.

As stated earlier, fatigue failure is generally considered to be a three-stage process as shown schematically in Fig. 11.44. *Stage I* involves the initiation of a crack from a defect and the subsequent growth of the crack along some favourably orientated direction in the microstructure. Eventually the crack will become sufficiently large that the microstructure has a reduced effect on the crack direction and the crack will propagate on average in a plane normal to the maximum principal stress direction. This is *stage II* growth which has attracted the greatest attention because it is easier to quantify than the initiation stage. When the crack has grown so that K_{IC} is approached the crack accelerates more rapidly until K_{IC} is exceeded and a final catastrophic failure occurs. This accelerated growth is classified as *stage III*.

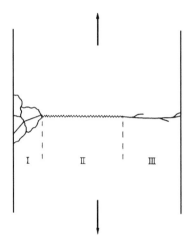

Fig. 11.44. Schematic representation of the three stages of fatigue crack growth.

The rate of growth of a fatigue crack is described in terms of the increase in crack length per load cycle, da/dN. This is related to the amplitude of the stress intensity factor, ΔK, during the cycle. If the amplitude of the applied stress remains constant then, as the crack grows, ΔK will increase. Such conditions produce growth-rate curves of the type shown in Fig. 11.45. Three distinct sections, which corresponds to the three stages of growth, can be seen.

There is a minimum value of ΔK below which the crack will not propagate. This is termed the *threshold value* or ΔK_{th} and is usually determined when the growth rate falls below 10^{-7} mm/cycle or, roughly, one atomic spacing. Growth rates of 10^{-9} mm/cycle can be detected but at this point we are measuring the average increase produced by a few areas of localised growth over the whole crack front. To remove any possibility of fatigue failure in a component it would be necessary to determine the maximum defect size, assume it was a sharp crack, and then ensure that variations in load do not produce ΔK_{th}.

Usually this would result in an over-strong component and it is necessary in many applications to assume that some fatigue crack growth will take place and assess the lifetime of the component before failure can occur. Only sophisticated detection techniques can resolve

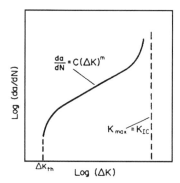

Fig. 11.45. Idealised crack growth rate plot for a constant load amplitude.

cracks in the initiation stage and generally it is assumed that the lifetime of fatigue cracks is the number of cycles endured in stage II.

For many materials stage II growth is described by the *Paris–Erdogan Law*[20].

$$\frac{da}{dN} = C(\Delta K)^m \tag{11.47}$$

C and m are material coefficients. Usually m lies between 2 and 7 but values close to 4 are generally found. This simple relationship can be used to predict the lifetime of a component if the stress amplitude remains approximately constant and the maximum crack size is known. If the stress amplitude varies, then the growth rate may depart markedly from the simple power law. Complications such as fatigue crack closure (effectively the wedging open of the crack faces by irregularities on the crack faces) and single overloads can reduce the crack growth rate drastically. Small changes in the concentration of corrosive agents in the environment can also produce very different results.

Stage III growth is usually a small fraction of the total lifetime of a fatigue crack and often neglected in the assessment of the maximum number of load cycles.

Since we are considering ΔK as the controlling parameter, only brittle materials or those with well-contained plasticity can be treated in this manner. When the plastic deformation becomes extensive we need another parameter. Attempts have been made to fit growth-rate curves to ΔJ the amplitude of the J integral. However while non-linear elastic and plastic behaviour may be conveniently merged in monotonic loading, in cyclic loading there are large differences in the two types of deformation. The non-linear elastic material has a reversible stress–strain relationship, while large hysteresis is seen when plastic material is stressed in the opposite sense. As yet the use of ΔJ has not been universally accepted but, on the other hand, no other suitable parameter has been developed.

11.3.8. Crack tip plasticity under fatigue loading

As a cracked body is loaded, a plastic zone will grow at the crack tip as described in §11.3.3. When the maximum load is reached and the load is subsequently decreased, the deformation of the plastic zone will not be entirely reversible. The elastic regions surrounding the plastic zone will attempt to return to their original displacement as the load is reduced. However, the plastic zone will act as a type of inclusion which the relaxing elastic material then loads in compression. The greatest plastic strain on the increasing part of the load cycle is near the crack tip, and is therefore subjected to the lightest compressive stresses when the load decreases. At a sufficiently high load amplitude the material near the crack tip will yield in compression. A "reverse" plastic zone is produced inside the material which has previously yielded in tension. Figure 11.46 shows schematically the configuration of crack tip plasticity and the variation in vertical stress, in plane stress conditions, at the minimum load of the cycle.

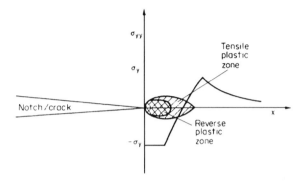

Fig. 11.46. Crack tip plasticity at the minimum load of the load cycle (plane stress conditions)

The material adjacent to the crack tip is therefore subjected to alternating plastic strains which lead to cumulative plastic damage and a weakening of the structure so that the crack can propagate. In metallic materials, striations on the fracture surface show the discontinuous nature of the crack propagation, and in many cases it can be assumed that the crack grows to produce a striation during each load cycle. Polymeric materials, however, can only show striations which occur after several thousands of load cycles.

11.3.9 Measurement of fatigue crack growth

In order to evaluate the fatigue properties of materials *SN* curves can be constructed as described in §11.1.1, or growth-rate curves drawn as shown in Fig. 11.45. Whilst non-destructive testing techniques can be used to detect fatigue cracks, e.g. ultrasonic detection methods to find flaws above a certain size or acoustic emission to determine whether cracks are propagating, growth-rate analysis requires more accurate measurement of crack length. Whilst a complete coverage of the many procedures available is beyond the scope of this text it is appropriate to introduce the most commonly used technique for metal fatigue studies, namely the D.C. potential drop method.

Essentially a large constant current ($\sim$ 30 amps) is passed through the specimen. As the crack grows the potential field in the specimen is disturbed and this disturbance is detected by a pair of potential probes, usually spot-welded on either side of the crack mouth. For single-edge notched (SEN) tensile and bend specimens theoretical solutions exist to relate the measured voltage to the crack length. In compact tension specimens (CTS) empirical calibrations are usually performed prior to the actual tests. Fig. 11.47 shows a block diagram of the potential drop technique. The bulk of the signal is "backed off" by the voltage source so that small changes in crack length can be detected. As the measured voltage is generally

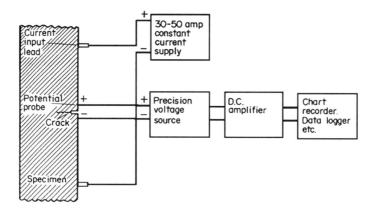

Fig. 11.47. Block diagram of the D.C. potential drop system for crack length measurement.

of the order of microvolts (steel and titanium) or nanovolts (aluminium), sensitive and stable amplifiers and voltage sources are required. A constant temperature environment is also desirable. If adequate precautions are taken, apparent increases in crack length of 10^{-9} mm can be detected in some materials.

If the material under test is found to be insensitive to loading frequency and a constant loading amplitude is required, the most suitable testing machine is probably one which employs a resonance principle. Whilst servo-hydraulic machines can force vibrations over a wider range of frequencies and produce intricate loading patterns, resonance machines are generally cheaper and require less maintenance. Each type of machine is usually provided with a cycle counter and an accurate load cell so that all the parameters necessary to generate the growth rate curve are readily available.

References

1. Goodman, J. *Mechanics Applied to Engineering*, vol. 1, 9th Ed. Longmans Green, 1930.
2. Gerber, W. "Bestimmung der zulossigen spannungen in eisen constructionen", *Z. Bayer Arch. Ing. Ver.*, **6** (1874).
3. Soderberg, C.R. "Factor of safety and working stresses" *Trans. ASME, J. App. Mech.*, **52** (1930).
4. Juvinall, R.C. *Engineering Considerations of Stress, Strain and Strength*. McGraw Hill, 1967.
5. Shigley, J.E. *Mechanical Engineering Design*, 3rd Edn. McGraw-Hill, 1977.
6. Osgood, C.C. Fatigue Design. 2nd Edn. Pergamon Press, 1982.
7. Miner, M.A. "Cumulative damage in fatigue", *Trans. ASME*, **67** (1945).
8. Coffin, L.F. Jnr. "Low cycle fatigue: a review", General Electric Research Laboratory, Reprint No. 4375, Schenectady, N.W., October 1962.
9. Basquin, H.O. "The exponential law of endurance tests", *Proc. ASTM*, **10** (1910).
10. Forsyth, P.J.E. *J. Inst. Metals*, **82** (1953).
11. Cottrell, A.H. and Hull, D. *Proc. Roy. Soc.* **A242** (1957).
12. Larson, F.R. and Miller, J. "A time−temperature relationship for rupture and creep stress", *Trans ASME*, **74** (1952).
13. Manson, S.S. and Haferd, A.M. "A linear time−temperature relation for extrapolation of creep and stress−rupture data", *NACA Tech.*, Note 2890 (1953).
14. Orr, R.L., Sherby, O.D. and Dorn, J.E. "Correlation of rupture data for metals at elevated temperatures", *Trans ASME*, **46** (1954).
15. Griffith, A.A. *Proc. Roy. Soc.*, A, **221** (1920).
16. Irwin, G.R. *J. Appl. Mech. Trans. ASME*, **24** (1957), 361.
17. Knott, J.F. *Fundamentals of Fracture Mechanics*. Butterworths, 1973.

18. Rice, J.R. *J. Appl. Mech. Trans. ASME*, **35** (1968), 379.
19. Begley, J.A. and Landes, D. *The J Integral as a Fracture Criterion*. ASTM STP 514 (1972).
20. Paris, P. and Erdogan, F. *J. Basic. Eng.*, **85** (1963), 265.

Examples

Example 11.1

The fatigue behaviour of a specimen under alternating stress conditions with zero mean stress is given by the expression:

$$\sigma_r^a \cdot N_f = K$$

where σ_r is the range of cyclic stress, N_f is the number of cycles to failure and K and a are material constants.

It is known that $N_f = 10^6$ when $\sigma_r = 300$ MN/m^2 and $N_f = 10^8$ when $\sigma_r = 200$ MN/m^2.

Calculate the constants K and a and hence the life of the specimen when subjected to a stress range of 100 MN/m^2.

Solution

Taking logarithms of the given expression we have:

$$a \log \sigma_r + \log N_f = \log K \tag{1}$$

Substituting the two given sets of condition for N_f and σ_r:

$$2.4771a + 6.0000 = \log K \tag{2}$$

$$2.3010a + 8.0000 = \log K \tag{3}$$

$$\therefore (3) - (2) \qquad \overline{-0.1761a + 2.0000 = 0}$$

$$a = \frac{2.0000}{0.1761}$$

$$= \mathbf{11.357}$$

Substituting in eqn. (2)

$$11.357 \times 2.4771 + 6.000 = \log K$$

$$= 34.1324$$

$$\therefore K = \mathbf{1.356 \times 10^{33}}$$

Hence, for stress range of 100 MN/m^2, from eqn (1):

$$11.357 \times 2.0000 + \log N_f = 34.1324$$

$$22.714 + \log N_f = 34.1324$$

$$\log N_f = 11.4184$$

$$N_f = \mathbf{262.0 \times 10^9 \ cycles}$$

Example 11.2

A steel bolt 0.003 m^2 in cross-section is subjected to a static mean load of 178 kN. What value of completely reversed direct fatigue load will produce failure in 10^7 cycles? Use the Soderberg relationship and assume that the yield strength of the steel is 344 MN/m^2 and the stress required to produce failure at 10^7 cycles under zero mean stress conditions is 276 MN/m^2.

Solution

From eqn. (11.9) of Soderberg

$$\sigma_a = \sigma_N \left[1 - \left(\frac{\sigma_m}{\sigma_y} \right) \right]$$

Now, mean stress σ_m on bolt

$$= \frac{178 \times 10^{-3}}{3 \times 10^{-3}}$$

$$= 59.33 \text{ MN/m}^2$$

$$\therefore \qquad \sigma_a = 276 \left(1 - \frac{59.33}{344} \right)$$

$$= 276(1 - 0.172)$$

$$= 276 \times 0.828$$

$$= 228.5 \text{ MN/m}^2$$

$$\therefore \qquad \text{Load} = 228.5 \times 0.003 \text{ MN}$$

$$= 0.6855 \text{ MN}$$

$$= \mathbf{685.5 \text{ kN}}$$

Example 11.3

A stepped steel rod, the smaller section of which is 50 mm in diameter, is subjected to a fluctuating direct axial load which varies from $+178$ kN to -178 kN.

If the theoretical stress concentration due to the reduction in section is 2.2, the notch sensitivity factor is 0.97, the yield strength of the material is 578 MN/m^2 and the fatigue limit under rotating bending is 347 MN/m^2, calculate the factor of safety if the fatigue limit in tension–compression is 0.85 of that in rotating bending.

Solution

From eqn. (11.12)

$$q = \frac{K_f - 1}{K_t - 1}$$

$$\therefore \qquad K_f = q(K_t - 1) + 1$$

$$= 0.97(2.2 - 1) + 1$$

$$= 2.16$$

But
$$\sigma_{\max} = \frac{178 \times 4}{\pi \times (0.05)^2}$$

$$= 90642 \text{ kN/m}^2$$

$$= \mathbf{90.64 \text{ MN/m}^2}$$

∴ $\sigma_{\min} = -90.64 \text{ MN/m}^2$ and $\sigma_{\text{mean}} = 0$

∴ Under direct stress conditions

$$\sigma_N = 0.85 \times 347$$

$$= \mathbf{294.95 \text{ MN/m}^2}$$

From eqn. (11.13)

$$\sigma_a = \frac{\sigma_N}{K_f F}\left(1 - \frac{\sigma_m \times F}{\sigma_y}\right)$$

∴ With common units of MN/m²:

$$90.64 = \frac{294.95}{2.16 \times F}\left(1 - \frac{0 \times F}{578}\right)$$

∴ $$F = \frac{294.95}{2.16 \times 90.64}$$

$$\mathbf{F = 1.5}$$

Example 11.4

The values of the endurance limits at various stress amplitude levels for low-alloy constructional steel fatigue specimens are given below:

σ_a (MN/m²)	N_f (cycles)
550	1 500
510	10 050
480	20 800
450	50 500
410	1 25 000
380	2 75 000

A similar specimen is subjected to the following programme of cycles at the stress amplitudes stated; 3 000 at 510 MN/m², 12 000 at 450 MN/m² and 80 000 at 380 MN/m², after which the sample remained unbroken. How many additional cycles would the specimen withstand at 480 MN/m² prior to failure? Assume zero mean stress conditions.

Solution

From Miner's Rule, eqn. (11.14), with X the required number of cycles:

$$\frac{n_1}{N_1} + \frac{n_2}{N_2} + \frac{n_3}{N_3} + \cdots \text{ etc} = 1.$$

∴ $$\frac{3\,000}{10\,050} + \frac{12\,000}{50\,500} + \frac{80\,000}{275\,000} + \frac{X}{20\,800} = 1$$

$$0.2985 + 0.2376 + 0.2909 + \frac{X}{20\,800} = 1$$

$$\frac{X}{20\,800} = 1 - 0.8270$$

$$\therefore X = \textbf{3598 cycles}.$$

Example 11.5

The blades in a steam turbine are 200 mm long and they elastically extend in operation by 0.02 mm. If the initial clearance between the blade tip and the housing is 0.075 mm and it is required that the final clearance be not less than 0.025 mm, calculate:

 (i) the maximum percentage creep strain that can be allowed in the blades,
(ii) the minimum creep strain rate if the blades are to operate for 10 000 hours before replacement.

Solution

$$\begin{array}{rcl}
\text{Permissible} \atop \text{creep extension} & = & \text{Initial} \atop \text{clearance} - \left(\text{Final} \atop \text{clearance} + \text{Elastic} \atop \text{extension} \right) \\[2mm]
& = & 0.075 - (0.025 + 0.02) \\[2mm]
& = & 0.03 \text{ mm} \\[2mm]
\therefore \text{ Max. percentage} \atop \text{creep strain} & = & \dfrac{0.03}{200} \times 100 \\[2mm]
& = & \textbf{0.015\%} \\[2mm]
\therefore \text{ Min. creep rate} & = & \dfrac{0.015}{10\,000} = \textbf{1.5} \times \textbf{10}^{-6}\textbf{\%/h}.
\end{array}$$

Example 11.6

The following secondary creep strain rates were obtained when samples of lead were subjected to a constant stress of 1.3 MN/m^2.

Temperature ($^\circ$C)	Minimum creep rate (ε_s^0) (s^{-1})
33	8.71×10^{-5}
29	4.98×10^{-5}
27	3.42×10^{-5}

Assuming that the material complies with the Arrhenius equation, calculate the activation energy for creep of lead. Molar gas constant, $R = 8.314$ J/mol K.

Solution
Construct a table as shown below:

$^\circ$C	K	$\dfrac{1}{T} \times 10^{-3}$	ε_s^0	$l_n \varepsilon_s^0$
33	306	3.27	8.71×10^{-5}	-9.3485
29	302	3.31	4.98×10^{-5}	-9.9156
27	300	3.33	3.42×10^{-5}	-10.2833

The creep rate is related to temperature by eqn. (11.19):

$$\varepsilon_s^0 = Ae^{-H/RT}$$

Hence we can plot $I_n \varepsilon_s^0$ against $\dfrac{1}{T}$ (as in Fig. 11.48)

From the graph.

$$\text{Slope} = 15.48 \times 10^3$$

But
$$H = \text{slope} \times R$$

$\therefore$
$$H = 15.48 \times 10^3 \times 8.314$$

$$= \textbf{128.7 kJ/mol}.$$

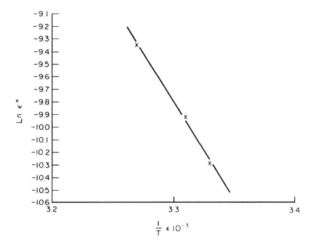

Fig. 11.48.

Example 11.7

An alloy steel bar 1500 mm long and 2500 mm^2 in cross-sectional area is subjected to an axial tensile load of 8.9 kN at an operating temperature of 600°C. Determine the value of creep elongation in 10 years using the relationship $\varepsilon_s^0 = \beta\sigma^n$ if, for 600°C, $\beta = 26 \times 10^{-12}$ h^{-1}(N/mm^2)$^{-6}$, and $n = 6.0$.

Solution

$$\text{Applied stress } \sigma = \frac{P}{A} = \frac{8900}{2500} = 3.56 \text{ N/mm}^2$$

$$\text{Duration of test} = 10 \times 365 \times 24 = 87\,600 \text{ hours}$$

$\therefore$ From eqn. (11.20)

$$\therefore \varepsilon = 26 \times 10^{-12} \times 87\,600 \times (3.56)^6$$

$$= 26 \times 10^{-12} \times 87\,600 \times 2036$$

$$= 4.637 \times 10^{-3}$$

Since the member is 1 500 mm long,

$$\text{total elongation} = 1500 \times 4.637 \times 10^{-3} = \textbf{6.96 mm}.$$

Example 11.8

Creep tests carried out on an alloy steel at 600°C produced the following data:

Stress (kN/m^2)	Minimum creep rate (% / 10 000 h)
10.2	0.4
13.8	1.2
25.5	10.0

A rod, 150 mm long and 625 mm² in cross-section, made of a similar steel and operating at 600°C, is not to creep more than 3.2 mm in 10 000 hours. Calculate the maximum axial load which can be applied.

Solution

$$\% \text{ Creep strain } = \frac{3.2}{150} \times 100 = 2.13\%/10\,000 \text{ h}$$

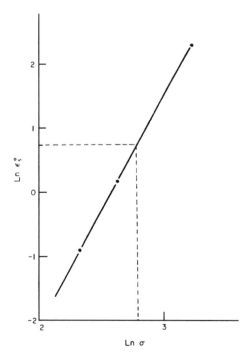

Fig. 11.49.

Since secondary creep rate is related to stress by the eqn. (11.20):

$$\varepsilon_s^0 = \beta\sigma^n,$$

a graph may be plotted of $l_n\varepsilon_s^0$ against $l_n\sigma$.

From the given data:

l_n stress	$l_n\varepsilon_s^0$
2.3224	−0.9163
2.6247	0.1823
3.2387	2.3026

Producing the straight line graph of Fig. 11.49.

For % creep strain rate 2.13%, $l_n\varepsilon_s^0 = 0.7561$.

∴ From graph, $l_n\sigma = 2.78$ and $\sigma = 16.12$ kN/m^2.

If the cross sectional area of the rod is 625 mm^2

then

$$\text{load} = 16\,120 \times 625 \times 10^{-6}$$

$$= \mathbf{10\ N.}$$

Example 11.9

The lives of Nimonic 90 turbine blades tested under varying conditions of stress and temperature are set out in the table below.

Stress (MN/m^2)	Temperature (°C)	Life (h)
180	750	3 000
180	800	500
300	700	5 235
350	650	23 820

Use the information given to produce a master curve based upon the Larson–Miller parameter, and thus calculate the expected life of a blade when subjected to a stress of 250 MN/m^2 and a temperature of 750°C.

Solution

(i) *To calculate C*: from eqn. (11.23), inserting *absolute* temperatures:

$$T_1(l_nt_r + C) = T_2(l_nt_r + C)$$

$$1023(l_n3000 + C) = 1073(l_n500 + C)$$

$$1023(8.0064 + C) = 1073(6.2146 + C)$$

$$8190.5 + 1023C = 6668.27 + 1073C$$

$$1522.23 = 50C$$

$$\mathbf{C = 30.44.}$$

(ii) *To determine P values*

Again, from eqn. (11.23):

$$P_1 = [T(l_n t_r + C)]$$

$$= 1023(l_n 3000 + 30.44)$$

$$= \mathbf{39\,330}$$

$$P_2 = 1073(l_n 500 + 30.44)$$

$$= 1073(6.2146 + 30.44)$$

$$= \mathbf{39\,330}$$

$$P_3 = 973(l_n 5235 + 30.44)$$

$$= 973(8.5632 + 30.44)$$

$$= \mathbf{37\,950}$$

$$P_4 = 923(l_n 23\,820 + 30.44)$$

$$= 923(10.0783 + 30.44)$$

$$= \mathbf{37\,398.}$$

Plotting the master curve as per Fig. 11.31 we have the graph shown in Fig. 11.50.
From Fig. 11.50, when the stress equals 250 MN/m^2 the appropriate parameter $P = 38\,525$
∴ For the required temperature of 750°C (= 1023° absolute)

$$38\,525 = 1023(l_n t_r + 30.44)$$

$$38\,525 = 1023 l_n t_r + 31\,144$$

$$l_n t_r = 7.219$$

$$\mathbf{t_r = 1365\ hours.}$$

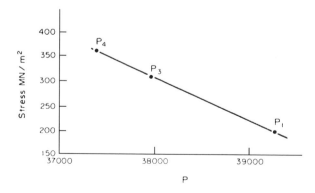

Fig. 11.50.

Example 11.10

The secondary creep rate in many metals may be represented by the equation

$$\varepsilon^0 = \beta\sigma^n.$$

A steel bolt clamping two rigid plates together is held at a temperature of 1000°C. If n is 3.0 and $\varepsilon^0 = 0.7 \times 10^{-9}\text{h}^{-1}$ at 28 MN/m^2, calculate the stress remaining in the bolt after 9000 h if the bolt is initially tightened to a stress of 70 MN/m^2.

Solution

From eqn. (11.20)

$$\varepsilon^0 = \beta\sigma^n$$

$$\therefore \qquad 0.7 \times 10^{-9} = \beta(28 \times 10^6)^3$$

$$\therefore \qquad \beta = \frac{0.7 \times 10^{-9}}{21\,952 \times 10^{18}}$$

$$= 3.189 \times 10^{-32}$$

Using eqn. (11.27) for stress relaxation

$$\frac{1}{\sigma^{n-1}} = \frac{1}{\sigma_i^{n-1}} + \beta E(n-1)t$$

$$\therefore \qquad \frac{1}{\sigma^2} = \frac{1}{(70 \times 10^6)^2} + 3.189 \times 10^{-32} \times 200$$

$$\times 10^9 \times 2 \times 900$$

$$\frac{1}{\sigma^2} = 10^{-16} \times 3.188$$

$$\sigma = 10^8 \times \sqrt{\frac{1}{3.188}}$$

$$= 10^8 \times 0.56$$

i.e. stress in bolt = **56 MN/m^2**

Example 11.11

A steel tie in a girder bridge has a rectangular cross-section 200 mm wide and 20 mm deep.

Inspection reveals that a fatigue crack has grown from the shorter edge and in a direction approximately normal to the edge. The crack has grown 23 mm across the width on one face and 25 mm across the width on the opposite face.

If K_{IC} for the material is 55 MN/m$^{3/2}$ estimate the greatest tension that the tie can withstand.

(Assume that the expression for K in a SEN specimen is applicable here.)

Solution

Since the crack length is not small compared with the width of the girder we need to calculate the compliance function.

Hence

$$a/W = 24/200 = 0.12$$

Then, from Table 11.1

$$Y = 1.99(0.12)^{1/2} - 0.41(0.12)^{3/2} + 18.70(0.12)^{5/2}$$

$$- 38.48(0.12)^{7/2} + 53.85(0.12)^{9/2}$$

$$= 0.689 - 0.017 + 0.093 - 0.023 + 0.004$$

$$= 0.745$$

Also, from eqn. (11.39),

$$K = \frac{PY}{BW^{1/2}}$$

At the onset of fracture $K = K_{IC}$

$$\therefore \qquad\qquad 55 \times 10^6 = \frac{P \times 0.745}{0.02 \times (0.2)^{1/2}}$$

Hence   failure load $P = \mathbf{660\ kN}$.

Example 11.12

A thin cylinder has a diameter of 1.5 m and a wall thickness of 100 mm. The working internal pressure of the cylinder is 15 MN/m^2 and K_{IC} for the material is 38 MN/m$^{3/2}$. Estimate the size of the largest flaw that the cylinder can contain. (Assume that for this physical configuration $K = \sigma\sqrt{\pi a}$.)

Non-destructive testing reveals that no flaw above 10 mm exists in the cylinder. If, in the Paris–Erdogan formula, $C = 3 \times 10^{-12}$ (for K in MN/m$^{3/2}$) and $m = 3.8$, estimate the number of pressurisation cycles that the cylinder can safely withstand.

Solution

Assume that the flaw is sharp, of length $2a$, and perpendicular to the hoop stress.
Then from §9.1.1[†] hoop stress

$$\sigma = \frac{Pd}{2t} = \frac{15 \times 1.5}{2 \times 0.1} = 112.5\ \text{MN/m}^2$$

From eqn. (11.37a), at the point of fracture

$$K = K_{IC} = \sigma\sqrt{\pi a}$$

$$38 = 112.5\sqrt{\pi a}$$

Hence $a = \mathbf{64.3\ mm}$

From eqn. (11.47) $\dfrac{da}{dN} = C(\Delta K)^m$

and for pressurisation from zero $\Delta K = K_{\max}$

$$\therefore \qquad\qquad \frac{da}{\Delta K^m} = C.\,dN$$

[†] E.J. Hearn, *Mechanics of Materials 1*, Butterworth-Heinemann, 1997.

$$N = \frac{1}{C} \int_a^{a_{max}} \frac{da}{(112.5\sqrt{\pi a})^{3.8}}$$

$$= 16.4 \left[\frac{-0.9}{a^{0.9}} \right]_{0.005}^{0.0643}$$

$$= 14.8(-11.88 + 1178)$$

$$= \mathbf{1156 \ cycles.}$$

Example 11.13

In a laboratory fatigue test on a CTS specimen of an aluminium alloy the following crack length measurements were taken.

Crack length (mm)	21.58	22.64	23.68	24.71	25.72	27.37	28.97	29.75
Cycles	3575	4255	4593	4831	5008	5273	5474	5514

The specimen has an effective width of 50.0 mm.
Load amplitude = 3 kN. Specimen thickness = 25.0 mm.
    Construct a growth rate curve in order to estimate the constants C and m in the Paris–Erdogan equation

$$da/dN = C(\Delta K)^m$$

Use the expression given in Table 11.1 to evaluate ΔK.
Use the three-point method to evaluate da/dN.
i.e. at point n;

$$\left(\frac{da}{dN} \right)_n = \frac{a_{n+1} - a_{n-1}}{N_{n+1} - N_{n-1}}$$

Solution

Equation (11.47) gives the Paris–Erdogan law as

$$da/dN = C(\Delta K)^m$$

$$\log(da/dN) = \log C + m\log(\Delta K)$$

From Table 11.1 we can calculate the amplitude of the stress intensity factor from the equation

$$\Delta K = \frac{\Delta P}{BW^{1/2}} \left[29.6 \left(\frac{a}{W} \right)^{1/2} - 185.5 \left(\frac{a}{W} \right)^{3/2} + 655.7 \left(\frac{a}{W} \right)^{5/2} \right.$$

$$\left. -1017 \left(\frac{a}{W} \right)^{7/2} + 638.9 \left(\frac{a}{W} \right)^{9/2} \right]$$

The crack growth rate is most easily found by using the three-point method. The crack growth rate at the point n is calculated as the slope of the straight line joining the $(n+1)$th and the $(n-1)$th points.

i.e.
$$\left(\frac{da}{dN} \right)_n = \frac{a_{n+1} - a_{n-1}}{N_{n+1} - N_{n-1}}$$

From these calculations we obtain the following results

Crack length	21.58	22.64	23.68	24.71	25.72	27.37	28.97	29.75
Cycles	3575	4355	4593	4831	5008	5273	5474	5514
ΔK (MN/m$^{3/2}$)	4.13	4.38	4.65	4.84	5.26	5.86	6.57	6.97
$\log_{10}(\Delta K)$	0.616	0.642	0.668	0.685	0.721	0.769	0.818	0.843
$\dfrac{da}{dN} \times 10^6$ m/cycle		2.06	4.34	4.91	6.02	6.97	8.38	
$\log_{10}$ (da/dN)		−5.68	−5.36	−5.31	−5.22	−5.15	−5.07	

A log–log plot of da/dN versus ΔK is shown in Fig. 11.51.

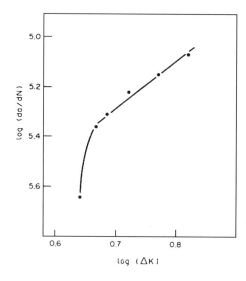

Fig. 11.51.

The first point is not close to the best straight line fit to the other points. The fatigue crack is normally initiated at a high stress amplitude in order to produce a uniform crack front. Although the stress amplitude is reduced gradually to the desired value the initial crack growth is through a crack tip plastic zone associated with the previous loading. The crack is then "blunted" by the larger plastic zone until the crack has grown through it. There is therefore a justification for ignoring this point.

A "least-squares" fit to the remaining points gives

$$\text{Slope} = 2.67 = m$$

$$\text{Intercept} = -7.57 = \log C$$

$$C = 2.69 \times 10^{-8}$$

The Paris–Erdogan equation (11.47) then becomes

$$\frac{da}{dN} = 2.69 \times 10^{-8}(\Delta K)^{2.67}$$

$$(\text{For } \Delta K \text{ in MN/m}^{3/2})$$

Problems

11.1 (B). (a) Write a short account of the microscopical aspects of fatigue crack initiation and growth.

(b) A fatigue crack is considered to have been initiated when the surface crack length has reached 10^{-3} mm. The percentage of cycle lifetime required to reach this stage may be calculated from the equation:

$$1000N_i = \sqrt{2.02} \times (N_f)^{\sqrt{2.02}}$$

Where N_i is the number of cycles to initiate the crack and N_f is the total number of cycles to failure.

 (i) Determine the cyclic lifetime of two specimens, one having a N_i/N_f ratio of 0.01 corresponding to a stress range $\Delta\sigma_1$ and the other having a N_i/N_f ratio of 0.99 corresponding to a stress range $\Delta\sigma_2$.

(ii) If the crack at failure is 1 mm deep, determine the mean crack propagation rate of $\Delta\sigma_1$ and the mean crack nucleation rate at $\Delta\sigma_2$.

[103 cycles, 5.66×10^6 cycles, 9.794×10^{-3} mm/cycle, 17.83×10^{-9} mm/cycle]

11.2 (B). (a) "Under fatigue conditions it may be stated that for less than 1000 cycles, life is a function of ductility and for more than 10,000 cycles life is a function of strength." By consideration of cyclic strain–stress behaviour, show on what grounds this statement is based.

(b) In a tensile test on a steel specimen, the fracture stress was found to be 520 MN/m^2 and the reduction in area 25%.

Calculate: (i) the plastic strain amplitude to cause fracture in 100 cycles; (ii) the stress amplitude to cause fracture in 10^6 cycles. [0.0268; 173.4 MN/m^2]

11.3 (B). An aluminium cantilever beam, 0.762 m long by 0.092 m wide and 0.183 m deep, is subjected to an end downwards fluctuating load which varies from a minimum value P_{min} of 8.9 kN to some maximum value P_{max}.

The material has a fatigue strength for complete stress reversal σ_N of 206.7 MN/m^2 and a static yield strength σ_Y of 275.6 MN/m^2.

By consideration of the Soderberg equation, derive an expression for P_{max} and show that it is equal to:

$$\frac{2I\sigma_N}{y(1+p)L} + \frac{(1-p)}{(1+p)}P_{min}$$

where $p = \sigma_N/\sigma_y$ and y is the distance of the extreme fibres from the neutral axis of bending and I is the second moment of area of the beam section. Determine the minimum value of P_{max} which will produce failure of the beam.

[159 kN]

11.4 (B). (a) Explain the meaning of the term "stress concentration" and discuss its significance in relation to the fatigue life of metallic components.

(b) A member made of steel has the size and shape indicated in Fig. 11.52.

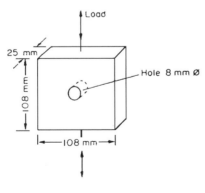

Fig. 11.52.

The member is subjected to a fluctuating axial load that varies from a minimum value of $P/2$ to a maximum value of P. Use the Soderberg equation to determine the value of P that will produce failure in 10^6 cycles.

Assume:

Yield strength of steel $= 420$ MN/m^2

Fatigue strength of steel $= 315$ MN/m^2 for 10^6 cycles

Notch sensitivity factor $= 0.9$

Static stress concentration factor $= 3.0$ [0.62 MN]

11.5 (B). (a). Explain briefly the concepts of survival probability and cumulative damage in respect of fatigue of structural components.

(b) The loading spectrum on an aluminium alloy component is given below for every 100 000 cycles. Also shown is the fatigue life at each stress level.

Determine an expected fatigue life based upon Miner's Hypothesis of damage.

Stress amplitude MN/m^2	Number of cycles in each 10^5 cycles	Fatigue life (N_f)
340	3000	80 000
290	8000	330 000
240	15 000	1 000 000
215	34 000	3 000 000
190	40 000	35 000 000

[11.21×10^5 cycles]

11.6 (B). A stress analysis reveals that at a point in a steel part the stresses are $\sigma_{xx} = 105$, $\sigma_{yy} = 35$, $\sigma_{zz} = -20$, $\tau_{xy} = 56$, $\tau_{yz} = 10$, $\tau_{zx} = 25$ MN/m^2. These stresses are cyclic, oscillating about a mean stress of zero. The fatigue strength of the steel is 380 MN/m^2. Determine the safety factor against fatigue failure. Assume that the fatigue could initiate at a small internal flaw located at the point in question, and assume a stress-concentration factor of 2. Ignore the effect of triaxial stresses on fatigue.   [1.35]

11.7 (B). (a) Briefly discuss how the following factors would affect the fatigue life of a component:

(i) surface finish,

(ii) surface treatment,

(iii) surface shape.

(b) An aluminium airframe component was tested in the laboratory under an applied stress which varied sinusoidally about a mean stress of zero. The component failed under a stress range of 280 MN/m^2 after 10^5 cycles and under a stress range of 200 MN/m^2 after 10^7 cycles. Assuming that the fatigue behaviour can be represented by:

$$A\sigma(N_f)^a = C$$

where a and C are constants, find the number of cycles to failure for a component subjected to a stress range of 150 MN/m^2.

(c) After the component has already endured an estimated 4×10^8 cycles at a stress range of 150 MN/m^2, it is decided that its failure life should be increased by 4×10^8 cycles. Find the decrease in stress range necessary to achieve this additional life.

You may assume a simple cumulative damage law of the form:

$$\sum \frac{N_i}{N_f} = 1$$

[15.2 MN/m^2]

11.8 (B). (a) Write an account of the effect of mean stress upon the fatigue life of a metallic component. Include within your account a brief discussion of how mean stress may be allowed for in fatigue calculations.

(b) A thin-walled cylindrical vessel 160 mm internal diameter and with a wall thickness of 10 mm is subjected to an internal pressure that varies from a value of $-P/4$ to P. The fatigue strength of the material at 10^8 cycles is 235 MN/m^2 and the tensile yield stress is 282 MN/m^2. Using the octahedral shear theory, determine a nominal allowable value for P such that failure will not take place in less than 10^8 cycles.   [36.2 MN/m^2]

11.9 (B). (a) The Manson–Haferd creep parameter method was developed on an entirely empirical basis, whilst those of Larson–Miller and Skerby–Dorn are based upon the well known Arrhenius equation. Compare all three extrapolation methods and comment on the general advantages and disadvantages of applying these methods in practice.

(b) The following table was produced from the results of creep tests carried out on specimens of Nimonic 80 A.

Stress (MN/m^2)	Temperature (°C)	Time to rupture (h)
180	752	56
180	502	1000
300	452	316
300	317	3160

Use the information given to produce a master curve based upon the Manson–Haferd parameter and thus estimate the expected life of a material when subjected to a stress of 250 MN/m^2 and a temperature of 400°C.

[1585 hours]

11.10 (B). (a) Briefly describe the generally desirable characteristics of a material for use at high temperatures.

(b) A cylindrical tube in a chemical plant is subjected to an internal pressure of 6 MN/m^2 which leads to a circumferential stress in the tube wall. The tube is required to withstand this stress at a temperature of 575°C for 9 years.

A designer has specified tubes of 40 mm bore and 2 mm wall thickness made from a stainless steel and the manufacturer's specification for this alloy gives the following information at $\sigma = 200$ MN/m^2.

Temp (°C)	500	550	600	650	700
ε/S	1.0×10^{-6}	2.1×10^{-6}	4.3×10^{-6}	7.7×10^{-6}	1.4×10^{-5}

Given that the effect of stress and temperature upon creep rate can be considered by the following equation:

$$\varepsilon° = A\sigma^6 e^{-\Delta H/RT}$$

and that failure of the tube will take place at a strain of 0.01, with the aid of a graph, calculate whether the tube will fulfil its design life function.

[No, $\varepsilon_9 = 2.07$]

11.11 (B). (a) By consideration of the Arrhenius equation, discuss the theoretical basis of the Larson–Miller parameter as applied to creep data and compare it with other well known alternative parameters.

(b) From the figures given below, determine the expected life at 650°C of an alloy steel when subjected to a stress of 205 MN/m^2.

Stress (MN/m^2)	Temperature (°C)	Life (h)
205	700	1000
205	720	315

[22 330 hours]

11.12 (B). A cylindrical polymer component is produced at constant pressure by expanding a smaller cylinder into a cylindrical mould. The initial polymer cylinder has length 1200 mm, internal diameter 20 mm and wall thickness 5 mm; and the mould has diameter 100 mm and length 1250 cm. Show that, neglecting end effects, the diameter of the cylindrical portion (which may be considered thin) will increase without any change of length until the material touches the mould walls. Hence determine the time taken for the plastic material to reach the mould walls under an internal pressure of 10 kN/m^2, if the uniaxial creep equation for the polymer is $d\varepsilon/dt = 32\sigma$, where t is the time in seconds and σ is the stress in N/mm^2

The Levy–von Mises equations which govern the behaviour of the polymer are of the form

$$d\varepsilon_1 = \frac{d\varepsilon_e}{\sigma_e}[\sigma_1 - \tfrac{1}{2}(\sigma_2 + \sigma_3)]$$

where σ_e is the equivalent stress and ε_e is the equivalent strain.

[1 second]

11.13 (B). (a) Explain, briefly, the meaning and importance of the term "stress relaxation" as applied to metallic materials.

(b) A pressure vessel is used for a chemical process operating at a pressure of 1 MN/m^2 and a temperature of 425°C. One of the ends of the vessel has a 400 mm diameter manhole placed at its centre and the cover plate is held in position by twenty steel bolts of 25 mm diameter spaced equally around the flanges.

Tests on the bolt steel indicate that $n = 4$ and $\varepsilon^\circ = 8.1 \times 10^{-10}$/h at 21 MN/m^2 and 425°C.
Assuming that the stress in a bolt at any time is given by the equation:

$$\frac{1}{\sigma_t^{n-1}} = \frac{1}{\sigma_i^{n-1}} + \beta E(n-1)t$$

and that the secondary creep rate can be represented by the relationship $\varepsilon^\circ = \beta \sigma^n$, with E for the bolt
steel $= 200$ GN/m^2, calculate:
 (i) the initial tightening stress in the bolts so that after 10 000 hours of creep relaxation there is still a safety factor
    of 2 against leakage,
(ii) the total time for leakage to occur.   [30.7 MN/m^2; 177.2 $\times 10^3$ hours]

11.14 (B). (a) The lives of Nimonic 90 turbine blades tested under varying conditions of stress and temperature
are set out in the table below:

Stress (MN/m^2)	Temperature (°C)	Life (h)
180	750	3000
180	800	500
300	700	5235
350	650	23 820

Use the information given to produce a master curve based upon the Larson–Miller parameter, and thus calculate
the expected life of a blade when subjected to a stress of 250 MN/m^2 and a temperature of 750°C.
   (b) Discuss briefly the advantages and disadvantages of using parametric methods to predict creep data compared
with the alternative method of using standard creep strain–time curves.   [1365 hours]

11.15 (B). (a) A support bracket is to be made from a steel to be selected from the table below. It is important
that, if overloading occurs, yielding takes place before fracture. The thickness of the section is 10 mm and the
maximum possible surface crack size that could have escaped non-destructive inspection is 10% of the section
thickness. Select a suitable tempering temperature from the list. Assume that this is equivalent to an edge crack in
a wide plate and assume an associated flaw shape parameter of 1.05.

Tempering temperature (°C)	Yield strength (MN/m^2)	K_{IC} (MN/m$^{3/2}$)
480	1207	98.9
425	1413	95.4
370	1586	41.8

                                                                                            [425 °C]

11.16 (B). A high-speed steel circular saw of 300 mm diameter and 3.0 mm thickness has a fatigue crack of
length 26 mm running in a radial direction from the spindle hole. Assume that this is an edge crack in a semi-infinite
plate. The proof strength of this steel is 1725 MN/m^2 and its fracture toughness is 23 MN/m$^{3/2}$. The tangential
stress component adjacent to the spindle hole during operation has been calculated as follows:

Periphery temperature relative to centre (°C)	Tangential stress (MN/m^2)
10	27
30	50
50	72

First estimate the size of the plastic zone in order to decide whether plane strain or plane stress conditions
predominate. Then estimate at what periphery temperature the saw is likely to fail by fast fracture.
                                                                        [0.0226 mm; 50°C]
   What relative safety factor (on tangential stress) can be gained by using a carbide-tipped lower-strength but
tougher steel saw of yield strength 1200 MN/m^2 and fracture toughness 99 MN/m$^{3/2}$, in the case of the same size
of fatigue crack?   [4.3]

11.17 (B). (a) When a photoelastic model similar to the one shown in Fig. 11.39 is stressed the fifth fringe is
found to have a maximum distance of 2.2 mm from the crack tip. If the fringe constant is 11 N/mm^2/fringe-mm

and the model thickness is 5 mm, determine the value of the stress-intensity factor (in $N/m^{3/2}$) under this applied load. Discuss any important errors which could be associated with this measurement. [1.293 $MN/m^{3/2}$]

(b) Suggest a way of checking to ensure that the stresses are purely mode I (i.e. those tending to "open" the crack) and that there is no superimposed mode II component (i.e. the tendency to shear the crack along its plane, as shown in Figure 11.34).

11.18 (B). (a) The stresses near the crack tip of a specimen containing a through-thickness crack loaded in tension perpendicular to the crack plane are given by the following equations

$$\sigma_{xx} = \frac{K_I \cos \frac{\theta}{2}}{\sqrt{2\pi r}} \left[1 - \sin \frac{\theta}{2} \cdot \sin \frac{3\theta}{2}\right]$$

$$\sigma_{yy} = \frac{K_I \cos \frac{\theta}{2}}{\sqrt{2\pi r}} \left[1 + \sin \frac{\theta}{2} \cdot \sin \frac{3\theta}{2}\right]$$

$$\sigma_{xy} = \frac{K_I \cos \frac{\theta}{2}}{\sqrt{2\pi r}} \left[\sin \frac{\theta}{2} \cdot \cos \frac{3\theta}{2}\right]$$

where K_I is the stress intensity factor, r is the distance from the crack tip and θ is the angle measured from the projected line of the crack in the uncracked region.

Using the proportions of Mohr's circle, or otherwise, show that the maximum shear stress near the crack tip is given by:

$$\tau_{max} = \frac{K_I \sin \theta}{2\sqrt{2\pi r}}$$

(b) Sketch the photoelastic fringe pattern which would be expected from a model of this loading case.

(c) In such a fringe pattern the fourth fringe occurs at a distance of 1.45 mm from the crack tip. If the material fringe constant is 10.5 kN/m^2/fringe/m and the model thickness is 5 mm determine the value of K_I under the given applied load. What error is associated with this measurement? [0.8 $MN/m^{3/2}$]

11.19 (B). (a) Write a short essay on the application of fracture mechanics to the problem of crack growth in components subjected to alternating loading conditions.

(b) After two years service a wide panel of an aluminium alloy was found to contain a 5 mm long edge crack orientated normal to the applied stress. The panel was designed to withstand one start-up/shut-down cycle per day for 20 years (assume 250 operating days in a year), the cyclic stress range being 0 to 70 MN/m^2.

If the fracture toughness of the alloy is 35 $MNm^{-3/2}$ and the cyclic growth rate of the crack is represented by the equation:

$$\frac{da}{dN} = 3.3 \times 10^{-9} (\Delta K)^{3.0}.$$

calculate whether the panel will meet its design life expectancy. (Assume $K_I = \sigma \sqrt{\pi a}$).

[No – 15.45 years]

11.20 (B). (a) Differentiate between the terms "stress concentration factor" and "stress intensity factor".

(b) A cylindrical pressure vessel of 7.5 m diameter and 40 mm wall thickness is to operate at a working pressure of 5.1 MN/m^2. The design assumes that failure will take place by fast fracture from a crack and to prevent this the total number of loading cycles must not exceed 3000.

The fracture toughness of the sheet is 200 $MN/m^{3/2}$ and the growth of the crack may be represented by the equation:

$$\frac{da}{dN} = A(\Delta K)^4$$

Where $A = 2.44 \times 10^{-14}$ and K is the stress intensity factor. Find the minimum pressure to which the vessel must be tested before use to guarantee against fracture in under 3000 cycles. [8.97 MN/m^2]

11.21 (B). (a) Write a short essay on application of fracture mechanics to the problem of crack growth in components subjected to alternating loading conditions.

(b) Connecting rods for an engine are to be made of S.G. iron for which $K_{IC} = 25$ $MNm^{-3/2}$. NDT will detect cracks or flaws of length 2a greater than 2 mm, and rods with flaws larger than this are rejected.

Independent tests on the material show that cracks grow at a rate such that

$$da/dN = 2 \times 10^{-15} (\Delta K_I)^3 \text{ m cycle}^{-1}$$

The minimum cross-sectional area of the rod is 0.01 m², its section is circular and the maximum tensile load in service is 1 MN.

Assuming that $K_I = \sigma\sqrt{\pi a}$ and the engine runs at 1000 rev/min calculate whether the engine will meet its design requirement of 20,000 h life. [$N_f = 4.405 \times 10^9$ and engine will survive]

11.22 (B). (a) By considering constant load conditions applied to a thin semi-infinite sheet and also the elastic energy in the material surrounding an internal crack of unit width and length $2c$, derive an expression for the stress when the crack propagates spontaneously.

(b) A pipeline is made from a steel of Young's modulus 2.06×10^{11} Nm^{-2} and surface energy 1.1 J m^{-2}.

Calculate the critical half-length of a Griffith crack for a stress of 6.2×10^6 Nm^{-2}, assuming that all the supplied energy is used for forming the fracture surface.

$$\left[\sqrt{\frac{2\gamma E}{\pi c}},\ 3.752\ \text{mm}\right]$$

11.23 (B). (a) Outline a method for determining the plane strain fracture toughness K_{IC}, indicating any criteria to be met in proving the result valid.

(b) For a standard tension test piece, the stress intensity factor K_I is given by:

$$K_I = \frac{P}{BW^{1/2}}\left[29.6\left(\frac{a}{W}\right)^{1/2} - 32.04\left(\frac{a}{W}\right)^{3/2}\right]$$

the symbols having their usual meaning.

Using the DC potential drop crack detection procedure the load at crack initiation P was found to be 14.6 kN. Calculate K_{IC} for the specimen if $B = 25$ mm, $W = 50$ mm and $a = 25$ mm.

(c) If $\sigma_1 = 340$ MN/m², calculate the minimum thickness of specimen which could be used still to give a valid K_{IC} value. [25.07 MNm$^{-3/2}$, 13.6 mm]

CHAPTER 12

MISCELLANEOUS TOPICS

12.1. Bending of beams with initial curvature

The bending theory derived and applied in *Mechanics of Materials 1* was concerned with the bending of initially straight beams. Let us now consider the modifications which are required to this theory when the beams are initially curved before bending moments are applied. The problem breaks down into two classes:

(a) initially curved beams where the depth of cross-section can be considered small in relation to the initial radius of curvature, and
(b) those beams where the depth of cross-section and initial radius of curvature are approximately of the same order, i.e. deep beams with high curvature.

In both cases similar assumptions are made to those for straight beams even though some will not be strictly accurate if the initial radius of curvature is small.

(a) Initially curved slender beams

Consider now Fig. 12.1, with Fig. 12.1 (a) showing the initial curvature of the beam before bending, with radius R_1, and Fig. 12.1 (b) the state after the bending moment M has been applied to produce a new radius of curvature R_2. In both figures the radii are measured to the neutral axis.

The strain on any element $A'B'$ a distance y from the neutral axis will be given by:

$$\text{strain on } A'B' = \varepsilon = \frac{A'B' - AB}{AB}$$

$$= \frac{(R_2 + y)\theta_2 - (R_1 + y)\theta_1}{(R_1 + y)\theta_1}$$

$$= \frac{R_2\theta_2 + y\theta_2 - R_1\theta_1 - y\theta_1}{(R_1 + y)\theta_1}$$

Since there is no strain on the neutral axis in either figure $CD = C'D'$ and $R_1\theta_1 = R_2\theta_2$.

$$\therefore \qquad \varepsilon = \frac{y\theta_2 - y\theta_1}{(R_1 + y)\theta_1} = \frac{y(\theta_2 - \theta_1)}{(R_1 + y)\theta_1}$$

and, since $\theta_2 = R_1\theta_1/R_2$.

$$\varepsilon = \frac{y\theta_1\left(\dfrac{R_1}{R_2} - 1\right)}{(R_1 + y)\theta_1} = \frac{y(R_1 - R_2)}{R_2(R_1 + y)} \qquad (12.1)$$

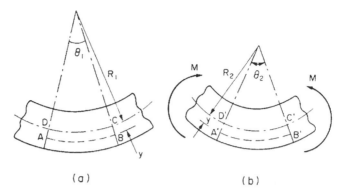

Fig. 12.1. Bending of beam with initial curvature (a) before bending, (b) after bending to new radius of curvature R_2.

For the case of slender, beams with y small in comparison with R_1 (i.e. when y can be neglected in comparison with R_1), the equation reduces to:

$$\varepsilon = y\frac{(R_1 - R_2)}{R_2 R_1} = y\left[\frac{1}{R_2} - \frac{1}{R_1}\right] \tag{12.2}$$

The strain is thus directly proportional to y the distance from the neutral axis and, as for the case of straight beams, the stress and strain distribution across the beam section will be linear and the neutral axis will pass through the centroid of the section. Equation (12.2) can therefore be incorporated into a modified form of the "simple bending theory" thus:

$$\frac{M}{I} = \frac{\sigma}{y} = E\left[\frac{1}{R_2} - \frac{1}{R_1}\right] \tag{12.3}$$

For initially straight beams R_1 is infinite and eqn. (12.2) reduces to:

$$\varepsilon = \frac{y}{R_2} = \frac{y}{R}$$

(b) Deep beams with high initial curvature (i.e. small radius of curvature)

For deep beams where y can no longer be neglected in comparison with R_1 eqn. (12.1) must be fully applied. As a result, the strain distribution is no longer directly proportional to y and hence the stress and strain distributions across the beam section will be non-linear as shown in Fig. 12.2 and the neutral axis will not pass through the centroid of the section. From eqn. (12.1) the stress at any point in the beam cross-section will be given by:

$$\sigma = E\varepsilon = \frac{Ey(R_1 - R_2)}{R_2(R_1 + y)} \tag{12.4}$$

For equilibrium of transverse forces across the section in the absence of applied end load $\int \sigma dA$ must be zero.

$$\therefore \qquad \int \frac{Ey(R_1 - R_2)}{R_2(R_1 + y)} dA = \frac{E(R_1 - R_2)}{R_2} \int \frac{y}{(R_1 + y)} \cdot dA = 0 \tag{12.5}$$

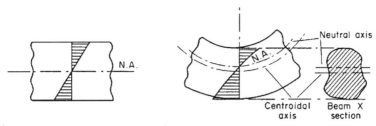

(a) Initially straight beam – Linear   (b) Initially curved beam – Non–linear stress
      stress distribution   distribution

Fig. 12.2. Stress distributions across beams in bending. (a) Initially straight beam linear stress distribution; (b) initially curved deep beam-non-linear stress distribution.

i.e.
$$\int \frac{y}{(R_1 + y)} \cdot dA = 0 \tag{12.6}$$

Unlike the case of bending of straight beams, therefore, it will be seen by inspection that the above integral no longer represents the first moment of area of the section about the centroid. Thus, *the centroid and the neutral axis can no longer coincide.*

The bending moment on the section will be given by:

$$M = \int \sigma \cdot dA \cdot y = \frac{E(R_1 - R_2)}{R_2} - \frac{y_2}{(R_1 + y)} \cdot dA \tag{12.7}$$

but
$$\int \frac{y^2}{(R_1 + y)} \cdot dA = \int \frac{y[(R_1 + y) - R_1]dA}{(R_1 + y)}$$

$$= \int y \cdot dA - R_1 \int \frac{y \cdot dA}{(R_1 + y)}$$

and from eqn. (12.5) the second integral term reduces to zero for equilibrium of transverse forces.

$$\therefore \qquad \int \frac{y_2}{(R_1 + y)} \cdot dA = \int y \cdot dA = A\bar{y} = Ah$$

where h is the distance of the neutral axis from the centroid axis, see Fig. 12.3. Substituting in eqn. (12.7) we have:

$$M = \frac{E(R_1 - R_2)}{R_2} \cdot hA \tag{12.8}$$

From eqn. (12.4)

$$\frac{\sigma}{y}(R_1 + y) = \frac{E}{R_2}(R_1 - R_2)$$

$$\therefore \qquad M = \frac{\sigma}{y}(R_1 + y)hA \tag{12.9}$$

i.e.
$$\frac{\sigma}{y} = \frac{M}{hA(R_1 + y)} \tag{12.10}$$

or
$$\sigma = \frac{My}{hA(R_1 + y)} = \frac{My}{hAR_0} \tag{12.11}$$

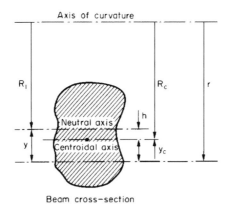

Fig. 12.3. Relative positions of neutral axis and centroidal axis.

On the opposite side of the neutral axis, where y will be negative, the stress becomes:

$$\sigma = -\frac{My}{hA(R_1 - y)} = -\frac{My}{hAR_i} \qquad (12.12)$$

These equations show that the stress distribution follows a hyperbolic form. Equation (12.12) can be seen to be similar in form to the "simple bending" equation[†].

$$\frac{\sigma}{y} = \frac{M}{I}$$

with the term $hA(R_1 + y)$ replacing the second moment of area I.

Thus in order to be able to calculate stresses in deep-section beams with high initial curvature, it is necessary to evaluate h and R_1, i.e. to locate the position of the neutral axis relative to the centroid or centroidal axis. This was shown above to be given by the condition:

$$\int \frac{y}{(R_1 + y)} \cdot dA = 0.$$

Now fibres distance y from the neutral axis will be some distance y_c from the centroidal axis as shown in Figs. 12.3 and 12.4 such that, in relation to the axis of curvature,

$$R_1 + y = R_c + y_c$$

with

$$y = y_c + h$$

∴ from eqn. (12.5)

$$\int \frac{(y_c + h)}{(R_c + y_c)} \cdot dA = 0$$

Re-writing

$$y_c + h = (R_c + y_c) - R_c + h = (R_c + y_c) - (R_c - h).$$

[†] Timoshenko and Roark both give details of correction factors which may be applied for standard cross-sectional shapes to be used in association with the simple straight beam equation. (S. Timoshenko, *Theory of Plates and Shells*, McGraw Hill, New York; R. J. Roark and W.C. Young, *Formulas for Stress and Strain*, McGraw Hill, New York).

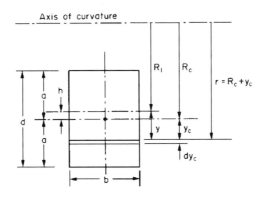

Fig. 12.4.

$$\int \frac{(y_c + h)}{(R_c + y_c)} = \int \frac{(R_c + y_c)}{(R_c + y_c)} \cdot dA - (R_c - h) \int \frac{1}{(R_c + y_c)} \cdot dA$$

$$= A - (R_c - h) \int \frac{1}{(R_c + y_c)} \cdot dA = 0$$

$$\therefore \qquad\qquad h = R_c - \frac{A}{\displaystyle\int \frac{dA}{(R_c + y_c)}} = R_c - \frac{A}{\displaystyle\int \frac{dA}{r}} \qquad\qquad (12.13)$$

and

$$R_1 = R_c - h = \frac{A}{\displaystyle\int \frac{dA}{(R_c + y_c)}} = \frac{A}{\displaystyle\int \frac{dA}{r}} \qquad\qquad (12.14)$$

Examples 12.1 and 12.2 show how the theory may be applied and Table 12.1 gives some useful equations for $\int \frac{dA}{r}$ for standard shapes of beam cross-section.

Note

Before applying the above theory for bending of initially curved members it is perhaps appropriate to consider the benefits to be gained over that of an approximate solution using the simple bending theory.

Provided that the curvature is not large then the simple theory is reasonably accurate; for example, for a radius to beam depth ratio R_c/d of as low as 5 the error introduced in the maximum stress value is only of the order of 7%. The error then rises steeply, however, as curvature increases to a figure of approx. 30% at $R_c/d = 1.5$.

(c) Initially curved beams subjected to bending and additional direct load

In many practical engineering applications such as chain links, crane hooks, G-clamps etc., the component cross-sections will be subjected to both bending and additional direct load, whereas the equations derived in the previous sections have all been derived on the assumption of pure bending only. It is therefore necessary in such cases to obtain a solution by the application of the principle of superposition i.e. by resolving the loading system into

Table 12.1. Values of $\int \dfrac{dA}{r}$ for curved bars.

Cross-section	$\int \dfrac{dA}{r}$
(a) Rectangle	$b \log_e \left(\dfrac{R_0}{R_i} \right)$ (N.B. The two following cross-sections are simply produced by the addition of terms of this form for each rectangular portion)
(b) T-section	$b_1 \log_e \left(\dfrac{R_i + d_1}{R_i} \right) + b_2 \log_e \left(\dfrac{R_0}{R_i + d_1} \right)$
(c) I-beam	$b_1 \log_e \left(\dfrac{R_i + d_1}{R_i} \right) + b_2 \log_e \left(\dfrac{R_0 - d_3}{R_i + d_1} \right) + b_3 \log_e \left(\dfrac{R_0}{R_0 - d_3} \right)$
(d) Trapezoid	$\left[\dfrac{(b_1 R_0 - b_2 R_i)}{d} \log_e \left(\dfrac{R_0}{R_i} \right) \right] - b_1 + b_2$
(e) Triangle	As above (d) with $b_2 = 0$ As above (d) with $b_1 = 0$
	$2\pi \{ (R_i + R) - [(R_i + R)^2 - R^2]^{1/2} \}$

its separate bending, normal (and perhaps shear) loads on the section and combining the stress values obtained from the separate stress calculations. Normal and bending stresses may be added algebraically and combined with the shearing stresses using two- or three-dimensional complex stress equations or Mohr's circle.

Care must always be taken to consider the direction in which the moment is applied. In the derivation of the equations in the previous sections it has been shown acting in a direction to increase the initial curvature of the beam (Fig. 12.1) producing tensile bending stresses on the outside (convex) surface and compression on the inner (concave) surface. In the practical cases mentioned above, however, e.g. the chain link or crane hook, the moment which is usually applied will tend to straighten the beam and hence reduce its curvature. In these cases, therefore, tensile stresses will be set up on the inner surface and these will add to the tensile stresses produced by the direct load across the section to produce a maximum tensile (and potentially critical) stress condition on this surface – see Fig. 12.5.

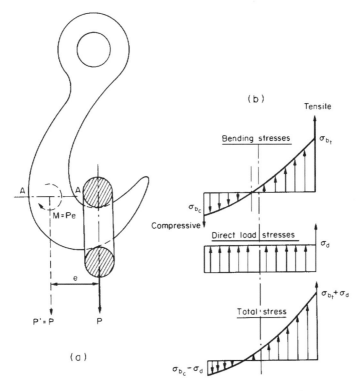

Fig. 12.5. Loading of a crane hook. (a) Load effect on section AA is direct load $P' = P$ plus moment $M = Pe$; (b) stress distributions across the section AA.

12.2. Bending of wide beams

The equations derived in *Mechanics of Materials 1* for the stress and deflection of beams subjected to bending relied on the assumption that the beams were narrow in relation to their depths in order that expansions or contractions in the lateral (z) direction could take place relatively freely.

For beams that are very wide in comparison with their depth – see Fig. 12.6 – lateral deflections are constrained, particularly towards the centre of the beam, and such beams become stiffer than predicted by the simple theory and deflections are correspondingly reduced. In effect, therefore, the bending of narrow beams is a plane stress problem whilst that of wide beams becomes a plane strain problem – see §8.22.

For the beam of Fig. 12.6 the strain in the z direction is given by eqn. (12.6) as:

$$\varepsilon_z = \frac{1}{E}(\sigma_z - v\sigma_x - v\sigma_y).$$

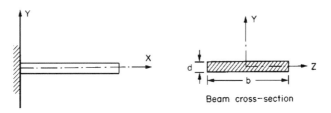

Fig. 12.6. Bending of wide beams ($b \gg d$)

Now for thin beams $\sigma_y = 0$ and, for total constraint of lateral (z) deformation at $z = 0$, $\varepsilon_z = 0$.

$$\therefore \qquad 0 = \frac{1}{E}(\sigma_z - v\sigma_x)$$

i.e. $\qquad \sigma_z = v\sigma_x$

Thus, the strain in the longitudinal x direction will be:

$$\varepsilon_x = \frac{1}{E}(\sigma_x - v\sigma_y - v\sigma_z)$$

$$= \frac{1}{E}(\sigma_x - 0 - v(v\sigma_x))$$

$$= \frac{1}{E}(1 - v^2)\sigma_x \qquad (12.15)$$

$$= \frac{(1 - v^2)}{E} \cdot \frac{My}{I} \qquad (12.16)$$

Compared with the narrow beam case where $\varepsilon_x = \sigma_x/E$ there is thus a reduction in strain by the factor $(1 - v^2)$ and this can be introduced into the deflection equation to give:

$$\frac{d^2y}{dx^2} = (1 - v^2)\frac{M}{EI} \qquad (12.17)$$

Thus, all the formulae derived in Book 1 including those of the summary table, may be used for wide beams *provided that they are multiplied by* $(1 - v^2)$.

12.3. General expression for stresses in thin-walled shells subjected to pressure or self-weight

Consider the general shell or "surface of revolution" of arbitrary (but thin) wall thickness shown in Fig. 12.7 subjected to internal pressure. The stress system set up will be three-dimensional with stresses σ_1 (hoop) and σ_2 (meridional) in the plane of the surface and σ_3 (radial) normal to that plane. Strictly, all three of these stresses will vary in magnitude through the thickness of the shell wall but provided that the thickness is less than approximately one-tenth of the major, i.e. smallest, radius of curvature of the shell surface, this variation can be neglected as can the radial stress (which becomes very small in comparison with the hoop and meridional stresses).

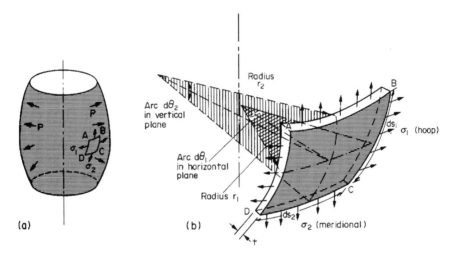

Fig. 12.7. (a) General surface of revolution subjected to internal pressure p; (b) element of surface with radii of curvature r_1 and r_2 in two perpendicular planes.

Because of this limitation on thickness, which makes the system statically determinate, the shell can be considered as a membrane with little or no resistance to bending. The stresses set up on any element are thus only the so-called *"membrane stresses"* σ_1 and σ_2 mentioned above, no additional bending stresses being required.

Consider, therefore, the equilibrium of the element ABCD shown in Fig. 12.7(b) where r_1 is the radius of curvature of the element in the horizontal plane and r_2 is the radius of curvature in the vertical plane.

The forces on the "vertical" and "horizontal" edges of the element are $\sigma_1 t \, ds_1$ and $\sigma_2 t \, ds_2$, respectively, and each are inclined relative to the radial line through the centre of the element, one at an angle $d\theta_1/2$ the other at $d\theta_2/2$.

Thus, resolving forces along the radial line we have, for an internal pressure p:

$$2(\sigma_1 t \, ds_1 \cdot \sin \frac{d\theta_1}{2} + \sigma_2 t \, ds_2 \cdot \sin \frac{d\theta_2}{2} = p \cdot ds_1 \cdot ds_2$$

Now for small angles $\sin d\theta/2 = d\theta/2$ radians

$$\therefore \qquad 2 \left(\sigma_1 t \, ds_1 \cdot \frac{d\theta_1}{2} + \sigma_2 t \, ds_2 \cdot \frac{d\theta_2}{2} \right) = p ds_1 \cdot ds_2$$

Also $ds_1 = r_2 \, d\theta_2$ and $ds_2 = r_1 \, d\theta_1$

$\therefore$ $\qquad\qquad\qquad \sigma_1 t \, ds_1 \cdot \dfrac{ds_2}{r_1} + \sigma_2 t \, ds_2 \dfrac{ds_1}{r_2} = p \cdot ds_1 \cdot ds_2$

and dividing through by $ds_1 \cdot ds_2 \cdot t$ we have:

$$\frac{\sigma_1}{r_1} + \frac{\sigma_2}{r_2} = \frac{p}{t} \tag{12.18}$$

For a general shell of revolution, σ_1 and σ_2 will be unequal and a second equation is required for evaluation of the stresses set up. In the simplest application, i.e. that of the sphere, however, $r_1 = r_2 = r$ and symmetry of the problem indicates that $\sigma_1 = \sigma_2 = \sigma$. Equation (12.18) thus gives:

$$\sigma = \frac{pr}{2t}$$

In some cases, e.g. concrete domes or dishes, the self-weight of the vessel can produce significant stresses which contribute to the overall failure consideration of the vessel and to the decision on the need for, and amount of, reinforcing required. In such cases it is necessary to consider the vertical equilibrium of an element of the dome in order to obtain the required second equation and, bearing in mind that self-weight does not act radially as does applied pressure, eqn. (12.18) has to be modified to take into account the vertical component of the forces due to self-weight.

Thus for a dome of subtended arc 2θ with a force per unit area q due to self-weight, eqn. (12.18) becomes:

$$\frac{\sigma_1}{r_1} + \frac{\sigma_2}{r_2} = \pm \frac{q \cos \theta}{t} \tag{12.19}$$

Combining this equation with one obtained from vertical equilibrium considerations yields the required values of σ_1 and σ_2.

12.4. Bending stresses at discontinuities in thin shells

It is normally assumed that thin shells subjected to internal pressure show little resistance to bending so that only membrane (direct) stresses are set up. In cases where there are changes in geometry of the shell, however, such as at the intersection of cylindrical sections with hemispherical ends, the "incompatibility" of displacements caused by the membrane stresses in the two sections may give rise to significant local bending effects. At times these are so severe that it is necessary to introduce reinforcing at the junction locations.

Consider, therefore, such a situation as shown in Fig. 12.8 where both the cylindrical and hemispherical sections of the vessel are assumed to have uniform and equal thickness membrane stresses in the cylindrical portion are

$$\sigma_1 = \sigma_H = \frac{pr}{t} \quad \text{and} \quad \sigma_2 = \sigma_L = \frac{pr}{2t}$$

whilst for the hemispherical ends

$$\sigma_1 = \sigma_2 = \sigma_H = \frac{pr}{t}.$$

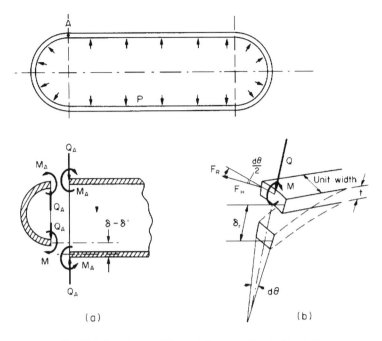

Fig. 12.8. Loading conditions at discontinuities in thin shells.

The radial displacements set up by these stress systems are; for the cylinder:

$$\delta = \frac{r}{E}(\sigma_H - v\sigma_L) = \frac{pr^2}{2tE}(2 - v)$$

and for the hemispherical ends:

$$\delta' = \frac{r}{E}(\sigma_H - v\sigma_H) = \frac{pr^2}{2tE}(1 - v).$$

There will thus be a difference in deformation radially of:

$$\delta - \delta' = \frac{pr^2}{2tE}[(2 - v) - 2(1 - v)]$$

$$= \frac{vpr^2}{2tE}$$

which can only be reacted by the introduction of shear forces and moments as shown in Fig. 12.8(a) where Q = shear force and M = moment, *both per unit length*.

Because of the total symmetry of the cylinder about its axis we may now consider bending of a small element of the cylinder of unit width as shown in Fig. 12.8 (b).

The shear stress Q produces inward bending of the elemental strip through a radial displacement δ_r and a compressive hoop or circumferential strain given by:

$$\varepsilon_H = \frac{\delta_r}{r}$$

with a corresponding hoop stress:

$$\sigma_H = \frac{E\delta_r}{r}$$

This stress sets up a force in the circumferential direction of

$$F_H = \sigma_H \times A = \frac{E\delta_r}{r} \times t \times 1.$$

This force has an outward radial component from both sides of the element of:

$$F_R = 2F_H \sin\frac{d\theta}{2} = 2F_H\frac{d\theta}{2} = \frac{2E\delta_r}{r}\frac{td\theta}{2}$$

$$= \frac{E\delta_r\ t\ d\theta}{r}$$

and since the strip is of unit width, $rd\theta = 1$

$$\therefore \qquad\qquad\qquad F_R = \frac{E\delta_r t}{r^2}$$

This force can be considered as a distributed load along the strip (since equal values will apply to all other unit lengths) and will act in opposition to the mis-match displacements caused by the membrane stresses.

It the strip were considered to be a simple beam then, the differential equation of bending would be:

$$\frac{EId^4 y}{dx^4} = -\frac{E\delta_r t}{r^2}$$

but, as for the case of the deformation of circular plates in 7.2, the restraint on distortion produced by adjacent strips needs to be allowed for by replacing EI by the plate stiffness constant or flexural rigidity

$$D = \frac{Et^3}{12(1 - v^2)} :$$

i.e.
$$D\frac{d^4 y}{dx^4} = -\frac{E\delta_r t}{r^2}$$

$$= -\left[D \times 12\frac{(1 - v^2)}{Et^3}\right]\frac{E\delta_r t}{r^2}$$

$$= -4D\beta^4\delta_r = -4D\beta^4 y \qquad\qquad (1)$$

where
$$\beta^4 = \frac{3(1 - v^4)}{r^2 t^2}\ \text{and}\ y = \delta_r.$$

The solution to eqn. (1) is of the form:

$$y = \delta_r = e^{\beta x}(A_1 \cos \beta x + A_2 \sin \beta x) + e^{-\beta x}(A_3 \cos \beta x + A_4 \sin \beta x) \qquad (2)$$

Now as $x \to \infty, \delta_r \to \infty$ and $A_1 = A_2 = 0$.

At $x = 0$, $M = M_A$ and $D\frac{d^2 y}{dx^2} = -M_A$.

At $x = 0$, $Q = Q_A$ and $D\dfrac{dy^3}{dx^3} = -Q_A$

Substituting these conditions into equation (2) gives:

$$A_3 = \frac{1}{2\beta^3 D}(Q_A - \beta M_A)$$

and

$$A_4 = \frac{M_A}{2\beta^2 D}$$

Substituting back into eqn. (2) we have:

$$y = \delta_r = \frac{e^{-\beta x}}{2\beta^3 D}[Q_A \cos \beta x - M_A \beta(\cos \beta x - \sin \beta x)] \tag{12.20}$$

which is the equation of a heavily damped oscillation, showing that significant values of σ_r, i.e. significant bending, will only be obtained at points local to the cylinder-end intersection. Any stiffening which is desired need, therefore, only to be local to the "joint".

In the special case where the material and the thickness are uniform throughout there will be no moment set up at the intersection A since the shear force Q_A will produce equal slopes and deflections in both the cylinder and the hemispherical end.

Bending stresses can be obtained from the normal relationship:

$$M = D\frac{d^2 y}{dx^2}$$

i.e by differentiating equation (12.20) twice and by substitution of appropriate boundary conditions to determine the unknowns. For cases where the thickness is not constant throughout, and M therefore has a value, the conditions are:

(a) the sum of the deflections of the cylinder and the end at A must be zero,
(b) the slope or angle of rotation of the two parts at A must be equal.

12.5. Viscoelasticity

Certain materials, e.g., rubbers and plastics, exhibit behaviour which combines the characteristics of a viscous liquid and an elastic solid and the term which is used to describe this behaviour is "viscoelasticity". In the case of the elastic solid which follows Hooke's law (a "Hookean" solid) stress is linearily related to strain. For so-called "Newtonian" viscous liquids, however, stress is proportional to strain rate. If, therefore, a tensile test is carried out on a viscoelastic material the resulting stress-strain diagram will depend significantly on the rate of straining $\dot{\varepsilon}$, as shown in Fig. 12.9. Further, whilst the material may well recover totally from its strained position after release of loading it may do so along a different line from the loading line and stress will not be proportional to strain even within this "elastic" range.

One starting point for the mathematical consideration of the behaviour of viscoelastic materials is the derivation of a linear differential equation which, in its most general form, can be written as:

$$A\sigma = B\varepsilon$$

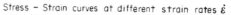

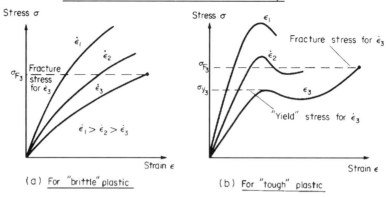

Fig. 12.9. Stress-strain curves at different strain rates $\dot{\varepsilon}$.

with A and B linear differential operators with respect to time, or as:

$$A_0\sigma + A_1\frac{d\sigma}{dt} + A_2\frac{d^2\sigma}{dt^2} + \ldots = B_0\varepsilon + B_1\frac{d\varepsilon}{dt} + B_2\frac{d^2\varepsilon}{dt^2} + \ldots \qquad (12.21)$$

In most cases this equation can be simplified to two terms on either side of the expression, the first relating to stress (or strain) the second to its first differential. This will be shown below to be equivalent to describing viscoelastic behaviour by mechanical models composed of various configurations of springs and dashpots. The simplest of these models contain one spring and one dashpot only and are due to Voigt/Kelvin and Maxwell.

(a) Voigt–Kelvin Model

The behaviour of Hookean solids can be simply represented by a spring in which stress is directly and linearly related to strain,

i.e. $$\sigma_s = E\varepsilon_s$$

The Newtonian liquid, however, needs to be represented by a dashpot arrangement in which a piston is moved through the Newtonian fluid. The constant of proportionality relating stress to strain rate is then the coefficient of viscosity η of the fluid.

i.e. $$\sigma_D = \eta\dot{\varepsilon}_D \qquad (12.22)$$

In order to represent a viscoelastic material, therefore, it is necessary to consider a suitable combination of spring and dashpot. One such arrangement, known as the *Voigt–Kelvin model*, combines the spring and dashpot in parallel as shown in Fig. 12.10.

The response of this model, i.e. the relationship between stress σ, strain ε and strain rate $\dot{\varepsilon}$ is given by:

$$\sigma = \sigma_s + \sigma_D$$

and since the strain is common to both parts of the parallel model $\varepsilon_S = \varepsilon_D = \varepsilon$

∴ $$\sigma = E\varepsilon + \eta\dot{\varepsilon} \qquad (12.23)$$

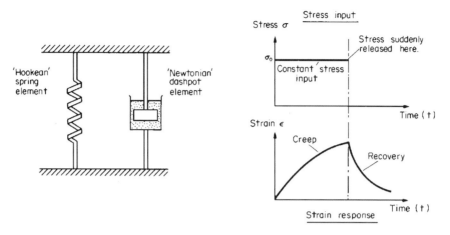

Fig. 12.10. Voigt–Kelvin spring/dashpot model with elements in parallel.

with the stress σ, in effect, shared between the two components of the model (the spring and the dashpot) as for any system of components in parallel.

The inclusion of the strain rate term $\dot{\varepsilon}$ makes the stress response time-dependent and this represents the principal difference in behaviour from that of elastic solids.

If a stress σ_0 is applied to the model, held constant for a time t and then released the strain response will be that indicated in Fig. 12.10. The first part of the response, i.e. the change in strain at constant stress is termed the *creep* of the material, the second part, when stress is removed, is termed the *recovery*.

For *stress relaxation*, i.e. relaxation of stress at constant strain

$$\varepsilon = \text{constant} \quad \text{and} \quad \frac{d\varepsilon}{dt} = 0$$

Equation (12.23) then gives

$$\sigma = E\varepsilon$$

indicating that, according to the Voigt–Kelvin model, the material behaves as an elastic solid under these conditions–clearly an inaccurate representation of viscoelastic behaviour in general.

For creep under constant stress $\sigma = \sigma_0$, however, eqn. (12.23) now gives;

$$\sigma_0 = E\varepsilon + \eta\frac{d\varepsilon}{dt}$$

from which it can be shown that

$$\varepsilon = \frac{\sigma_0}{E}[1 - e^{-Et/\eta}] \tag{12.24}$$

In the special case where $\sigma = \sigma_0 = 0$, the so-called "*recovery*" stage, this reduces to:

$$\varepsilon = \varepsilon_0 e^{-Et/\eta} = \varepsilon_0 e^{-t/t'} \tag{12.25}$$

and this equation indicates that the strain recovers exponentially with time, with t' a characteristic time constant known as the "*retardation time*".

(b) Maxwell model

An alternative model for viscoelastic behaviour proposed by Maxwell again uses a combination of a spring and dashpot but this time in series as shown in Fig. 12.11.

Whereas in the Voigt–Kelvin (parallel) model the stress is shared between the components, in the Maxwell (series) model the stress is common to both elements.

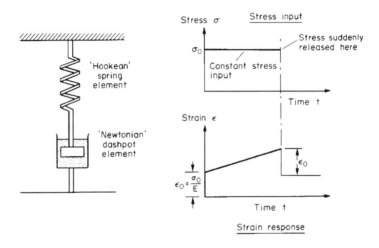

Fig. 12.11. Maxwell model with elements in series.

The strain, however, will be the sum of the strains of the two parts, i.e., the strain of the spring ε_S plus the strain of the dashpot ε_D

$$\therefore \qquad \varepsilon = \varepsilon_S + \varepsilon_D$$

Differentiating:

$$\dot{\varepsilon} = \dot{\varepsilon}_S + \dot{\varepsilon}_D \qquad (1)$$

Now $\sigma_S = E\varepsilon_S$ $\therefore$

$$\dot{\varepsilon}_S = \frac{\dot{\sigma}_S}{E}$$

and $\sigma_D = \eta\dot{\varepsilon}_D$ $\therefore$

$$\dot{\varepsilon}_D = \frac{\sigma_D}{\eta}$$

Now, for the series model,

$$\sigma_S = \sigma_D = \sigma$$

$\therefore$ substituting in (1) we obtain the basic response equation for the Maxwell model.

$$\dot{\varepsilon} = \frac{\dot{\sigma}}{E} + \frac{\sigma}{\eta} \qquad (12.26)$$

The response of this model to a stress σ_0 held constant over a time t and released, is shown in Fig. 12.11.

Let us now consider the response of the Maxwell model to the "standard" relaxation and recovery stages as was carried out previously for the Voigt–Kelvin model.

For stress relaxation $d\varepsilon/dt = 0$, and from eqn. (12.26)

$$0 = \frac{\dot{\sigma}}{E} + \frac{\sigma}{\eta}$$

i.e.

$$\frac{d\sigma}{\sigma} = -\frac{E}{\eta} \cdot dt$$

If, at $t = 0$, $\sigma = \sigma_0$, the initial stress, this equation can be integrated to yield

$$\sigma = \sigma_0 e^{-Et/\eta} = \sigma_0 e^{-t/t''} \tag{12.27}$$

This is analogous to the strain "recovery" equation (12.25) showing that, in this case, stress relaxes from its initial value σ_0 exponentially with time dependent upon the relaxation time t''.

For the creep recovery stage from a constant level of stress, $d\sigma/dt = 0$ and eqn. (12.26) gives

$$\dot{\varepsilon} = \frac{\sigma}{\eta} \tag{12.28}$$

the basic equation of pure Newtonian flow. Generally, however, the creep behaviour of viscoelastic materials is far more complex and, once again, the model does not adequately represent both recovery and relaxation situations. More accurate model representations can only be obtained, therefore, by suitable combinations of the Voigt–Kelvin and Maxwell models (see Figs. 12.12 and 12.13).

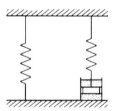

Fig. 12.12. The "standard linear solid" model.

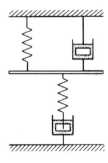

Fig. 12.13. Maxwell and Voigt–Kelvin models in series.

(c) Linear and non-linear viscoelasticity

Both the Voigt–Kelvin and Maxwell models represent so-called *linear viscoelasticity* (which must not be interpreted as meaning that stress is proportional to strain as indicated earlier). Linear viscoelasticity is said to occur when, as a result of a series of creep tests at constant stress levels, the ratios of strain to stress are plotted against time either in the form:

$$\varepsilon = \sigma f(t) \text{ or } \varepsilon = f_1(\sigma) f_2(t).$$

The strain to stress ratio in such tests is termed the *creep compliance*.

Neither the Voigt–Kelvin nor the Maxwell model, will fully represent the behaviour of polymers although the combination of the two, in series, as shown in Fig. 12.13, will give a reasonable approximation of polymer linear viscoelastic behaviour. Unfortunately, however, the range of strain over which linear viscoelasticity is exhibited by polymers is very small.

Non-linear viscoelasticity occurs when the creep compliance–time curve follows an equation of the form:

$$\varepsilon = f'(\sigma, t)$$

This form of viscoelasticity can only be modelled using non-linear springs and dashpots, and the analysis of such systems can become extremely complex.

A convenient approximate solution[1,2] for the design of components constructed from polymers employs the use of "*isochronous*" *stress–strain curves* and a "*secant modulus*" $E_s(t)$. If a series of creep tests are carried out to produce a set of strain-time curves at various stress levels a number of constant time sections can be taken through the curves to enable isochronous (constant time) stress–strain diagrams to be plotted in Fig. 12.14. Such results may be obtained under tensile, compressive or shear loading. Alternatively these data may be obtained from manufacturers' data sheets. One of these isochronous curves can then be selected on the basis of the known lifetime requirement of the component and used for the determination of the secant modulus.

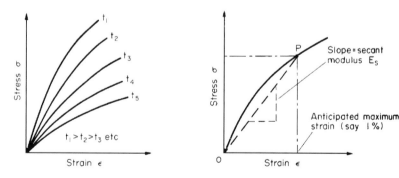

Fig. 12.14. Use of isochronous curves for design.

Defining a point P on the isochronous curve, be it either the expected maximum stress or strain (usually taken as 1%), allows a straight line to be drawn from P to the origin O, the slope of which gives the secant modulus. As stated above, this modulus may be as a result of tension, compression or shear and the appropriate value can then be used to replace E and G in the standard elastic formulae derived in other chapters of this text. If such formulae also contains Poisson's ratio v this must also be replaced by its equivalent under creep conditions, the so-called "*creep contraction*" or "*lateral strain ratio*" $v(t)$. See Example 12.2.

References

1. Benham, P. P. and McCammond, D., "Approximate creep analysis for thermoplastic beams and struts", *J.S.A.*, **6**, 1, 1971.
2. Benham, P. P. and McCammond, D., "A study of design stress analysis problems for thermoplastics using time dependent data", *Plastics and Polymers*, Oct. 1969.

Examples

Example 12.1

The gantry shown in Fig. 12.15 is constructed from 100 mm × 50 mm rectangular cross-section and, under service conditions, supports a maximum load P of 20 kN. Determine the maximum distance d at which P can be safely applied if the maximum tensile and compressive stresses for the material used are limited to 30 MN/m^2 and 100 MN/m^2 respectively.

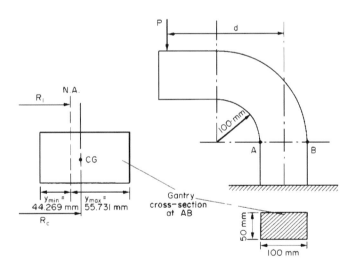

Fig. 12.15.

How would this value change if the cross-section were circular, but of the same cross-sectional area?

Solution

For the gantry and cross-section of Fig. 12.15 the following values are obtained by inspection:

$$R_c = 150 \text{ mm} \quad R_i = 100 \text{ mm} \quad R_o = 200 \text{ mm} \quad b = 50 \text{ mm}.$$

∴ From Table 12.1(a)

$$\int \frac{dA}{r} = b \log_e \left(\frac{R_o}{R_i} \right) = 50 \log_e \left(\frac{200}{100} \right)$$

$$= 34.6574 \text{ mm}$$

∴
$$R_1 = \frac{A}{\int \dfrac{dA}{r}} = \frac{50 \times 100}{34.6574} = 144.269 \text{ mm}$$

∴
$$h = R_c - R_1 = 150 - 144.269 = 5.731 \text{ mm}.$$

Direct stress (compressive) due to $P = \dfrac{P}{A} = \dfrac{20 \times 10^3}{(100 \times 50)10^{-6}} = 4 \text{ MN/m}^2$

Thus, for maximum tensile stress of 30 MN/m² to be reached at B the bending stress (tensile) must be $30 + 4 = 34$ MN/m².

$$\text{Now } y_{max} = 50 + 5.731$$

$$= 55.731 \text{ at } B$$

and bending stress at $B = \dfrac{My}{hA(R_1 + y)} = 34 \text{ MN/m}^2,$

∴
$$\frac{(40 \times 10^3 d) \times 55.731 \times 10^{-3}}{(5.731 \times 10^{-3})(50 \times 100 \times 10^{-6})(200 \times 10^{-3})} = 34 \times 10^6$$

∴
$$d = 174.69 \text{ mm}.$$

For maximum compressive stress of 100 MN/m² at A the compressive bending stress must be limited to $100 - 4 = 96$ MN/m² in order to account for the additional direct load effect.

∴ At A, with $y_{min} = 50 - 5.731 = 44.269$

$$\text{bending stress} = \frac{(20 \times 10^3)44.269 \times 10^{-3}}{(5.731 \times 10^{-3})(50 \times 100 \times 10^{-6})(100 \times 10^{-3})} = 96 \times 10^6$$

∴
$$d = 310.7 \text{ mm}.$$

The critical condition is therefore on the tensile stress at B and the required maximum value of d is **174.69 mm**.

If a circular section were used of radius R and of equal cross-sectional area to the rectangular section then $\pi R^2 = 100 \times 50$ and $R = 39.89$ mm.

∴ From Table 12.1 assuming R_c remains at 150 mm

$$\int \frac{dA}{r} = 2\pi \{(R_i + R) - \sqrt{(R_1 + R)^2 - R^2}\}$$

$$= 2\pi \{150 - \sqrt{150^2 - 39.894^2}\}$$

$$= 2\pi \times 5.4024 = 33.944 \text{ mm}.$$

∴
$$R_1 = \frac{A}{\int \dfrac{dA}{r}} = \frac{50 \times 100}{33.944} = 147.301 \text{ mm},$$

with $h = R_c - R_1 = 150 - 147.3 = 2.699$ mm.

∴ For critical tensile stress at B with $y = 39.894 + 2.699 = 42.593$.

$$\frac{My}{hA(R_1 + y)} = 34 \text{ MN/m}^2.$$

∴

$$\frac{(20 \times 10^3 \times d)(42.593 \times 10^{-3})}{2.699 \times 10^{-3} \times (\pi \times 39.894^2 \times 10^{-6})(150 + 39.894)} = 34 \times 10^6$$

$$d = 102.3$$

i.e. Use of the circular section reduces the limit of d within which the load P can be applied.

Example 12.2

A constant time section of 1000 h taken through a series of strain–time creep curves obtained for a particular polymer at various stress levels yields the following isochronous stress–strain data.

$\sigma(\text{kN/m}^2)$	1.0	2.25	3.75	5.25	6.54	7.85	9.0
$\varepsilon(\%)$	0.23	0.52	0.85	1.24	1.68	2.17	2.7

The polymer is now used to manufacture:

(a) a disc of thickness 6 mm, which is to rotate at 500 rev/min continuously,
(b) a diaphragm of the same thickness which is to be subjected to a uniform lateral pressure of 16 N/m² when clamped around its edge.

Determine the radius required for each component in order that a limiting stress of 6 kN/m² is not exceeded after 1000 hours of service. Hence find the maximum deflection of the diaphragm after this 1000 hours of service.

The lateral strain ratio for the polymer may be taken as 0.45 and its density as 1075 kN/m³.

Solution

(a) From eqn. (4.11) the maximum stress at the centre of a solid rotating disc is given by:

$$\sigma_{r_{max}} = \sigma_{\theta_{max}} = (3 + \nu)\frac{\rho\omega^2 R^2}{8}$$

For the limiting stress condition, therefore, with Poissons ratio ν replaced by the lateral strain ratio:

$$6 \times 10^3 = 3.45 \times 1075 \times \frac{(500 \times 2\pi)^2}{60} \times \frac{R^2}{8}$$

From which $R^2 = 0.00472$

and $R = 0.0687$ m = **68.7 mm**.

(b) For the diaphragm with clamped edges the maximum stress is given by eqn. (22.24) as:

$$\sigma_{r_{max}} = \frac{3qR^2}{4t^2}$$

∴ $$6 \times 10^3 = \frac{3 \times 16 \times R^2}{4 \times (6 \times 10^{-3})^2}$$

From which $R = 0.134$ m = **134 mm**.

The maximum deflection of the diaphragm is then given in Table 7.1 as:

$$\delta_{max} = \frac{3qR^4}{16Et^3}(1 - v^2)$$

Here it is necessary to replace Young's modulus E by the secant modulus obtained from the isochronous curve data and Poisson's ratio by the lateral strain ratio.

The 1000 hour isochronous curve has been plotted from the given data in Fig. 12.16 producing a secant modulus of 405 kN/m^2 at the stated limiting stress of 6 kN/m^2; this being the slope of the line from the origin to the 6 kN/m^2 point on the isochronous curve.

∴ $$\delta_{max} = \frac{3 \times 16 \times (134 \times 10^{-3})^4 \times (1 - 0.45^2)}{16 \times 405 \times 10^3 \times (6 \times 10^{-3})^3}$$

$$= 0.0088 \text{ m} = \textbf{8.8 mm}.$$

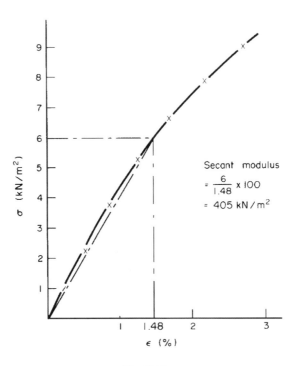

Fig. 12.16.

Problems

12.1 (B). The bracket shown in Fig. 12.17 is constructed from material with 50 mm × 25 mm rectangular cross-section and it supports a vertical load of 10 kN at *C*. Determine the magnitude of the stresses set up at *A* and *B*.

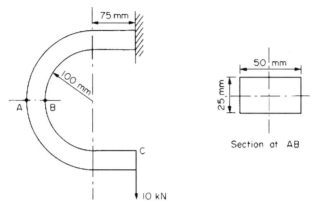

Fig. 12.17.

What percentage error would be obtained if the simple bending theory were applied?

$$[-161.4 \text{ MN/m}^2, + 169.4 \text{ MN/m}^2, 19\%, 13.6\%]$$

12.2 (B). A crane hook is constructed from trapezoidal cross-section material. At the critical section *AB* the dimensions are as shown in Fig. 12.18. The hook supports a vertical load of 25 kN with a line of action 40 mm from *B* on the inside face. Calculate the values of the stresses at points *A* and *B* taking into account both bending and direct load effects across the section. $\qquad$ $[129.2 \text{ MN/m}^2, -80.3 \text{ MN/m}^2]$

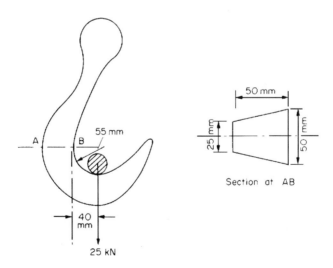

Fig. 12.18.

12.3 (B). A G-clamp is constructed from I-section material as shown in Fig. 12.19. Determine the maximum stresses at the central section *AB* when a clamping force of 2 kN is applied.

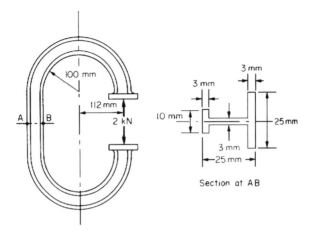

Fig. 12.19.

How do these values compare with those which would be obtained using simple bending theory applied to a straight beam of the same cross-section?   [267 MN/m², −347.5 MN/m², 240 MN/m², 380.7 MN/m²]

12.4 (B). Part of the frame of a machine tool can be considered to be of the form shown in Fig. 12.20. A decision is required whether to construct the frame from T or rectangular section material of the dimensions shown.

Compare the critical stresses set up at section *AB* for each of the cross-sections when the frame is subjected to a peak load of 5 kN and discuss the results obtained in relation to the decision required.

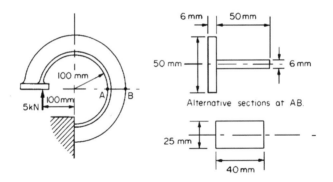

Fig. 12.20.

Plot diagrams of the stress distribution across *AB* for each cross-sectional shape.
          [122.7 MN/m², −198 MN/m², 193.4 MN/m², −143.2 MN/m²]

12.5 (B). (a) By consideration of the Maxwell model, derive an expression for the internal stress after time *t* of a polymer held under constant strain conditions and hence show that the relaxation time is equal to η/G where η is the coefficient of viscosity and G is the shear modulus.

(b) A shear stress of 310 MN/m² is applied to a polymer which is then held under fixed strain conditions. After 1 year the internal stress decreases to a value of 207 MN/m². Calculate the value to which the stress will fall after 2 years, assuming the polymer behaves according to the Maxwell model.   [$\tau = \tau_0 e^{-Gt/\eta}$; 138 MN/m²]

12.6 (B). (a) Spring and dashpot arrangements are often used to represent the mechanical behaviour of polymers. Analyse the mathematical stress strain relationship for the Maxwell and Kelvin–Voigt models under conditions of (i) constant stress, (ii) constant strain, (iii) recovery, and draw the appropriate strain–time, stress–time diagrams, commenting upon their suitability to predict behaviour of real polymers.

(b) Maxwell and Kelvin–Voigt models are to be set up to simulate the behaviour of a plastic. The elastic and viscous constants for the Kelvin–Voigt model are 2×10^9 N/m^2 and 100×10^9 Ns/m^2 respectively and the viscous constant for the Maxwell model is 272×10^9 Ns/m^2. Calculate a value for the elastic constant for the Maxwell model if both models are to predict the same strain after 100 seconds when subjected to the same stress.

$$[15.45 \times 10^9 \text{ N/m}^2]$$

12.7 (B). The model shown in Fig. 12.21 is frequently used to simulate the mechanical behaviour of polymers:

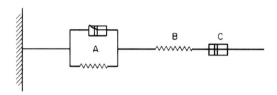

Fig. 12.21.

(a) With reference to Figure 12.21, state what components of total strain the elements A, B and C represent.
(b) Sketch a typical strain–time graph for the model when the load F is applied and then removed. Clearly label those parts of the graph corresponding to the strain components ε_1, ε_2 and ε_3.
(c) A certain polymer may be modelled on such a system by using the following constants for the elements:

Dashpot A:	viscosity $= 10^6$ Ns/m^2
Dashpot B:	viscosity $= 100 \times 10^6$ Ns/m^2
Spring A:	shear modulus $= 50 \times 10^3$ N/m^2
Spring B:	shear modulus $= 10^9$ N/m^2

This polymer is subjected to a direct stress of 6×10^3 N/m^2 for 30 seconds ONLY.
Determine the strain in the polymer after 30 seconds, 60 seconds and 2000 seconds.

$$[3.17 \times 10^{-2}, 0.75 \times 10^{-2}, 0.06 \times 10^{-2}]$$

12.8 (C). For each of the following typical engineering components and loading situations sketch and dimension the components and allocate appropriate loadings. As a preliminary step towards finite element analysis of each case, select and sketch a suitable analysis region, specify complete boundary conditions and add an appropriate element mesh. Make use of symmetry and St. Venant's criteria wherever possible.

(a) A shelf support bracket welded to a vertical upright.
(b) An engine con-rod with particular attention paid to shoulder fillet radii for weight reduction purposes (see Fig. 6.1)
(c) A washing machine agitator cross-section (see Fig. 5.14), bar-tube fillet radii and relative thicknesses of particular concern.
(d) The extruded alloy section of Fig. 1.21. Model to be capable of consideration of varying lines of action of applied force.
(e) A circular pipe flange used to connect two internally pressurised pipes. Model to be capable of including the effect of bolt tensions and external moments on the joint. You may assume that the pipe is free to expand axially.
(f) A C.T.S. (compact test specimen) for brittle fracture compliance testing. Stress distributions at the crack tip are required.
(g) A square storage hopper fabricated from thin rectangular plates welded together and supported by means of welded angle around the upper edge. It may be assumed that the hopper is full with an equivalent hydrostatic pressure p throughout. The supporting frame can be assumed rigid.
(h) A four-point beam bending test rig with plastic beam mounted on steel pads over steel knife edges. The degree of indentation of the plastic and deformation of the steel pad are required.
(i) Thick cylinder with flat ends and sharp fillet radii subjected to internal pressure. The model should be capable of assessing the effect of different end plate thicknesses.
(j) A pressurised thick cylinder containing a 45° nozzle entry. Stress concentrations at the nozzle entry are required.

APPENDIX 1

TYPICAL MECHANICAL AND PHYSICAL PROPERTIES FOR ENGINEERING METALS

Material	Young's modulus of elasticity E (GN/m^2)	Shear modulus G (GN/m^2)	"Elastic" limit σ_y (MN/m^2)	Shear yield strength τ_y (MN/m^2)	Tensile strength (MN/m^2)	Ultimate strength in shear (MN/m^2)	Percentage elongation (%)	Density (Kg/m^3)	Linear coefficient of thermal expansion ($\times10^{-6}$/°C)
Aluminium alloy	69	26	230	–	390	240	23	2770	23
Brass	102	38	–	–	350	–	40	8350	18.9
Bronze	115	45	210	–	310	–	20	7650	18
Cast iron: Grey	90	41	–	–	210	–	8	7640	10.5
Malleable	170	83	248	166	370	330	12	7640	12
Low carbon (mild) steel	207	80	280	175	480	350	25	7800	11.7
Nickel-chrome steel	208	82	1200	650	1700	950	12	7800	11.7
Titanium	107	40	480	–	551	–	–	4507	9.5
Magnesium	45	17	262	–	379	165	–	1791	28.8

TYPICAL MECHANICAL PROPERTIES OF NON-METALS

Material	Young's modulus of elasticity E (GN/m²)	Tensile strength (MN/m²)	Compressive strength (MN/m²)	Elongation (maximum) %
Acetals	–	69	124	75
Cellulose acetate	1.4	41	207	20
Cellulose nitrate	1.4	48	138	40
Epoxy (glass filler)	–	145	234	–
Hard rubber	3.0	48	–	–
Melamine	8.0	55	227	0.7
Nylon filaments	4.1	340	–	–
Polycarbonate – unreinforced Makralon	2.3	70	83	100
Reinforced Makralon	6.0	90	–	8
Polyester (unfilled)	2.0	41	–	2
Polyethylene H.D.	–	28	22	100
Polyethylene L.D.	–	10	–	800
Polypropylene	–	34	510	250
Polystyrene	3.4	20	76	1.2
Polystyrene – impact resistant	1.4	38	41	80
P.T.F.E.	–	34	248	70
P.V.C. (rigid)	3.4	50–60	69	40
P.V.C. (plasticised)	–	20	0.7	200
Rubber (natural-vulcanised)	–	7–34	–	–
Silicones (elastomeric)	–	1.5–6	–	–
Timber	9.0	70	–	–
Urea (Cellulose filler)	10.0	62	241	0.7

* Data taken in part from *Design Engineering Handbook on Plastics* (Product Journals Ltd).

OTHER PROPERTIES OF NON-METALS*

Material	Chemical resistance					Max useful temp. (C)
	Organic	Acids		Alkalis		
	Solvents	Weak	Strong	Weak	Strong	
Acetal	×	×	00	×	×	90
Acrylic	Varies	×	×−0	×	×	90
Nylon 66	×	×	00	×	×	150
Polycarbonate	Varies	×	0	×−0	00	120
Polyethylene LD	×	×	×−00	×	×	90
Polyhethylene HD	×	×	×−0	×	×	120
Polypropylene	×	×	×−0	×	×	150
Polystyrene	Varies	×	×−0	×	×	95
PTFE	×	×	×	×	×	240
PVC	Varies	×	×−0	×	×	80
Epoxy	×	×	×	×	0	430
Melamine	×	×	00	×	0	100−200
Phenolic	×	×−0	0−00	0−00	00	200
Polyester/glass	×−0	0	00	0	00	250
Silicone	×−0	×−0	00	0−00	00	180
Urea	×−0	×−0	00	0−00	00	90
× − Resistant, 0 − Slightly attacked, 00 − markedly attacked						

*Data taken from Design Engineering Handbook on Plastics. (Product Journals Ltd)

INDEX